CARBON NANOTUBES

CARBON NANOTUBES
Properties and Applications

Liam Anderson
Zoe Chapman
Leah Patterson

Kruger Brentt
Publishers

2026

Kruger Brentt Publishers UK. LTD.
Company Number 9728962

Regd. Office: 68 St Margarets Road, Edgware, Middlesex HA8 9UU

ISBN: 978-1-78715-645-6

For information on all our publications visit our website at http://krugerbrentt.com/

PREFACE

Carbon Nanotubes: Properties and Applications provides a comprehensive exploration of one of the most exciting and versatile materials in modern science and engineering. Carbon nanotubes (CNTs), with their unique structural, electrical, thermal, and mechanical properties, have opened new horizons in nanotechnology, materials science, electronics, energy storage, and biotechnology. This book aims to offer both foundational knowledge and practical insights for students, researchers, and professionals seeking to understand and harness the potential of CNTs.

The text begins with an introduction to the structure and classification of carbon nanotubes, including single-walled and multi-walled varieties. It then delves into their remarkable properties—mechanical strength, electrical conductivity, thermal stability, and chemical reactivity—highlighting how these attributes distinguish CNTs from conventional materials and enable novel applications.

Subsequent chapters focus on the synthesis methods of CNTs, such as chemical vapor deposition, arc discharge, and laser ablation, and discuss challenges related to purity, scalability, and controlled growth. The book also examines functionalization techniques that enhance CNT compatibility with polymers, metals, and biological systems, making them suitable for diverse industrial and research applications.

A significant portion of the book is dedicated to the applications of CNTs across multiple domains. Topics include electronics and semiconductors, energy storage and conversion (batteries, supercapacitors, fuel cells), composites and structural materials, sensors, drug delivery, and environmental remediation. Case studies and real-world examples illustrate how CNTs are transforming technology and research.

Special attention is given to safety, environmental, and ethical considerations, including potential toxicity, occupational exposure, and sustainability challenges associated with large-scale CNT production and use.

We hope that Carbon Nanotubes: Properties and Applications will serve as an essential resource for students, researchers, and industry professionals, providing both a solid theoretical foundation and practical guidance for exploring and applying the remarkable potential of carbon nanotubes.

Liam Anderson

Zoe Chapman

Leah Patterson

CONTENTS

Chapter-1

Carbon Nanotubes - A scientometric study

1. Introduction

In contrast to our previous study (Barth & Marx, 2008) dealing with a currently decreasing research field (high-temperature superconductors) we analyzed here a topic which has raised a strongly increasing interest among researchers: research activities around carbon nanotubes (CNTs or NTs). Carbon nanotubes (often named only nanotubes) are graphite sheets rolled up into cylinders with diameters of the order of a few nanometers and up to some millimeters in length with at least one end capped with a hemisphere of the fullerene structure. There are two main types of nanotubes: the single-walled nanotubes (SWCNTs or SWNTs) and the multi-walled nanotubes (MWCNTs or MWNTs), in particular the doublewalled nanotubes (DWCNTs or DWNTs). MWCNTs consist of a single sheet of graphite rolled in around itself (like a rolled up newspaper) or consist of multiple layers of graphite arranged in concentric cylinders (like a Russian Doll).

Nanotubes exhibit some remarkable properties: They feature extraordinary strength, show efficient conductivity of heat, and unique electrical properties (metallic conductivity and semiconductivity). These properties make them potentially useful in a wide range of applications like in materials science, electronics, and nanotechnology. The one-atom thick single graphite layers building up the nanotube cylinders are named graphene, the newest member of this structural family. This species was presumed not to exist in the free state before it was discovered in the year 2004.

The large number of articles with respect to nanotubes has brought about that scientists being active in this research field have increasingly problems to overview their discipline. On the other hand, modern information systems offer databases and analysis tools providing remedy. However, due to lack of access and experience, many scientists do not take advantage of them. In this analysis we demonstrate the potential of such tools with respect to different kinds of meta-information. The data presented here are not expected to reveal surprising insights for experts working in this research field. However, they provide a quantification of (1) the productivity of the active players and (2) of the impact of their works. Moreover, the data could also be interesting for scientists working in neighboring research fields.

2. Methodology and Information Sources

The data presented in this study are based on the Science Citation Index (SCI) including the Conference Proceedings Citation Index, Science (CPCI-S) under the Web of Science (WoS). The SCI under the WoS stretches back to 1900 and the CPCI-S covers conference proceedings published since 1992. The WoS is accessible under the Web of Knowledge (WoK), the search platform provided by Thomson Reuters (Thomson Reuters, 2009), the former Institute for Scientific Information (ISI). The research field analyzed here stretches throughout most natural sciences disciplines being covered by the multi-disciplinary SCI and CPCI-S. However, the WoS source journals selected by the Thomson Reuters staff as contributing to the progress of science do not cover all publications being relevant here, in particular with respect to application and technology.

Therefore, the literature file of the Chemical Abstracts Service (CAS), a division of the American Chemical Society (ACS), has been consulted as an alternative information source. The CAS literature file is available via the online service STN International (STN International, 2009). The literature file CAplus is seen as the most extensive source of substance related publications (either articles or patents) in the fields of chemistry, materials science, and physics. Specific functions of the STN search system for carrying out statistical investigations have made it possible to perform extensive scientometric studies. Additional information is accessible via STN AnaVist, an analysis tool developed by STN International. However, the competent use of such databases and search systems requires some experience and awareness of the possibilities and pitfalls: e.g. about the coverage of the research disciplines by the various databases, the appropriate search and analyze functions available under the different search systems or the significance and the limitations of citation analysis (bibliometry).

3. Overall Productivity: Nanotubes vs Fullerenes and Graphene

The WoS offers two search modes: The General Search and the Cited Reference Search (the latter is not relevant for this study). The General Search mode reveals publications which appeared in WoS source journals (in particular articles, reviews, and meeting abstracts - no books, no popular scientific publications, no conference proceedings unless they appear in source journals or in the CPCI-S). The number of articles published in the WoS source journals has become a standard measure for scientific productivity (output in terms of the number of publications). The number of publications per year can easily be plotted as a function of the publication years using the WoS analyze function.

At the date of search (01-07-2009) the SCI including the CPCI-S revealed altogether 57128 publications related to nanotubes. The terms "nanotub*" or "nano tub*" (* = wildcard allowing to include the plural or synonyms like tubulus) were searched in the title and the abstract search fields. The search has not been restricted to WoS specific document types. Additionally searching the relevant abbreviations in common use to distinguish single walled, double walled, and multi walled nanotubes (SWCNT, SWNT, DWCNT, DWNT, MWCNT, MWNT) increased the total number of papers only marginally (57208). Due to the fact that the abbreviations hardly appear without additionally mentioning the full term and the potential ambiguity, the abbreviations were not taken into consideration here. Figure 1 shows the time curve of the articles of the entire nanotubes research field. The time evolution of the related fullerene and graphene literature are shown for comparison. The total number of articles covered by the SCI is included as a measure for the growth of the overall scientific literature.

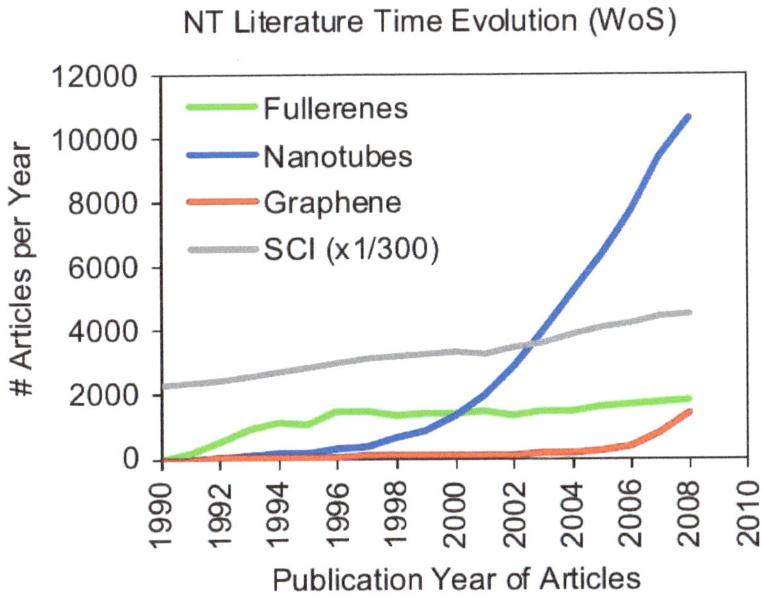

Fig. 1. Time dependent number of articles dealing with fullerenes, nanotubes, and graphene. The total number of articles covered by the SCI is shown as a rough measure for the growth of scientific literature. Source: SCI and CPCI-S under WoS.

According to Figure 1, the productivity (total number of articles per year) of the research activities dealing with nanotubes steadily increased, reaching about 11000 papers published in the year 2008 (compared to "only" 2000 fullerene papers published in the same year). The output increased by a factor of ten since 2000, which is far above the growth of the overall scientific literature in the same time period (about a factor of 1.25 with respect to the literature appearing in the source journals covered by the SCI).

In contrast to the nanotubes productivity, the time evolution of the fullerene literature shows a distinct saturation since about five years after discovery (Braun, 1992). Obviously, the nanotubes are one of the hottest research topics within the last decades with an undamped evolution. Since 2000 they have supplanted the (firstly) more promising fullerenes. The graphene related articles (meanwhile about 4500) show a rather similar increase, obviously starting a follow-up boom beside the ongoing fullerenes and nanotubes research activities.

Google and Google Scholar have become powerful search engines for web resources. Searching the world wide web for "nanotube(s)" with Google results in the large number of 3.1 million entries, while Google Scholar results in 0.27 million hits. The Google searches were carried out without any limitations concerning format, language, or time.

4. Productivity: Authors and Research Organizations

The almost 60,000 articles dealing with nanotubes and selected under the WoS were analyzed using the WoS analyze function. The most productive authors and research organizations, the countries of authors, and the leading journals were determined and are given in the Tables 1-4 below (date of search: 01-07-2009).

Table 1. Top authors with at least 200 nanotubes articles based on the SCI and the CPCI-S under WoS.

Rank	Author	Country	# Articles
1	Iijima, S	Japan	333
2	Bando, Y	Japan	307
3	Ajayan, PM	USA	294
4	Dresselhaus, MS	USA	291
5	Chen, Y	PR China	288
6	Roth, S	Germany	269

7	Lee, YH	South Korea	262
8	Chen, J	PR China	259
9	Wang, J	USA	258
10	Wang, Y	PR China	258
11	Zhang, Y	PR China	258
12	Zhang, J	PR China	248
13	Kataura, H	Japan	246
14	Goldberg, D	Japan	239
15	Li, Y	PR China	238
16	Terrones M	England	232
17	Wang, X	PR China	230
18	Liu, J	USA	225
19	LI, J	PR China	212
20	Smalley, RE	USA	209
21	Liu, Y	PR China	208
22	Zhang, L	PR China	208

Among the authors is a clear dominance of researchers from the East Asian countries and the US. Only a single author from a European country (S. Roth, Germany) is found in the group of the top ten authors in Table 1. Please note: Except for reprint authors, the SCI author addresses (and countries) are not allocated to the corresponding author names. In addition, some authors changed their affiliation. Asian names with only one forename initial comprise namesakes. Hence, the countries of authors given in Table 1 are not fully definite.

The top research organizations with respect to the number of articles dealing with nanotubes are shown in Table 2. Among the top positions are many Chinese research organizations and universities. The "weight" of the Chinese nanotubes research is confirmed further with Peoples Republic of China on rank two of the top countries of authors given in Table 3 further below.

Table 2. Top research organizations with at least 400 nanotubes articles based on the SCI and the CPCI-S under WoS.

Rank	Research Organization	# Articles	% Articles
1	Chinese Acad Sci	2840	5.0
2	Tsing Hua Univ	903	1.6
3	Russian Acad Sci	881	1.5
4	Peking Univ	684	1.2
5	Tohoku Univ	630	1.1
6	Rice Univ	628	1.1
7	Univ Sci & Technol China	616	1.1
8	Univ Cambridge	609	1.1
9	MIT	607	1.1
10	Univ Tokyo	607	1.1
11	Osaka Univ	573	1.0
12	Nanjing Univ	556	1.0
13	Zhejiang Univ	551	1.0
14	Natl Univ Singapore	514	0.9
15	NASA	513	0.9
16	Univ Illinois	497	0.9
17	CNRS	496	0.9
18	Seoul Natl Univ	479	0.8
19	Univ Calif Berkeley	476	0.8

20	Penn State Univ	460	0.8
21	Natl Inst Mat Sci	453	0.8
22	Natl Inst Adv Ind Sci & Technol	449	0.8
23	Natl Tsing Hua Univ	445	0.8
24	Sungkyunkwan Univ	427	0.7
25	Rensselaer Polytech Inst	414	0.7
26	Nanyang Technol Univ	406	0.7
27	Georgia Inst Technol	401	0.7

Please note: (1) In contrast to author names and journal titles, author addresses are not fully standardized in literature databases. Hence, the data offer only a rough picture of the leading research organizations and do not provide an exact ranking. (2) Many publications have been assigned to more than one country of author resulting in a substantial overlap.

5. Productivity: Countries and Continents

The ranking of the countries of authors having published articles dealing with nanotubes is given in Table 3.

Table 3. Top countries of authors with at least 500 nanotubes articles based on the SCI and the CPCI-S under WoS.

	Country of Author	# Articles	% Articles
1	USA	15845	27.7
2	Peoples R China	13386	23.4
3	Japan	6683	11.7
4	South Korea	3660	6.4
5	Germany	3423	6.0
6	France	2469	4.3

7	England	2391	4.2
8	Taiwan	1743	3.1
9	Russia	1691	3.0
10	Italy	1394	2.4
11	India	1335	2.3
12	Spain	1177	2.1
13	Canada	1094	1.0
14	Australia	994	1.7
15	Singapore	965	1.7
16	Switzerland	756	1.3
17	Belgium	694	1.2
18	Brazil	662	1.2
19	Poland	599	1.0
20	Israel	589	1.0
21	Sweden	523	0.9
22	Mexico	508	0.9

The countries of authors ranking given in Table 3 is based on articles dealing with any type of nanotubes. A ranking based on articles dealing only with multi-walled nanotubes reveals that the top position is taken up by the Peoples Republic of China. A broader view is given in Figure 2, showing the share of the continents of the authors publishing nanotubes related articles being covered by the WoS source journals.

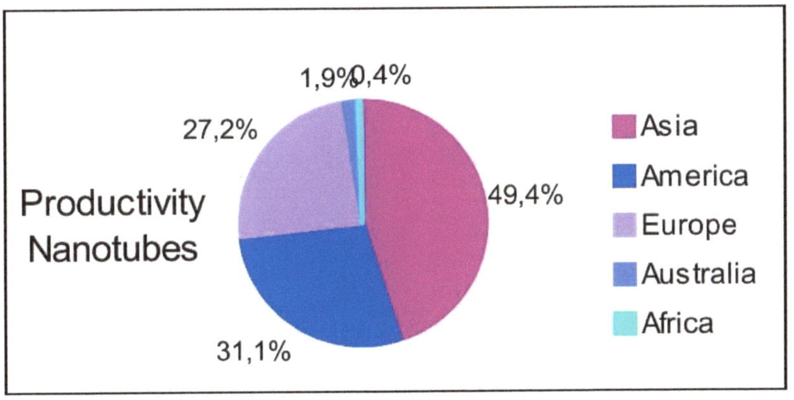

Fig. 2. Share of the continents of the authors of nanotubes related articles (with Europe incl. Russia and Australia incl. New Zealand).

6. Research Output: Journal Articles and Patents

The scientific literature dealing with nanotubes has been published in about 500 different WoS source journals. Almost half of this literature appeared in only 25 key-journals (according to Bradford's law of scattering). The leading journals each with their number of nanotubes articles are given in Table 4. Among these journals about one fourth are journals around nanotechnology or material science. For scientists, such information could be valuable with respect to the selection of the appropriate journals for publishing their own results.

Table 4. Top WoS source journals with more than 500 nanotubes articles based on the SCI and the CPCI-S under WoS.

Rank	SCI Source Journal	# Articles	% Articles
1	Physical Review B	2428	4.3
2	Applied Physics Letters	2242	3.9
3	Nanotechnology	1671	2.9
4	Carbon	1652	2.9
5	Chemical Physics Letters	1295	2.3
6	Nano Letters	1286	2.3

7	Journal of Physical Chemistry B	1164	2.0
8	Journal of Physical Chemistry C	1106	1.9
9	Journal of the American Chemical Society	919	1.6
10	AIP Conference Proceedings	915	1.6
11	Physical Review Letters	880	1.5
12	Journal of Nanoscience and Nanotechnology	865	1.5
13	Advanced Materials	825	1.4
14	Abstracts of Papers of the American Chemical Society	801	1.4
15	Journal of Applied Physics	779	1.4
15	Proceedings of the Society of Photo-Optical Instrumentation Engineers (SPIE)	717	1.3
17	Chemistry of Materials	618	1.1
18	Diamond and Related Materials	610	1.1
19	Langmuir	536	0.9

A search in the CAS literature file CAplus (based on the corresponding WoS search query) revealed 53772 publications and 13184 patents. The lower number of publications in CAplus compared to the multidisciplinary WoS (57128) results from the narrower field coverage of CAS mainly focusing on chemistry. The publications searched in CAplus have been taken for establishing a nanotubes research landscape (see Figure 4 further below) - this is not possible with the WoS records. At first we had a closer look at the patents and show in Figure 3 the time evolution of the articles and patents as covered by CAS. The patents have been further analyzed with respect to technical applications in broad areas of activity. The country specific number of patents are presented in Table 5 as a heat map.

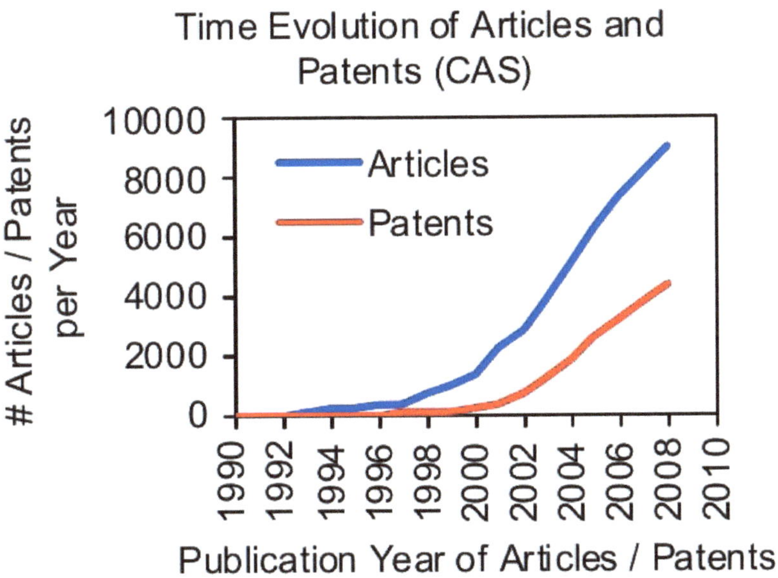

Fig. 3. Time dependent number of articles and patents dealing with nanotubes and covered by the CAS literature file. Source: CAplus under STN International.

Table 5. Number of patents with respect to nanotubes published by the top 20 countries (country of author) in the order of ranking with respect to the research areas.

Japan	66	368	722	52	919	704	1206	4037
USA	251	552	825	90	627	388	865	3598
S. Korea	19	196	342	44	588	276	385	1850
P. R. China	26	216	235	24	410	202	564	1677
Taiwan	3	66	82	7	182	72	125	537
Germany	17	72	112	20	62	37	107	427
France	12	23	46	4	44	46	67	242
United Kingdom	6	15	28	2	21	11	30	113
Canada	2	12	9	0	13	12	22	70

Switzerland	2	6	14	0	11	13	14	60
Netherlands	1	8	15	3	12	3	10	52
Singapore	2	8	6	0	11	8	16	51
Russia	3	6	8	0	9	5	16	47
Austria	1	7	5	0	10	3	17	43
Italy	0	5	12	4	4	2	11	38
Australia	7	8	5	1	5	4	8	38
Israel	1	5	8	0	10	5	6	35
Sweden	1	3	12	0	6	3	9	34
Belgium	0	7	6	0	7	1	10	31
Brazil	2	2	4	1	4	8	4	25
	Health Care	Sensors	Electronics & Logic	Mass Data Storage	Displays & Field Emiss.	Energy Storage	Materials & Reactions	Total Sum

The ranking of nations can be depicted from the last column in Table 5 which shows the total number of patents for each country. The data show a clear dominance of East Asian countries with Japan (1), S. Korea (3), P. R. China (4), and Taiwan (5). USA is second in the ranking while most European countries show a lower rank: Germany (6), France (7), and United Kingdom (8). The same tendency can be seen with respect to the research areas analyzed. In Health Care, Sensors, and Electronics & Logic, however, the ranking between Japan and the United States is flipped. The highest number of patents has been published in the area of Materials & Reactions.

7. Content Analysis: Research Landscape

STN International has recently launched a new interactive analysis tool called STN AnaVist (STN AnaVist, 2009) which focuses mainly on patent analysis (Fischer & Lalyre, 2006). This tool can also be applied to journal articles exploring basic research topics. STN AnaVist has been used to refine the nanotubes literature analysis. The selection of the articles to be analyzed has to be done in the CAS literature file. Currently, the STN AnaVist system limits do not allow to process more than 20,000 articles. Therefore, the literature analysis had to be restricted to the time period 2008-2009.

One of the functions offered by STN AnaVist implies the creation of so-called research landscapes: "Significant keywords and concepts are derived from document titles and abstracts. These keywords are used to determine the similarity between documents. An algorithm uses document similarity scores to position each document relative to one another in a two-dimensional space, with each document positioned at one point. This process is repeated until all documents have been clustered and each assigned to a single x, y coordinate pair. A graphical map is generated. The z coordinates, determining the height of each 'peak', are calculated based on the density of the documents in an area." (STN International, 2009). The research landscape of the past-2007 articles dealing with nanotubes is shown in Figure 4.

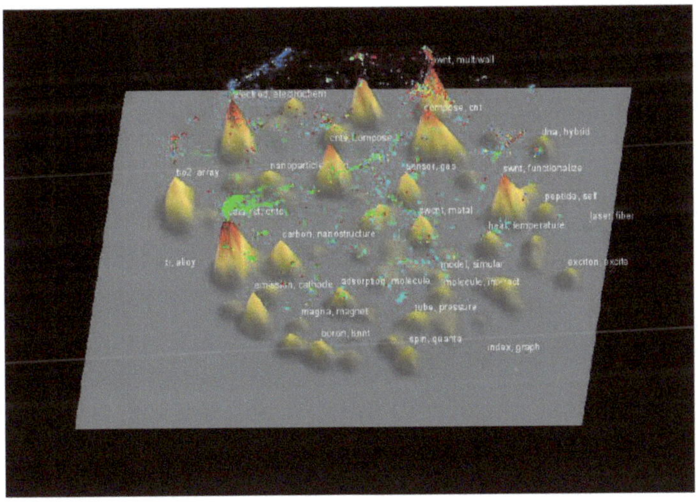

Fig. 4. Research landscape of the nanotubes literature published in the time period 2008-2009 established using STN AnaVist (cnts = carbon nanotubes, swnt = single-walled nanotubes, mwnt = multi-walled nanotubes).

The nanotubes landscape in Figure 4 shows two different layers one upon the other: (1) the mountains with the related keyword pairs based on the text analysis as mentioned and (2) the colored patches above based on the CAS classification categories (technical indicators). The keyword analysis reveals major studies on composites, catalysis, and electrochemistry. The mountains are characterized by the two top clustering words, e. g. "catalyst, cnts" at the left of the landscape or "compose, cnt" top right. The colored dots correspond to documents with different technical indicators (green: vapor deposition process, red: nanocomposites, blue: biosensors, cyan: electric conductivity, and yellow: overlaps). Please note: STN AnaVist is an interactive tool that allows to focus and to zoom according to the specific needs. The research landscape, for example, can be twisted for better analysis.

8. Most Highly-Cited Papers

The number of citations is often taken as a measure of the attention an article, a researcher, an institute or even a country has attracted. Although citation numbers are no ultimate scale of the final importance and quality of articles, they reflect strengths and shortcomings of research activities and are therefore frequently used for research evaluation. Being cited means that a given article appears as a reference in the article of another author for additional reading. The number of citations of a specific paper is thus a rough measure of the importance or usefulness of the paper within the scientific community. The top ten most highly-cited nanotubes articles until present are given in Table 6. Note that such lists actually imply no real ranking because the various papers accumulated their citations over different time periods.

Table 6. The top ten most highly-cited nanotubes papers based on the SCI and the CPCI-S under WoS (date of search: 01.07.09).

Author(s)	Title	Journal	# Citations
Iijima S	Helical microtubules of graphitic carbon	Nature 354, 56-58 (1991)*	8545
Thess A, Lee R, Nikolaev P, et al.	Crystalline ropes of metallic carbon nanotubes	Science 273, 483-487 (1996)	2872
Pan ZW, Dai ZR, Wang ZL	Nanobelts of semiconducting oxides	Science 291, 19471949 (2001)	2680

Xia YN, Yang PD, Sun YG, et al.	One-dimensional nanostructures: synthesis, characterization, and applications	Advanced Materials 15, 353-389 (2003)	2621
Tans SJ, Verschueren ARM, Dekker C	Room-temperature transistor based on a single carbon nanotube	Nature 393, 49-52 (1998)	2539
Baughman RH, Zakhidov AA, de Heer WA	Carbon nanotubes - the route toward applications	Science 297, 787-792 (2002)	2299
Morales AM, Lieber CM	A laser ablation method for the synthesis of crystalline semiconductor nanowires	Science 279, 208-211 (1998)	2137
Kong J, Franklin NR, Zhou CW, et al.	Nanotube molecular wires as chemical sensors	Science 287, 622-625 (2000)	2103
Iijima S, Ichihashi T	Single-shell carbon nanotubes of 1 – nm diameter	Nature 363, 603-605 (1993)	1975
Kitagawa S, Kitaura R, Noro S	Functional porous coordination polymers	Angewandte Chemie (Int. Edit.) 43, 2334-2375 (2004)	1907

Please note: The lijima paper on rank 1 illustrates the problems when selecting literature by using search terms: Primarily, this paper has not been included in the answer set, because the term "nanotube" does not appear within the title or the abstract of this early paper.

The graph displaying the time evolution of the citations of a single article is called its citation history. Each article develops its own life span as it is being cited. With time, the citations per year (citation rate) normally evolve following a similar pattern: The citations generally do not increase substantially until one year after the publication. They reach a summit after about three years, the peak position depending somewhat on the research discipline. Subsequently, as the articles are displaced by newer ones, their impact decreases, accumulating citations at a lower level. Finally, most of the articles are barely cited or forgotten.

Normally, articles receive only a few citations in the first years after the publication. Those which are decisive for research usually garner hundreds or even thousands of citations and keep being cited for a longer time, often for decades. The five most highly-cited nanotubes articles given in Table 6 are no exceptions. Figure 5 shows the citation history of these highimpact nanotubes papers.

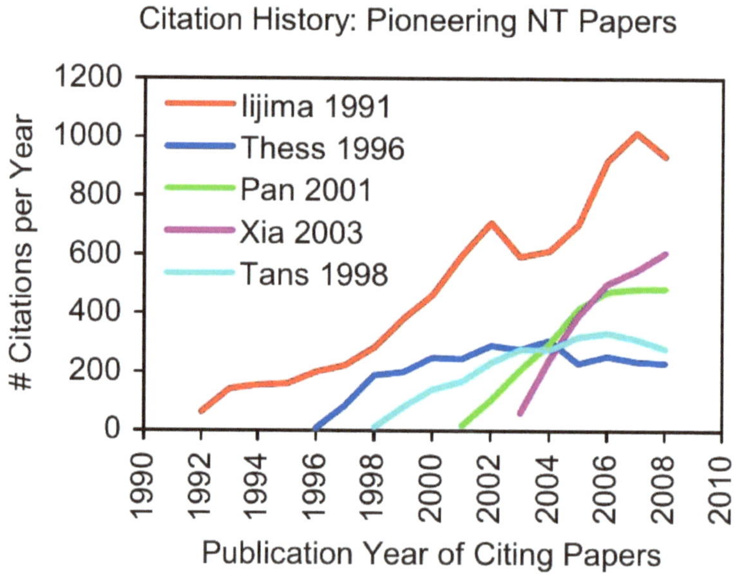

Fig. 5. Time dependent number of citations (citation history) of the five most highly-cited nanotubes papers (see Table 6).

Recently, an article published in Carbon, one of the leading nanotubes journals (see Table 4) discussed the nanotubes history and raised the question, to whom the credit for the discovery should be given (Monthioux & Kuznetsov, 2006). The first evidence for nanosized carbon tubes is believed to have been published 1952 in the Russian Journal of Physical Chemistry by Radushkevich and Lukyanovich (Radushkevich & Lukyanovich, 1952). The article contains clear images of carbon tubes with about 50 nanometers in diameter based on Transition Electron Microscopy (TEM). However, due to the Russian language and the limited access of Western scientists to Russian publications during the time of the Cold War the discovery was largely unnoticed. Another important article prior to 1991 was published 1976 by Oberlin et al. (Oberlin et al., 1976).

The citation history of these two articles is shown in Figure 6. The Radushkevich paper received altogether 73 citations (many from Russian scientists) and was barely cited until present (19 citations in 2008). The Oberlin article received altogether 600 citations (72 citations prior to 1991) showing a remarkable impact increase (75 citations in 2008). These two articles seem to belong to the few papers in science being nearly ignored for decades and then are "rediscovered" (so-called sleeping beauties).

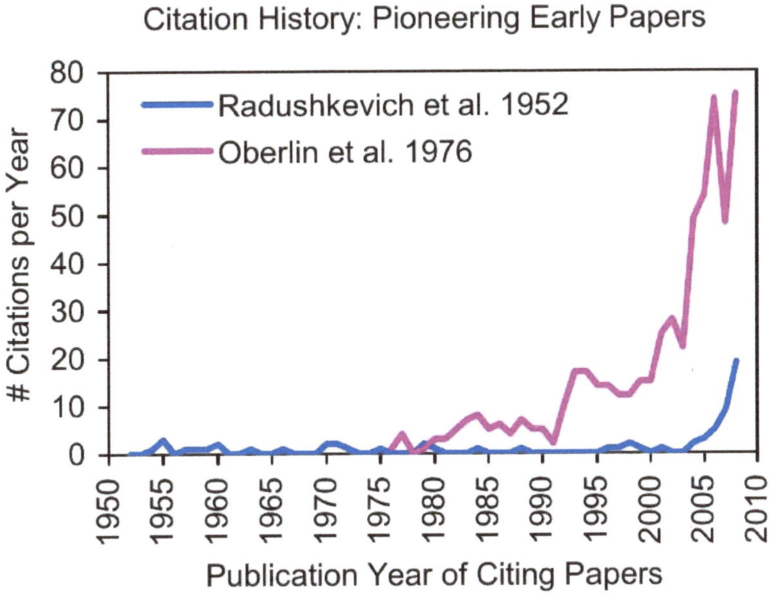

Fig. 6. Time dependent number of citations (citation history) of the Radushkevich & Lukyanovich and the Oberlin paper.

The "weight" of research fields can be indicated by their Hirsch index (h-index, h-number). The h -index is defined as the number of articles in WoS source journals that have had h citations or more. For example, a researcher with an h-index of 40 will have published 40 articles that have received at least 40 citations each. The h-index can be seen as a rough indicator for measuring the output and impact of a researcher in a single number (Hirsch, 2005). This index reflects a researcher's contribution based on a broad body of publications rather than on a few high-impact papers. The h-index may well be extended to journals (Bornmann et al., 2009) as well as to research fields (Banks, 2006). The h-index for the overall literature related to nanotubes, fullerenes, graphene, and further research fields is given in Table 7. The data can be taken as a rough measure of the relative "weight" of these fields in terms of both publications and citations.

Table 7. Number of articles related to nanotubes, fullerenes, graphene, and further research fields and their corresponding h-index (date of search: 01-07-2009).

Rank	Research Field	# Articles	h-index
1	Nanotubes	57128	305
2	Fullerenes	24552	204
3	Graphene	4639	112
4	Quantum Dots	38457	245
5	Nanowires	27939	201
6	Amorphous Silicon	18060	145
7	Porous Silicon	8824	134

9. Citation Graphs

Using the HistCite® software of Eugene Garfield, we established the citation graph shown in Figure 7, which is based on the most highly-cited nanotubes articles accessible in WoS at the date of search (22.09.09). This graph visualizes the citation network within the high impact nanotubes literature published so far: The nodes represent the papers with at least 1000 citations (31 papers). Lower citation limits increase the number of nodes considerably without changing the node pattern substantially (see Figure 8 further below). The specific papers represented by the circles of the citation graph of Figure 7 are given in Table 8 in short form only. Like STN AnaVist®, also HistCite® is used in an interactive mode and can deliver much more information than given in the limited citation graphs shown here.

The citation graph of Figure 7 shows the lijima 1991-paper (see Table 6) as the starting point of the nanotubes research field activities at the top. The graph reveals a pronounced cluster of high-impact publications based on nanotubes within the time period 1995-2003. The more recently published articles are increasingly rare because of the time delay of citations. These publications had so far not sufficient time for accumulating enough citations to reach the high limit of 1000 citations specified in Figure 7.

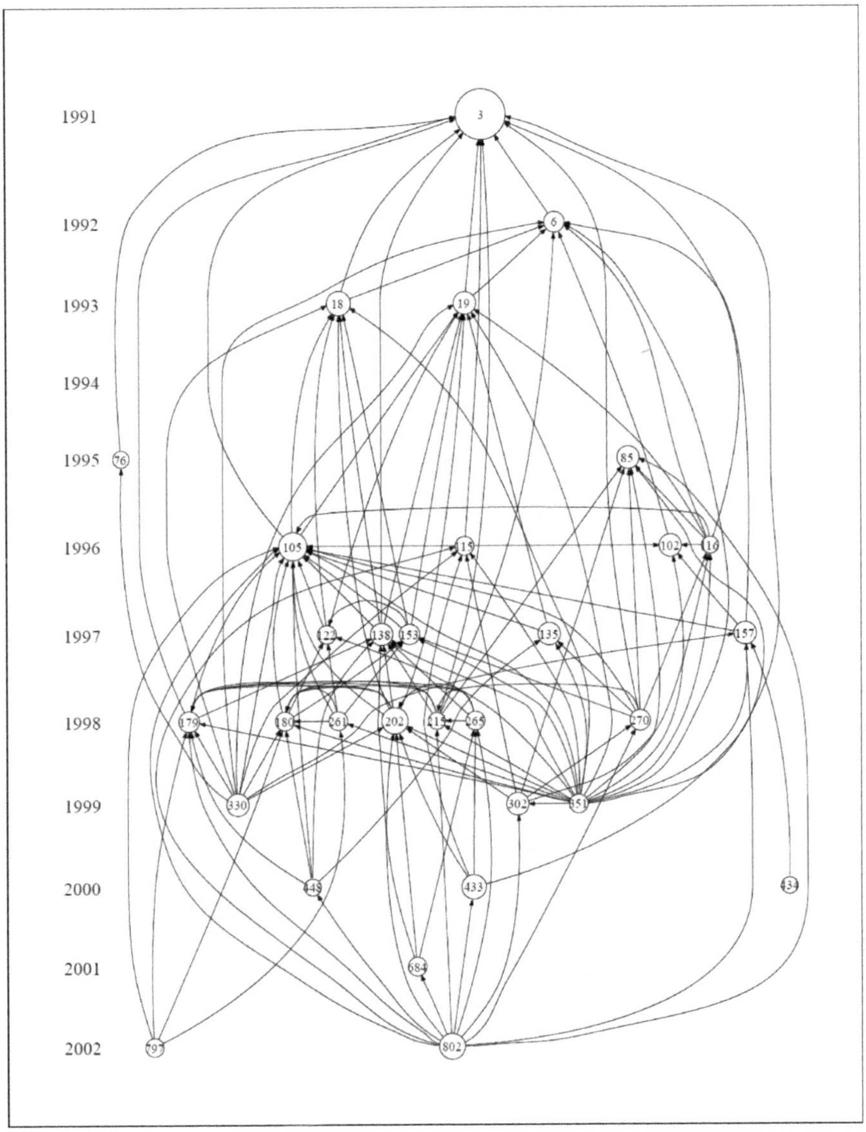

Fig. 7. Citation graph based on the most highly-cited nanotubes articles searched in the WoS. The nodes represent the currently 31 nanotubes papers with at least 1000 citations until the date of search. The circle diameter is proportional to the overall number of citations and the arrows indicate the citation direction. The numbers within the circles originate from the consecutive numbering of the nanotubes papers by the software (see Table 8) and do not represent their citation counts. Source: HistCite® and SCI under WoS.

Table 8. List of the specific papers of the HistCite® based citation graph of Figure 7 represented by the numbered circles. LCS = local citation score: citations collected within the ensemble of the selected nanotubes articles, GCS = global citation score: overall number of citations as given by the WoS under "times cited".

#	No.	Papers	LCS	GCS
1	3	IIJIMA S, 1991, NATURE, V354, P56	377	8787
2	6	EBBESEN TW, 1992, NATURE, V358, P220	150	1628
3	18	IIJIMA S, 1993, NATURE, V363, P603	161	2005
4	19	BETHUNE DS, 1993, NATURE, V363, P605	125	1686
5	76	CHOPRA NG, 1995, SCIENCE, V269, P966	48	1133
6	85	DEHEER WA, 1995, SCIENCE, V270, P1179	105	1831
7	102	Treacy MMJ, 1996, NATURE, V381, P678	125	1779
8	105	Thess A, 1996, SCIENCE, V273, P483	264	2917
9	115	Dai HJ, 1996, NATURE, V384, P147	68	1245
10	116	Li WZ, 1996, SCIENCE, V274, P1701	75	1066
11	122	Rao AM, 1997, SCIENCE, V275, P187	96	1259
12	135	Dillon AC, 1997, NATURE, V386, P377	75	1854
13	138	Tans SJ, 1997, NATURE, V386, P474	131	1708
14	153	Journet C, 1997, NATURE, V388, P756	111	1519
15	157	Wong EW, 1997, SCIENCE, V277, P1971	102	1756
16	179	Wildoer JWG, 1998, NATURE, V391, P59	141	1497
17	180	Odom TW, 1998, NATURE, V391, P62	121	1194
18	202	Tans SJ, 1998, NATURE, V393, P49	143	2606
19	215	Frank S, 1998, SCIENCE, V280, P1744	57	1143
20	261	Chen J, 1998, SCIENCE, V282, P95	120	1280

21	265	Martel R, 1998, APPL PHYS LETT, V73, P2447	68	1322
22	270	Ren ZF, 1998, SCIENCE, V282, P1105	78	1496
23	302	Fan SS, 1999, SCIENCE, V283, P512	75	1708
24	330	Hu JT, 1999, ACCOUNT CHEM RES, V32, P435	11	1682
25	351	Ajayan PM, 1999, CHEM REV, V99, P1787	45	1153
26	433	Kong J, 2000, SCIENCE, V287, P622	89	2181
27	434	Yu MF, 2000, SCIENCE, V287, P637	35	1036
28	448	Collins PG, 2000, SCIENCE, V287, P1801	63	1060
29	684	Bachtold A, 2001, SCIENCE, V294, P1317	36	1232
30	797	O'Connell MJ, 2002, SCIENCE, V297, P593	60	1243
31	802	Baughman RH, 2002, SCIENCE, V297, P787	48	2432

The citation graph of Figure 8 shows more clearly the pattern of the citation network based on the highly-cited papers, in particular within the starting period after 1991. Again, the past-2002 papers are increasingly rare because of the time delay of the citations.

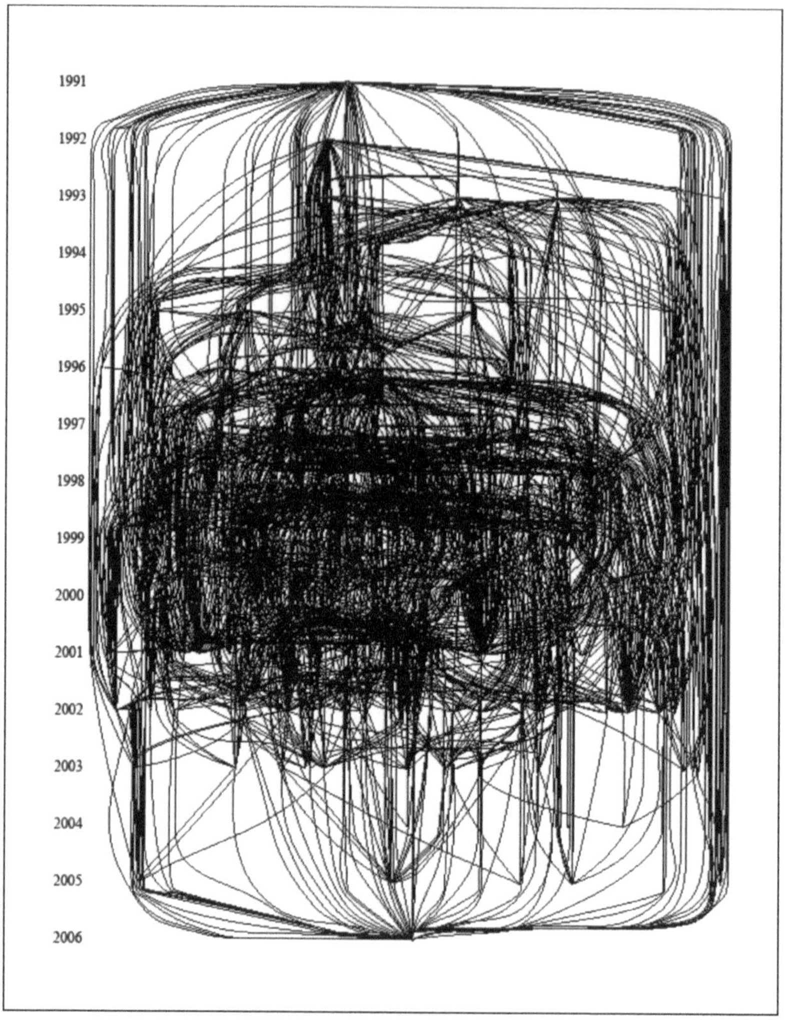

Fig. 8. Citation graph based on the highly-cited nanotubes articles searched in the WoS. The nodes represent the currently 226 nanotubes papers with at least 300 citations until the date of search (rather than only the 31 papers with at least 1000 citations as in Figure 7). Source: HistCite® and SCI under WoS.

10. Summary

A currently most popular research discipline has been reviewed by scientometric methods. The time dependent overall number of nanotubes articles is given, revealing a still strongly increasing research activity in this field. The nanotubes articles were analyzed with respect to the most productive authors, research organizations, countries of authors, and leading journals. Among the authors is a clear dominance of researchers from the East Asian countries and the US. Among the top positions are many Chinese research organizations and universities. The time evolution of the nanotubes patents shows a similar increase as the evolution of the articles. The number of patents of the various organizations related to broad areas of research and technology were determined and visualized as a heat map. The research landscape of the nanotubes research field was established using the new analysis tool STN AnaVist. The STN AnaVist based landscape shows a pronounced clustering with respect to specific keywords and technical indicators. The time evolution of the most highlycited nanotubes papers shows an unusually long-lasting impact. Two early papers may be characterized as "sleeping beauties". Finally, a citation graph has been established revealing the historiography of the nanotubes research with regard to the high impact papers.

11. References

Banks, M.G. (2006). An extension of the Hirsch-index: indexing scientific topics and compounds. Scientometrics 69, 161-168.

Barth, A. & Marx, W. (2008). Mapping High-Temperature Superconductors - A Scientometric Approach. Journal of Superconductivity and Novel Magnetism 21, 113128. (PDF), URL : http://arxiv.org/abs/cond-mat/0609114

Bornmann, L.; Marx, W. & Schier, H. (2009). Hirsch-type index values for organic chemistry journals: A comparison of new metrics with the Journal Impact Factor. European Journal of Organic Chemistry 10, 1471-1476.

Braun, T. (1992). The epidemic spread of fullerene research. Angew. Chem. Int. Edit. 31, 588589.

Fischer, G. & Lalyre, N. (2006). Analysis and visualisation with host-based software - The features of STN®AnaVist ™, World Patent Information.

Hirsch, J.E. (2005). An index to quantify an individual's scientific research output. Proceedings of the National Academy of Sciences of the USA (PNAS) 102, 16569-16572.

Monthioux, M. & Kuznetsov V. L. (2006). Who should be given the credit for the discovery of carbon nanotubes? Carbon 44, 1621-1623.

Oberlin, A.; Endo, M. & Koyama, T. (1976). Filamentous growth of carbon through benzene decomposition. Journal of Crystal Growth 32, 335-349.

Radushkevich, L.V. & Lukyanovich, V.M. (1952). Russian Journal of Physical Chemistry 26, 8895.

STN International (2009), URL: www.stn-international.de/

Thomson Reuters (2009), URL: www.isinet.com/isi/

STN AnaVist (2009).

Chapter-2

Vapour Deposition of CNTs Using Structural Nanoparticle Catalysts

1. Introduction

The extraordinary mechanical, electrical and optical properties of CNTs have stimulated extensive research since their discovery by Sumio Iijima of the NEC Corporation in the early 1990s (Iijima, 1991). Possible applications for carbon nanotubes range from nanoelectronics, quantum wire interconnects, sensors and field emitters to composites (Meyyappan et al., 2005). These applications require reliable synthesis techniques capable of generating large quantities of high purity material. In addition, applications in nanoelectronics and photonics may require controlled growth at precise lithographically patterned areas. Understanding how to control the synthesis of SWNTs is vital in order to deterministically integrate such nanostructures into various technologies. Of the various methodologies developed for carbon nanotube (CNT) synthesis, those based upon catalyst-assisted chemical vapour deposition (CCVD) appear to be best suited to satisfy these requirements. Traditionally, 3*d* valence transition metal nanoparticles, such as Fe, Co and Ni, are used to catalyse CNT growth. These catalysts are thought to work in accordance with the model for carbon filament growth, derived by Baker et al. (Baker, 1989; Baker et al., 1972; 1973), from concepts of vapour-solid-liquid theory (Wagner & Ellis, 1964). In this model, hydrocarbons adsorbed on the metal nanoparticle are catalytically decomposed resulting in atomic carbon dissolving into the liquid catalyst particle, and when a supersaturated state is reached, carbon precipitates in a tubular, crystalline form. Recently, several groups have reported successful growth of CNTs from noble metal (Lee et al., 2005; Takagi et al., 2006; Yoshihara et al., 2008; Yuan et al., 2008; Zhou et al., 2006), ceramic (Liu et al., 2008b; Steiner et al., 2009) and semiconducting nanoparticles (Takagi et al., 2007; Uchino et al., 2009; 2008; 2005b), all of which are regarded as unable to catalyse the dissociation of hydrocarbons. In addition to this, in their bulk form, these materials do not have a catalytic function to produce graphite. This implies that given enough energy, carbon atoms on a nanoparticle are capable of a structural reorganisation into CNTs. This leads to a new in-terpretation of the role of the catalyst in nanotube growth in which only a nanoscale curvature and carbon adsorption sites are necessary This work examines the recent developments in non-traditional CCVD of CNTs with a view to determine the essential role of the catalyst in nanotube growth. Section 2 provides a brief overview of the techniques reliant on the structural reorganization of carbon to form CNTs. An in-depth analysis of CNT synthesis based upon ceramic (Section 3), noble metal (Section 4) and semiconducting nanoparticle catalysts (Section 5) is presented. Various approaches to germanium catalyst preparation are compared in terms of growth density and quality of synthesized nanotubes. Scanning electron microscopy measurements indicate that a technologically relevant density is achievable using non conventional catalysts. Raman measurements have identified the synthesized nanotubes as single walled and, in terms of graphitization and structure, of a high quality. Extensive atomic force microscopy characterisation of the catalyst has been undertaken in order to ascertain the influence of morphology on the ability of the catalyst to yield CNT growth. A model for CNT growth consistent with the experimental results is proposed in Section 6. Finally, a summary of challenges and future directions for investigations is presented in Section 7.

2. Catalyst Free Synthesis of CNTs

Silicon carbide has been the most widely used non metallic catalyst. The synthesis techniques involving this catalyst have produced high densities of carbon nanostructures by annealing either SiC particles (Botti et al., 2004; Takikawa et al., 1998), amorphous SiC films (Botti et al., 2001; Kusonoki et al., 1997) or hexagonal SiC ($6H-SiC$) (Derycke et al., 2002; Kusonoki et al., 1997) in a vacuum. In these methodologies, the nanotube formation can be explained by the mechanism proposed by Kusonoki et al. (1997). Owing to the low vacuum in the chamber when annealing, the SiC oxidises forming SiO_2. As a consequence, the carbon atoms are free to bond with other atoms. If they bond to neighbouring carbon atoms, graphite fragments are formed containing dangling bonds. Thermodynamics drives the folding of graphitic fragments so that the dangling bonds of opposite edges are saturated. The as-formed nanotube segments act as seeds for the attachment of new carbon atoms, leading to CNT growth. However, these techniques require high temperature annealing at approximately 1650°C.

An alternate approach involves using carbon nanoparticles as a catalyst. This technique depends on the structural reorganization of carbon aggregates into nanotubes upon annealing. Botti et al. (2002) report a dense array of CNTs grown on silicon by spraying amorphous hydrogenated carbon nanoparticles on a Si substrate. Other similar approaches have been reported in the literature (Koshio et al., 2002; Larciprete et al., 2002; 2003).

Figure 1 shows transmission electron microscope images of carbon nanostructures deposited on a carbon implanted Ge nanoparticle sample. By inspection of Figure 1(a) it can be determined that this is made up of a collection of small multi-walled carbon nanotubes (MWNTs). As there were no catalyst particles detected at either ends of the nanotubes, it is therefore believed that these nanotubes were nucleated by the structural reorganization of amorphous carbon deposits without the influence of the Ge nanoparticles. This hypothesis is further reinforced by Figure 1(b), which shows a double walled nanotube (DWNT) in the early stages of growth. Again, there is no catalyst detected, and it appears that the nanotube was nucleated from the nearby carbon deposits. Both nanostructures demonstrate a good degree of graphitization and furthermore there are few impurities present, such as amorphous carbon, in the materials.

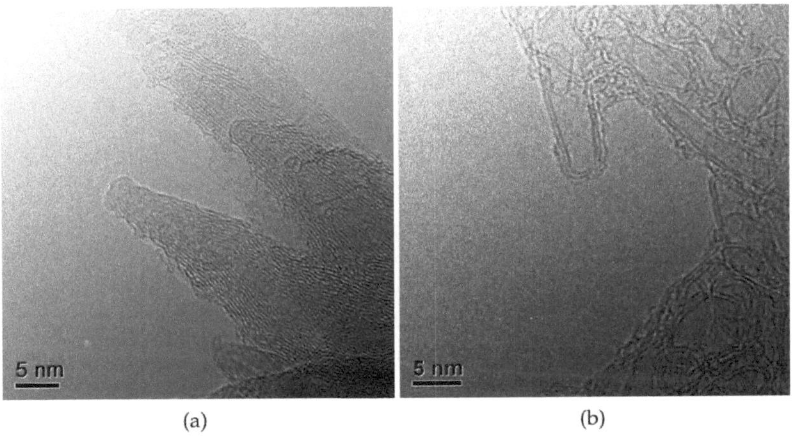

(a) (b)

Fig. 1. Transmission electron micrographs of self-assembled carbon nanostructures deposited on a carbon implanted Ge nanoparticle sample. The TEM sample was prepared by scraping the substrate surface with a surgical blade and collecting the material on a holey-carbon TEM grid. (a) Image of MWNTs and (b) a DWNT in the early stages of growth formed by the structural reorganization of carbon.

These techniques, although not strictly classed as chemical vapour deposition of CNTs, provide some insight into the behaviour of carbon aggregates at elevated temperatures without the influence of an external catalyst with a function to produce graphite. These results also demonstrate that regardless of the catalyst, the formation of CNTs involves two important processes: (i) the diffusion of carbon and (ii) the nucleation of a graphitic cap or fragment followed by the further incorporation of carbon into the growing nanotube. It has been reported that the diffusion process on a nanoparticle surface or across its interior is a rate limiting step (Bartsch et al., 2005; Hofmann et al., 2005), while the chirality of the growing CNT is decided upon the formation of the graphitic cap (Reich et al., 2006; Yazyev & Pasquarello, 2008).

3. Ceramic Nanoparticle Catalysts

Ceramic materials, such as Al_2O_3, have typically been used as a buffer layer to disperse metallic catalyst particles and enhance their catalytic properties in CNT growth (Takagi et al., 2007). However, the simplistic view that the support only plays a catalytically passive role in the formation of carbon nanotubes requires some examination. Rummeli et al. (2007) demonstrated that under typical conditions for CVD growth of CNTs, nanoparticles of difficult-to-reduce metal-oxides are exceedingly good at promoting ordered carbon (graphene) growth. As was expected, there was no observation of ordered carbon formation in bulk/film samples. This difference was attributed to the presence of surface defect sites on the nanoparticle oxides, and it was argued that in the substrate-based CNT synthesis routes, the interface between the catalyst particle and the surface behaves as an annular defect site. These sites would then promote the formation of cylindrical graphene structures, or nanotubes.

A recent study by Liu et al. (2008b) reports the formation of dense CNT layers catalysed by Al_2O_3 nanoparticles. Raman spectra of the synthesized nanotubes indicated that the nanotubes synthesized were predominantly single walled and of a good quality. Interestingly, the authors surmise that the mechanism of formation is different from the traditional vapour-liquid-solid mechanism as the nanoparticles are likely to be in the solid state during growth. This finding reinforces the argument of Rummeli et al. (2007), and additionally indicates that the growth of single-walled carbon nanotubes (SWNTs) on flat Al_2O_3 substrates may be possible by nanostructuring their surfaces. This hypothesis was partially confirmed by Liu et al. (2009) using a nanostructured SiO_2 substrate to grow SWNTs.

Another ceramic catalyst reported in the literature is ZrO_2 (Steiner et al., 2009). In this publication, dense growth of either MWNTs or SWNTs was possible, depending on the carbon feedstock used. ZrO_2 was typically deposited on either Al_2O_3 capped SiO_2 supports or Si substrates with an oxynitride support through a chloride salt solution. Samples were pretreated in H_2 prior to the introduction of the carbon feedstock. ZrO_2 is known to not be reduced by H_2 (Mctaggart, 1961) and additionally, carbothermic reduction of ZrO_2 does not yield Zr metal, but results in the formation of ZrC (Berger et al., 1999). In-situ x-ray photoelectron spectroscopy (XPS) revealed that the state of the catalyst after H_2 pretreatment showed two phases; a stoichiometric and an oxygen deficient phase of zirconia. The role of the H_2 pretreatment in this work, while shown not to result in the formation of Zr metal, is thought to introduce surface defect sites into zirconia nanoparticles that aid in enhancing catalytic ability. It should be noted that, as observed by in-situ XPS, CNT growth seems to begin only after the introduction of both the hydrocarbon and hydrogen. It is speculated that the introduction of hydrogen aids in the transformation of the hydrocarbon into other organic precursors which can then be uptaken and catalysed into CNTs.

4. Noble Metal Nanoparticle Catalysts

Nanosized iron-group metals (Fe, Co, Ni) are known for their ability to catalyse SWNT growth in chemical vapour deposition. It has been generally accepted that these metals and their alloys consistently show the highest catalytic activity (Awasthi et al., 2005; Melechko et al., 2005). This is attributed to the solubility of carbon in the metal-solid solution (Deck & Vecchio, 2006). However, noble metals such as Au, Ag or Cu have both negligible carbon solubility and negligible carbide formation, and have recently been identified as catalysts for the growth of CNTs. Takagi et al. (2006) have found that the yield of SWNTs from noble metals is comparable to that of iron-group metals. Moreover, noble metals, in particular Cu, are thought to favour CVD growth of CNT nanotubes at low temperatures with a narrow chirality distribution (Yazyev & Pasquarello, 2008).

Bulk Au is considered a noble metal, as it is highly unreactive and catalytically inactive. Au is the only metal with an endothermic chemisorbtion requirement, and in addition it has d-states so low in energy that the interaction with oxygen $2p$-states is net repulsive. Nevertheless, in its nanoparticle form, Au is capable of catalysing a wide variety of reactions. These include the oxidation of CO (Hvolbaek et al., 2007), the selective hydrogenation of acetylene (Jia et al., 2000), hydrogenation of halogen compounds, reduction of nitrogen oxides and photocatalytic hydrogen production (Haruta, 1997). The origin of this effect is believed to be the increase in the fraction of low-coordinated Au atoms as the size of the Au cluster is reduced. In some cases, the catalytic nature of supported Au clusters can be explained by assuming the Ausupport perimeter interface acts as a site for activating at least one of the reactants.

The first demonstration of CNT growth from Au nanoparticle catalysts by Lee et al. (2005), involved the decomposition of acetylene over nanoparticles supported on $SiO_2 - Al_2O_3$. This support showed a good propensity for the decomposition of acetylene and demonstrated strong interactions between the Au nanoparticles and its surface. The synthesized products were predominantly MWNTs, with average diameters of ≤ 20 nm. The first reports of the formation of SWNTs from small Au nanoparticle catalysts were by Takagi et al. (2006) and Bhaviripudi et al. (2007). XPS measurements in both publications showed that CNT growth was only possible from contaminant-free catalyst nanoparticles, once the residual shell of gold oxides or gold chlorides were reduced by H_2. The findings of Liu et al. (2008a) corroborate this finding. Interestingly, neither paper detected any radial breathing modes in the low Raman shift region, corresponding to large diameter nanotubes.

Figure 2(a) shows a scanning electron microscope image of carbon nanotubes synthesized from a Au nanoparticle catalyst. In this experiment, colloidal gold nanoparticles were spin coated on SiO_2(300 nm) capped Si substrates. Atomic force microscopy (AFM) characterisation of the catalyst revealed that the nanoparticles were approximately 1.4 nm in diameter, with a very narrow particle size distribution, shown in Figure 2(b). The measurements indicated a density of 2500 ± 790 (mean ± standard deviation) particles $/\mu m^2$, which corresponds to an interparticle separation of approximately 20 ± 3 nm. The samples were then pretreated in an H_2 atmosphere for 10 minutes at temperatures ranging from $850 - 1050°C$, followed by a growth step in a mixture of CH_4 and H_2 at 850°C. The highest area density was found for samples pretreated at 1000°C. Raman spectroscopy, with an excitation wavelength of 632.8 nm , was performed in order to evaluate the synthesized nanotubes; a typical spectrum is shown in Figure 2(c). The spectrum exhibits the radial breathing mode feature, indicating that the synthesized nanotubes are predominantly single walled.

From AFM measurements taken to analyse the influence of the pretreatment step on catalyst morphology, we can determine that the initial density of particles is reduced as pretreatment increases. The initial density of 2500 ± 790 particles $/\mu m^2$ was reduced to 420 and 290 particles $/\mu m^2$ after a pretreatment at 900°C and 1000°C, respectively. Broadening of the nanoparticles size distribution is accompanied by a reduction of the modal height as pretreatment temperatures increase. This reduction in density and modal height of the catalyst nanoparticles can be attributed to the evaporation of Au from the substrate in conjuction with the possible diffusion of the Au into the substrate. A similar effect was reported by Bhaviripudi et al. (2007), and may be the cause of the difficulty in synthesizing large diameter nanotubes in other reports. The broadening of the particle size distribution is thought to be owed to particle coalescence by ripening and migration. This change in morphology during pretreatment may even modify the nature of the catalyst surface, thus modifying the catalytic properties of the nanoparticles and consequently the morphology of the carbon products produced (Pisana et al., 2007; Wang et al., 2007).

Typically, the vapour-liquid-solid mechanism is used to explain the mechanism of carbon uptake, supersaturation and precipitation in the catalyst. However, owing to the low solubity of carbon in bulk Au, this must be reviewed for the synthesis of CNTs from Au nanoparticles. By studying the formation of carbon nanowires (CNWs) from Au catalysts, Takagi et al. (2008) inferred that nanosized Au shows some carbon solubility and that Au can form Au-C nanoalloy droplets and produce CNWs by the VLS mechanism. When the catalyst size approaches ≤ 5 nm , the carbon solubility and the nanowire nucleation energy increases dramatically, leading to a structural change in the synthesized carbon products from CNW to SWNT. This finding is in agreement with simulations by Yazyev & Pasquarello (2008), who determined that monatomic carbon in Au can diffuse uniformly across the nanoparticle, even at low temperatures. However, it should be noted that no direct evidence of this effect was demonstrated and that similar studies (Yoshihara et al., 2008) could not determine whether carbon atoms were supplied to the nanotube from the Au-C liquid phase or through surface diffusion. Metallic Cu , long considered to be a contaminant in the growth of SWNTs, has also been reported as an efficient catalyst for SWNT formation in several studies (Takagi et al., 2006; Yuan et al., 2008; Zhou et al., 2006). Figure 3(a) shows a SEM image of CNTs synthesized from a Cu catalyst. In this experiment, Cu nanoparticles were formed by the thermal decomposition of $Cu(NO_3)_2$ in air at 400°C, deposited from a 1 mM isopropanol solution on a SiO_2 support. The particle size distribution of the catalyst, as determined by AFM, is shown in Figure 3(b). The mean particle size was found to be 1.5 ± 0.4 nm (mean ± standard deviation), with a particle density 350 ± 50 particles $/\mu m^2$. The samples were then pretreated in an H_2 atmosphere for 10 minutes at 900°C, followed by a growth step in a mixture of CH_4 and H_2 at the same temperature. Raman spectroscopy showed that the synthesized carbon products were predominantly high quality SWNTs (Figure 3(c)).

It has been reported that the carbon solubility in a metallic catalyst should be in the range of $0.5-1.5$wt% carbon in order to efficiently form CNTs (Deck & Vecchio, 2006). Therefore, it is surprising that Cu can be catalytically active, as its carbon solubility is extremely low. However, Zhou et al. (2006) argue that the low solubility of carbon in Cu results in an increased rate of carbon precipitation. Additionally, Cu has a lower catalytic ability for the dissociation of alkanes than traditional catalysts, resulting in a slower supply of carbon in the CVD process. Thus, matching the supply of carbon to the formation rate of nanotubes will result in the production of high quality SWNTs (Lu & Liu, 2006). This argument is supported by Yazyev & Pasquarello (2008), who state that the stability and the diffusion barriers of diatomic carbon on Cu allow one to restrict the diffusion pathways to the nanoparticle surface by choosing an appropriate gas-phase carbon source, resulting in the preferred formation of high quality SWNTs.

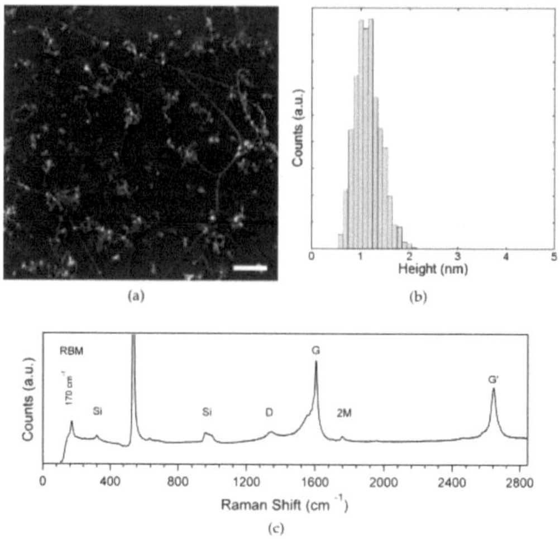

Fig. 2. (a) SEM image of CNTs synthesized from Au nanoparticles pretreated in H_2 at 1000°C. Scale bar corresponds to 250 nm . (b) Particle size distribution of the as-deposited Au catalyst on a SiO_2 substrate. (c) Typical Raman spectrum from CNTs synthesized from the Au catalyst at the optimum growth condition.

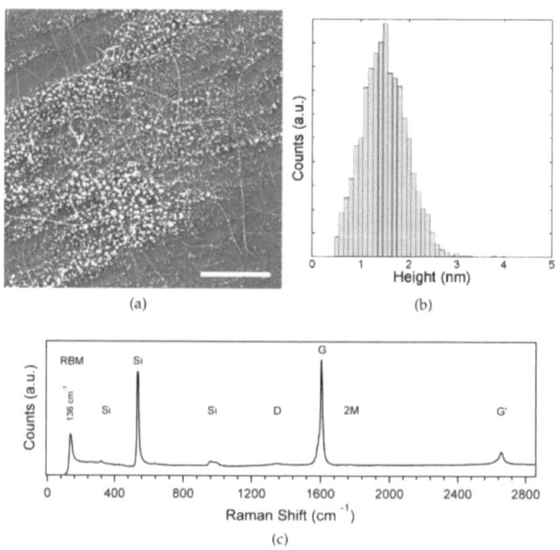

Fig. 3. (a) SEM image of CNTs synthesized from Cu nanoparticles at 900°C. Cu nanoparticles were fabricated by the thermal decomposition of $Cu(NO_3)_2$, deposited from solution. Scale bar corresponds to 1μ m. (b) Particle size distribution of the decomposed Cu catalyst on a SiO_2 substrate. (c) Typical Raman spectrum from CNTs synthesized from the Cu catalyst at the optimum growth condition.

Interestingly, Zhou et al. (2006) reported a higher ratio of metallic SWNTs in Cu catalysed samples, determined from Raman analysis. This characteristic was not detected in our experiments. However, only one laser excitation line was used and the sampling size was too small to draw any significant conclusions. Simulations by Yazyev & Pasquarello (2008) also found that the nucleation of graphitic fragments bound to the Cu nanoparticle catalyst favours the formation of metallic nanotubes. In addition, the low melting point and low carbon diffusion barriers suggest that CVD synthesis could take place at much lower temperatures. In these conditions, the chirality preference would be further enhanced.

5. Semiconductor Nanoparticle Catalysts

Results presented in the previous sections demonstrate that hydrocarbon dissociation and graphite formation abilities are not essential in a catalyst to synthesize CNTs. This leads to a new interpretation of the role of the catalyst particle in CNT growth, where only a nanoscale curvature is needed to act as a template for nanotube formation. This assertion is supported by the reports of CNT formation from semiconductor nanoparticles (Takagi et al., 2007; Uchino et al., 2009; 2008; 2005b), from which no catalytic functions were expected.

The first reports of CNT growth from semiconducting catalysts were by Uchino et al. (2005b). In this experiment, carbon-doped SiGe islands, deposited by CVD on Si , form nanoscale clusters through various mechanisms which act as a seed for SWNT growth. These results were supported by the work of Takagi et al. (2007), who showed that CNT growth from Ge, Si and SiC nanoparticles was possible. More recently, there have been various reports of CNT growth from SiO_2 nanoparticles (Huang et al., 2009; Liu et al., 2009), which are thought to be promising catalysts owing to their ability to maintain a narrow size distribution at CNT growth temperatures.

In this section, research on the use of germanium for carbon nanotube growth is reviewed. Four different techniques to synthesize CNTs based upon Ge nanoparticle catalysts are investigated. These are based on SiGe islands, Ge Stranski-Krastanow dots, Ge nanoparticles formed by ion implantation and colloidal Ge nanoparticles. It is shown that in all cases high quality SWNTs can be grown.

5.1 SiGe Islands

A 50 nm thick $Si_{0.7}Ge_{0.3}$ layer was deposited by CVD on Si(001) wafers after the growth of a thin Si buffer layer. To accommodate the stress resulting from the lattice mismatch between Si and Ge , the SiGe layer forms islands on top of a thin wetting layer. The heights of the islands ranged from 20 to 50 nm . Subsequently, the islands were implanted with carbon ions (energy 30 keV , dose of 3×10^{16} cm^{-2}). This heavy ion implantation is thought to induce damage and form an amorphous layer at the surface (Uchino et al., 2005a). The substrates were then dipped in buffered HF solution to remove the native oxide. Chemical oxidation was performed using a 30% hydrogen peroxide (H_2O_2) solution at room temperature. This step was followed by a pretreatment step in a mixture of Ar and H_2 for 10 minutes at 900°C, followed by the CNT growth step in a mixture of CH_4 and H_2 at 850°C.

Figure 4(a) shows a SEM image of the as-synthesized products on SiGe islands. In this image, two distinct types of nanostructures are visible. The short and thick nanofibres, approximately 20 nm in diameter and 1μ m in length, are formed during the pretreatment step. These nanostructures were identified as SiO_x nanowires by TEM, Raman and photoluminescence measurements, and are formed by the carbothermic reduction of SiO_2 (Lee et al., 2004; Li et al., 2004). These fibres were easily removed by an HF vapour etch, as shown in Figure 4(b). The second type of nanostructure forms during the growth step and comprises straight and thin fibres of less than 10 nm diameter and approximately 5μ m in length. Raman measurements, shown in Figures 4(c) and 4(d), confirm that these fibres are SWNTs. Despite considerable effort, the disorder induced D-band feature that is normally seen at 1350 cm^{-1} (Dresselhaus et al., 2005) could not be detected. This indicates that the nanotubes have a low defect density, and thus could be described as high quality. In this experiment, nanoscale Ge clusters are formed following the chemical oxidation and annealing of the SiGe layers. The oxidation behaviour of SiGe layers has been studied to a great extent (Liou et al., 1991; Paine et al., 1991). Si is known to have a stronger thermodynamic tendency to be oxidised in comparison to Ge. Therefore, the dry oxidation of SiGe alloys, with a low Ge content, results in the formation of SiO_2 and the segregation of Ge clusters from the growing oxide (Sass et al., 2002). It should be noted that this effect is less pronounced with wet oxidation, and the oxide layer typically contains a mixture of Si – O and Ge – O bonds. However, upon annealing in a reducing atmosphere, the Ge – O bonds are preferentially broken owing to a lower stability, resulting in the formation of nanoscale Ge clusters (Paine et al., 1993). These clusters are thought to act as the catalyst for the growth of CNTs in this methodology.

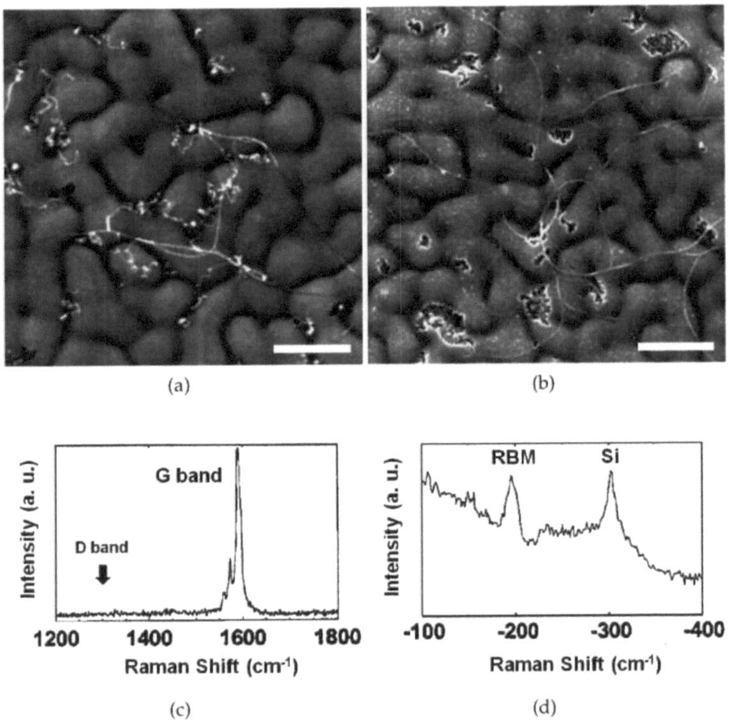

Fig. 4. SEM images of as grown CNTs and SiO_x nanowires synthesized from C implanted SiGe islands (a) before and (b) after HF vapour etching, showing that only carbon nanotubes remain. Scale bar corresponds to 500 nm . Typical Raman spectra of the as-grown CNTs showing (c) G-band characteristic and (d) anti-Stokes spectra showing the radial breathing mode characteristic.

5.2 Ge Stranski-Krastanow Dots

Figure 5(a) shows a TEM image of a bundle of SWNTs grown from Ge Stranki-Krastanow dots. In this experiment, Ge Stranski-Krastanow dots are formed by CVD deposition of Ge atop a thin Si buffer layer. This step forms Ge dots in the form of cones with diameters from 20 to 250 nm and heights between 10 and 25 nm . Subsequently, the islands were implanted with carbon ions (energy 30 keV , dose of 3×10^{16} cm^{-2}). The substrates were then dipped in buffered HF solution to remove the native oxide and subjected to a chemical oxidation using a 30% hydrogen peroxide (H_2O_2) solution at room temperature. This step was followed by a pretreatment step in a mixture of Ar and H_2 for 10 minutes at 900°C, followed by the CNT growth step in a mixture of CH_4 and H_2 at 850°C. Raman measurements on the synthesized CNTs (Figure 5(b)) clearly show the radial breathing mode feature and tangential G band mode expected for SWNTs. The radial breathing modes indicate that the diameters of the synthesized CNTs are in the range 1.6 to 2.1 nm , which are slightly larger than those on SiGe islands (Uchino et al., 2005b). Again, the disorder induced D-band could not be detected, indicating that these CNTs are of a high quality.

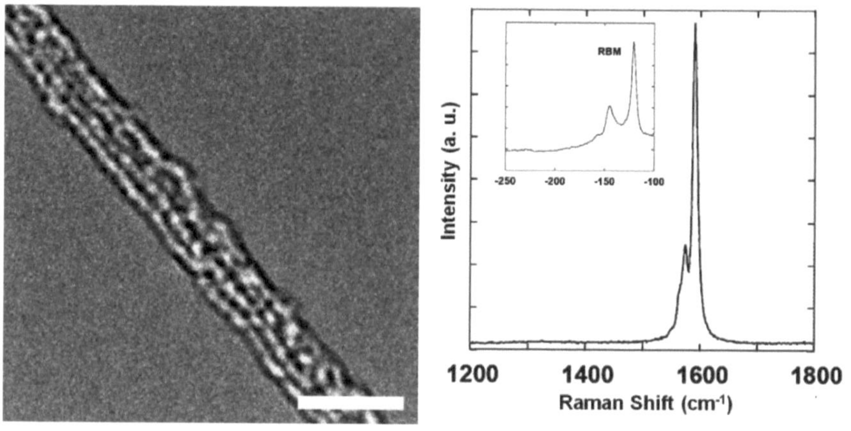

Fig. 5. (a) TEM image of a bundle of SWNTs synthesized from C implanted Ge StranskiKrastanow dots. Scale bar corresponds to 10 nm . The TEM sample was prepared by scraping the substrate surface with a surgical blade and collecting the material on a holey-carbon TEM grid. (b) Typical Raman spectra of the as-grown CNTs showing the G-band characteristic. Inset shows anti-Stokes spectra displaying the radial breathing mode characteristic.

It is believed that the mechanism of formation is very similar to that of the CNTs grown from SiGe islands. Upon chemical oxidation of the Ge Stranski-Krastanow dots, a thin layer of SiGe oxide is formed. Following a subsequent anneal in a reducing atmosphere, Ge clusters are nucleated and it is believed that these act as catalysts in this growth technique. Sass et al. (2002) reported that after the oxidation of Ge islands on Si(001), recovery of the original dot structure was not possible. Instead, a reduction of the GeO_2 around the single crystalline core of the non-oxidized Ge dot materials results in only Ge-enriched clouds, surrounded by a matrix of non-reductible material. This assertion is supported by SEM images taken after each stage in the process (not shown), which indicate a definite change of morphology after the chemical oxidation and reduction steps (Uchino et al., 2008). In fact, there have been reports of the formation of ultra-high density Ge nanoparticles, with diameters of about 4 nm , from the oxidation/reduction of Ge/Si surfaces (Nakamura et al., 2004), which further supports this hypothesis.

5.3 Ge Nanoparticles fabricated by Ion Implantation

In order to further investigate the role of Ge nanoparticles in the growth of carbon nanotubes, Ge nanoparticles were fabricated directly by Ge ion implantation into a layer of thermally oxidised SiO_2, and subsequently annealed at 600°C. This step was followed by a HF vapour etch to remove the SiO_2 and expose the Ge nanoparticles. AFM characterisation revealed that a uniform layer of nanoparticles had been synthesized with a mean density of 460 ± 30 particles $/\mu m^2$ and a modal height of 1.8 nm . The particle size distribution is shown in Figure 6(a). This result shows good agreement with others in the literature, for instance Min et al. (1996) reported the formation of Ge nanocrystals by a similar process, with an average size of 1.9 ± 0.8 nm. Selected samples were then implanted with C . The C implanted samples showed a lower particle density (70 ± 18 particles $/\mu m^2$), a lower modal height (0.7 nm), and a narrower size distribution, shown in Figure 6(b). This change in morphology is attributed to a sputtering effect caused by the ion implantation.

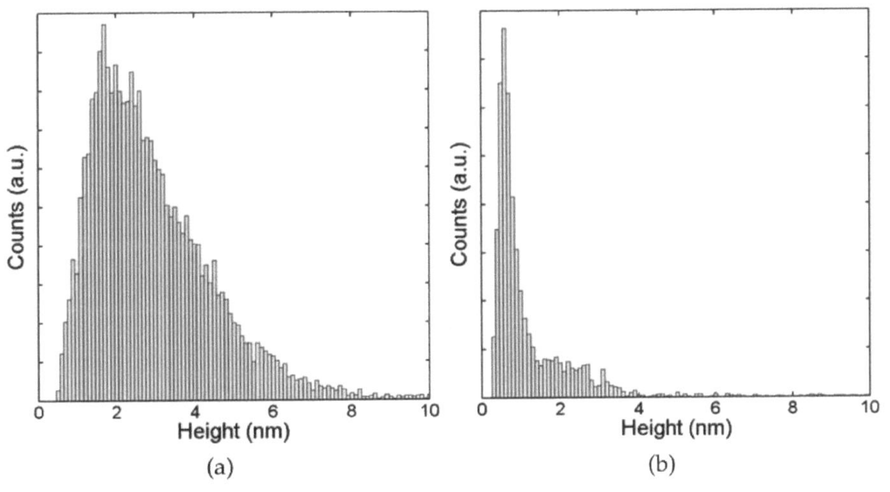

Fig. 6. (a) AFM particle size distributions of Ge nanoparticles synthesized by Ge ion implantation (20keV, 5×10^{15} cm^{-2}) into a 30 nm thick SiO_2 layer and annealed at 600°C for 40 min in N_2 followed by an HF vapour etch to expose the nanoparticles. (b) AFM particle size distribution for samples given a C implant (30keV, 3×10^{16} cm^{-2})

Figure 7(a) and 7(b) shows a typical SEM image after CNT growth for a sample without and with C implantation, respectively. Both images show that a good density of CNTs are achievable using this methodology. Representative Raman spectra for samples without and with C implantation are shown in Figure 7(c) and 7(d), respectively. All samples (insets) clearly show the radial breathing mode, indicating that single walled nanotubes are present. In the case of CNTs grown without C implantation, a small D-band peak is visible around 1320 cm^{-1}, which can be attributed to disorder in the nanotubes. In contrast, samples grown from the C implanted samples show no D-band peak. This indicates that the SWNTs synthesized from C implanted Ge nanocrystals have a low defect density, and are thus high quality.

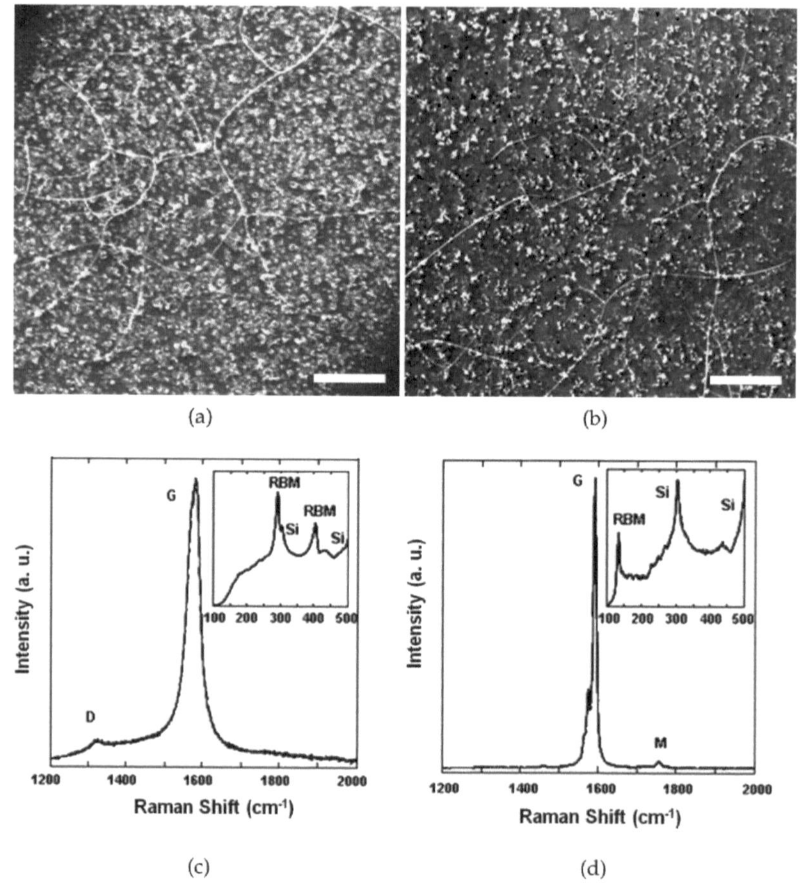

Fig. 7. SEM images of CNTs grown from Ge nanoparticles fabricated by ion implantation from (a) nonimplanted samples and (b) carbon implanted samples. Scale bar corresponds to 500 nm . Typical Raman spectra of the as-grown CNTs from (c) nonimplanted and (d) carbon implanted Ge nanocrystals. Inset shows Stokes spectra displaying the radial breathing mode characteristic.

At the optimum growth condition, there is no statistically significant benefit in terms of area density from the C implant. However, results at other growth conditions show that successful CNT growth can be achieved for a wider range of temperatures when the C implant is formed. AFM measurements taken to analyse the influence of the pretreatment temperature revealed that samples without a C implant show a strong reduction in particle density with increasing pretreatment temperature. In contrast, samples with a C implant show a much smaller decrease in particle density with increasing pretreatment temperature. This suggests that the C implant might increase the Ge melting point through formation of a $Ge_{1-y}C_y$ alloy. This hypothesis is supported by the phase diagram of the Ge-C system presented by Scace & Slack (1959), which indicates that the presence of a small percentage of C has a strong effect in raising the melting point of Ge . Furthermore, Kanazawa et al. (2001) demonstrate that $Ge_{1-y}C_y$ alloys can be successfully formed by C implantation into Ge .

5.4 Colloidal Germanium Nanoparticles

Figure 8(a) shows an AFM image of Ge nanoparticles deposited from a 1 mM colloidal solution on a SiO_2 support by spin coating. The colloidal solution was synthesized by means of an inverse micelle method. The synthesis of a $Ge[N(SiCH_3)_2]_2$ precursor for the formation of Ge nanocrystals was based on the works of Lessing et al. (1977a;b), with some minor modifications. The precursor (50mg) was dissolved in 7 ml of trioctylamine and injected into hot (340 °C), molten hexadecylamine (HDA, 1 g). The injection method was inspired from various publications and has been standardised for this purpose (Nair et al., 2002, and references therein). The residue was then dissolved in toluene, re-precipitated with methanol and suspended in tryoctlamine. This process yielded nanoparticles with a narrow size distribution, shown in Figure 8(b). The synthesized nanoparticles had a mean size of 1.5 ± 0.4 nm, and the density of the spin coated layer was 430 ± 60 particles $/\mu m^2$.

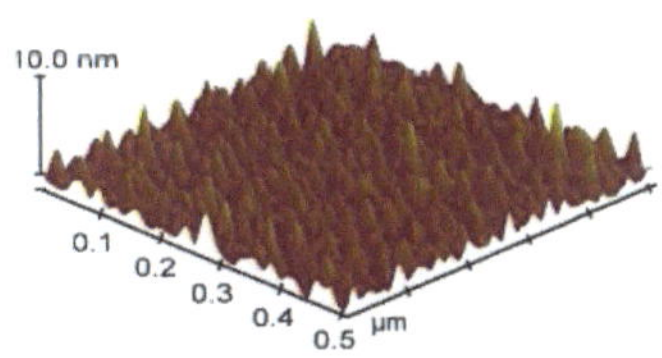

(a)

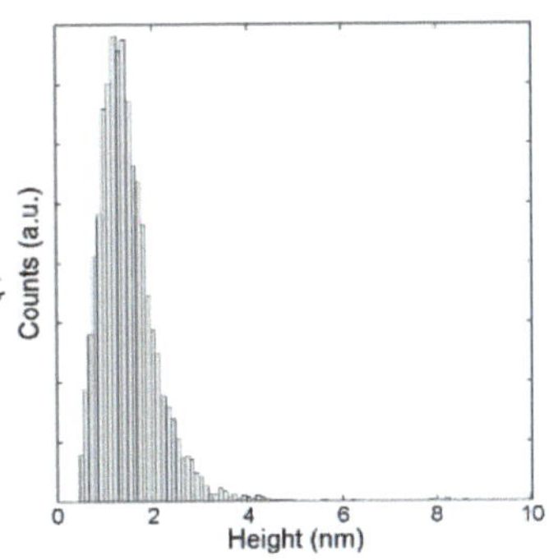

(b)

Fig. 8. (a) AFM image and (b) particle size distributions of colloidal Ge nanoparticles, deposited on SiO_2 by spin coating of a 1 mM solution. After deposition, samples were cleaned in a 100 WO_2 plasma for 30 minutes.

Samples were then subjected to a 100 WO_2 plasma for 30 minutes in order to remove the organic residue left from deposition. Subsequently, samples were pretreated in an H_2 atmosphere for 10 minutes at temperatures ranging from 850 – 1050°C, followed by a growth step in a mixture of CH_4 and H_2 at 850°C. The highest area density was found for samples pretreated at 900°C, shown in Figure 9(a). Raman spectroscopy showed that the synthesized carbon products were predominantly high quality SWNTs (Figure 9(c)). Synthesis on sapphire (Al_2O_3) substrates showed that a slightly higher uniformity and area density of CNTs was achievable on this support medium, shown in Figure 9(b). This is attributed to the ability of the Al_2O_3 support to provide an interface for graphite formation (Rummeli et al., 2007).

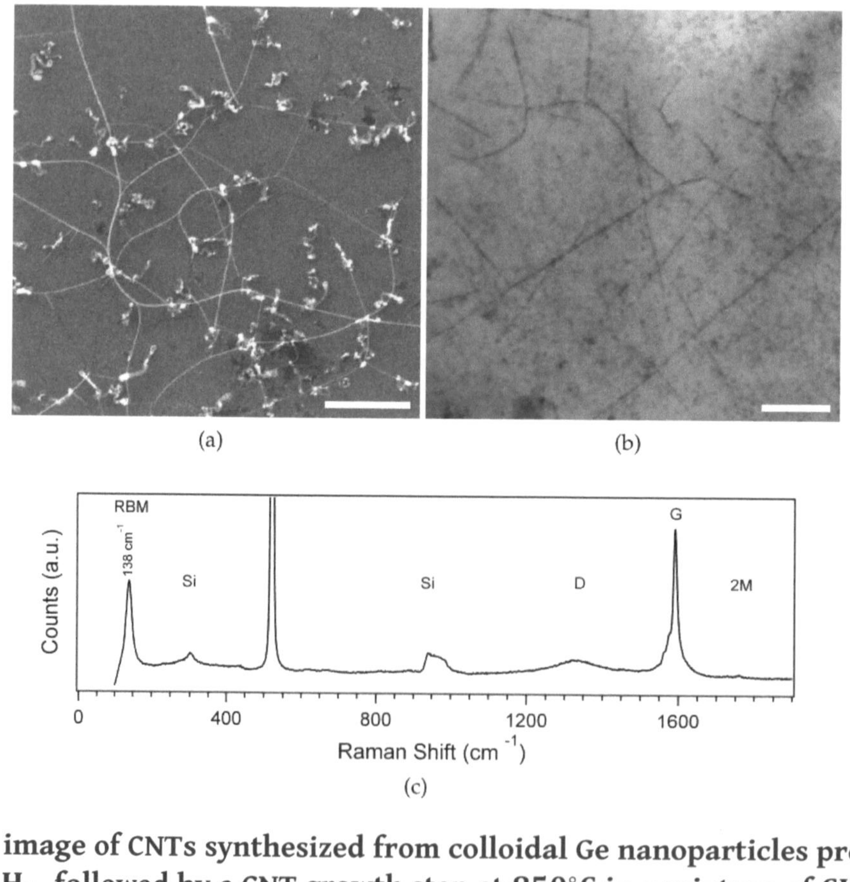

Fig. 9. SEM image of CNTs synthesized from colloidal Ge nanoparticles pretreated at 900 °C in H_2, followed by a CNT growth step at 850°C in a mixture of CH_4 and H_2 on a (a) SiO_2 support and a (b) sapphire (Al_2O_3) support. Scale bar corresponds to 500 nm in both images. (c) Typical Raman spectrum from CNTs synthesized from the colloidal Ge catalyst at the optimum growth condition.

In comparison to the non C implanted Ge nanoparticles fabricated by ion implantation, the process window for this methodology was considerably wider, with little change in CNT area density in samples pretreated at temperatures from 900 to 1000°C. AFM studies of the catalyst after H_2 pretreatment revealed that there was no statistically significant reduction in particle density with increasing pretreatment temperature. In addition, rather than a reduction in mean particle size, a very slight increase in size was found with increasing pretreatment temperature. This effect is believed to be due to the organic cap on the nanoparticles acting as a stabilising agent as the furnace ramps up to temperature. It should be noted that at the synthesis temperatures, the organic cap on the nanoparticles is expected to be fully reduced.

6. Growth Mechanism Discussion

Despite enormous strides in the synthesis of carbon nanotubes, the mechanism for growth is still a highly debated issue. As discussed previously, it is generally accepted that the model for carbon filament growth (Baker, 1989; Baker et al., 1972; 1973), derived from concepts of vapour-liquid-solid theory, also applies to carbon nanotube growth. This belief arises from the visual observation (by TEM) of catalyst particles on the ends of nanotubes, as was the case with carbon filaments. In this model, hydrocarbons adsorbed on the metal nanoparticle are catalytically decomposed resulting in atomic carbon dissolving into the liquid catalyst particle, and when a supersaturated state is reached, carbon precipitates in a tubular, crystalline form.

However, the results presented in this work suggest that this belief holds several observational inconsistencies that do not support this mechanism for CVD production. The successful CNT growths from catalyst free, noble metal catalysts and semiconducting catalysts imply that hydrocarbon dissociation ability is not essential in a catalyst. It should be noted that the catalytic behaviour of Cu and Au may be explained by electron donation to the support (Vander Wal et al., 2001), creating *d*-vacancies which may cause hydrocarbon dissocation. However, the ability of catalyst-free and semiconducting catalysts to seed CNT growth cannot be explained by the same mechanism. Reilly & Whitten (2006) argue that a more likely scenario is that a free radical condensate (FRC) provides carbon species through a leaving group, such as hydrogen (or oxygen). FRCs naturally form during hydrocarbon pyrolysis by the breaking of carbon-hydrogen or carbon-carbon bonds with each fragment keeping one electron to form two radicals. The presence of a radical in a hydrocarbon molecule permits rapid rearrangement of carbon bonds. In this case, the catalyst particle's role is to simply provide an interface where carbon rearrangement can occur and act as a template for growth.

Typically, metal catalysts with no *d*-vacancies, such as Cu and Au , do not offer sites to dissolve carbon, such that neither saturation nor precipitation is possible. However, despite the low carbon solubility, these catalysts have demonstrated an ability to catalyse CNT formation. Additionally, catalysts with a high melting point such as Al_2O_3 or ZrO_2 are thought to be solid at CNT synthesis temperatures. However, if nanoparticles of these metals are small enough (≤ 5 nm), the increasing fraction of low-coordinated atoms may lead to surface saturation followed by carbon precipitation, as reported by Takagi et al. (2006). Considering that carbon penetration inside small nanoparticles is unlikely (Raty et al., 2005), the growth of CNTs is most likely a process primarily controlled by surface diffusion (Chadderton & Chen, 1999; Ding et al., 2005). Indeed, it is believed by several groups that the rate-limiting factor in CNT synthesis is the surface diffusion of carbon across the catalyst (Bartsch et al., 2005; Hofmann et al., 2005). Additionally, this factor could explain the influence of the carbon source on the ability of a catalyst to synthesize CNTs. Yazyev & Pasquarello (2008) reported different activation energies for the surface diffusion of C dimers and adatoms on noble metal catalysts, and argued that appropriate choice of a diatomic or monatomic carbon gas-phase source could significantly accelerate diffusion.

The specificity of the growth of nanotubes on nanoparticles with regard to the growth of carbon filaments is their nanometer dimensions. Other mechanisms are therefore required to explain the nucleation of CNTs from nanoparticle catalysts. One such model is the Yarmulke mechanism proposed by Dai et al. (1996). In the Yarmulke mechanism, a graphene cap is assembled on the particle surface with its edges strongly chemisorbed to the catalyst. The graphene cap acts to reduce the high total surface energy of the particle caused by its high curvature, owing to the fact that the basal plane of graphite has an extremely low surface energy. As additional carbon atoms are added, the hemifullerene cap formed on the particle surface lifts off, creating a hollow tube with constant diameter which grows away from the particle (Nikolaev et al., 1999). This model was supported by molecular dynamics simulations by Shibuta & Maruyama (2003). Recent works in high-resolution in-situ TEM observation of the catalytic growth of CNTs have verified this mechanism (Helveg et al., 2004; Hofmann et al., 2007). These studies have also shown that cap stabilisation and nanotube growth involve reshaping of the catalyst nanoparticle.

7. Conclusions and Future Directions

In this review, the role of the catalyst in the selective growth of SWNTs by CVD has been studied. Evidence of self-assembled carbon nanostructures was presented, indicating that upon annealing, C undergoes a structural reorganization to form graphitic structures. In addition, SWNT growth was shown to be possible from ceramic, noble metal and semiconducting catalysts. This demonstrates that hydrocarbon dissociation and graphitization ability are not essential in a catalyst, and it was argued that carbon species are supplied through a free radical condensate. The results presented show that the commonly utilised model of carbon filament growth is inadequate to describe SWNT growth from non traditional catalysts. A new interpretation of the role of the catalyst was presented where only a nanoscale curvature is necessary to grow CNTs. A mechanism for SWNT growth was suggested where the surface saturation and diffusion of *C* on the catalyst nanoparticle lead to the formation of a graphitic cap followed by the further incorporation of C into a growing nanotube.

Although there have been tremendous advances in the fabrication of CNTs, the integration of these nanostructures into successful applications and large-scale production processes depend on the understanding of several fundamental issues, which are yet to be addressed. A few of these issues are briefly discussed below.

The role played by the support in the CVD of CNTs is not yet fully understood. The simplistic view that the support only plays a catalytically passive role in the formation of CNTs requires examination. The work of Rummeli et al. (2007) demonstrated that under typical CVD growth conditions, nanoparticles of difficult-to-reduce metal oxides were capable of promoting ordered carbon growth. The authors attributed this to the presence of surface defect sites on the nanoparticle oxides. However, the interface between the catalyst nanoparticle and the support is thought to act as a annular defect site. This would indicate that the nature of the support-nanoparticle interface may be very important to the behaviour of the catalyst.

There appears to be a consensus in the literature concerning the correlation of catalyst size and SWNT diameter. Several groups have observed a direct dependence of the two quantities (Jeong et al., 2005; Wei et al., 2001). However, whether there is a correlation between the atomic structure/facets of the nanoparticle catalyst and the chirality of the synthesized nanotube has not yet been determined. If a good correlation is found between the facets and the chirality of the nanotube, how can the physical structure of the nanoparticles be retained as the sample is heated to the synthesis temperature?

The low-temperature synthesis of CNTs is of significant technological importance. This is essential for synthesis on glass substrates or other applications requiring plastic substrates. However, it is not yet clear whether this is possible and what the effect on the material quality would be. The integration of carbon nanotubes in IC manufacturing would also require large area growth uniformities, however most research is undertaken in small, low-throughput batch reactors, typically on small pieces of wafers. This is expected to change as the market for CNT-based products begins to emerge, creating the need for large-scale commercial reactors.

8. References

Awasthi, K., Srivastava, A. & Srviastava, O. N. (2005). Synthesis of Carbon Nanotubes, J. Nanosci. Nanotechnol. 5: 1616-1636.

Baker, R. T. K. (1989). Catalytic Growth of Carbon Filaments, Carbon 27: 315.

Baker, R. T. K., Barber, M. A., Harris, P. S., Feates, F. S. & White, R. J. (1972). Nucleation and Growth of Carbon Deposites from the Nickel Catalyzed Decomposition of Acetylene, J. Catal. 26: 51.

Baker, R. T. K., Harris, P. S., Thomas, R. B. & Waite, R. J. (1973). Formation of Filamentous Carbon from Iron, Cobalt and Chromium Catalyzed Decomposition of Acetylene, J. Catal. 30: 86.

Bartsch, K., Biedermann, K., Gemming, T. & Leonhardt, A. (2005). On the Diffusion Controlled Growth of Multiwalled Carbon Nanotubes, J. Appl. Phys. 97: 114301

Berger, L.-M., Gruner, W., Langholf, E. & Stolle, S. (1999). On the Mechanism of Carbothermal Reduction Processes of TiO_2 and ZrO_2, Int. J. Refract. Met. Hard Mater. 17: 235-243.

Bhaviripudi, S., Mile, E., Steiner, S. A., Zare, A. T., Dresselhaus, M. S., Belcher, A. M. & Kong, J. (2007). CVD Synthesis of Single-Walled Carbon Nanotubes from Gold Nanoparticle Catalysts, J. Am. Chem. Soc. 129: 1516-1517.

Botti, S., Asilyan, C. R. L., Dominicis, L. D., Fabbri, F., Orlanducci, S. & Fiori, A. (2004). Carbon Nanotubes Grown by Laser-annealing of SiC Nano-particles, Chem. Phys. Lett. 400: 264-267.

Botti, S., Asilyan, L. S., Ciardi, R., Fabbri, F., Lortei, S., Santoni, A. & Orlanducci, S. (2001). Catalyst-Free Growth of Carbon Nanotubes by Laser Annealing of Amrphous SiC Films, Chem. Phys. Lett. 396: 1-5.

Botti, S., Ciardi, R., Terranova, M. L., Piccirillo, S., Sessa, V., Rossi, M. & Vittori-Antisari, M. (2002). Self-Assembled Carbon Nanotubes Grown Without Catalyst From Nanosized Carbon Particles Adsorbed on Silicon, Appl. Phys. Lett. 80: 1441.

Chadderton, L. T. & Chen, Y. (1999). Nanotube Growth by Surface Diffusion, Phys. Lett. A 263: 401-405.

Dai, H., Rinzler, A. G., Nikolawv, P., Thess, A., Colbert, D. T. & Smalley, R. E. (1996). Single-wall Nanotubes Produced by Metal-Catalyzed Disproportionation of Carbon Monoxide, Chem. Phys. Lett. 260: 471-475.

Deck, C. P. & Vecchio, K. (2006). Prediction of Carbon Nanotube Growth Success by the Analysis of Carbon-Catalyst Binary Phase Diagrams, Carbon 43: 2654-2663.

Derycke, V., Martel, R., Radosavljevic, M., Ross, F. M. & Avouris, P. (2002). Catalyst Free Growth of Ordered Single Walled Carbon Nanotube Networks, Nano Lett. 2: 10431046.

Ding, F., Rosen, A. & Bolton, K. (2005). Dependence of SWNT Growth Mechanism on Temperature and Catalyst Particle Size: Bulk versus Surface Diffusion, Carbon 43: 2215-2217.

Dresselhaus, M. S., Dresselhaus, G., Sait, R. & Jorio, A. (2005). Raman Spectroscopy of Carbon Nanotubes, Phys. Rep. 409: 47.

Haruta, M. (1997). Size- and Support-Dependency in the Catalysis of Gold, Catal. Today 36: 156-166.

Helveg, S., Lopez-Cartes, C., Sehested, J., Hansen, P. L., Clausen, B. S., Rostrup-Nielsen, J. R., Abild-Pedersen, F. & Norskov, J. K. (2004). Atomic-scale Imaging of Carbon Nanofibre Growth, Nature 427: 426-429.

Hofmann, S., Csányi, G., Ferrari, A. C., Payne, M. C. & Robertson, J. (2005). Surface Diffusion: The Low Activation Energy Path for Nanotube Growth, Phys. Rev. Lett. 95: 036101.

Hofmann, S., Sharma, R., Ducati, C., Du, G., Mattevi, C., Cepek, C., Cantoro, M., Pisana, S., Parvez, A., Cervantes-Sodi, F., Ferrari, A. C., Dunin-Borkowski, R., Lizzit, S., Petaccia, L., Goldoni, A. & Robertson, J. (2007). In situ Observations of Catalyst Dynamics during Surface-Bound Carbon Nanotube Nucleation, Nano Lett. 7: 602-608.

Huang, S., Cai, Q., Chen, J., Qian, Y. & Zhang, L. (2009). Metal-Catalyst-Free Growth of SingleWalled Carbon Nanotubes on Substrates, J. Am. Chem. Soc. 131: 2094-2095.

Hvolbaek, B., Janssens, T. V. W., Clausen, B. S., Falsig, H., Christensen, C. H. & Norkov, J. K. (2007). Catalytic Activity of Au Nanoparticles, Nano Today 2: 14-18.

Iijima, S. (1991). Helical Microtubules of Graphitic Carbon, Nature 354: 56.

Jeong, G.-H., Suzuki, S., Kobayashi, Y., Yamazaki, A., Yoshimura, H. & Homma, Y. (2005). Effect of Nanoparticle Density on Narrow Diameter Distribution of Carbon Nanotubes and Particle Evolution During Chemical Vapour Deposition, J. Appl. Phys. 98: 124311.

Jia, J., Haraki, K., Kondo, J. N., Domen, K. & Tamaru, K. (2000). Selective Hydrogenation of Acetylene over Au/Al_2O_3 Catalyst, J. Phys. Chem. B 104: 1153-1156.

Kanazawa, Y., Katayama, K., Nozawa, K., Saitoh, T. & Kubo, M. (2001). Preparation of $Ge_{1-y}C_y$ Alloys by C Implantation into Ge Crystal and Their Raman Spectra, Jpn. J. Appl. Phys. 40: 5880.

Koshio, A., Yudasaka, M. & Iikima, S. (2002). Metal-free Production of High-Quality MultiWall Carbon Nanotubes, in which the Innermost Nanotubes Have a Diameter of 0.4 nm, Chem. Phys. Lett. 356: 595-600.

Kusonoki, M., Rokkak, M. & Suzuki, T. (1997). Epitaxial Carbon Nanotube Fil Self Organised by Sublimation Decomposition of Silicon Carbide, Appl. Phys. Lett. 71: 2620.

Larciprete, R., Lizzit, S., Botti, S., Cepek, C. & Goldini, A. (2002). Structural Reorganisation of Carbon Nanoparticles into Single-Walled Nanotubes, Phys. Rev. B.: Condes. Matter Mater. Phys. 66: 121402.

Larciprete, R., Lizzit, S., Cepek, C., Botti, S. & Goldini, A. (2003). Thermal Reactions at the Interface between Si and C Nanoparticles: Nanotube Self-Assembling and Transformation into SiC, Surf. Sci. 532: 886-891.

Lee, K.-H., Yang, H. S., Baik, K. H., Bang, J., Vanfleet, R. R. & Sigmund, W. (2004). Direct Growth of Amorphous Silica Nanowiresby Solid State Transformation of SiO_2 Films, Chem. Phys. Lett. 383: 380.

Lee, S. Y., Yamada, M. & Miyake, M. (2005). Synthesis of Carbon Nanotubes Over Gold Nanoparticle Supported Catalysts, Carbon 43: 2654-2663.

Lessing, J. G. V., Fouche, K. F. & Retief, T. T. (1977a). Redox Extractions from Molten AlkaliMetal Cyanides. Part 1. Mechanism for Extractions with Liquid Zinc Alloys, Dalton Trans. 20: 2020.

Lessing, J. G. V., Fouche, K. F. & Retief, T. T. (1977b). Redox Extractions from Molten Alkalimetal Cyanides. Part 2. Mechanism for Extractions with Liquid Tin Alloys, Dalton Trans. 20: 2004.

Li, S.-H., Zhu, X.-F. & Zhao, Y.-P. (2004). Carbon-Assisted Growth of SiO_x Nanowires, J. Phys Chem. B 108: 17032.

Liou, H. K., Mei, P., Gennser, U. & Yang, E. S. (1991). Effects of Ge Concentration on SiGe Oxidation Behavior, Appl. Phys. Lett. 59: 1200.

Liu, B., Ren, W., Gao, L., Li, S., Pei, S., Liu, C., Jiang, C. & Cheng, H.-M. (2009). Metal-CatalystFree Growth of Single-Walled Carbon Nanotubes, J. Am. Chem. Soc. 131: 2082-2083.

Liu, H., Takagi, D., Ohno, H., Chiashi, S., Chokan, T. & Homma, Y. (2008a). Effect of Ambient Gas on the Catalytic Properties of Au in Single Walled Carbon Nanotube Growth, Jpn. J. Appl. Phys. 47: 1966-1970.

Liu, H., Takagi, D., Ohno, H., Chiashi, S., Chokan, T. & Homma, Y. (2008b). Growth of Single Walled Carbon Nanotubes from Ceramic Particles by Alcohol Chemical Vapor Deposition, Appl. Phys. Express 1: 014001.

Lu, C. & Liu, J. (2006). Controlling the Diameter of Carbon Nanotubes in Chemical Vapor Deposition Method by Carbon Feeding, J. Phys. Chem. B 110: 20254.

Mctaggart, F. K. (1961). Reduction of Zirconium and Hafnium Oxides, Nature 191: 1192.

Melechko, A. V., Merlukov, V. I., McKnight, T. E., Guillorn, M. A., Lowndes, D. H. & Simpson, M. L. (2005). Vertically Aligned Carbon Nanofibers and Related Structures: Controlled Synthesis and Directed Assembly, J. Appl. Phys. 97: 041301.

Meyyappan, M., Yamada, T., Sarrazin, P. & Li, J. (2005). Carbon Nanotubes: Science and Applications, CRC Press.

Min, K. S., Shcheglov, K. V., Yang, C. M., Atwater, H. A., Brongersma, M. L. & Polman, A. (1996). The Role of Quantum-Confined Excitons vs Defects in the Visible Luminescence of SiO_2 Films Containing Ge Nanocrystals, Appl. Phys. Lett. 68: 2511.

Nair, P. S., Radhakrishnan, T., Revaprasadu, N., Kolawolea, G. & OŠBrien, P. (2002). Cadmium Ethylxanthate: A Novel Single-Source Precursor for the Preparation of CdS Nanoparticles, J. Mater. Chem. 12: 2722.

Nakamura, Y., Nagadomi, Y., Sugie, K., Miyata, N. & Ichikawa, M. (2004). Formation of Ultrahigh Density Ge Nanodots on Oxidized Ge/Si(111) Surfaces, J. Appl. Phys. 95: 5014.

Nikolaev, P., Bronikowski, M. J., Bradley, R. K., Rohmund, F., Colbert, D. T., Smith, K. A. & Smalley, R. E. (1999). Gas-Phase Catalytic Growth of Single-Walled Carbon Nanotubes from Carbon Monoxide, Chem. Phys. Lett. 313: 91-97.

Paine, D. C., Caragianis, C., Kim, T. Y., Shigesato, Y. & Ishahara, T. (1993). Visible Photoluminescence from Nanocrystalline Ge Formed by H_2 Reduction of $Si_{0.6}Ge_{0.4}O_2$, Appl. Phys. Lett. 62: 2842.

Paine, D. C., Caragianis, C. & Schwartzman, A. F. (1991). Oxidation of $Si_{1x}Ge_x$ Alloys at Atmospheric and Elevated Pressure, J. Appl. Phys. 70: 5076.

Pisana, S., Cantoro, M., Parvez, A., Hofmann, S., Ferrari, A. C. & Robertson, J. (2007). The Role of Precursor Gases on the Surface Restructuring of Catalyst Films During Carbon Nanotube Growth, Physica E37: 1.

Raty, J.-Y., Gygi, F. & Galli, G. (2005). Growth of Carbon Nanotubes on Metal Nanoparticles: A Microscopic Mechanism from Ab-Initio Molecular Dynamics, Phys. Rev. Lett. 95: 096103.

Reich, S., Li, L. & Robertson, J. (2006). Control the Chirality of Carbon Nanotubes by Epitaxial Growth, Chem. Phys. Lett. 421: 469.

Reilly, P. T. A. & Whitten, W. B. (2006). The Role of Free Radical Condensates in the Production of Carbon Nanotubes During the Hydrocarbon CVD Process, Carbon 44: 1653-1660.

Rummeli, M. H., Kramberger, C., Gruneis, A., Ayala, P., Gemming, T., Buchner, B. & Pichler, T . (2007). On the Graphitization Nature of Oxides for the Formation of Carbon Nanostructures, Chemistry of Meterials 19: 4105-4107.

Sass, T., Zela, V., Gustafsson, A., Pietzonka, I. & Seifert, W. (2002). Oxidation and Reduction Behavior of Ge-Si Islands, Appl. Phys. Lett. 81: 3455.

Scace, R. I. & Slack, G. A. (1959). Solubility of Carbon in Silicon and Germanium, J. Chem. Phys. 30: 1551.

Shibuta, Y. & Maruyama, S. (2003). Molecular Dynamics Simulation of Formation Process of Single-Walled Carbon Nanotubes by CCVD Method, Chem. Phys. Lett. 382: 381-386.

Steiner, S. A., Baumann, T. F., Bayer, B. C., Blume, R., Worsley, M. A., MoberlyChan, W. J., Shaw, E. L. ., Schlogl, R., Hart, A. J., Hofmann, S. & Wardle, B. L. (2009). Nanoscale Zirconia as a Nonmetallic Catalyst for Graphitization of Carbon and Growth of Single- and Multiwall Carbon Nanotubes, J. Am. Chem. Soc. 131: 12144-12154.

Takagi, D., Hibino, H., Suzuki, S., Kobayashi, Y. & Homma, Y. (2007). Carbon Nanotube Growth from Semiconductor Nanoparticles, Nano Lett. 7: 2272-2275.

Takagi, D., Homma, Y., Hibino, H., Suzuki, S. & Kobayashi, Y. (2006). Single Walled Carbon Nanotube Growth from Highly Activated Metal Nanoparticles, Nano Lett. 6: 26422645.

Takagi, D., Kobayashi, Y., Hibino, H., Suzuki, S. & Homma, Y. (2008). Mechanism of GoldCatalyzed Carbon Material Growth, Nano Lett. 8: 832-835.

Takikawa, H., Miyano, R., Yatsuki, M. & Sakakibara, T. (1998). Carbon Nanotubes on SiC Powder Surface Grown by a Vacuum Heating Process, Jpn. J. Appl. Phys., 2: L187.

Uchino, T., Ayre, G. N., Smith, D. C., Hutchison, J. L., de Groot, C. H. & Ashburn, P. (2009). Growth of Single-Walled Carbon Nanotubes Using Germanium Nanocrystals Formed by Implantation, J. Electrochem. Soc. 156: K144-K148.

Uchino, T., Bourdakos, K. N., Ayre, G. N., de Groot, C. H., Ashburn, P. & Smith, D. C. (2008). CMOS Compatible Synthesis of Carbon Nanotubes, Mater. Res. Soc. Symp. Proc. 1081: 1081-P01-09.

Uchino, T., Bourdakos, K. N., de Groot, C. H., Ashburn, P., Kiziroglou, M. E., Dilliway, G. D. & Smith, D. C. (2005a). Catalyst Free Low Temperature Direct Growth of Carbon Nanotubes, Proceedings of 2005 5th IEEE Conference on Nanotechnology 5: 1-4.

Uchino, T., Bourdakos, K. N., de Groot, C. H., Ashburn, P., Kiziroglou, M. E., Dilliway, G. D. & Smith, D. C. (2005b). Metal Catalyst-Free Low-Temperature Carbon Nanotube Growth on SiGe Islands, Appl. Phys. Lett. 86: 233110.

Vander Wal, R. L., Tichich, T. M. & Curtis, V. E. (2001). Substrate-support Interactions in Metal-Catalyzed Carbon Nanofiber Growth, Carbon 39: 2277-2289.

Wagner, R. S. & Ellis, W. C. (1964). Vapor-Liquid-Solid Mechanism of Single Crystal Growth, Appl. Phys. Lett. 4: 89-90.

Wang, W. P., Wen, H. C., Jian, S. R., Juang, J. Y., Lai, Y. S., Tsai, C. H., Wi, W. F., Chen, K. T. & Chou, C. P. (2007). The Effects of Hydrogen Plasma Pretreatment on the Formation of Vertically Aligned Carbon Nanotubes, Appl. Surf. Sci. 253: 9248.

Wei, Y. Y., Eres, G., Merlukov, V. I. & Lowndes, D. H. (2001). Effect of Catalyst Film Thickness on Carbon Nanotube Growth by Selective Area Chemical Vapour Deposition, Appl. Phys. Lett. 78: 1394.

Yazyev, O. V. & Pasquarello, A. (2008). Effect of Metal Elements in Catalytic Growth of Carbon Nanotubes, Phys. Rev. Lett. 100: 156102.

Yoshihara, N., Ago, H. & Tsuji, M. (2008). Growth Mechanism of Carbon Nanotubes over Gold-Supported Catalysts, Jpn. J. Appl. Phys. 47: 1944-1948.

Yuan, D., Ding, L., Chu, H., Feng, Y., McNicholas, T. P. & Liu, J. (2008). Horizontally Aligned Single-Walled Carbon Nanotube on Quartz from a Large Variety of Metal Catalysts, Nano Lett. 8: 2576-2579.

Zhou, W., Han, Z., Wang, J., Zhang, Y., Jin, Z., Sun, Z., Zhang, Y., Yan, C. & Li, Y. (2006). Copper Catalyzing Growth of Single-Walled Carbon Nanotubes on Substrates, Nano Lett. 6: 2987-2990.

Chapter-3

Properties of Carbon Nanotubes under External Factors

1. Introduction

Carbon nanotubes are quasi-one dimensional nanostructures with many unique properties. These materials are considered as promising building blocks in nanoelectronics, spintronics, and nano-electromechanics. Potential applications of carbon nanotubes have strongly relied on their extraordinary ability of property changes upon external factors, such as mechanical alterations or applied external fields. The success in application of carbon nanotubes has significantly depended on predictions from theoretical calculations and modeling. In particular, first principal simulation methods have proven to be useful techniques in predicting the fundamental nature of changes in carbon nanotube properties and how these properties can be altered under external factors. There is a considerable amount of research on simulations of carbon nanotubes under one type of external factors, such as mechanical defects, mechanical deformations, applied external electric fields, or applied external magnetic fields. However, the development and design of current and new carbon nanotube based nanodevices has led to the search of additional ways to modify and tailor their properties. To achieve this flexibility, researchers have begun to study the influence of more than one external factor simultaneously. This is a relatively new area which opens up more possibilities for new devices. Thus, the knowledge of physical property changes in carbon nanotubes under several external factors is crucial for fundamental science and technological prospects. Our primary focus in this chapter is first principal simulations of electronic and magnetic properties of single wall carbon nanotubes under two external factors. The discussion involves understanding the combined effect of radial deformation and mechanical defects, such as a Stone-Wales defect, nitrogen impurity, and mono-vacancy in the carbon network. Furthermore, we will discuss the nanotube property changes due to the application of radial deformation and external electric field. To provide the basis for further arguments, we start with a review of important results of carbon nanotubes under one external factor such as mechanical deformations, mechanical defects, or external fields. After we provide the description of the calculation method and model, we arrive at the major issue of radially deformed carbon nanotubes with defects or external electric field. We start with the analysis of the structural changes in defective carbon nanotubes under two types of radial deformations. One is deformation achieved by squeezing of the nanotube between two hard narrow surfaces, which have width smaller than the nanotube cross section. The other one is achieved by squeezing of the nanotube between two hard surfaces with width larger than the nanotube cross section. As a result, we show the characteristic energies of deformation and defect-formation. Further, we show first principal calculations of band gap modulations, band structure evolutions, and changes in magnetic properties of carbon nanotubes in terms of types of defects and degrees of radial deformation. A special consideration is given to the mechanically induced magnetic properties. Finally, we present the combined effect of radial deformation and external electric field on the nanotube electronic structure and band gap modulations. The chapter reveals greater capabilities for carbon nanotube applications in new devices.

2. Carbon Nanotubes under One External Factor

In many earlier theoretical calculations, carbon nanotubes were considered to be with perfect structure, where ideal graphite plane was rolled up into a cylindrical tube. Depending on the particular way of graphene rolling, nanotubes can be metallic or semiconducting. The rolling direction is characterized by a chiral index (n, m). Achiral zigzag $(n, 0)$ and armchair (n, n) nanotubes are distinguished from the rest (chiral nanotubes). Armchair tubes are always metallic, while zigzag tubes can be semiconducting or metallic (Reich et al., 2004). It has been revealed that these nanostructures have interesting mechanical and electronic properties strongly depended on chirality, which immediately drew interest to their application in micro-electro-mechanics, spintronics, and nanoelectronics. However, experimental measurements have shown that the nanotube structure is not perfect (Fan et al., 2005; Hashimoto et al., 2004; Suenaga et al., 2007). External factors, such as mechanical deformations, mechanical defects, or applied external fields can induce various changes in nanotube properties In addition, researchers have realized that these external factors can be used in new methods to control and/or tune carbon nanotube properties for practical applications.

2.1 Mechanical Deformation of Carbon Nanotubes

Carbon nanotubes have high elastic modulus, strength, flexibility, and low mass density, which makes them excellent candidates for polymer composite materials and nano-devices (Ashraf et al., 2006; Fennimore et al., 2003; Itkis et al., 2008). They are rigid in the axial direction while they are flexible in the radial direction. In actual experiments, nanotubes undergo structural modifications such as bending, twisting, stretching, and radial deformation (Barboza et al., 2008; Cohen-Karni et al., 2006; Falvo et al., 1997). It has been suggested that the most common mechanical deformation is radial squeezing. Such deformation takes place in bundles of nanotubes, under applied hydrostatic pressure, in interaction of nanotubes with surfaces, nano-indentation experiments, and atom force microscopy (AFM) squeezing of nanotubes (Hertel et al., 1998; Minary-Jolandan & Yu 2008; Tang et al., 2000). In many experiments the carbon nanotubes are manipulated by an AFM tip. This has a major effect not only on the mechanical, but also on the electronic and transport properties of nanotubes. For instance, pressing suspended nanotube over trench by an AFM tip causes the local deformation in the nanotube structure (Fig. 1). Subsequently, the deformed region affects the electron flow between the source and drain electrodes, and as a result, the conductance of carbon nanotubes decreases several times as compared to the theoretically predicted one (Leonard et al., 2005; Maiti et al., 2002).

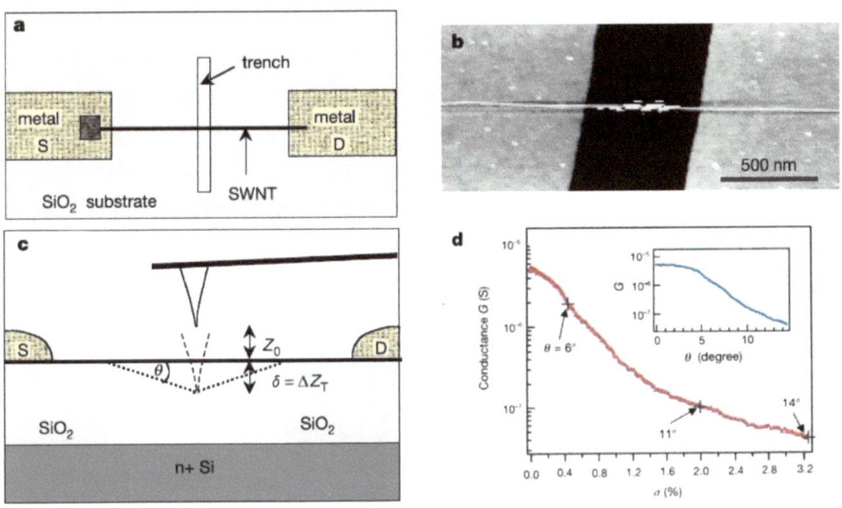

Fig. 1. Electromechanical measurements of a partly suspended nanotube over a trench. a) Device view from above. The substrate has a trench of 500 nm in width and 175 nm in depth. A pair of metal electrodes (S - source and D - drain) is bridged by nanotube suspended over the trench. b) An atom force microscopy image of suspended nanotube. c) Side view of experimental setup when the nanotube is pushed into a trench by atom force microscopy tip. d) Experimentally measured conductance of nanotube as a function of strain. All figures are from (Tombler et al., 2000).

Cross-sectional changes of carbon nanotube geometry under radial deformation are characterized by two applied strains ϵ_{xx} and ϵ_{yy} along x and y axis, respectively, and they are defined as follows:

$$\epsilon_{xx} = \frac{D_0 - a}{D_0} \tag{1}$$

and

$$\epsilon_{yy} = \frac{D_0 - b}{D_0} \tag{2}$$

where D_0 is the diameter of the undeformed nanotube, a and b are the major and minor axes of the ellipse (Fig. 2a-b). For discussion purposes here we use percentage (%) values of the applied strains ϵ_{xx} and ϵ_{yy}. Depending on the degree of radial deformation, the radial cross section of the nanotube can take a different form. When the applied strain is small, the nanotube has an elliptical-like shape. However, if the applied strain continues to increase, then nanotube cross section undergoes Peanut-like or Flat-like forms. The Peanut-like form corresponds to squeezing of the nanotube between two hard surfaces with a cross section smaller than its radial one, while the Flat-like form corresponds to radial squeezing between two hard surfaces with a cross section larger than its radial one (Fig. 2c-d). The Peanut and Flat deformations can also occur during the experimental measurements when the size of AFM tip is smaller or larger than the cross section of the nanotube.

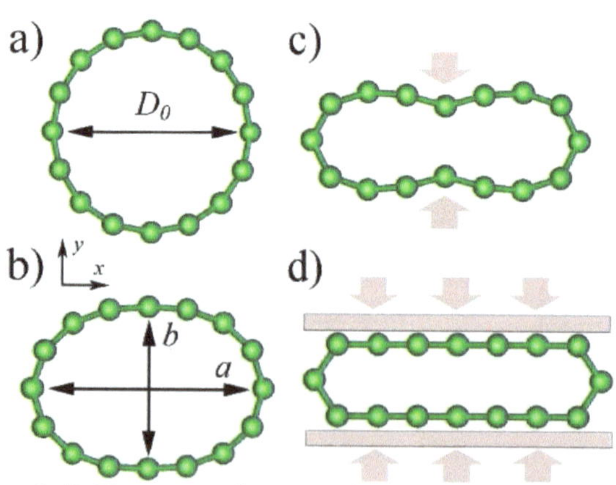

Fig. 2. Cross section view of (8, 0) nanotubes under radial deformation. a) Perfect nanotube. b) Nanotube under small radial deformation with elliptical shape of the cross section. c) Peanut-like deformed nanotube. d) Flat-like deformed nanotube.

First principal calculations have shown that the band gap of zigzag semiconductor nanotubes decreases with the increase of radial deformation while for zigzag metallic nanotube it increases and then decreases (Fig. 3). At the same time, if the mirror symmetry of armchair carbon nanotubes is not broken, the radial deformation does not have any effect on the band gap (Gulseren et al., 2002; Park et al., 1999). Therefore, the application of radial deformation causes chirality dependent semiconductor-metal and metal-semiconductor transitions in carbon nanotubes. There are several factors responsible for these transitions, such as $\sigma - \pi$ hybridization of the atomic orbitals located on the high curvature regions, interaction between low curvature regions, breaking mirror-symmetry of carbon nanotubes, and geometry deformations (Giusca et al., 2007, Gulseren et al., 2002; Mazzoni & Chacham 2000; Mehrez et al., 2005; Shan et al., 2005; Vitali et al., 2006).

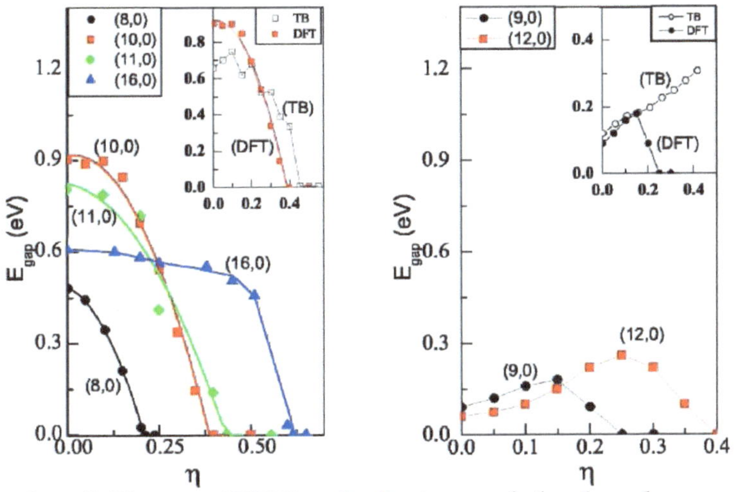

Fig. 3. Density Functional Theory (DFT) calculations of the band gap as a function of the cross sectional flattering for variety of semiconductor (left panel) and metallic (right panel) carbon nanotubes. The insets on both panels compare tight-binding (TB) and DFT results for $(10,0)$ and $(9,0)$ nanotubes, respectively from (Shan et al., 2005).

2.2 Mechanical Defects in Carbon Nanotubes

The presence of mechanical defects in carbon nanotube structure drastically affects all their properties such as transport, mechanical, magnetic, electronic, and optical. Defects can occur during growth, purification, alignment, or device application processes. They can also be introduced intentionally to modify the carbon nanotube properties. Common defects are Stone-Wales defects (Fig.4b), substitutional impurities (Fig. 4c), and vacancies (Fig. 4d). A Stone-Wales defect takes place in a carbon network when one C-C bond is rotated by 900 degree (Stone & Wales 1986). Such rotation creates two pairs of a pentagon and a heptagon around the rotated C-C bond. It is also known as a 5-7-7-5 defect (Nardelli et al., 2000). The pentagon/heptagon pairs create a local disturbance which serves as a reactive center for adsorption of various atoms, nanoparticles, and molecules (Bettinger 2005; Picozzi et al., 2004; Yang et al., 2006). Increasing the concentration of Stone-Wales defects leads to semiconductor-metal transitions in zigzag nanotubes. It also increases the density of states around the Fermi level for armchair nanotubes due to new defect states, thus, making the tubes more metallic (Crespi & Cohen 1997).

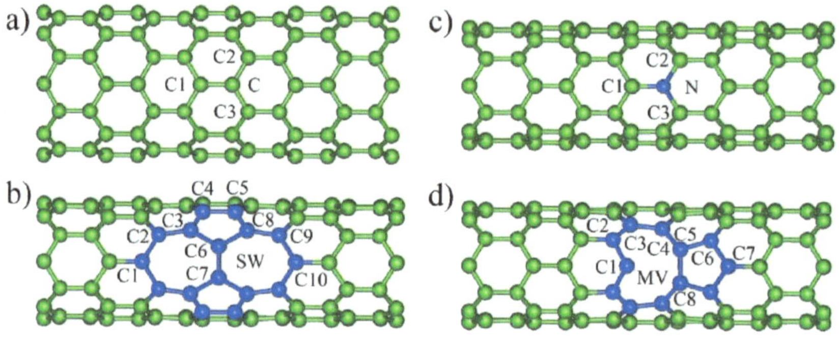

Fig. 4. Side view of $(8,0)$ nanotube structure a) without deformation, and with different defects at $\epsilon_{yy} = 20\%$ deformation for b) a Stone-Wales defect, c) nitrogen substitutional impurity, and d) mono-vacancy. The atoms of the defective sites are labeled.

A substitutional impurity in nanotube can happen when a carbon atom is substituted by a foreign one (Fig. 4c). It has been considered that the common doping atoms in carbon nanotubes are boron and nitrogen, which can appear during laser-ablation and arcdischarge synthesizing processes or during substitutional reaction methods (Droppa et al., 2002; Gai et al., 2004; Golberg et al., 1999). Since B and N atomic radii have similar size to the C atom, they create a small perturbation in the nanotube structure in comparison to the perfect one. However, other atoms such as Li, K, Br, and Ni, which can create a larger disturbance, are also considered for new nano-device application of nanotubes (Lee et al., 1997; Santos et al., 2008). The B and N serve as acceptors and donors of electrons in the nanotubes, respectively, since B has one electron less and N has one electron more than the carbon atom. Theoretical calculations have revealed that such impurities lead to the shift of the Fermi level, additional impurity states in the band structure, and a break of the nanotube mirror symmetry (Czerw et al., 2001; Moradian & Azadi 2006; Yi & Bernholc 1993). Fig. 5b shows evidence of new impurity levels in the density of states of (10,0) nanotube around the Fermi level under nitrogen and boron doping. The impurity effect on electronic properties becomes more evident if their concentration increases. Nitrogen doping has drawn substantial interest due to the one extra electron provided by a nitrogen atom, which causes magnetic effects in nanotubes, transforms semiconductor nanotube to metallic one, and serves as a reactive center in the interaction of nanotubes with other molecules (Nevidomskyy et al., 2003).

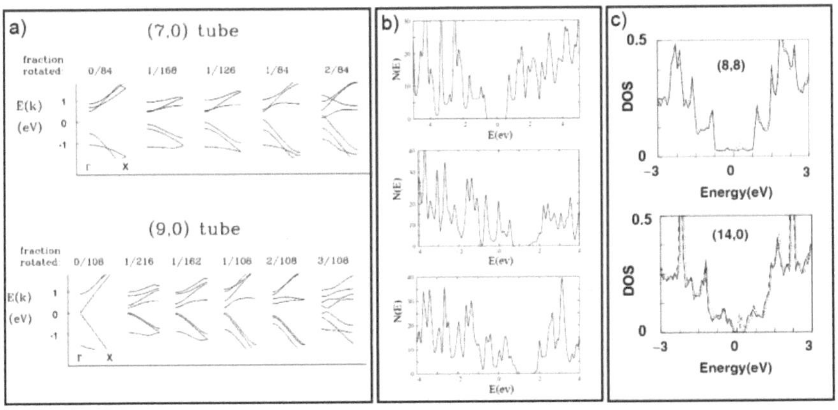

Fig. 5. Band structure and density of states of defective nanotubes. a) Band structure evolutions of (7, 0) and (9, 0) nanotubes upon increasing the concentration of Stone-Wales defect from (Crespi et al., 1997). b) Density of states (top) of (10,0) nanotube, (middle) where one of carbon atom is substituted by a nitrogen atom, and (bottom) where a carbon atom is substituted by a boron atom in the supercell of 120 carbon atom from (Moradian & Azadi 2006). The nitrogen and boron defect concentration is 0.83%. c) Density of states for (8,8) and (14, 0) nanotubes without/with mono-vacancy defect in their structure (solid/dashed line respectively) from (Lu & Pan 2004).

Furthermore, removing a carbon atom leaves a vacancy defect with three dangling carbon bonds in the nanotube network (Rossato et al., 2005). These three dangling bonds are unstable and undergo recombination to make a chemical bond between two of them forming a pentagon ring and one remaining dangling bond - in a nonagon ring (Fig. 4d) (Berber, S. & Oshiyama 2006). The existence of vacancies has been demonstrated experimentally (Hashimoto et al., 2004). It occurs during synthesis of nanotubes or under intentional ion or electron irradiation techniques (Krasheninnikov et al., 2001; Krasheninnikov et al., 2005). The formation of vacancy and its orientation depends on the radius and chirality of nanotubes. Such a disturbance causes significant changes in their transport, magnetic, mechanical, and optical properties (Orellana & Fuentealba 2006; Tien et al., 2008; Yuan & Liew 2009). Fig. 5c shows the density of state of (8,8) and (14,0) carbon nanotubes without a vacancy (solid line) and with vacancy defect (dashed line). The vacancy defect gives the sharp peak in the density of states above the Fermi level, which consists of localized states from the C dangling bond. Such imperfections can reduce the mean-free path and mobility of the free carrier in nanotubes. This leads to the decrease in the nanotube conductivity. On the other hand, the dangling bond can participate in the interaction with other molecules as well as it can be the center of functionalization of nanotubes. Theoretical studies have shown that carbon nanotubes with vacancies demonstrate ferromagnetic ordering due to the dangling bond of the nonagon ring (Orellana & Fuentealba 2006). Moreover, the magnetic moment depends on the defect concentration. It opens up opportunities for the application of carbon nanotubes in spintronics, logical, and memory storage devices.

2.3 Carbon Nanotubes under External Fields

The successful application of carbon nanotubes in nano-devices such as filed-effecttransistors (Tans et al., 1998), rectifiers (Yao et al., 1999), p-n junctions (Lee et al., 2004), or sensors (Ghosh et al., 2003) requires the knowledge of changes in electronic properties of nanotubes under external electric or magnetic fields. However, it is also important to realize that carbon nanotube properties can be tuned by such fields to reach a desirable functionality. The cylindrical structure of nanotubes suggests two high symmetry directions for application of those external fields: parallel and perpendicular to the nanotube axis (Fig. 6). Theoretical and experimental studies have shown that external electric and magnetic fields have a drastic effect on electronic and magnetic properties of carbon nanotubes (Lai, et al., 2008; Li. & Lin 2006; Minot et al., 2004). The general pattern of external electric and magnetic fields is an electronic band structure modulation with semiconductor-metal or metal-semiconductor transition, as well as lifting the degeneracy of sub-bands (Fig. 6).

As Fig. 6 shows, the energy band gap decreases, and the degeneracy of the energy bands around E_F disappears with the increase of the strength of applied fields. Application of the external electric field parallel to the nanotube axis has attracted a lot of attention because of the excellent field emission properties of carbon nanotubes (Jonge et al., 2004). Such properties have effectively been used in prototype devices such as flat panel display (Choi et al., 1999), x-ray tube (Sugie et al., 2001), and scanning x-ray source (Zhang et al., 2005). Nevertheless, the search for new devices is ongoing. Currently, researchers are exploring the perpendicular direction of field applications as new opportunities of controlling electronic properties of nanotubes.

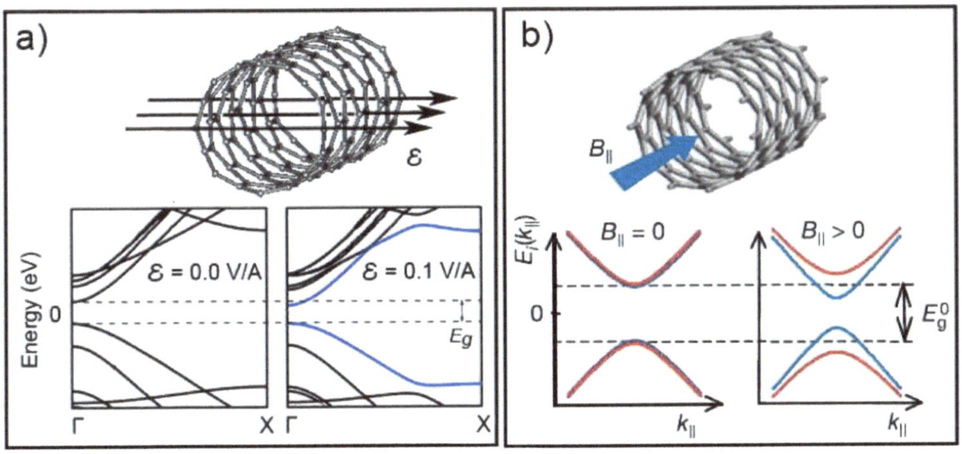

Fig. 6. a) Prospective view of $(8,0)$ nanotube under transverse external electric field (top) and band structure changes under $\mathcal{E} = 0$ V/Å and $\mathcal{E} = 0.1$ V/Å external electric field (bottom). b) Prospective view of nanotube in the presence of magnetic field along nanotube axis (top) and band structure changes under magnetic field from (Minot et al., 2004). The magnetic and electric fields lift of the degeneracy of the band structure.

The application of external magnetic field leads to metal-insulator transition in nanotubes. Moreover, the carbon nanotubes can have diamagnetic or paramagnetic characteristics which depend on their chirality and the orientation of magnetic field (Lu 1995). When the magnetic field is applied along the nanotube axis (Fig.6b), oscillations in the magnetoresistance or in the band gap can be observed due to the Aharonov-Bohm effect, which is a periodic behavior of the wave-function phase factor in the cylindrical geometry (Bachtold et al., 1999; Lee et al., 2000). The Zeeman effect has been found to cause metallization of carbon nanotubes at certain values of the magnetic flux (Jiang et al., 2000). However, perpendicularly applied magnetic field has a different effect on the electronic structure of nanotubes. It opens the band gap at a weak strength and closes that band gap at a high strength of magnetic field for zigzag nanotube. When the magnetic length $l = \sqrt{c\hbar/eH}$ is smaller than the nanotube circumference, the formation of Landau level formations at high magnetic fields closes the band gap, while those levels do not appear in small magnetic fields (Ajiki & Ando 1996). The application of external electric or magnetic fields has also been used in experimental alignment of carbon nanotube samples (Kordas et al., 2007; Kumar et al., 2003).

3. Calculations Method and Model

3.1 Method

First principal simulations are considered to be a state of the art approach for nanoscale design. Such simulations allow investigating a large variety of systems, while these can be very difficult to recreate in actual experiments. They are a powerful tool to explore the functional properties of prospective nanodevices. Among atomistic simulation methods, density functional theory (DFT) is the most successful quantum mechanical technique for calculating the electronic structures of many materials. DFT results are found in a satisfactory agreement with numerous experimental data. Our main method of investigation is based on self-consistent DFT calculations within the local density approximation (LDA) for the exchange-correlation functional implemented within the Vienna Ab Initio Simulation Package (VASP) (Kresse & Furthmuller 1996). The code uses plane-wave basis set and it is a periodic supercell approach. The core electrons of the atoms are treated by ultrasoft Vanderbilt pseudopotentials or by the projector-augmented wave method (Kresse & Joubert 1999). This DFT method is rescalable for large systems, and it requires relatively small plane-wave basis set for each atom in the calculation. It also allows the application of external electric fields in the system (Neugebauer & Scheffler 1992). Over the last decay, DFT-LDA has been a useful model in studying large graphitic structures. It provides an accurate description of the electronic structure along the graphite planes and nanotube axis, where the electronic structure is characterized by sp^2 orbitals of the carbon atoms (Kolmogorov & Crespi 2005; Shtogun et al., 2007; Woods et al., 2007). Based on its past success, we utilize DFT-LDA for our studies. The calculation criteria for relaxation of the atom positions have been 10^{-5}eV for the total energy and 0.005eV/Å for the total force. Spin-polarization effects are also included in the study for defective carbon nanotubes.

3.2 Model

As a model for investigating the combined effect of radial deformation and defects, we study a zigzag (8,0) single wall carbon nanotube. To avoid the interaction between nearestneighbor defects we construct a supercell consisting of four unit cells along the z-axis of the nanotube which has the length of 17.03Å after relaxation. The radial deformation is simulated by applying stress to the cross section of the nanotube along the y-direction from both sides. Consequently, the nanotube is compressed along the y-direction and extended along the x-direction (Fig. 2a-b). The nanotubes are squeezed by two types of strain. The first one is deforming the nanotube between two hard surfaces with a cross section smaller than the nanotube one, which we call "Peanut" deformation

(Fig. 2c). The second one represents squashing the nanotube between two hard surfaces with a cross section larger than the nanotube one, which we call "Flat" deformation (Fig. 2d). Such deformations are modeled by fixing the y-coordinates of the top and bottom rows at low curvature in Peanut and Flat structures, while the rest of the atoms are free to relax. We show that the (8,0) nanotubes can undergo large radial deformations - up to $\epsilon_{yy} = 75\%$ and $\epsilon_{yy} = 65\%$ for Peanut and Flat deformation, respectively (Shtogun & Woods 2009a).

After the radial deformation, defects, such as a Stone-Wales defect, nitrogen impurity, or mono-vacancy are introduced on the high curvature of the nanotube. Experimentally this can be achieved, for example, by deforming the tube using an atom force microscope and then intentionally introducing defects using electron or ion irradiation methods. The radially deformed carbon nanotube with or without defects has an elastic response even under high degree of squeezing. When the applied stress is removed, the nanotube comes back to its original circular cross-sectional form regardless of the defect presence. To study the combined effect of radial deformation and an external electric field we focus on zigzag semiconducting (8,0) and metallic (9,0) single wall carbon nanotubes. The constructed supercell consists of one unit cell along z-axis with length of 4.26Å for the (8,0) and 4.25Å for the (9,0) nanotubes after relaxation. The external applied electric field is generated by a saw-tooth electric potential (Neugebauer & Scheffler1992) in three different directions: $\mathcal{E}_x$ and $\mathcal{E}_y$ along x and y-axes, respectively, and $\mathcal{E}_{45^0}$ along 45^0 from the x-axis (Shtogun & Woods 2009b).

4. Carbon Nanotubes under Two External Factors

Many devices associated with different applications are based on the unique properties of carbon nanotubes and the opportunity of controlling them. The search for new methods to modulate nanotube properties has led to the introduction of deformation, mechanical defects, and external fields. Reaching the desirable properties of nanotubes under one external factor, however, can be difficult in device engineering. To overcome such difficulties, researchers now consider how more than one external factors can alter the carbon nanotube properties. Thus, it is important to build fundamental understanding of the combination of more than one external stimulus by starting with applying two stimuli simultaneously.

Studies have shown that the combination of radial deformation and impurity or vacancy, for example, can facilitate tuning electronic properties of nanotubes (Fagan et al., 2003; Fagan et al., 2004). As we have shown above, the radial deformation causes the closing of band gap in semiconductor nanotubes but how fast this band gap is closed can be controlled by various defects such as impurities and vacancies (Fagan et al., 2004). Radial deformation produces two regions with high and low curvatures (Fig.2b). As a result, the atoms on the high curvature tend to have sp^3 configuration, and they are more strained than the ones on the low curvature with sp^2 configuration. It makes it easier to create the defect on the high curvature in comparison to the low one. The flexibility in localization of the defect in radially deformed nanotubes gives more control over nanotube properties.

Band gap modulation can also be achieved by combining external electric fields and defects in nanotubes. This is of particular interest in application of nanotubes in field-effect transistors since the current can be turned off or on by applying external electric field. Such combination provides an opportunity to identify presence of defects in carbon nanotubes. Rotating nanotube in external electric field provides different responses in different directions (Tien et al., 2005). Fig. 7 shows the presence of a vacancy defect in (10,0) nanotubes with an applied external electric field for various directions. We can see that the value of the band gap can be controlled by the presence of vacancy, strength, and direction of electric field. It is interesting that usually semiconducting tubes are used in nanoelectronics. Metallic nanotubes have been discarded because their resistance is not sensitive to gate voltage or transverse electric field (Li et al., 2003).

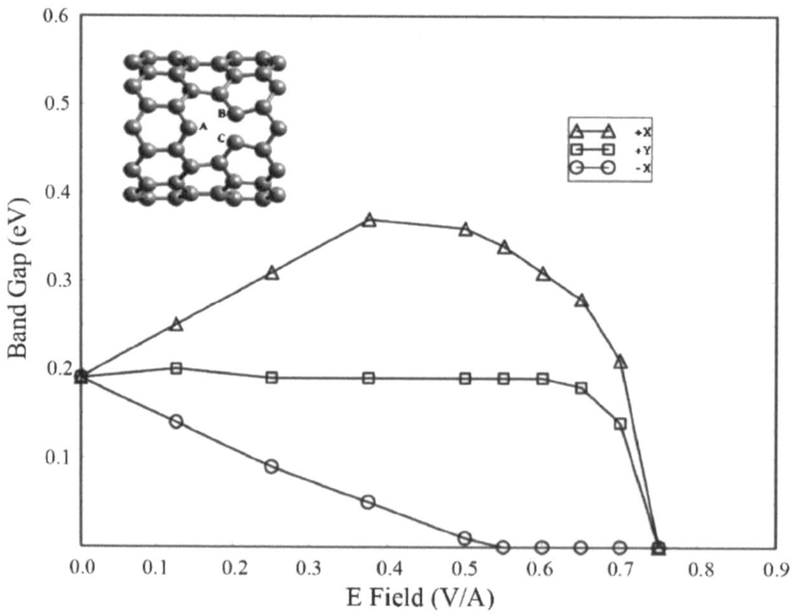

Fig. 7. Energy band gap evolution as a function of external electric field in three directions such as $+x$-axis, $-x$-axis, and $+y$-axis for vacancy defect in $(10, 0)$ nanotube with 79 carbon atoms from (Tien et al., 2005). The insert shows the optimized structure of vacancy defect.

New views on metallic nanotube application are considered due to the combined effect of external electric field and impurities, which allows tuning the resistance by two orders of magnitude. The origin of this change comes from the backscattering of conduction electron by impurities, and it strongly depends on the strength and orientation of the applied electric field (Son et al., 2005).

Nowadays, most magnetic materials are based on d and f elements. However, as we mentioned above, sp^2 and sp^3 hybridization due to mechanical defects (vacancy and impurity) can also induce magnetism in carbon nanotubes. This is of great importance for nanotube applications in magnetic and spintronic devices. Magnetism can be further influenced by the combined effect of mechanical deformation and defects. The strong magneto-mechanical coupling has been shown in defective nanotubes under mechanical deformation (Zheng & Zhuang 2008). It can increase or decrease the magnetic moment and cause changes in the spin transport. To further investigate this, we have explored energetic, structural, magneto-mechanical, and electronic changes in carbon nanotubes under combined effect of radial deformation and external electric field along with radial deformation and different defects.

4.1 Structural Changes in Radially Deformed and Defective Carbon Nanotubes

In this section we analyze the changes in the carbon nanotube structure under radial deformation and various defects under two types of deformation - Peanut and Flat. The main focus is on the bonds on the high curvature and the bonds comprising the defects (Fig. 4). When the perfect nanotube is radially deformed, the bonds along the nanotube axis (C-C1) decrease while the curvature bonds (C-C2) increase to reduce the strain on the high curvature (Fig. 8a). Under moderate radial deformation ($\epsilon_{yy} < 50\%$) there is no significant difference between Peanut and Flat deformation. However, at high deformation ($\epsilon_{yy} > 50\%$) the difference becomes more evident since the Flat deformation applies much more distortion on the high curvature rather than the Peanut does. As a result, the bonds on the high curvature change more in Flat than Peanut deformation, while there are no changes along nanotube axis (Firg. 8a). It is worth mentioning that in the region of ($\epsilon_{yy} \sim 20 \div 25\%$), the (8,0) nanotube shows interesting changes in the structure, which can be seen in the jump of the bond length profile. Later, there will be more discussion of this jump due to the admixture of different orbitals from the C atoms.

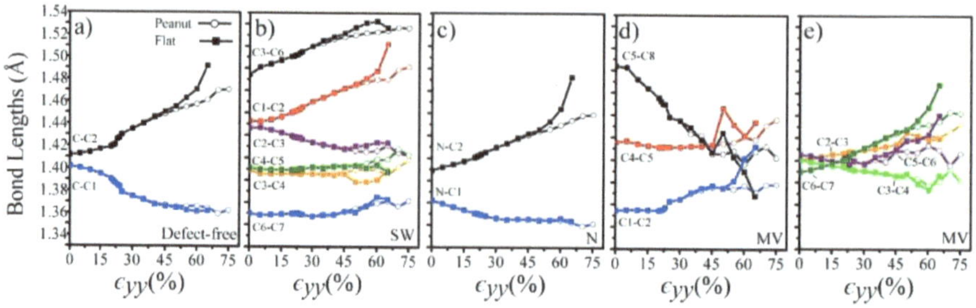

Fig. 8. (8,0) nanotube bond length evolution as a function of applied strain ϵ_{yy} for a) no defect, b) with a Stone-Wales defect, c) with a nitrogen impurity, d) and e) with a monovacancy.

Energetically favorable configuration of Stone-Wales defect happens when two heptagon rings align with the nanotube axis (Fig. 4b). Fig. 8b) shows the evolution of various bond changes as a function of ϵ_{yy}. Both, Peanut and Flat, deformations produce almost the same changes in the defect bond lengths with the exception of the C1-C2 bond. In fact, the bonds in the heptagon rings (C1-C2, C2-C3, and C3-C6) undergo more changes than the bonds in pentagon ones (C3-C4, C4-C5, and C6-C7). Such changes are explained by the location of the heptagon rings on the high curvature because the pentagon rings are distant from it.

Moreover, the C1-C2 functional behavior looks similar to the one for the C-C2 in the defectfree nanotube.

The presence of a nitrogen impurity on the high curvature of the nanotube does not create much disturbance in its structure (Fig. 4c). Since the atomic size of the nitrogen atom is larger than the carbon one, it leads to the decrease in $N - C1$ and $N - C2$ bond lengths by 0.01Å and 0.03Å as compared to the $C - C1$ and $C - C2$ in defect-free nanotube for all ϵ_{yy}, respectively (Fig. 8c). Another evidence of small disturbance due to the nitrogen atom can be seen from the functional behavior of the $N - C1$ and $N - C2$ bonds, which exhibits similar dependence on ϵ_{yy} as the corresponding C - C bonds in the defect-free nanotube (Fig. 8a).

Removing a carbon atom from the carbon network creates a vacancy leaving three dangling bonds behind. Such configuration is not stable, and it undergoes reconstruction with the formation pentagon and nonagon along the axis of zigzag nanotube (Fig. 4d). On the contrary to the previous cases (defect-free, Stone-Wales, and nitrogen impurity) where the bond lengths show continuous behavior, vacancy has more dramatic dependence as a function of deformation ϵ_{yy} (Fig. 8d-e). In particular, when the distance between the two low curvatures becomes comparable with graphite interplanar distance, there are some relatively "sudden" jumps and dips in bond lengths for Flat deformation. We again observe that most changes in the bond lengths happen on the high curvature. For instance, the bond lengths of C1-C2 and C6-C7 increase by 0.06Å and 0.08Å, respectively, while the C5-C8 decreases by 0.11Å for the Flat deformation.

As a result, the presence of mechanical defects causes the structural changes in nanotubes where nitrogen impurity creates the lowest perturbation in nanotube network whiles the mono-vacancy - the largest one.

4.2 Characteristic Energies of Deformation and Defect Formation in Carbon Nanotubes

The structural changes described above lead to changes in deformation and defect formation energies. We introduce the characteristic energies of radial deformation and defect formations. The energy required to perform radial deformation is defined as:

$$E^f_{\epsilon_{yy}} = E_{\epsilon_{yy}} - E \qquad (3)$$

where $E_{\epsilon_{yy}}$ is the total energy for the deformed nanotube and E is the total energy for the undeformed perfect nanotube. Fig. 9 shows the deformation energy for (8,0) and (9,0) carbon nanotubes as a function of ϵ_{yy}. The deformation energy has non-linear dependence and it rises with the increasing ϵ_{yy} showing that higher deformation requires more energy to overcome strain energy on the high curvature.

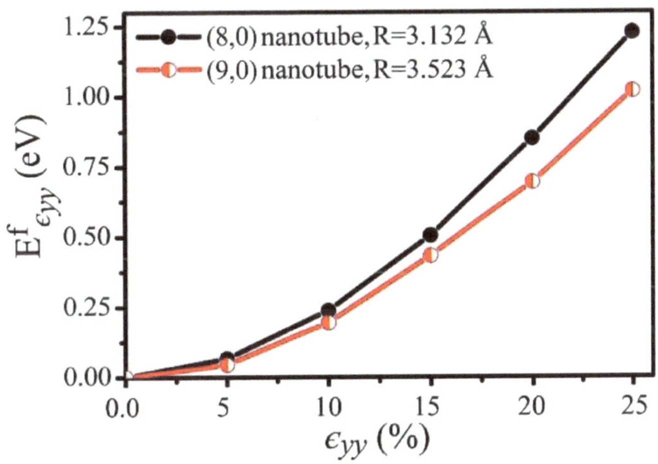

Fig. 9. The deformation energy as a function of radial deformation ϵ_{yy} for one unit cell of $(8,0)$ and $(9,0)$ nanotubes.

On the other hand, the deformation energy for (8,0) nanotube is steeper than for (9,0) due to the larger curvature effect in (8,0) nanotube as compared to the (9,0) one.

The defect formation energies for the Stone-Wales defect, nitrogen impurity, and vacancy at the high curvature of the radially deformed nanotube are calculated, respectively as:

$$E^f_{\epsilon_{yy}/SW} = E_{\epsilon_{yy}/SW} - E_{\epsilon_{yy}} \tag{4}$$

$$E^f_{\epsilon_{yy}/N} = E_{\epsilon_{yy}/N} - E_{\epsilon_{yy}} - \mu_N + \mu_C \tag{5}$$

$$E^f_{\epsilon_{yy}/MV} = E_{\epsilon_{yy}/MV} - E_{\epsilon_{yy}} + \mu \tag{6}$$

where $E_{\epsilon_{yy}/SW,N,MV}$ are the total energies for the deformed nanotube with the appropriate defect (SW - Stone-Wales, N - nitrogen impurity, and MV - mono-vacancy), $\mu_{N,C}$ are the chemical potentials for a free nitrogen and carbon atoms, respectively, and μ is the chemical potential of a carbon atom in the nanotube (energy per C atom in the supercell).

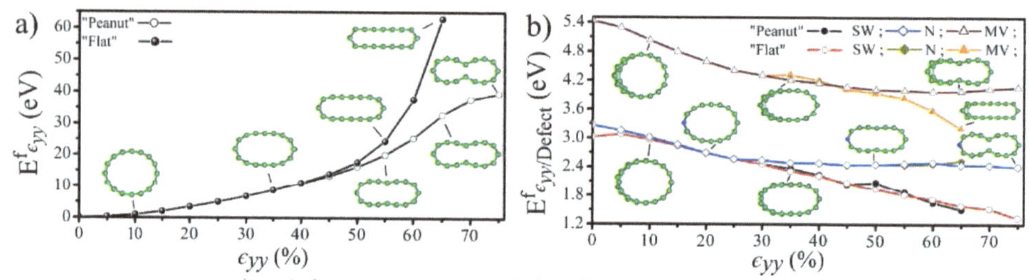

Fig. 10. The energy of a) deformation and b) defect formation as a function of ϵ_{yy} for Peanut and Flat deformations in (8,0) nanotube. Cross-sectional views of the deformed and defective (8,0) nanotubes are also provided.

Applied radial deformation of up to $\epsilon_{yy} < 35\%$ transforms the circular (8,0) nanotube to an elliptical-like one (Fig. 2b). Further increase of $\epsilon_{yy} > 35\%$ results in Peanut or Flat structures.

As it can be seen from Fig. 10a, the deformation energy becomes more distinctive between Peanut and Flat deformations for higher radial stress $-\epsilon_{yy} > 45\%$. What actually happens for high deformation is that for the Peanut deformation only one row set on the top and the bottom of low curvatures are brought together. Consequently, there are no significant changes on the high curvature and elongation along a-axis (Fig. 2c). The Flat deformation, however, has five rows of atoms from top and bottom on the low curvatures brought together (Fig. 2d). As a result, it leads to major changes on high curvature and appropriate elongation along a-axis (Fig. 2d). Meanwhile, radial deformation energy shows steeper increase in Flat that in Peanut deformation. In addition, there is a repulsion between two low curvatures of nanotube at $\epsilon_{yy} > 50\%$, where separation distance becomes smaller than graphite interplanar distance. The repulsion is stronger for Flat rather than Peanut deformation because fewer atoms are brought together in Peanut as compared to Flat deformation.

Fig. 10b also shows variations in the defect formation energy as a function of the deformation. All energies $E^{f}_{\epsilon_{yy}/SW,N,MV}$ decrease for all types of defects as ϵ_{yy} increases.

Therefore, the defect formation occurs much easier in radially deformed carbon nanotubes than in the undeformed ones. This has been shown for Si doping of carbon nanotubes (Fagan et al., 2004). Analysis of defect formation energies shows that the lowest amount of energy is required to produce Stone-Wales defect, more energy is needed to make nitrogen impurity and the most energy-expensive is the vacancy formation at high curvature of carbon nanotubes. There is a small difference between Peanut and Flat deformations for Stone-Wales defect and nitrogen impurity, while this difference is significant for monovacancy. It relates to the fact that Stone-Wales and nitrogen impurity create small disturbance in nanotube structure for both deformations where the $C-C$ bonds in monovacancy exhibit more dramatic changes for the Flat that Peanut deformations (Fig. 8d-e). Thus it is easier to break the three bonds to form a mono-vacancy after squashing between two wide surfaces as compared to the case of two narrow ones (Fig. 2c-d). As a result, the higher the deformation ϵ_{yy} is, the easier it is to make defects in carbon nanotube structure.

4.3 Electronic and Magnetic Properties of Carbon Nanotubes under Combined Effect of Radial Deformation and Defects

One of the main questions we consider here is what electronic response carbon nanotubes show under the combined effect of radial deformation and different defects. We address this issue by analyzing the modulation of the band gap E_g of defect-free and defective (8,0) nanotube under radial deformations ϵ_{yy} (Fig. 11). Initially, the defect-free (8,0) nanotube is a semiconductor with $E_g = 0.55\text{eV}$. Applying radial deformation leads to decreasing of E_g until its closure at $\epsilon_{yy} = 23\%$ where semiconductor-metal transition occurs. As we observed earlier, this semiconductor-metal transition is reflected in discontinuous change in the C-C1 and $CC2$ bonds (Fig. 8a). On the other hand, it corresponds to a strong $\sigma - \pi$ hybridization on high curvature which causes structural and semiconductor-metal transformations in (8,0) nanotube. Interestingly, such transition arises before (8,0) nanotube undergoes Peanut or Flat deformations. Further increase of radial deformation does not open second bang gap in the (8,0) nanotube.

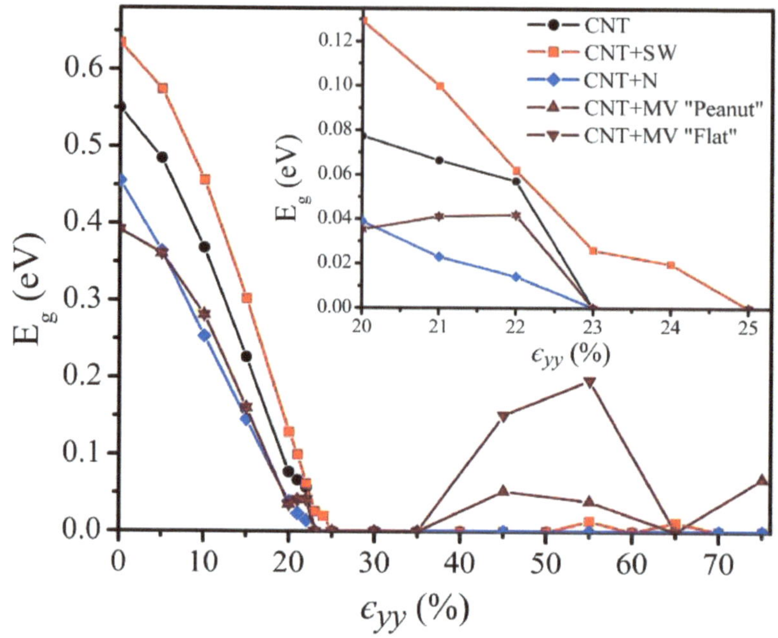

Fig. 11. Energy band gap E_g of perfect and defective $(8,0)$ nanotubes as a function of radial deformation ϵ_{yy}. Insert shows the semiconductor-metal transition in $(8,0)$ nanotube in detail.

The presence of defects at high curvature tunes the value of band gap in nanotubes. For instance, Stone-Wales defect increases E_g, while E_g decreases for nitrogen impurity and mono-vacancy as compared to defect-free nanotube at $\epsilon_{yy} = 0\%$. Applying radial deformation to the defective nanotube has shown similar trends as in defect-free nanotubes (Fig. 11). Moreover, the semiconductor-metal transition happens at the same degree of deformation $\epsilon_{yy} = 23\%$ for nitrogen impurity and mono-vacancy. However, it occurs at $\epsilon_{yy} = 25\%$ for StoneWales defect. Further increasing deformation opens the band gap in nanotube with monovacancy and Stone-Wales defect again, while nanotube with nitrogen impurity remains metallic for all deformations. In addition, the band gap can be modulated by Peanut or Flat deformations at high deformation. Still, we can see that strong $\sigma - \pi$ hybridization is a major factor in semiconductor-metal transition in defective nanotubes but defect-deformation coupling is significant for further opening of the band gap at high deformations.

To elucidate the mechanism of semiconductor-metal transformation in defect-free and defective nanotubes we carry out comprehensive analysis of the electronic structure changes for various cases. The applied radial deformation removes the degeneracy in sub-bands of defect-free nanotubes (Fig. 12a). Increasing ϵ_{yy} causes the lowest conduction band and the highest valence band to move towards the Fermi level. They cross the Fermi level at the point different than Γ for $\epsilon_{yy} = 23\%$. The further deformation leads to the shift of crossing point with E_F towards the X point. For sufficiently large $\epsilon_{yy} > 50\%$, multiple crossings at the Fermi level from higher conduction and lower valence bands occur, which improve metallization of nanotubes and increase the density of states at the Fermi level. Initial analysis reveals that all atoms of carbon network equally contribute to the energy band. However, this input changes with applied deformation. The atoms from high curvature have the main contribution to the energy bands around the Fermi level. For $\epsilon_{yy} > 50\%$ the atoms from the low curvature in the Peanut deformation (top and bottom rows) also contribute to the energy bands at the Fermi level.

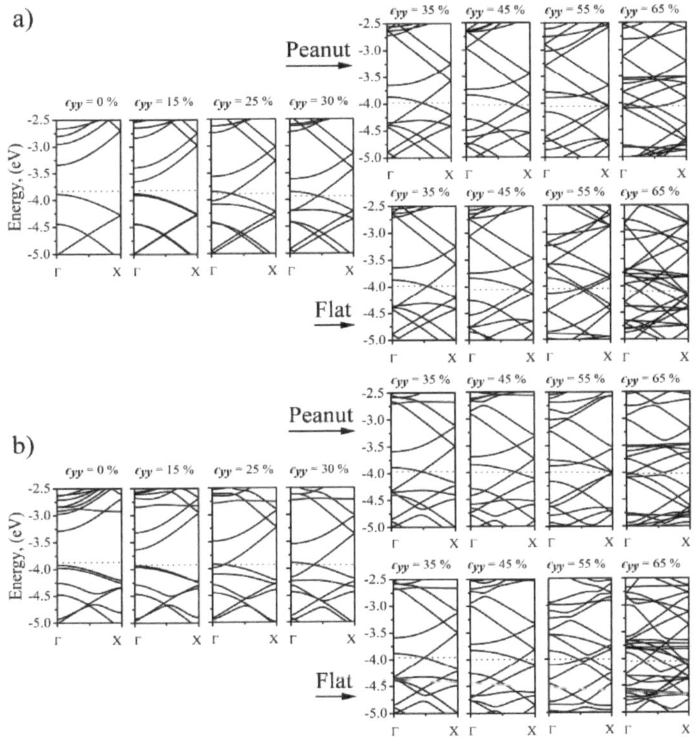

Fig. 12. The energy band structure of a) defect-free $(8,0)$ nanotube, b) $(8,0)$ nanotubes with Stone-Wales defect at high curvature site for several degrees of ϵ_{yy}. The red level is associated with the Stone-Wales defect.

The band structure evolution of the deformed (8,0) nanotube with a Stone-Wales defect shows similar behavior as compared to the defect-free one (Fig. 12b). However, Stone-Wales defect increases initial value of band gap and introduces flat defect level in the conduction band at $E \approx -2.9$eV (in red). Under applied radial deformations $\epsilon_{yy} > 0\%$, this flat level starts to move away from the Fermi level while increasing $\sigma - \pi$ hybridizations brings the lowest conduction and the highest valence bands closer together. Small band gap $E_g \approx 0.01$eV opens up for the Peanut deformation at $\epsilon_{yy} = 55\%$ and $\epsilon_{yy} = 65\%$. There is no such opening for Flat deformation, and the band gap remains closed. The analysis of conduction and valence bands reveals that the lowest conduction bands consist of the orbital from defect atoms where $\sigma - \pi$ hybridizations is significant.

The applied radial deformation induces nonuniform charge distribution on the nanotube surface (Fig. 13). For a defect-free nanotube, the charge accumulation appears on the high curvature. Even more irregular allocation is caused by Peanut and Flat deformations (Fig. 13a-c). Stone-Wales defect creates local disturbance in the nanotube structure, which reduces charge accumulation on the defect itself, except for the C6-C7 bond even though it is on the high curvature. Most of the charge is gathered around the atoms surrounding Stone-Wales defect (Fig. 13d-f).

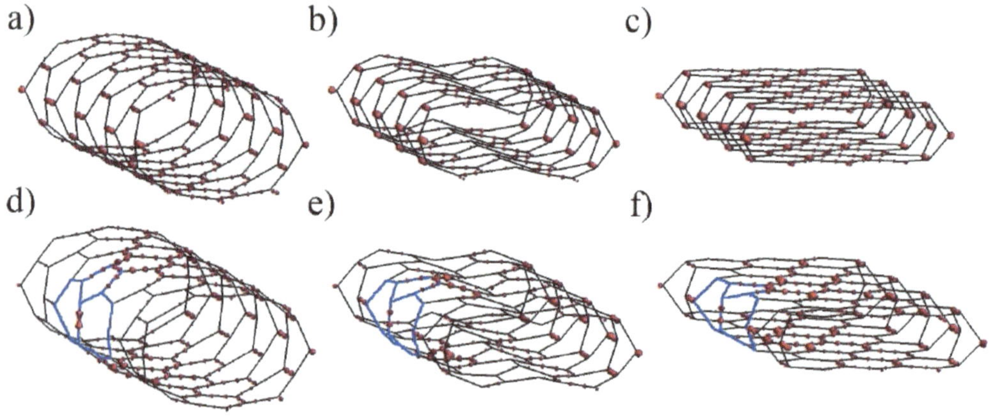

Fig. 13. Total charge density plots with isosurface value is $0.0185e/\text{Å}^3$ for a) defect-free $(8,0)$ nanotube with $\epsilon_{yy} = 25\%$, b) Peanut structure with $\epsilon_{yy} = 65\%$, c) Flat structure with $\epsilon_{yy} = 65\%$, and d) $(8,0)$ nanotube with Stone-Wales defect with $\epsilon_{yy} = 25\%$, e) Peanut structure with $\epsilon_{yy} = 65\%$, f) Flat structure with $\epsilon_{yy} = 65\%$.

The effects of nitrogen impurity and mono-vacancy in radially deformed nanotubes have been investigated in terms of Density of States (DOS) for different ϵ_{yy}. Since the nitrogen atom supplies one extra electron in nanotube, and mono-vacancy provides dangling bond, the calculations include spin-polarized effect to account for magnetic effect in nanotubes. The possibility of magnetism in nanotube has been predicted by other authors (Lim et al., 2007; Orellana & Fuentealba 2006; Wu & Hagelberg 2009). The density of states of nitrogen doped nanotubes is responsive to the concentration of impurities (Orellana & Fuentealba 2006). Increasing the impurity concentration more than 1% provides a flat level in the band gap. In our calculation we do not observe any flat level because the nitrogen concentration is 0.78% (Fig. 14a). As the radial deformation ϵ_{yy} increases, the band gap decreases, and the conduction and valance bands move towards the Fermi level. For $\epsilon_{yy} >$ 45%, the DOS for Peanut and Flat deformation raises around the Fermi level which suggests that the nanotube shows more metal-like behavior.

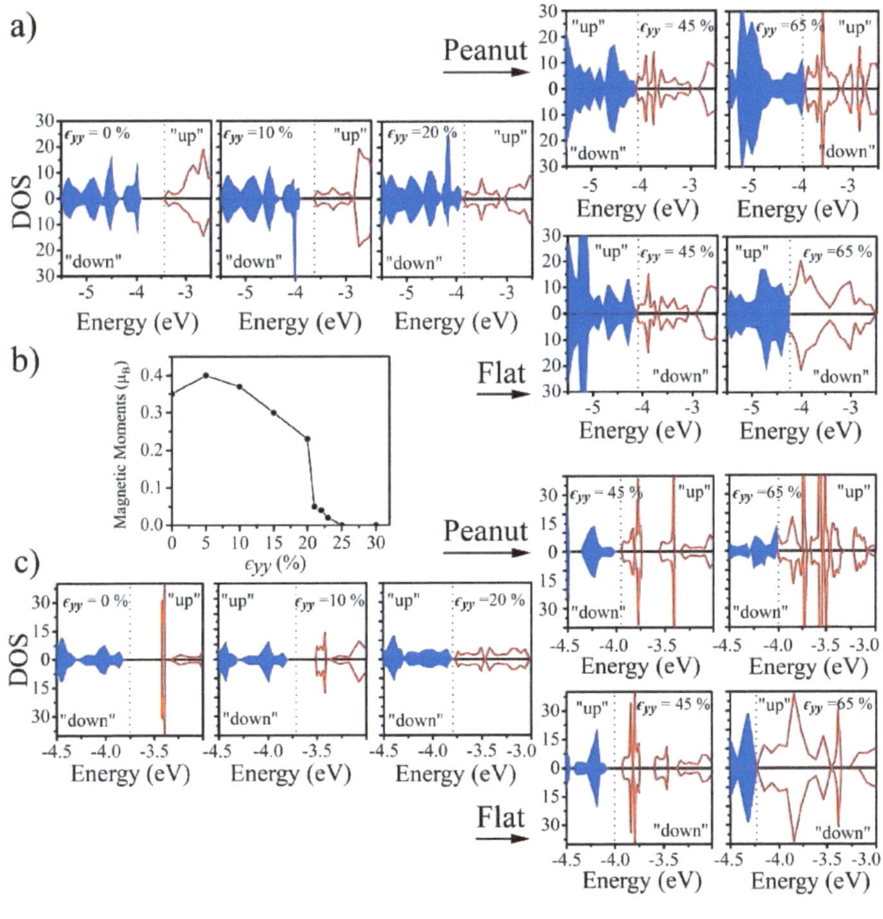

Fig. 14. a) Total Density of States for spin "up" and "down" carriers of (8, 0) nanotube with a substitutional nitrogen impurity on its high curvature at different ϵ_{yy} for Peanut and Flat structures. b) The magnetic moment of the (8, 0) nanotube with the nitrogen impurity as a function of ϵ_{yy}. c) Total Density of States for spin "up" and "down" carriers of (8, 0) nanotube with mono-vacancy defect on its high curvature at different ϵ_{yy} for Peanut and Flat structures.

The nitrogen impurity contributes significantly to conduction band in the vicinity of the Fermi level which is partially occupied. The highest valence band consists of contribution from atoms on the high curvature. The extra nitrogen e^-induces spin-polarization effects (magnetism) in the nanotube electronic structure at $\epsilon_{yy} = 0\%$ and sharp features in the DOS for spin "up" around ~ -2.8eV (Fig. 14a). As ϵ_{yy} increases, the spin-polarized effect is overtaken by $\sigma - \pi$ hybridization. Eventually, it completely disappears at $\epsilon_{yy} > 25\%$ except for at $\epsilon_{yy} = 65\%$ for Flat deformation. Thus, applied radial deformation provides the opportunity to control not only electronic but also magnetic properties of nanotubes. To elucidate this effect further, we investigate changes in local magnetic moment m under ϵ_{yy} (Fig. 14b). For small deformations m increases with a maximum at $\epsilon_{yy} = 5\%$ and then it reduces to $m = 0$ at $\epsilon_{yy} = 25\%$ when the nanotube becomes almost a metal ($\epsilon_{yy} = 23\%$). However, complicated magnetomechanical coupling produces large magnetic moment $m = 0.92\mu_B$ (not shown on Fig. 14b) at $\epsilon_{yy} = 65\%$. It gives evidence of new methods of tuning nanotube properties and complex interrelation between defects and deformations.

Finally, we consider a radially deformed (8,0) nanotube with a mono-vacancy (Fig. 14c). The presence of mono-vacancy is manifested in a sharp localized peak around -3.41 eV in the DOS conduction region for the spin "up" and "down" at $\epsilon_{yy} = 0\%$. This state mainly consists of contributions from the mono-vacancy atoms. As the deformation progresses, the peak intensity decreases and becomes broader, and peak location moves closer to the Fermi level due to $\sigma - \pi$ hybridization from the nanotube atoms on high curvature along with the C atoms of mono-vacancy. Such process repeats again for $\epsilon_{yy} > 40\%$, where sharp peaks around the Fermi level occur and become wider and sharper with increasing ϵ_{yy}. The major contribution to these peaks comes from dangling bond atom - C1 and the C atoms of the pentagon ring of mono-vacancy (Fig. 4d). DOS for spin "up" and "down" demonstrates that there is no spinpolarized effect in radially deformed nanotubes with mono-vacancy for all ϵ_{yy} except $\epsilon_{yy} = 65\%$ in Peanut deformation where a small magnetic moment $m = 0.04\mu_B$ has been found. These results are in partial agreement with previously reported findings that zigzag nanotubes with mono-vacancy do not show magnetic response for given orientation of defect (Berber & Oshiyama 2006). However, the appearance of magnetism at $\epsilon_{yy} = 65\%$ can originate from several reasons such as uncoordinated C atom with a localized unpaired spin, concentration of vacancies, nanotube chirality, sp^3 pyramidization, and related bond length changes, as well as vacancy location with respect to the nanotube structure (Ma et al., 2004; Orellana & Fuentealba 2006; Zheng & Zhuang 2008). Another explanation of the magnetism in radially deformed nanotubes with mono-vacancy is the magnetic flat-band theory together with σ - π hybridization (Fujita et al., 1996; Nakada et al., 1996). The presence of flat bands around the Fermi level leads to increased electron-electron interaction and results in the occurrence of magnetic polarization which is common for edge states in zigzag nanoribbons (Fujita et al., 1996; Nakada et al., 1996). We can consider a mono-vacancy on the high curvature and large deformations such edge states. Our results suggest that there is a complex relation among mechanical, electric, and magnetic coupling in radially deformed nanotubes with the mono-vacancy.

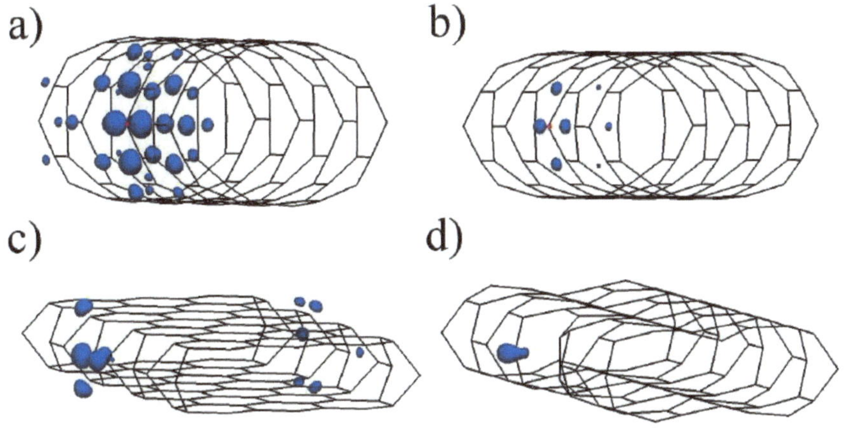

Fig. 15. Spin density isosurface plots for a) $(8, 0)$ nanotube with nitrogen impurity for $\epsilon_{yy} = 10\%$ at $0.004\mu_B/\text{Å}^3$, b)$(8, 0)$ nanotube with nitrogen impurity for $\epsilon_{yy} = 20\%$ at $0.004\mu_B/\text{Å}^3$, c) Flat $(8, 0)$ nanotube with nitrogen impurity for $\epsilon_{yy} = 65\%$ at $0.0122\mu_B/\text{Å}^3$, d) Peanut $(8, 0)$ nanotube with mono-vacancy for $\epsilon_{yy} = 65\%$ at $0.018\mu_B/\text{Å}^3$.

Further insight in the nanotube magnetism can be gained by exploring the spin density distribution for various deformations ϵ_{yy} (Fig. 15). Initially extra nitrogen $e^-$is localized on the defect site but increasing deformation obliterates this localization. As an example, the unpaired $e^-$is much more delocalized at $\epsilon_{yy} = 20\%$ in comparison to the $\epsilon_{yy} = 10\%$ case, where the e^-spin density localized around the impurity site is much more intense. Similar behavior has been reported for semiconducting nanotubes, where the nitrogen $e^-$is localized around the impurity, while for metallic nanotube it is completely delocalized (Nevidomskyy et al., 2003). However, the e^-spin localization around the impurity appears again at $\epsilon_{yy} = 65\%$ in Flat deformation, where the nanotube is a metal, but the magnetic moment has the highest value of - $m = 0.92\mu_B$. Moreover, similar localization of the spin density happens around the dangling bond in the nanotube with mono-vacancy $\epsilon_{yy} = 65\%$. This suggests that the uncoordinated C atom with unpaired spin is vital in the magnetic properties of defective and deformed nanotubes.

4.4 Electronic Properties of Carbon Nanotubes under Radial Deformation and an External Electric Field

Next, we explore the electronic properties of radially deformed nanotubes in an external radial electric field. We focus on the modulation of the band gap by an external electric field in nanotubes with various degrees of deformation. As we have already seen from earlier discussion, applying radial deformation causes the metal semiconductor transition in semiconductor nanotubes and causes opening and closing of a band gap in metallic nanotubes because of the $\sigma - \pi$ hybridization from the high curvature regions. Fig. 16a-b show these dependences in semiconductor (8,0) and semi-metal (9,0) carbon nanotubes as a function of ϵ_{yy}. The same behavior can be achieved by the application of an external electric field to undeformed (8,0) and (9,0) nanotubes. E_g dependences in Fig. 16c-d and Fig. 16a-b look similar. However, the semiconductor-metal transition happens faster for the (8,0) nanotube over a shorter field strength interval as compared to the (9,0) nanotube. In this case, E_g for (8,0) nanotube is closed at $\mathcal{E} = 0.4$ V/Å, while for (9,0) tube the closure occurs at $\mathcal{E} = 0.8$ V/Å. Therefore, we can obtain the same electronic characteristics of nanotubes from radial deformation and the applied external electric field. In addition, it provides the flexibility of controlling nanotube properties at different circumstances.

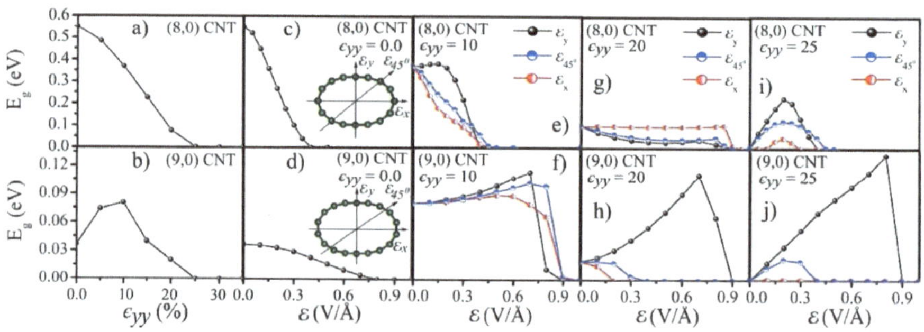

Fig. 16. E_g as a function of ϵ_{yy} for a) (8,0) and b) (9,0) nanotubes; E_g as a function of $\mathcal{E}$ for c) (8, 0) at $\epsilon_{yy} = 0\%$; d)(9, 0) at $\epsilon_{yy} = 0\%$; e)(8, 0) at $\epsilon_{yy} = 10\%$; f (9, 0) at $\epsilon_{yy} = 10\%$; g)(8, 0) at $\epsilon_{yy} = 20\%$; h)(9, 0) at $\epsilon_{yy} = 20\%$; i) (8, 0) at $\epsilon_{yy} = 25\%$; j) (9, 0) at $\epsilon_{yy} = 25\%$. The insert indicates electric field directions as follows: $\mathcal{E}_x$ is along x-axis, $\mathcal{E}_y$-along y-axis, and $\mathcal{E}_{45}$ ^{0}-along 45^0 from x-axis in $x - y$ plane.

The symmetry of the nanotube is decreased under applied radial deformation. Consequently, the x and y directions of the cross section are not equivalent any more (Fig. 2b). It has been shown before that electronic structure of the defective nanotube depends on the orientation of external electric field (Tien et al., 2005). Here, we analyze the E_g

evolution of the deformed (8,0) and (9,0) nanotubes under $\mathcal{E}$ along x, y-axes and the 45^0 direction in the $x - y$ plane (insert in Fig. 16c-d).

Radially deformed (8,0) nanotube at $\epsilon_{yy} = 10\%$ still has a relatively large band gap - $E_g = 0.37$ eV , which decreases under applied electric field and vanishes at $\mathcal{E} \sim 0.45$ V/Å for all three directions. Each E_g dependence has distinct profile for $\mathcal{E} < 0.4$ V/Å.$_g$ initially increases and then starts to decrease for y-direction. However, it sharply decreases for x-direction, and it shows intermediate, almost linear, decline for 45^0-direction (Fig.16e). At $\epsilon_{yy} = 20\%$ the applied electric field has a small effect on E_g over relatively long range of electric field strength for all directions until it closes E_g at $\mathcal{E} \sim 0.8$ V/Å (Fig. 16g). The (8,0) nanotube becomes metallic at $\epsilon_{yy} = 25\%$ but applied electric field opens E_g with its maximum at $\mathcal{E} \sim 0.2$ V/Å for all directions. The largest E_g is for the electric field along y-direction and the lowest - for x direction (Fig. 16i).

On the other hand, the radially deformed (9,0) nanotube at $\epsilon_{yy} = 10\%$ has the largest $E_g = 0.08$ eV , which continues to increase under applied electric field in all directions until $\mathcal{E} \sim 0.6 \div 0.7$V/Å, where E_g is sharply closed (Fig. 16f). This behavior is similar for E_g in the (8,0) nanotube at $\epsilon_{yy} = 20\%$. More drastic behavior of the band gap is observed at $\epsilon_{yy} = 20\%$ for electric field along y-direction. The original value of the band gap increases by ~ 4 times at $\mathcal{E} = 0.7$ V/Å which is followed by its rapid closure at $\mathcal{E} = 0.9$ V/Å. At the same time, the weak electric field $\mathcal{E}$ along x - and 45^0 - directions closes the band gap. The same dependence has been found at $\epsilon_{yy} = 25\%$. The band gap reaches maximum for $\mathcal{E} = 0.8$ V/Å while there is no effect of applied electric field on the band gap along x-direction.

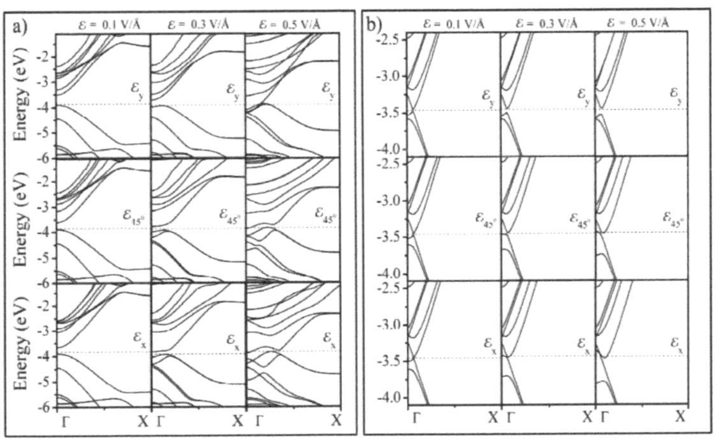

Fig. 17. Energy band structure evolutions for a) $(8,0)$ nanotube at $\epsilon_{yy} = 10\%$ and b) $(9,0)$ nanotubes at $\epsilon_{yy} = 25\%$ and electric field strength $\mathcal{E} = 0.1, 0.3, 0.5$ V/Å

along x, y-axis, and 45 ^{0} from x-axis in $x - y$ plane, respectively. The Fermi level is shown as a dashed line.

Better understanding of band gap modulations comes from the analysis of the electronic band structure for (8,0) and (9,0) nanotubes at $\epsilon_{yy} = 10\%$ and $\epsilon_{yy} = 25\%$ respectively, for several orientations (Fig. 17). The electronic level of conduction and valance bands are associated with the quantum angular momentum J which defines the rule for admixture of allowed states (Vukovic et al., 2002). It has been shown that electronic states can mix according to the selection rule $\Delta J = 0, \pm 2$ for radial deformation of nanotubes (Park et al., 1999). This admixture and increasing $\sigma - \pi$ hybridization on high curvature bring the lowest conduction and the highest valance bands - closer to the Fermi level. However, the applied electric field mixes the electronic states in accordance with $\Delta J = \pm 1$ and also decreases the values of the band gap to its closure at E_F (Kim & Chang 2001; Li et al., 2006). The combination of radial deformation and external electric field causes the band structure modulation according to $\Delta J = 0, \pm 1, \pm 2$ selection rule and $\sigma - \pi$ hybridization on the high curvature of nanotubes. As we have seen from the band gap behavior, the strength and orientation of applied electric field with respect to the nanotube cross section are crucial for electronic properties of nanotubes. This has been reflected in the band structures for (8,0) and (9,0) nanotubes (Fig. 17). Most changes happen with the lowest unoccupied single degenerate conduction state for both (8,0) and (9,0) nanotubes. Therefore, application of electric field along x-direction (through the high curvature) has revealed the largest $\sigma - \pi$ hybridization on the high curvature. Thus, depending on the interplay between the radial deformation and electric field strength/orientation, the electronic properties of nanotubes can reveal semiconductor-metal or metal semiconductor transformation (Fig. 17).

5. Conclusion

In this chapter we discuss electronic and magnetic carbon nanotube properties under two external factors: radial deformation and various defects along with radial deformation and external electric fields. Our analysis is based on density functional theory calculations. We show that defects, such as Stone-Wales, nitrogen impurity, and mono-vacancy, create a local distortion in the carbon network. This disturbance is lowest for nitrogen, larger for StoneWales defect, and largest for mono-vacancy. The analysis of the characteristic energies reveals that it takes more energy to deform a smaller diameter nanotube in comparison to larger one. The larger degree of radial deformation makes it easier to introduce a defect on the high curvature of the nanotube. The Stone-Wales defect requires the lowest formation energy, while it is higher for nitrogen impurity, and it is the largest for mono-vacancy. We find that applied radial deformation causes semiconductor-metal transition in defective nanotubes at approximately the same value of applied strain ($\epsilon_{yy} = 23 \div 25\%$) regardless of the type of defect.

The electronic structure calculations show that the main factors responsible for semiconductor-metal transition in defective nanotubes are the $\sigma - \pi$ hybridization from the high curvature and the presence of defect states around the Fermi level region. We observe that interplay between radial deformation and type of defects such as Stone-Wales defect and mono-vacancy can open the band gap again, after it has been closed due to the same factors. However, there is not such opening for the nitrogen impurity and defect-free nanotubes regardless of the type of deformation. On the other hand, mutual effect of radial deformation and defects leads to magneto-mechanical coupling in the nanotube structure because of the presence of extra an electron from the nitrogen atom and dangling bond from the mono-vacancy. The magnetic properties of nanotubes can be controlled by the degree of radial deformation. The origin of magnetic properties of nanotubes is also discussed.

The combined effect of radial deformation and external electric field is given in terms of band gap modulations, electronic band structure changes, strength of applied electric field, electric field orientation, and selection rules. We show various responses in semiconductor and metallic carbon nanotubes. The electronic structure is sensitive to the strength and electric field orientation with respect to the cross section of nanotubes. The interplay between selection rules of mixing different states in the radial deformation and applied electric field along with $\sigma - \pi$ hybridization controls and tunes the band gap of nanotubes. Finally, we discuss semiconductor-metal transition in semiconducting nanotubes and opening of the band gap in metallic nanotubes. This chapter provides a comprehensive study of electronic and magnetic properties of nanotubes under two external factors. It discusses additional opportunities in controlling and tuning carbon nanotube properties for their expanded applications in current and new nano-devices.

6. References

Ajiki, H. & Ando, T. (1996). Energy bands of carbon nanotubes in magnetic fields. J. Phys. Soc. Jap., Vol. 65, No. 2, 505-514, ISSN 00314-9015

Ashraf, B.; Hubert, P. & Vengallatore, S. (2006). Carbon nanotube-reinforced composites as structural materials for microactuators in microelectromechanical systems. Nanotechnology, Vol. 17, No. 19, 4895-4903, ISSN 0957-4484

Bachtold, A.; Strunk, C.; Salvetat, J.-P.; Bonard, J.-M.; Forro, L.; Nussbaumer, T. & Schonenberg, C. (1999). Aharonov-Bohm oscillations in carbon nanotubes. Nature, Vol. 397, No. 6721, 673-675, ISSN 0028-0836

Barboza, A.; Gomes, A.; Archanjo, B.; Araujo, P.; Jorio, A.; Ferlauto, A.; Mazzoni, M.; Chacham, H. & Neves, B. (2008). Deformation induced semiconductor-metal transition in single wall carbon nanotubes probed by electric force microscopy. Phys. Rev. Lett., Vol. 100, No. 25, 256804-1-4, ISSN 1079-7114

Berber, S. & Oshiyama, A. (2006). Reconstraction of mono-vacancies in carbon nanotubes: atomic relaxation vs. spin polarization. Physica B, Vol. 376-377, No. 1, 272-275, ISSN 0921-4526

Bettinger, H. (2005). The reactivity of defects at the sidewalls of single-walled carbon nanotubes: the Stone-Wales defect. J. Phys. Chem., Vol. 109, No. 15, 6922-6924, ISSN 1089-5647

Choi, W.; Chung, D.; Kang, J.; Kim, H.; Jin, Y.; Han, I.; Lee, Y.; Jung, J.; Lee, N.; Park, G. & Kim, J. (1999). Fully sealed, high-brightness carbon-nanotube field-emission display. Appl. Phys. Lett., Vol. 75, No. 20, 3129-3131, ISSN 0003-6951

Chopra, N.; Benedict, L.; Crespi, V.; Cohen, M.; Louie, S. & Zettl, A. (1995). Fully collapsed carbon nanotubes. Nature, Vol. 377, No. 6544, 135-138, ISSN 0028-0836

Cohen-Karni, T.; Segev, L.; Srur-Lavi, O.; Cohen, S. & Joselevich, E. (2006). Torsional electromechanical quantum oscillation in carbon nanotubes. Nature Nanotechnology, Vol. 1, No. 1, 36-41, ISSN 1748-3387

Crespi, H.; Cohen, L. & Rubio, A. (1997). In situ band gap engineering of carbon nanotubes. Phys. Rev. Lett., Vol. 79, No. 11, 2093-2096, ISSN 1079-7114. Figure 5a is used with permission from American Institute of Physics, Copyright 1997.

Czerw, R.; Terrones, M.; Charlier, J.-C.; Blase, X.; Foley, B.; Kamalakaran, R.; Grobert, N.; Terrones, H.; Tekleab, D.; Ajayan, P.; Blau, W.; Ruhle, M. & Carroll, D. (2001). Identification of electron donor stets in N-doped carbon nanotubes. Nano Lett., Vol. 1, No. 9, 457-460, ISSN 1530-6984

Droppa, R.; Hammer, P.; Carvalho, A.; Santos, M. & Alvarez, F. (2002). Incorporation of nitrogen in carbon nanotubes. J. Non-Cryst. Sold., Vol. 299-302, No. 2, 874-879, ISSN 0022-3093

Fagan, S.; da Silva, L. & Mota, R. (2003). Ab initio study of radial deformation plus vacancy on carbon nanotubes: energetics and electronic properties. Nano Lett., Vol. 3, No. 3, 289-291, ISSN 1530-6984

Fagan, S.; Mota, R.; da Silva, A. & Fazzio, A. (2004). Substitutional Si doping in deformed carbon nanotubes. Nano Lett., Vol. 4, No. 5, 975-977, ISSN 1530-6984

Fan, Y.; Goldsmith, B. & Collins, P. (2005). Identifying and counting point defects in carbon nanotubes. Nature Materials, Vol. 4, No. 12, 906-911, ISSN 1476-1122

Favlo, M.; Clary, G.; Taylor II, R.; Chi, V.; Brooks, F.; Washburn, S. & Superfine, R. (1997). Binding and buckling of carbon nanotubes under large strain. Nature, Vol. 389, No. 6651, 582-584, ISSN 0028-0836

Fennimore, A.; Yuzvinsky, T.; Han, W.-Q.; Fuhrer, M.; Cumings, J. & Zetti, A. (2003). Rotational actuators based on carbon nanotubes. Nature, Vol. 424, No. 6947, 408410, ISSN 0028-0836

Fujita, M.; Wakabayashi, K.; Nakada, K. & Kusakabe, K. (1996). Peculliar localized states at zig-zag graphite edge. J. Phys. Soc. Jap., Vol. 65, No. 7, 1920-1923, ISSN 0031-9015

Hashimoto, A.; Suenaga, K.; Gloter, A.; Urita, K, & lijima, S. (2004). Direct evidence for atomic defects in graphene layers. Nature, Vol. 430, No. 7002, 870-873, ISSN 00280836

Hertel, T.; Walkup, R. & Avouris, P. (1998). Deformation of carbon nanotubes by surface van der Waals forces. Phys. Rev. B, Vol. 58, No. 20, 13870-73, ISSN 1098-0121

Gai, P.; Odile, S.; Mcguire, S.; Rao, A.; Dresselhaus, M.; Dresselhaus, G. & Colliex, C. (2004). Structural systematic in boron-doped single wall carbon nanotubes. J. Mater. Chem., Vol. 14, No. 4, 669-675, ISSN 0959-9428

Gao, G.; Cagin, T. & Goddard, W. (1998). Energetics, structure, mechanical and vibrational properties of single-walled carbon nanotubes. Nanotechnoly, Vol. 9, No. 3, 184-191, ISSN 0957-4484

Ghosh, A.; Sood, A. & Kumar, N. (2003). Carbon nanotube flow sensors. Science, Vol. 299, No. 5609, 1042-1044, ISSN 003-8075

Giusca, C.; Tison, Y. & Silva, S. (2007). Atomic and electronic structure in collapsed carbon nanotubes evidenced by scanning tunneling microscopy. Phys. Rev. B, Vol. 76, No. 3, 035429-1-6, ISSN 1098-0121

Golberg, D.; Bando, Y.; Han, W.; Kurashima, K. & Sato, T. (1999). Single-walled B-doped carbon and BN nanotubes synthesized from single walled carbon nanotubes through a substitutional reaction. Chem. Phys. Lett., Vol. 308, No. 3-4, 337-342, ISSN 0009-2614

Gulseren, O.; Yildirim, T.; Ciraci, S. & Kilic, C. (2002). Reversible band-gap engineering in carbon nanotubes by radial deformation. Phys. Rev. B, Vol. 65, No. 15, 155410-1-7, ISSN 1098-0121

Itkis, M.; Yu, A. & Haddon, R. (2008). Single-walled carbon nanotube thin film emitterdetector integrated optoelectronic device. Nano Lett., Vol. 8, No. 8, 2224-2228, ISSN 1530-6984

Jiang, J.; Dong, J. & Xing, D. (2000). Zeeman effect on the electronic spectral properties of carbon nanotubes in an axial magnetic field. Phys. Rev. B, Vol. 62, No. 19, 1320913215, ISSN 1098-0121

Jonge, N.; Allioux, M.; Doytcheva, M.; Kaiser, M.; Teo, K.; Lacerda, R. & Milne, W. (2004). Characterization of the field emission properties of individual carbon nanotubes. Appl. Phys. Lett., Vol. 85, No. 9, 1607-1609, ISSN 0003-6951

Kim, Y.-H. & Chang K. (2001). Subband mixing rules in circumferentially perturbed carbon nanotubes: effects of transverse electric fields. Phys. Rev. B, Vol. 64, No. 15, 153404-1-4, ISSN 1098-0121

Kolmogorov, A. & Crespi, V. (2005). Registry-dependent interlayer potential for graphitic systems. Phys. Rev. B, Vol. 71, No. 23, 235415-1-6, ISSN 1098-0121

Kordas, K.; Mustonen, T.; Toth, G.; Vahakangas, J.; Uusimaki, A.; Jantunen, H.; Gupta, A.; Rao, K.; Vajtai, R. & Ajayan, P. (2007). Magnetic-field induced efficient alignment of carbon nanotubes in aqueous solutions. Chem. Mater., Vol. 19, No. 4, 787-791, ISSN 0897-4756

Krasheninnikov, A.; Nordlund, K.; Sirvio, M.; Salonen, E. & Keinonen, J. (2001). Formation of ion-irradiation-induced atomic-scale defects on walls of carbon nanotubes. Phys. Rev. B, Vol. 63, No. 24, 245405-1-6, ISSN 1098-0121

Krasheninnikov, A.; Banhart, F.; Li, J.; Foster, A. & Nieminen, R. (2005). Stability of carbon nanotubes under electron irradiation: role of tube diameter and chirality. Phys. Rev. B, Vol. 72, No. 12, 125428-1-6, ISSN 1098-0121

Kresse, G. & Furthmuller, J. (1996). Efficient interactive schemes for ab initio total-energy calculations using a plane-wave basis set. Phys. Rev. B, Vol. 54, No. 16, 11169-11186, ISSN 1098-0121

Kresse, G. & Joubert, D. (1999). From ultrasoft pseudopotentials to the projector augmentedwave method. Phys. Rev. B, Vol. 59, No. 3, 1758-1775, ISSN 1098-0121

Kumar, M.; Lee, S.; Kim, T.; Kim, T.; Song, S.; Yang, J.; Nahm, K. & Suh, E.-K. (2003). DC electric field assisted alignment of carbon nanotubes on metal electrodes. Solid State Electron., Vol. 47, No. 11, 2075-2080, ISSN 0038-1101

Lai, P.; Chen, S. & Lin, M. (2008). Electronic properties of single-walled carbon nanotubes under electric and magnetic fields. Physica E, Vol. 40, No. 6, 2056-2058, ISSN 13869477

Lee, J.; Gipp, P. & Heller, C. (2004). Carbon nanotube $p-n$ junction diodes. Apll. Phys. Lett., Vol. 85, No. 1, 145-147, ISSN 0003-6951

Lee, J.-O.; Kim, J.-R.; Kim, J.-J.; Kim, J.; Kim, N.; Park, J. & Yoo, K.-H. (2000). Observation of magnetic-field-modulated energy gap in carbon nanotubes. Solid State Comm., Vol. 115, No. 9, 467-471, ISSN 0038-1098

Lee, R.; Kim, H.; Fischer, J.; Thess, A. & Smalley, R. (1997). Conductivity enhancement in single-walled carbon nanotube bundles doped with K and Br. Nature, Vol. 388, No. 6639, 255-257, ISSN 0028-0836

Leonard, F.; Jones, F.; Talin, A. & Dentinger P. (2005). Robustness of nanotube electronic transport to conformational deformations. Appl. Phys. Lett., Vol. 86, No. 9, 093112-13, ISSN 0003-6951

Li, T. & Lin, M. (2006). Electronic properties of carbon nanotubes under external fields. Phys. Rev. B, Vol. 73, No. 7, 075432-1-7, ISNN 1098-0121

Li, Y.; Rotkin, S. & Ravaioli, U. (2003). Electronic response and band structure modulation of carbon nanotubes in a transverse electrical field. Nano Lett., Vol. 3, No. 2, 183-187, ISSN 1530-6984

Lim, S.; Li, R.; Ji, W. & Lin, J. (2007). Effect of nitrogenation on single-walled carbon nanotubes within density functional theory. Phys. Rev. B, Vol. 76, No. 19, 195406-116, ISSN 1098-0121

Lu, A. & Pan, B. (2004). Nature of single vacancy in achiral carbon nanotubes. Phys. Rev. Lett., Vol. 92, No. 10, 105504-1-4, ISSN 1079-7114. Figure 5c is used with permission from American Institute of Physics, Copyright 2004.

Lu, J. (1995). Novel magnetic properties of carbon nanotubes. Phys. Rev. Lett., Vol. 74, No. 7, 1123-1126, ISSN 1079-7114

Ma, Y.; Lehtinen, P. Foster, A. & Nieminen. (2004). Magnetic properties of vacancies in graphene and single-walled carbon nanotubes. New J. Phys., Vol. 6, No. 1, 68-1-15, ISSN 1367-2630

Maiti, A.; Svizhenko, A. & Anantram, M. (2002). Electronic transport through carbon nanotubes: effects of structural deformation and tube chirality. Phys. Rev. Lett., Vol. 88, No. 12, 126805-1-4, ISSN 1079-7114

Mazzoni, M. & Chacham, H. (2000). Bandgap closure of a flattened semiconductor carbon nanotube: a first principal study. Appl. Phy. Lett., Vol. 76, No. 12, 1561-1563, ISSN 0003-6951

Mehrez, H.; Svizhenko, A.; Anantram, M.; Elstner, M. & Frauenheim, T. (2005). Analysis of band-gap formation in squashed armchair carbon nanotubes. Phys. Rev. B, Vol. 71, No. 15, 155421-1-7, ISSN 1098-0121

Minary-Jolandan, M. & Yu, M.-F. (2008). Reversible radial deformation up to the complete flattening of carbon nanotubes in nanoindentation. J. Appl. Phys., Vol. 103. No. 7, 073516-1-5, ISSN 0021-8979

Minot, E.; Yaish, Y.; Sazonova, V & McEuen, P. (2004). Determination of electron orbital magnetic moments in carbon nanotubes. Nature, Vol. 428, No. 6982, 536-539, ISSN 0028-0836. Figure 6b is used with permission from Macmillan Publisher Ltd: Nature, Copyright 2004.

Moradian, R. & Azadi, S. (2006). Boron and nitrogen-doped single-walled carbon nanotube. Physica E, Vol. 35, No. 1, 157-160, ISSN 1386-9477. Figure 5b is used with permission from Elsevier, Copyright 2005.

Nakada, K.; Fujita, M.; Dresselhaus, G. & Dresselhaus, M. (1996). Edge states in graphene ribbonce: nanometer size effects and edge shape dependence. Phys. Rev B, Vol. 54, No. 24, 17954-17961, ISSN 1098-0121

Nardelli, M.; Fattebert, J.-L.; Orlikowski, D.; Roland, C.; Zhao, Q. & Bernholc, J. (2000). Mechanical properties, defects and electronic behavior of carbon nanotubes. Carbon, Vol. 38, No. 11-12, 1703-1711, ISSN 0008-6223

Neugebauer, J. & Scheffler, M. (1992). Adsorbate-substrate and adsorbate-adsorbate interactions of Na and K adlayers on Al(111). Phys. Rev. B, Vol. 46, No. 24, 16067106080, ISSN 1098-0121

Nevidomskyy, A.; Csanyi, G. & Payne, M. (2003). Chemically active substitutional nitrogen impurity in carbon nanotubes. Phys. Rev. Lett., Vol. 91, No. 10, 105502-1-4, ISSN 1079-7114

Orellana, W. & Fuentealba, P. (2006). Structural, electronic and magnetic properties of vacancies in single-walled carbon nanotubes. Surf. Sci., Vol. 600, No. 18, 4305-4309, ISSN 0039-6028

Park, C.; Kim, Y. & Chang, K. (1999). Band-gap modification by radial deformation in carbon nanotubes. Phys. Rev. B, Vol. 60, No. 15, 10656-10659, ISSN 1098-0121

Picozzi, S.; Santucii, S. & Lozzi, L. (2004). Ozone adsorption on carbon nanotubes: the role of Stone-Wales defects. J. Chem. Phys., Vol. 120, No. 15, 7147-7152, ISSN 0021-9606

Reich, S.; Thomsen, C. & Maultzsch, J. (2004). Carbon Nanotubes. EILEY-VCH Verlag GmbH & Co. KGaA, ISBN 3-527-40386-8

Rossato, J.; Baierle, R.; Fazzio, A. & Mota, R. (2005). Vacancy formation process in carbon nanotubes: first-principal approach. Nano Lett., Vol. 5, No. 1, 197-200, ISSN 15306984

Santos, E.; Ayuela, A.; Fagan, S.; Filho, J. Azevedo, D.; Souza, A. & Sanchez-Portal, D. (2008). Switching on magnetism in Ni-doped graphene: density functional calculation. Phys.Rev. B, Vol. 78, No. 19. 195420-1-5, ISSN 1098-0121

Shan, B.; Lakatos, G.; Peng, S. & Cho, K. (2005). First-principal study of band-gap change in deformed nanotubes. Appl. Phys. Lett., Vol. 87, No. 17, 173109-1-3, ISSN 0003-6951. Figure 3 is used with permission from American Institute of Physics, Copyright 2005.

a Shtogun, Y. & Woods, L. (2009). Electronic and magnetic properties of deformed and defective single wall carbon nanotubes. Carbon, Vol. 47, No. 14, 3252-3262, ISSN 0008-6223

b Shtogun, Y. & Woods, L. (2009). Electronic structure modulations of radially deformed single wall carbon nanotubes under transverse external electric fields. J. Phys. Chem. C, Vol. 113, No. 12, 4792-4796, ISSN 1932-7447

Shtogun, Y, Woods, L. & Dovbeshko, G. (2007). Adsorption of adenine and thymine and their radicals on single-wall carbon nanotubes. J. Phys. Chem. C, Vol. 111, No. 49, 18174-18181, ISSN 1932-7447

Son, Y.; Ihm, J.; Cohen, M.; Louie, S. & Choi H. (2005). Electrical switching in metallic carbon nanotubes. Phys. Rev. Lett., Vol. 95, No. 21, 216602-1-4, ISSN 1079-7114

Stone, A. & Wales, D. (1986). Theoretical studies of icosahedral C_{60} and some related species. Chem Phys. Lett., Vol. 128, No. 5-6, 501-503, ISSN 0009-2614

Suenaga, K.; Wakabayashi, H.; Koshino, M.; Sato, Y.; Urita, K. & Iijima, S. (2007). Imaging active topological defects in carbon nanotubes. Nature Nanotechnology, Vol. 2, No. 6, 358-360, ISSN 1748-3387

Sugie, H.; Tanemura, M.; Filip, V.; Iwata, K.; Takahashi, K. & Okuyama, F. (2001). Carbon nanotubes as electron source in an x-ray tube. Appl. Phys. Lett., Vol. 78, No. 17, 25782580, ISSN 0003-6951

Tang, J.; Qin, L.-C.; Sasaki, T.; Yudasaka, M.; Matsushita, A. & IIjima, S. (2000). Compressibility and polygonization of single-wall carbon nanotubes under hydrostatic pressure. Phys. Rev. Lett., Vol. 85, No. 9, 1887-1889, ISSN 1079-7114

Tans, S.; Verschueren, A. & Dekker, C. (1998). Room-temperature transistor based on a single carbon nanotube. Nature, Vol. 393, No. 6680, 49-52, ISSN 0028-0836

Tien, L.; Tsai, C.; Li, F. & Lee, M. (2005). Band-gap modification of defective carbon nanotubes under a transverse electric field. Phys. Rev. B, Vol. 72, No. 24, 245417-1-6, ISSN

1098-0121. Figure 7 is used with permission from American Institute of Physics, Copyright 2005.

Tien, L.-G.; Tsai, C.-H.; Li, F.-Y. & Lee, M.-H. (2008). Influence of vacancy defect on electronic properties of armchair single wall carbon nanotube. Diam. Relat. Mater., Vol. 17, No. 4-5, 563-566, ISSN 0925-9635

Tombler, T.; Zhou, C.; Alexseyev, L.; Kong, J.; Dai, H.; Liu, L.; Jayanthi, C.; Tang, M. & Wu, S. (2000). Reversible electromechanical characteristics of carbon nanotubes under local-probe manipulation. Nature, Vol. 405, No. 6788, 769-772, ISSN 0028-0836. Figure 1 is used with permission from Macmillan Publisher Ltd: Nature, Copyright 2000.

Vitali, L.; Burghard, M.; Wahl, P.; Scheider, M. & Kern, K. (2006). Local pressure-induced metallization of a semiconducting carbon nanotube in a crossed junction. Phys. Rev. Lett., Vol. 96, No. 8, 086804-1-4, ISSN 1079-7114

Vukovic, T.; Milosevic, I. & Damjanovic, M. (2002). Carbon nanotubes band assignation, topology, Bloch states, and selection rules. Phys. Rev. B, Vol. 65, No. 4, 045418-1-9, ISSN 1098-0121

Wang, C.; Zhou, G.; Liu, H.; Wu, J.; Qiu, Y.; Gu, B.-L. & Duan, W. (2006). Chemical functionalization of carbon nanotubes by carboxyl groups on Stone-Wales defects: a density functional theory study. J. Phys. Chem. B, Vol. 110, No. 21, 10266-10271, ISSN 1089-5647

Woods, L.; Badescu, S. & Reinecke, T. (2007). Adsorption of simple benzene derivatives on carbon nanotubes. Phys. Rev. B, Vol. 75, No. 15, 155415-1-9, ISSN 1098-0121

Wu, J. & Hagelberg, F. (2009). Magnetism in finite-sized single-walled carbon nanotubes of zizzag type. Phys. Rev. B, Vol. 79, No. 11, 115436-1-9, ISSN 1098-0121

Yang, S.; Shin, W. & Kang, J. (2006). Ni adsorption on Stone-Wales defect sites in single-wall carbon nanotubes. J. Chem. Phys., Vol. 125, No. 8, 084705-1-5, ISSN 0021-9606

Yao, Z.; Postma, H.; Balents, L. & Dekker, C. (1999). Carbon nanotube intramolecular junction. Nature, Vol. 402, No. 6759, 273-276, ISSN 0028-0836

Yi, J.-Y. & Bernholc, J. (1993). Atomic structure and doping of microtubules. Phys. Rev. B, Vol. 47, No. 3, 1708-1711, ISSN 1098-0121

Yuan, J. & Liew, K. (2009). Effects of vacancy defect reconstruction on the elastic properties of carbon nanotubes. Carbon, Vol. 47, No. 6, 1526-1533, ISSN 0008-6223

Zhang, J.; Yang, G.; Cheng, Y.; Gao, B.; Qiu, Q. & Lee, Y. (2005). Stationary scanning x-ray source based on carbon nanotube field emitters. Appl. Phys. Lett., Vol. 86, No. 18, 184104-1-3, ISSN 0003-6951

Zheng, G. & Zhuang, H. (2008). Magneto-mechanical coupling behavior of defective singlewalled carbon nanotubes. Nanotechnology, Vol. 19, No. 32, 325701-1-9, ISSN 09576984

Chapter-4

Electronic Structure of Fluorinated Carbon Nanotubes

1. Introduction

Chemical functionalization of carbon nanotubes (CNTs), i.e., attachment of individual atoms/molecules or their aggregates to CNTs, can extend the field of application of these nanosystems in nanoelectronics, sensorics, hydrogen power engineering, bioengineering, medicine, etc. (Dresselhaus & Dresselhaus, 2001; Burghard, 2005). In this respect, the fluorination of CNTs is of special interest because the fluorination results in a considerable decrease in the chemical inertness of the initial systems. Therefore, the fluorination is considered as a promising technological process for the first stage of the CNT chemical functionalization (Mickelson et al., 1998; Mickelson et al., 1999; Khabashesku et al., 2002; Lee, 2007). However, carbon materials react with fluorine over a wide range of external conditions. Consequently, the chemical composition, as well as the atomic and electronic structures of fluorinated carbon nanotubes (F-CNTs), depends substantially on the structure and properties of the initial materials and fluorination conditions, such as the reaction temperature and duration, the presence of catalysts, pressure, and concentration of fluorinating reactants (Touhara & Okino, 2000). All these factors need comprehensive investigations of fluorination products by different experimental methods. Fitting of fluorination conditions that are necessary to perform the preset controllable restructuring of the CNT electronic structure remains at present the most topical problem of the carbon nanomaterial chemistry. Therefore, their atomic and electronic structures have been studied by using a limited number of experimental techniques. First and foremost, these are different microscopic techniques ensuring their visualization (Mickelson et al., 1998; Mickelson et al., 1999; Khabashesku et al., 2002; Lee, 2007; Touhara & Okino, 2000), Hamwi et al., 1997; Yudanov et. al., 2002; Hayashi et al., 2002; Lee et al., 2003), x-ray diffraction methods characterizing the degree of F-CNT crystallinity (Hamwi et al., 1997; Yudanov et. al., 2002) UV spectroscopy (Mickelson et al., 1999), Raman spectroscopy (Mickelson et al., 1998; Mickelson et al., 1999; Khabashesku et al., 2002; Lee, 2007; Touhara & Okino, 2000; Hamwi et al., 1997; Lee et al., 2003) and C 1s and F 1s core-level x-ray photoelectron spectroscopy (XPS) (Touhara & Okino, 2000; Hamwi et al., 1997;

Lee et al., 2003; An et al, 2002) which provide information on the nature of chemical bonding between carbon C and fluorine F atoms in F-CNTs. The investigations performed made it possible to determine the main conditions for the fluorination of single-walled carbon nanotubes (SWCNTs) and multi-walled carbon nanotubes (MWCNTs) to characterize their atomic structure and to obtain preliminary information on the specific features of their electronic structure due to the chemical bonding between C and F atoms. However, the data obtained in those works even with due regard for the results of the first theoretical calculations (Kudin et al., 2001; Park et al., 2003; Lebedev et al., 2003; Van Lier et al., 2005; Ewels et al., 2006) do not provide clear perception of the CNT fluorination mechanism and properties of fluorination products, which is necessary for expansion of practical use of F-CNTs.

Near-edge x-ray absorption fine-structure spectroscopy (NEXAFS) is at present one of the most efficient experimental methods (Stöhr, 1992; Chen, 1997). NEXAFS spectroscopy uses the relation between the spectral characteristics of the near-edge fine structure of x-ray absorption spectra and the parameters of the local atomic and electronic structures of the material under investigation in the vicinity of absorbing atoms. There are a few works in which NEXAFS spectroscopy has been successfully used to characterize pristine and oxidized CNTs (Kuznetsova et al., 2001; Tang et al., 2002; Schiessling et al., 2003; Banerjee et al., 2004; Banerjee et al., 2005; Hemray-Benny et al., 2006; Zhou et al., 2007). Total electron yield (TEY) mode is the most popular for recording of NEXAFS spectra (Gudat & Kunz, 1972). At the same time, TEY mode with probing depth approximately equal to the average MWCNT diameter (several tens of monolayers) does not allow one to get information about characteristics of investigated fluorinated multi-walled carbon nanotubes (F-MWCNTs) in their near-surface region where fluorination process can differ from that at greater depths. However, these data are of genuine fundamental interest and is indeed necessary for optimization of the MWCNT fluorination techniques. Such information for near-surface region only several graphene layers thick can be obtained, particularly, by XPS providing C 1*s* and F 1*s* core-level spectra of F-MWCNTs under investigation. XPS is an effective surfacesensitive method for investigation of material electronic structure (Huefner, 2003). In this case, the probing depth is adequate to the depth the registered electrons are emitted from, which depends on the electron kinetic energy. Therefore, the probing depth can vary from several tenth of nanometer (two to three atomic monolayers) to nanometers, depending on the energy of excitation photons. Presently, several works with XPS used to characterize fluorinated SWCNTs (F-SWCNTs) (Lee et al., 2007; Lee et al., 2003; An et al., 2002), fluorinated multiwalled nanoparticles (Okotrub et al., 2000), and F- CNTs (Shul'ga et al., 2007) are known.

In this chapter, the results of combined investigation of electronic structure of F-CNTs are presented. The high-resolution NEXAFS and XPS spectroscopies were used. The main goal of the present project is to elucidate the nature of chemical bonding between the carbon and fluorine atoms in the SWCNTs and MWCNTs, to evaluate a chemical state of fluorine and carbon atoms in these CNTs, to analyze modification of electronic structure of CNTs under fluorination, to characterize mechanism of CNT fluorination. The defluorination process of F-CNTs on thermal annealing has been investigated.

2. Experimental Details

MWCNTs were produced by arc-discharge evaporation of graphite rods (extremely pure, spectral purity, 99.9992wt%) in the helium atmosphere (500 Torr) at current density of 175 A/ cm^2 and voltage of 23 V . Material with MWCNTs was withdraw from the cathodic deposit center and had a columnar structure with the column length of up to 10 mm (Kiselev et. al., 1999). The carbon-based material containing nanotubes was only grinded and screened by using a sieve with 0.25 mm mesh without any chemical treatment.

The powdered samples of MWCNTs were fluorinated in a nickel reactor at the temperature $T_F = 420°C$ in a flow of molecular fluorine produced by electrolyzing acidic potassium trifluoride $KF \cdot 2HF$ containing up to 3wt% of HF .

SWCNTs were synthesized by the electroarc method using nickel-yttrium catalyst. Amorphous carbon and metal catalyst were removed from the initial condensation products (containing 15-20 weight % of SWCNTs) by multiple alternating air oxidations at temperatures up to 550°C and by rinsing with hydrochloric acid (Krestinin et al., 2003a). After purifying, nanotubes in the form of SWCNT powder were obtained, their main substance content being about 80 − 85 weight %.

Diameters of the purified nanotubes varied within a narrow range around the average value of about 1.5 nm ; the nanotubes existed in the SWCNT powder in a highly aggregated state in the form of bundles, micro-crystal films, and carpets with polycrystalline structures (Krestinin et al., 2003b). SWCNTs were isolated from the main impurity by dispersing SWCNT powder in aqueous solution of surfactants. As a result, high-purity SWCNT were obtained in the form of paper (SWCNT paper) where the main substance content was over 98-99 wt%, and also there was a small dope of ultra highly-dispersed particles of graphitized soot less than 20 nm in size.

Direct fluorination was carried out in a stainless steel reactor. The SWCNT sample in an aluminum boat was placed into the reactor, which was evacuated at room temperature until the residual vapor pressure became about 10^{-1}mbar. Next, undiluted fluorine (the amount of impurities in fluorine did not exceed 0.1 volume %) was inserted into the reactor to achieve the pressure of 0.8 bar, and the reactor was heated for one hour to 222°C (sample 1) and 190°C (sample 2). Thereafter, the samples were kept at the given temperature for four hours more. Fluorine content was estimated by the increase in the sample weight and appeared to be 35% for sample 1 (SWCNT+F35%) and 40% for sample 2 (SWCNT+F40%).

In this work, the spectra of "white" graphite fluoride (WGF) ($C_F = 62.4\text{wt}\%$), a highly oriented pyrolytic graphite (HOPG) crystal (Brzhezinskaya et al., 2008, 2009) and nanodiamond (Brzhezinskaya et al., 2010) were used as the reference spectra.

The C 1*s* and F 1*s* x-ray absorption spectra of the pristine and fluorinated MWCNTs, HOPG, and WGF were measured using monochromatic synchrotron radiation and the facilities of the Russian-German beamline (RGBL) at the BESSY II (Berlin) (Molodtsov et al., 2009). The samples for x-ray absorption and photoelectron measurements were prepared in air. For other details of sample preparation see (Brzhezinskaya et al., 2009).

The NEXAFS spectra were obtained by recording the TEY of the x-ray photoemission (Gudat & Kunz, 1972) in the mode of measurement of the drain current of the sample by varying the energy of incident photons. All x-ray absorption and photoelectron spectra were measured under ultrahigh vacuum ($\sim 2 \times 10^{-10}$Torr).

The energy resolutions ΔE of the monochromator in the range of the F1*s* x-ray absorption edge ($hv \sim 680$eV) and the C1*s* x-ray absorption edge ($hv \sim 285$eV) were equal to ~ 150 and 70 meV , respectively. The hv in the range of the fine structure of the F and C x-ray absorption spectra was calibrated against the energy positions of the first narrow peak in the F 1s x-ray absorption spectrum of K_2TiF_6(683.9eV) and the C1*s* x-ray absorption spectrum of HOPG (285.45eV).

In recording x-ray photoelectron spectra, hv varied from 345 to 1130 eV . Energy resolution ΔE of the monochromator with the 200μ m exit slit was $145 - 750$meV within the above mentioned range of hv. Photoelectron spectra of all the samples were measured in the normal photoemission registration mode by using the Phoibos 150 spherical analyzer from Specs whose resolution in recording the spectra remained equal to 200 meV . The analyzer was calibrated against energy based on photoelectron spectra of Au atom $4f7/2,5/2$ electrons. The monochromator was calibrated by recording the basic photoelectron lines of C 1*s* spectra exited by radiation reflected from the diffraction grating in the first and second diffraction orders.

For more experimental details see (Brzhezinskaya et al., 2009; 2010).

3. Experimental Results and Discussion

3.1 NEXAFS spectra

Fig. 1 presents $CK(1s)$ absorption spectra of initial SWCNTs, MWCNTs, HOPG crystal and nanodiamond measured under the same experimental conditions (Brzhezinskaya et al., 2008, 2009, 2010). All the spectra were normalized to the same level of continuous $C1s$ absorption at photon energy of 315 eV . The number and energy positions of $A - F$ fine structure features of the initial MWCNTs, SWCNTs and HOPG C $1s$ spectra are similar. This fact confirms the crucial role a single graphene layer plays in formation of the structure of carbon $1s$ absorption spectra in HOPG, MWCNTs and SWCNTs (Brzhezinskaya et al., 2008, 2009; Krestinin et al., 2009). It also confirms that the SWCNT conduction band structure is slightly affected by nanotube curvature (Louie, 2001). In the scope of this philosophy, we associate the A and B-C peaks in the HOPG spectrum with dipole-allowed electron transitions from $1s$ carbon states to the unoccupied π and σ symmetry states in the conduction band that are formed from carbon $\pi 2p_z$ and $\sigma 2p_{x,y}$ states oriented perpendicular to and parallel to the carbon layer plane (graphene), respectively. Such states are mostly associated with free electron states of individual carbon atoms. The states are actually quasimolecular as they are similar to those of benzene molecule C_6H_6 and are localized within one carbon hexagon (Brzhezinskaya et al., 2008; Comelli et al., 1988). The broad A' band that determines high level of continuous absorption between the A and B resonance peaks is associated with electron transitions to unoccupied π type states that are less localized in graphene layer than the state responsible for π resonance peak A. The broad $D - F$ absorption bands are associated with electron transitions to the unoccupied σ states in the graphite conduction band arising due to interaction between carbon hexagons in graphene layer.

Comprehensive investigation of spectra has revealed slight differences between nanotube and HOPG spectra, as well as between SWCNT and MWCNT spectra. Those are mostly the differences in the low-energy spectral region ($hv = 284 - 295\text{eV}$). All structural features become a little broader in nanotube spectra as compared to HOPG spectrum, the broadening being different for SWCNTs and MWCNTs. This is most clearly seen for π band A : its full width at half maximum (FWHM) increases from 1.15 eV (for HOPG spectrum) to 1.3 eV for SWCNT spectra and up to 1.45 eV for MWCNT spectra. Moreover, for SWCNT and MWCNT spectra, extra weak structural features a and a, b, c arise in the gaps between absorption bands A and $B - C$, respectively.

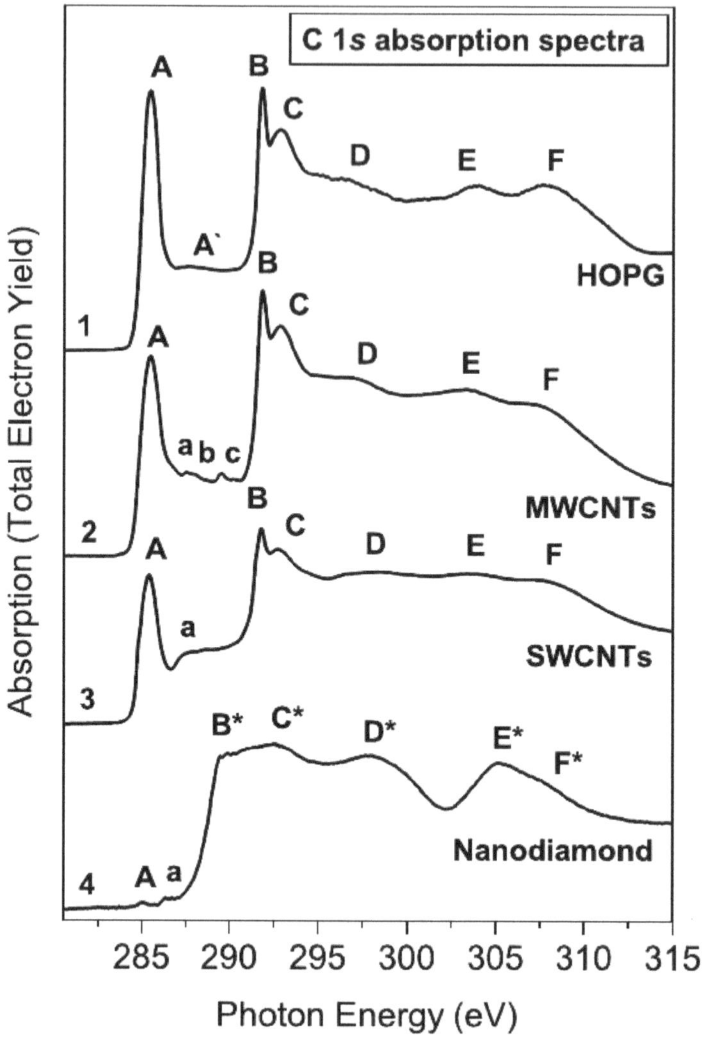

Fig. 1. C 1*s* absorption spectra of HOPG (1), MWCNTs (2), SWCNTs (3) and nanodiamond (4)

Naturally, the observed broadening of π band A in CNTs carbon spectrum as compared to HOPG spectrum can be associated with splitting of the graphite $\pi 2p_z$ conduction sub-band, which is caused by the reduction in local symmetry of nanotubes carbon atoms due to the fact that graphene layers become curved. However, for single-walled nanotubes, one should have expected greater broadening of band A since the SWCNT diameters ($\sim$ 1.5 nm) are an order of magnitude less than those of MWCNTs ($\sim$ 10 – 30 nm). In reality, as shown above, the situation is quite opposite.

In interpreting the differences in the C 1s absorption spectra recorded by measuring the total electron yield, it is necessary to take into account that the nanotube spectra under consideration are obtained by averaging a great number of individual SWCNTs in the bundles or in MWCNTs within the area limited by the size of the focused beam spot ($0.2 \times 0.1\ mm^2$) on the sample surface and by the probing depth (15-20 nm) (Chen, 1997). In this case, the differences in chemical states of carbon atoms in different tubes or parts of individual tubes result in energy dispersion of C1*s* electron transitions to π symmetry electron states and, hence, can contribute significantly to broadening of π band *A*. Thus, the greater broadening of π band *A* in MWCNT spectrum may be attributed to greater variations in the carbon atom chemical states in MWCNTs as compared to SWCNTs. Greater variations in diameters and concentration of defects in the MWCNTs as compared to SWCNTs may be the reasons for this. It is also possible that just the higher degree of order of carbon atoms and their electron states in the SWCNT bundles provides high energy of interaction (about 0.5 eV per 1 nm of length) between individual SWCNTs in the bundles.

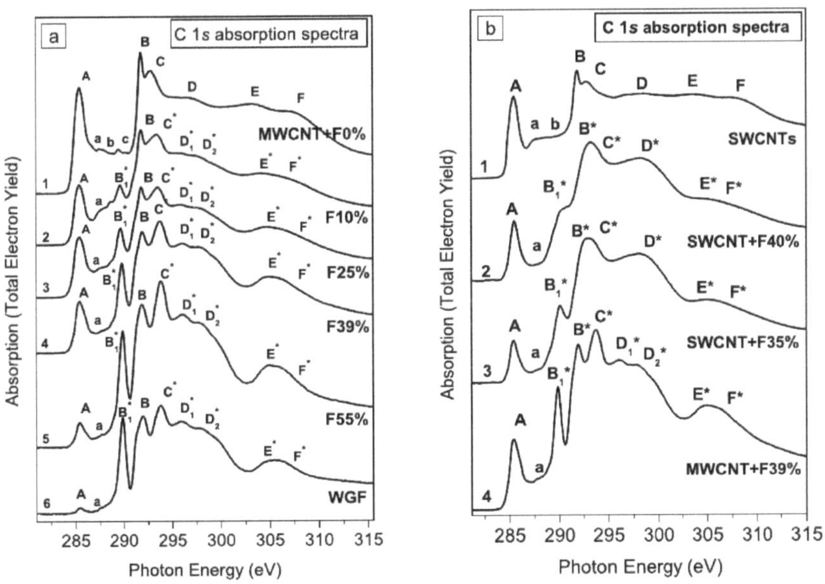

Fig. 2. (a) C 1*s* absorption spectra of (1) MWCNTs, (2-5) F-MWCNTs containing (2) 10%, (3) 25%, (4) 39%, (5) 55% and (6) WGF (62.4%); and (b) C 1 s absorption spectra of SWCNTs (1), F-SWCNTs containing 40% (2) and 35% (3) and MWCNTs+F39% (4)

In its turn, the rise of weak structures *a* and *a*, *b*, c between π and σ peaks in the nanotubes spectrum is typically attributed to the C1*s* electrons transitions to free 2*p* states of carbon atoms localized on nanotube surface areas oxidized during synthesis (Banerjee et al., 2004;

Kuznetsova et al., 2001). It is quite possible that the weak structures under discussion are partly associated with the $1s{-}{>}\ 2p$ absorption transitions in carbon atoms at the fullerene-like edges of nanotubes.

Comparing the spectra of pristine and fluorinated CNTs (Fig. 2), we can see that their fine structures differ significantly. What is most important is that the C 1s NEXAFS spectra of the F-MWCNTs with the highest fluorine content ($C_F = 55$wt%) (MWCNTs+F55%) and WGF [Fig. 2(a), curves 5 and 6] are very similar and differ substantially from the spectra of HOPG and MWCNTs (Fig. 1, curves 1 and 2). The most significant differences involve a drastic decrease in the intensity of the resonance A and the band A' associated with the free π states and the appearance of the high-energy σ band $E^* - F^*$. It is evident that the observed changes are governed by the fluorination of the HOPG and MWCNTs. The resonance A is retained in the spectra of the fluorinated samples but has a considerably lower intensity with respect to the other spectral portion associated with the transitions of the $1s$ electrons to the free states of the conduction band. The band B is also retained in the spectra, whereas the other specific features in the spectra of HOPG and MWCNTs have no clear analogs in the spectra of the fluorinated samples. New specific features $B_1{}^* - F^*$ in the spectra of the F-MWCNTs and WGF should be unambiguously treated as a result of the transitions of the C$1s$ electrons to the free states of the new phase formed in the MWCNTs and HOPG due to their fluorination. It should be emphasized that in the given case, only one particular phase can be formed. In the presence of several fluorocarbon phases, the C$1s$ spectra would be virtually structureless and similar in this respect to the spectrum of amorphous carbon (Rosenberg et al., 1986). Hence, the spectra of the F-MWCNTs and WGF are a superposition of the dominant C 1s spectrum of the fluorocarbon phase and the low-intensity spectrum of the pristine sample (MWCNTs and HOPG). In other words, despite the maximum possible fluorine content in the F-MWCNTs and WGF (55.0 and 62.4wt%, respectively), they are incompletely fluorinated and contain regions composed of the pristine HOPG and MWCNTs. This is most probably explained by the fact that a limiting depth of fluorination exists under these conditions of the synthesis. This fluorination depth is less than the probing depth ($d \sim 15$ nm) of the sample when the TEY method is employed for recording the NEXAFS spectra (Stohr, 1992; Chen, 1997). Since the sample contains MWCNTs of different sizes ($L \sim 0.5 - 2.0\mu$ m and $D \sim 10 - 30$ nm), including nanotubes with a diameter $D > 15$ nm, the measured C$1s$ spectrum should involve the contributions of the fluorinated and pristine carbon phases.

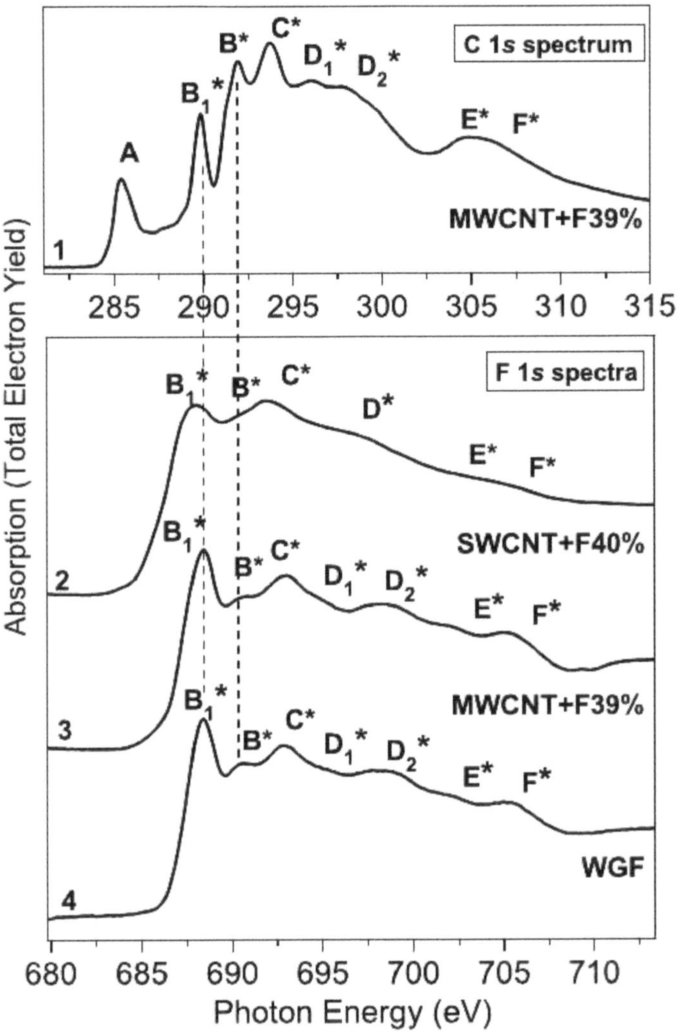

Fig. 3. Comparison of C 1s (1) and F 1s (2) NEXAFS spectra of MWCNTs+F39% and F 1s NEXAFS spectra of SWCNTs +F40% (3) and WGF (4). The F 1*s* absorption spectra are reduced to the energy scale of the C 1*s* absorption spectrum with the use of energy spacing $\Delta E($ F1s $-$ C1$s) = 398.4$eV (measured by x-ray photoemission spectroscopy) between the F1*s* and C1*s* core levels.

Since the C 1*s* NEXAFS spectra of the F-MWCNTs and WGF do not contain the intense π resonance A and the π band A', we can make the inference that the C atoms are involved in the chemical bonding upon formation of the fluorocarbon phase. This implies that upon fluorination, the F atoms are attached to the graphene layer perpendicular to it rather than replace the C atoms in the layer. In this case, the F atoms form bonds with the C atoms with the participation of the $C2p_z$ states. As a result, the C atom has a spatial coordination rather than the planar coordination.

One more indirect evidence for the change in the coordination of the C atom in HOPG and MWCNTs during fluorination can be the appearance of a single σ band $E^* - F^*$ in the x-ray absorption spectra of the fluorinated samples ($hv \sim 303 - 310\text{eV}$). A similar high-energy band, together with the absence of the π resonance, is considered as a characteristic difference between the C 1s NEXAFS spectrum of diamond (Fig. 1, curve 4) with the sp^3 tetrahedral coordination of the C atoms and the spectrum of HOPG with the sp^2 trigonal coordination. This conclusion is in good agreement with the results obtained from the investigation into the mechanism of the formation of graphite fluoride C_x F by Raman spectroscopy (Gupta et al., 2003).

It is interesting to compare the C1s x-ray absorption spectra of the F-MWCNTs with different fluorine contents ($0\% - 55\%$) [Fig. 2(a)] because they correspond to different stages of the MWCNT fluorination. These spectra demonstrate a monotonic character of variations in the fine structure with an increase in the degree of MWCNT fluorination. As a result, a gradual disappearance of the π structure (the resonance A and the band A'), as well as the formation of the characteristic resonances and in the region of the σ structure in addition to a single high-energy band $E^* - F^*$, is observed. These changes manifest the uniform character of the interaction between the F and C atoms in all stages of the fluorination process.

In analyzing C 1s absorption spectra of fluorinated SWCNTs with the almost equal fluorine content of 35 and 40wt% [Fig. 2(b), curves 2 and 3] it is necessary to notice, first of all, similarity of their general spectral shape manifesting itself in coincidence of the fine structure features number and energy positions. The only slight difference between those spectra is somewhat better resolution of band $B_1{}^*$ on the spectra of the SWCNTs with lower fluorine content. Obviously, similarity of fine structures of the spectra of two fluorinated SWCNT samples reflect the similarity of their atomic and electronic structures; therefore, this confirms the reproducibility of SWCNTs fluorination results for the fluorination process described above.

Comparing the spectra of pristine and fluorinated SWCNTs, we can see that their fine structures differ significantly. Indeed, after fluorination, the only fine structure feature still remaining in the nanotube spectra is the depressed π resonance of A, all other features $B_1{}^* - F^*$ being new. Evidently, the presence of band A typical of the pristine nanotube spectrum in the F-SWCNT spectra concurrently with the new absorption bands is the evidence of incomplete fluorination of single-walled nanotube bundles within the probing depth of 1520 nm . Probably, it is caused by the fact that the SWCNT bundle fluorination depth is much shorter than the probing depth in the absorption spectra under consideration.

The above-described variations in the F-SWCNT carbon absorption spectra fine structures are quite similar but not identical to the variations observed in the structures of F-MWCNT spectrum [Fig. 2(b), curve 4] with respect to the pristine MWCNT spectrum (Fig. 1, curve 2). Indeed, Fig. 3 demonstrates clearly that, having the general form quite similar to that of spectrum of F-MWCNTs with almost the same content of fluorine (39wt%), the F-SWCNT spectra (F35 wt% and F40 wt%) are less structured and manifest broader absorption bands. Bands $B^* - C^*$ and D^* in F-SWCNT spectra were assumed to be the result of superposition of individual absorption bands B^* and $C^*, D_1{}^*$ and $D_2{}^*$ in F-MWCNT spectra.

Getting back to the differences between F-SWCNT and F-MWCNT C 1*s* spectra, we can briefly characterize them as significant broadening and blurring of prominent and wellresolved features of F-MWCNT spectrum as compared with the F-SWCNT spectra, the general spectrum shape character remaining almost unchanged. Assuming that only one fluorine-carbon phase arises in fluorination in both SWCNTs and MWCNTs, we can attribute the observed differences in their spectra to different levels of structural ordering in the phase formed. This could be due to different temperatures of nanotubes fluorination by molecular fluorine: 420°C for multi-walled (Brzhezinskaya et al., 2008, 2009) and 190°C and 222°C for single-walled nanotubes. Indeed, the differences in structural ordering of fluorinated regions of single-walled nanotubes should be accompanied by differences in the strength of chemical carbon-fluorine bonding and, hence, by dispersion of charge (chemical) states of those atoms. In its turn, this should cause noticeable dispersion of energies of identical electron transitions and, hence, broadening and casual blurring of fine structure features of the carbon and fluorine absorption spectra of F-SWCNTs.

However, the nature of differences in C 1*s* absorption spectra of F-SWCNTs and FMWCNTs is, probably, more complicated. In the case of F-MWCNTs with diameters of approximately 10 to 30 nm , the spectra under consideration are related to the major (bulk) fluorine-carbon phase due to quite large probing depth ($15-20$ nm) (Brzhezinskaya et al., 2009). This phase is highly ordered because of strong covalent carbon-fluorine bonding; it dominates in the F-MWCNTs over the less structured near-surface phase that coexists with the major phase within $2-3$ graphene monolayers (Brzhezinskaya et al., 2009). The two phases revealed in F-MWCNTs are characterized by chemical bonding with different electron transfers between carbon and fluorine atoms, which gives rise to different carbon atomic states in those phases. Relying on these facts, we assume that, in the case of FSWCNTs (with diameters ~ 1.5 nm), similar phases should coexist in all tubes of the bundles and, consequently, their C 1*s* absorption spectra should manifest structural features of both phases concurrently. So, blurring of fine structure features in the F-SWCNT spectra as compared to F-MWCNT spectrum may be attributed to superposition of the spectral structure features characteristic of the two fluorine-carbon phases whose carbon atoms are in different chemical states. Theoretical calculations (Ewels et al., 2006) also point to the possible presence of two fluorine super-lattices that give rise to fluorine-carbon phases with the C_4 F and C_2 F stechiometries on the SWCNTs lateral surface in the temperature range of $200 - 250°C$.

The above conclusion is confirmed by the results of comparison of the F1*s* x-ray absorption spectra of F-MWCNTs and F-SWCNTs (Fig. 3). All the F 1*s* x-ray absorption spectra of the FMWCNTs and WGF exhibit a similar fine structure (Brzhezinskaya et al., 2009). The main difference between the spectra is that the measured intensity of the spectra increases monotonically with respect to the background intensity with increasing fluorine content in the F-MWCNTs. This behavior of the F 1*s* x-ray absorption spectra of the F-MWCNTs with an increase in the fluorine content in the samples can be interpreted only as a result of the formation of a fluorocarbon phase, which is identical in all stages of the fluorination process. This conclusion, of course, should correlate with the probing depth ($\sim$ 15 nm) of the nanotube samples during the recording of the x-ray absorption spectra in this work. The F 1s x-ray absorption spectra are also characterized by rich and clearly pronounced fine structure. This circumstance indicates a well-defined atomic structural order in the fluorinated samples. In other words, the fluorine atoms, such as the carbon atoms, have an identical coordination, which reflects chemical bonding of the C and F atoms.

The above-made conclusion on the covalent character of fluorine-carbon bonds in FMWCNTs and retention of laminar character of the carbon atomic structure up to maximal fluorine concentrations ensuring the binary (CF) stoichiometry of fluorination products agrees well with wellknown results of graphite fluorination. It was found out for graphite that when it is maximally fluorinated, fluorinecarbon phase poly(carbon monofluoride) $(CF)_n$ predominates (Gupta et al., 2003; Sato et al., 2003) which is characterized by the retention of laminar graphitelike structure, although the layers are no longer planar but become corrugated because fluorine atoms are attached to them. What is important is that this graphite fluoride is characterized by covalent binding with sp^3 hybridization of carbonatom valent states. This fact is confirmed also by the similarity of absorption spectra of FMWCNTs and WGF reference samples. All this confirms the above conclusion that MWCNTs retain their basic structure at high fluorination levels, such as 55wt%, where almost every carbon atom throughout all CNT layers is in the sp^3 configuration.

F 1*s* absorption spectra of F-SWCNTs with almost the same fluorine content of 35 and 40 wt % are in general similar: numbers of their spectral features are identical, their energies are close to each other (Brzhezinskaya et al., 2010). The only slight difference between them is somewhat better fine structure resolution of the spectra of tubes with less fluorine content; this is clearly seen when comparing the B_1 * bands.

It is very important that both F 1*s* and C 1*s* absorption spectra of F-SWCNTs are, in general, very similar to respective F-MWCNT spectra (Fig. 3, curve 4). The main difference is that the fluorine spectra for F-MWCNTs are better structured (as the carbon spectra are). This fact clearly confirms our conclusion made above that the major covalent fluorine/carbon phases arising in fluorination of single- and multi-walled nanotubes are similar to each other. Hence, we can assume that SWCNT fluorination occurs mainly by bonding fluorine atoms to carbon atoms on nanotube sidewalls (similar to MWCNT fluorination); this is accompanied by generation of $\sigma(\mathrm{C}-\mathrm{F})$ bonds by covalent mixing of valence F2*p* and $\mathrm{C}2p_z\pi$ electronic states without any destruction of the tubular structure of carbon layers (Brzhezinskaya et al., 2008, 2009). Such fluorine bonding transforms carbon triangular arrangement of initial nanotubes into almost tetrahedron one of fluorinated nanotubes, which becomes possible only if sp^2 hybridization of carbon atom valence states in pristine nanotubes transforms into sp^3 hybridization of F-SWCNTs. Such transformation of atomic structure should cause corrugation of nanotube sidewalls.

It is interesting to analyze the fluorinated sample unoccupied electronic state spectrum by using C 1*s* and F 1*s* X-ray absorption spectra that reflect energy distribution of free electron states formed from the 2*p* states of the carbon and fluorine atoms, respectively, because of the dipole nature of absorption transitions. For this purpose, the carbon and fluorine atom spectra for F-MWCNTs with the 35wt% fluorine content (curves 1 and 3) are given in Fig. 3 on the photon energy scales that are brought into coincidence by using the ΔE (F1*s* – C1*s*) = 398.4eV energy shift between the fluorine and carbon 1*s* states in the system. In this work, energy positions of the core levels were measured the by X-ray photoemission technique at the exciting photon energy of 1030 eV .

Comparing the spectra, we can clearly see that both spectra manifest fine structures similar in the number and energy positions of their key features $B_1{}^* - F^*$ (the *A* band in the carbon spectrum is associated with non-fluorinated tubes). Vertical dashed lines in Fig. 3 indicate that energy positions of bands B_1 and $B^* - C^*$ in the carbon and fluorine absorption spectra of F-SWCNTs and F-MWCNTs are close to each other. Such correlation in the F-CNT carbon and fluorine spectra means that carbon and fluorine 1*s* electron transitions to the unoccupied electronic states formed from C 2*p* and F 2*p* electronic states are responsible for the key features of the spectra compared. It indicates the hybrid (covalent) F 2*p* + C2*p* nature of the free electronic states in the systems discussed, which complies with the conclusion about sp^3 hybridization of the carbon atom valence states, which was made as a result of independent analyses of carbon and fluorine spectra.

3.2 XPS spectra

X-ray photoelectron spectra of SWCNTs and MWCNTs were measured by using exciting photons with energies from 385 to 1030 eV . All the measurements were carried out under the same experimental conditions. For both SWCNTs and MWCNTs, the A line in the C 1s spectra is a narrow peak at $BE = 284.8\text{eV}$ which is asymmetrically shaped at the higher binding energy side and whose full-width at half maximum (FWHM) is 0.7 eV .

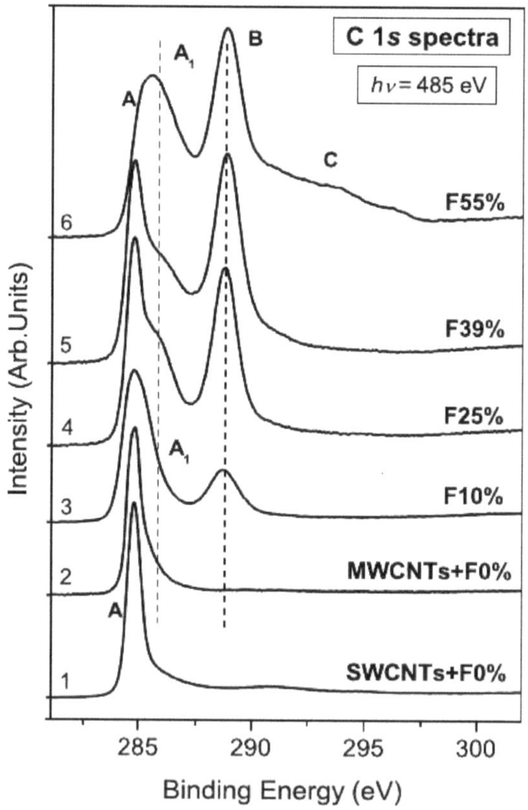

Fig. 4. C 1*s* x-ray photoelectron spectra of (1) SWCNTs, (2) MWCNTs and F-MWCNTs with fluorine concentrations: $10\%(3), 25\%(4), 39\%(5), 55\%(6)$.

Fig. 4 demonstrates that C 1s photoelectron spectrum structure changes radically as pristine MWCNTs are transformed into F-MWCNTs. Shoulder A_1 shows up near the main line A at $BE = 285.7\text{eV}$, and the second line B arises at $BE = 288.8\text{eV}$. As the fluorine concentration in FMWCNTs is growing, the intensities of the spectrum new features increase, while their energy positions do not change. When the concentration reaches $C_F = 39\text{wt}\%$, intensity B becomes almost equal to the main C1s line intensity A. Spectrum measured at the maximum fluorine concentration ($C_F = 55\text{wt}\%$) differs significantly from the other spectra: band A almost fully disappears against the background of the intense band into which shoulder A is transformed, whereas band B becomes the most intense structure feature having an intense, extended and poorly structured tail C on the high-energy side.

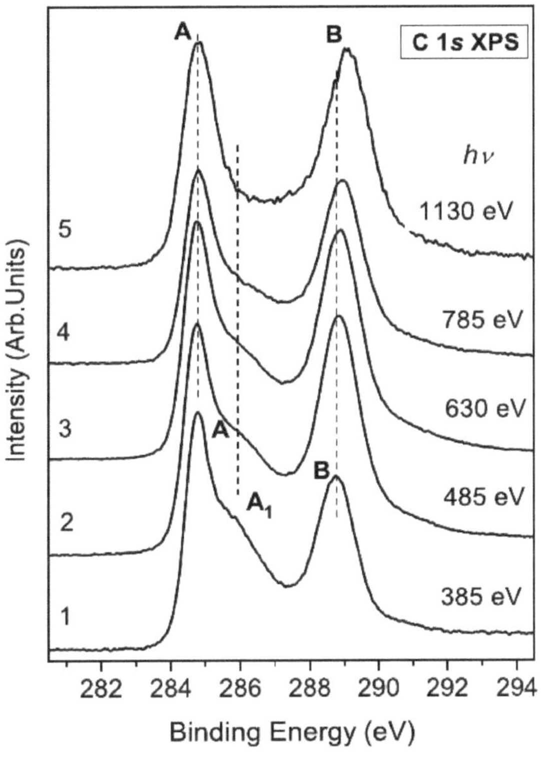

Fig. 5. C 1s x-ray photoelectron spectra of MWCNTs+F39% measured at photon energies of 385eV(1), 485eV(2), 630eV(3), 785eV(4) and 1130eV(5).

As shown in our work (Brzhezinskaya et al., 2009), appearance of new features in the FMWCNT C 1s spectra structure as compared with those of MWCNTs can be attributed to CNT fluorination; therefore, it can be regarded as a result of carbon-fluorine chemical bonding. This interaction results in formation of fluorine-carbon phases and is accompanied by charge transfer from carbon atoms to fluorine atoms due to the fact that fluorine electronegativity is higher. Charge states (oxidation rates) of carbon atoms in at least two of the fluorine-carbon phases are different and can be characterized by chemical shifts of their C 1s photoelectron lines (peaks A_1 and B) from the energy position of this line in pristine MWCNT spectrum (peak A). Energy positions of bands A_1 and B relative to peak A are significantly different (0.9 eV and 4.0 eV , respectively). This large difference ($> 3\text{eV}$) indicates considerable differences in the natures of C-F atom interactions in the analyzed fluorine-carbon phases that are formed in F-MWCNTs.

Relative intensities of bands A and B do not change much with hv increasing up to 1030 eV , while the A_1 band intensity decreases, and the corresponding shoulder gets barely visible against band A background. It is reasonable to connect those observations with the probing depth increase from ~ 0.6 nm ($1-2$ graphene layers) to 2 nm (5-6 layers). The weak relationship between the peaks A and B relative intensities indicates that the ratio of the initial and fluorinated MWCNTs phases for the samples with $C_F = 10 - 39\text{wt}\%$ changes only slightly with depth within several graphene layers. This statement is consistent with the conclusion that the fluorination process in MWCNTs under investigation remains uniform within the depth limits of ~ 15 nm (Brzhezinskaya et al., 2008).

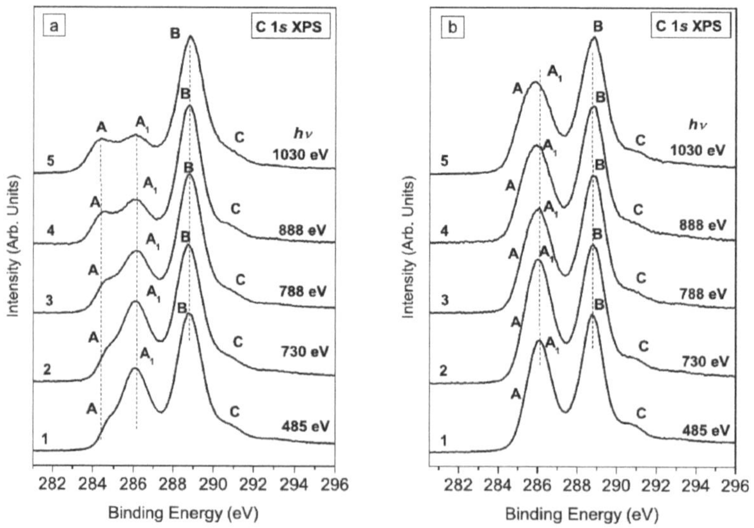

Fig. 6. C 1s x-ray photoelectron spectra of (a) SWCNTs +F35% and (b) SWCNTs +F40% measured at photon energies of 385eV(1), 730eV(2), 788eV(3), 888eV(4), 1030eV(5).

A significant decrease in the band A_1 intensity with the probing depth increase from 0.6 to 2 nm shows unambiguously that phase 2 is of the near-surface character and is formed by fluorination of only a few outer graphene layers. This conclusion is consistent with the specific shape of C 1s photoelectron spectrum obtained for MWCNTs +F55%.

The energy gap between bands A and B grows by $\sim$ 0.5eV as photon energy changes from $h\nu = 385\text{eV}$ to $h\nu = 1130\text{eV}$. It means that chemical shift of the C 1 s level grows from 4.0 to 4.5 eV with increasing probing depth for spatial fluorocarbon phase. Such an increase in the chemical shift is observed for $h\nu > 700\text{eV}$ only. It may be a result of a small change in the chemical state of phase 1 carbon atoms located deeper in the samples. In this case, the chemical shift of 4.5 eV for C 1s level is practically equal to the 4.45 eV chemical shift in the C 1s absorption spectra (Brzhezinskaya et al., 2008).

The feature A_1 behavior is similar in all F-MWCNTs spectra; Fig. 5 represents MWCNT +F39% to illustrate this fact. The band A_1 intensity gradually decreases with increasing probing depth ($d = 0.4 - 2$ nm and $h\nu = 345 - 1130\text{eV}$). It is reasonable to connect the gradual band A_1 weakening with formation of fluorocarbon phase in the very near-surface layer of the tubes. This layer is one to two graphene monolayers thick as this is just the depth ($0.4 - 0.8$ nm), where from C1s photoelectrons generated by absorption of photon with $h\nu = 345 - 630\text{eV}$ and having kinetic energies of $E_k = 60 - 345\text{eV}$ can be emitted without energy losses (Huefner, 2003; Brown et al., 1978; Martin et al., 1985).

The spectra of F-SWCNTs (Fig. 6) were measured at photon energies from 385 to 1030 eV . According to the universal curve of electron mean free path versus kinetic energy (Brzhezinskaya et al., 2009), the escape depth of C 1s photoelectrons excited by photons with energies of 385 to 1030 eV varies from about 0.5 to 1.5 nm ; thus, the probing depth lays actually within the subsurface layer about one nanotube diameter wide. As compared with F-MWCNTs (Fig. 4), for F-SWCNTs, C 1s photoelectron spectrum structure changes even more radically as initial SWCNTs are being transformed into F-SWCNTs. In the C 1s photoelectron spectrum of SWCNTs +F35%, there are 4 peaks, $A(BE = 284.6\text{eV}), A_1(BE = 286.1\text{eV}), B(BE = 288.8\text{eV})$ and $C(BE = 290.9\text{eV})$. Notice that at small photon energies ($< 888\text{eV}$) band A manifests itself as a shoulder next to peak A_1 on the lower binding energy side. In the C 1s spectra of SWCNTs +F40% there are only three explicit features: bands $A_1(BE = 286.1\text{eV}), B(BE = 288.8\text{eV})$, and shoulder $C(BE = 290.9\text{eV})$. The presence of band A is indicated by only the large width and slight low energy asymmetry of band A_1. Coincidence of bands A_1 positions and bands B positions in the F-SWCNTs and F-MWCNTs indicates their connection with two fluorine-carbon phases that are similar for the single- and multi-walled nanotubes. This fact indicates their relation to two different fluorine-carbon phases that are respectively similar in single- and multi-walled nanotubes. As shown above, a single fluorine-carbon phase uniformly distributed within the sample probing depth is formed in fluorination of both MWCNTs and SWCNTs. Peak B in C 1s photoelectron spectra of FMWCNTs is attributed to this phase. Then the same peak B should correspond to this phase in the F-SWCNT C 1s photoelectron spectra.

Energy positions of bands A_1, B and C relative to peak A are significantly different (1.3 eV, 4.0 eV and 6.1 eV , respectively) for SWCNTs+F35%. For SWCNTs+F40%, energy shifts of peaks A_1, B and C are the same when measured relative to peak A energy position for SWCNTs +F35%. Note that peak A cannot be clearly identified in the SWCNTs +F40% spectra [Fig. 6(b)]. This large difference ($\sim$ 3 and 6 eV) indicates considerable differences in the natures of C-F atom interactions in the analyzed fluorine-carbon phases formed in FSWCNTs.

It is interesting that in the C 1s spectra of SWCNTs+F35% [Fig. 6(a)] measuring with photon energy increasing from 485 to 1030 eV and, hence, with probing depth increasing from $\sim$ 0.5 nm to $\sim$ 1.5 nm, band A_1 relative intensity becomes more than twice lower than that of band B. Probably, this indicates a significant decrease in the fluorination level of inner nanotubes located inside the bundle, namely, all the nanotubes that do not participate in forming the bundle surface layer. This conclusion correlates well with the earlier revealed increase in the SWCNT fluorination level accompanied by improvement of their dispersion, i.e., by the decrease in transverse size of the bundles formed by the nanotubes (Krestinin et al., 2009). However, the decrease in SWCNTs +F40% band A_1 intensity is not so dramatic [Fig. 6(b)]. Generally, shapes of SWCNTs+F40% [Fig. 6(b)] and MWCNTs+F55% (Fig. 4, curve 5) C 1s spectra are similar, except for the band C shape. Thus, the first fluorine-carbon phase (phase 1, peak B) is apparently formed by fluorine atoms covalent bonding with the graphene layer. Contrary to phase 2 (peak A_1), this phase is uniformly distributed within the sample probing depth ($0.5-1.5$ nm for XPS). Moreover, this fluorine-carbon phase contributes most to C 1s absorption spectra of fluorinated nanotubes at larger probing depths (Brzhezinskaya et al., 2010). In addition, just phase 1 is associated with formation of the C_2 F super-lattice on the surfaces of both F-SWCNTs and F-MWCNTs.

This is confirmed by most theoretical calculations (Bettinger et al., 2001; Kudin et al., 2001; Seifert et al., 2000).

Chemical shift in the second phase (phase 2, peak A_1) of the C1s electron binding energy (1.2 eV) is significantly smaller than that in the first phase (4.0 eV). Hence, unlike phase 1, phase 2 formation is accompanied by significantly lower carbon-fluorine electron density transfer resulting in a weaker chemical bond. Phase 2 is most probably associated with formation of C_4 F superlattice on the surfaces of both F-SWCNTs and F-MWCNTs. This idea is consistent with the calculation results (Ewels et al., 2006) showing that the most stable phase is (1,2)(C_2 F) phase followed by the (1,4)(C_4 F) phase. When the probing depth increases from $\sim$ 0.5 nm to $\sim$ 1.5 nm, the band A_1 intensity decreases relatively to the band B intensity. For F-SWCNTs, such a transformation of band A_1 is likely to confirm the above conclusion about incomplete SWCNT fluorination inside the bundles. Therefore, phase 2 is developed only on SWCNTs located in the bundle near-surface layers, at least SWCNTs located on the bundle outer periphery having the greatest chance to become completely fluorinated. It is known that individual SWCNTs are arranged in the bundles almost coaxially and practically without tangles, the number of tubes per bundle being about 300 . The F-SWCNT defluorination experimental results discussed below also point to a weaker carbon-fluorine bond.

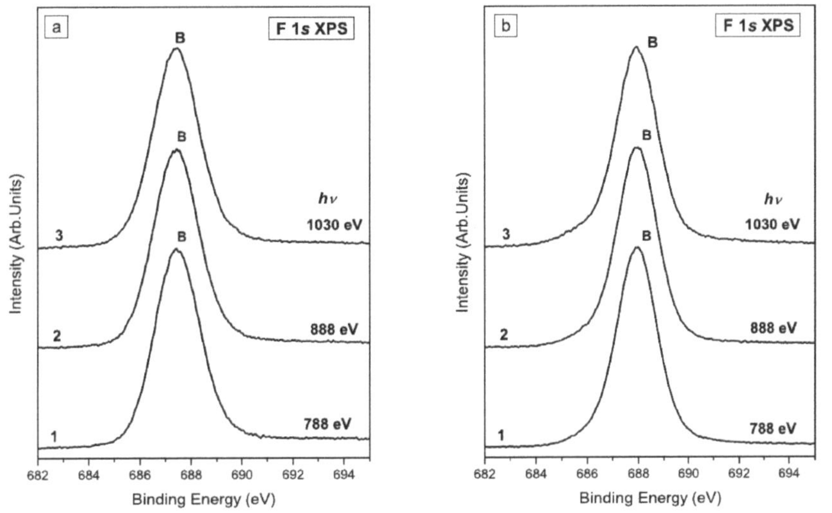

Fig. 7. F $1s$ x-ray photoelectron spectra of (a) SWCNTs $+F40\%$ and (b) MWCNTs $+F39\%$ measured at photon energies of $788eV(1), 888eV(2), 1030eV(3)$.

In addition to new peaks A_1 and B in the F-SWCNT C $1s$ photoelectron spectra there is one more feature C whose intensity is significantly lower than that of peaks $A-B$. Feature C is approximately 6 eV apart peak A itself in the case of SWCNTs $+F35\%$ or from peak A assumed position in the case of SWCNTs+F40%. In this case, chemical shift of feature C carbon atom $1s$ electron binding energy ($\sim$ 6eV) is much larger than the shift corresponding to phase 1(4.0eV). Hence, this peak is obviously associated with F-SWCNT carbon atoms with higher oxidation levels as compared with carbon atoms of phase 1. Relative intensity of peak C is independent of the excited photon energy (Fig. 6). It is possible that, contrary to MWCNTs, an extra fluorine-carbon phase 3 is formed as a result of SWCNT fluorination. As compared with phase 1 , formation of phase 3 is accompanied by even higher carbonfluorine electron density transfer resulting in formation of a stronger chemical bond. Phase 3 is probably formed due to covalent bonding of fluorine atoms to carbon atoms on the SWCNT open edges. Theoretical calculations performed for nanographite and STM measurements (Saito et al., 1999; Maruyama et al., 1994; Kobayashi et al., 2005) also point to such possibility. Notice that destruction of nanotube fullerene-like edges is a particular feature of just the SWCNT purification process (Krestinin et al., 2003). If this is true, fluorine atoms should be able to get attached to SWCNT graphene mesh not only by sharing carbon atom $2\Pi p_z$ orbitals but also by using dangling $2\sigma p_{x,y}$ bonds at the SWCNT graphene mesh edges at the points where fullerene-like structures were previously attached. This would result in formation of even stronger carbon-fluorine chemical bond of the CF type due to covalent mixing of the F $2p$ and C $2\sigma p_{x,y}$ electron states.

(Saito et al., 1999) showed, the fluorine atoms get first attached to the edge carbon atoms and only later they begin bonding to carbon atoms located along the nanotube perimeter. It was shown that doped F is more stable at the edge than in the interior region. The dopant's electron is strongly localized around the F site, and even two F atoms can interact with a single carbon atom, i.e., $\mathrm{CF}_x(x > 1)$ clusters are formed. There are two ways for fluorine atoms to get attached to the edges of nanotubes: by forming either Kleins bearded zigzag edge (Klein, 1994) (two fluorine atoms are attached to a single carbon atom at the edge) or Fujitas normal zigzag edge (Fujita et al., 1996) (one fluorine atom is attached to one carbon atom). This conclusion is confirmed by the peak C behavior in the process of SWCNTs +F40% thermal defluorination (see below), which shows that, contrary to peaks B and A_1, peak C can not be totally reduced. Seifert et al., 2000 calculated that equimolar $(\mathrm{CF})_n$ nanotubes are stable. However, the authors believe that formation of stable $(\mathrm{CF})_n$ nanotubes may be hindered by the height of the energy barrier preventing fluorine atoms penetration into CNTs. But in our case, when the most part of fullerene-like nanotube edges are already destructed, it is easier for fluorine to penetrate into SWCNTs.

This difference in carbon atom chemical states in the fluorine-carbon phases compared should manifest itself in different spatial structures, particularly, in the increase of interatomic spacing when moving from phase 1 to phase 2 and then to phase 3 .

F $1s$ spectra with $hv = 788 - 1088\mathrm{eV}$ were also measured for F-SWCNTs and F-MWCNTs (Fig. 7). Here the photoelectron kinetic energies were about $E_k \sim 100 - 400\mathrm{eV}$, which corresponds to the probing depth of about $d \sim 0.5 \div 1.0$ nm. F$1s$ spectra of SWCNTs +F35% and SWCNTs +F40% are the same. F $1s$ photoelectron spectra of F-MWCNTs keep their energy positions and, contrary to the C$1s$ spectra, do not exhibit any significant changes in their shapes up to $C_F = 39\mathrm{wt}\%$. F $1s$ photoelectron spectra of F-CNTs [(Fig. 7(a)] consist of the main peak B only. The binding energy corresponding to the F $1s$ peak of F-SWCNTs is $BE = 687.4\mathrm{eV}$; it is shifted by 0.5 eV towards smaller BE as compared with MWCNTs +F39% ($BE = 687.9\mathrm{eV}$) [Fig. 7(b)] (Brzhezinskaya et al., 2009). At the same time, its FWHM is 2.1 eV , which is 0.2 eV larger than that of F $1s$ peak of F-MWCNTs (1.9 eV).

F 1*s* spectra are less informative as compared to C 1*s* spectra. The effective charge of fluorine atom in F-SWCNTs, as in the most of other compounds, is negative and varies in a limited range of 0 to -1 when the atom's chemical state changes. A small change in the valence electron density in a fluorine atom is accompanied, due to its participation in chemical bonding with carbon atoms, by a weak change in screening of F1*s* core level electrons by valence electrons, which manifests itself in small values of chemical shifts of F1*s* electron bonding energies.

Thus, the fluorine spectrum is a superposition of contributions from all the F1*s* electrons involved in chemical bonds with carbon atoms. In the case of F-SWCNTs, there is one extra fluorine-carbon phase with much stronger fluorine-carbon bond (as compared with the other two phases observed also in F-MWCNTs) and, hence, with greater carbon-fluorine electron charge transfer. C 1*s* photoelectron spectra showed that the phase 3 contribution is smaller than that from phases 1 and 2, and, hence, the shift of the F-SWCNTs F 1*s* peak as a whole towards the lower binding energies is also insignificant (0.5 eV).

3.3 Results of defluorination experiments

Investigation of CNT defluorination process makes it possible to estimate the type and power of carbon-fluorine bonding and to find out whether the graphene mesh is distorted. Thermal annealing is one of the possible methods of F-CNT defluorination. Our data obtained for F-MWCNT annealing (Brzhezinskaya et al., 2009) showed that annealing started at temperatures almost equal to the fluorination temperature (T_F). As was shown above the results of fluorination process is more complicated for F-SWCNTs (Brzhezinskaya et al., 2010). Therefore, the results of defluorination experiments are presented here for FSWCNTs only (for F-MWCNTs see Brzhezinskaya et al., 2009). The thermal annealing process was controlled by both NEXAFS and XPS.

The first changes in the C 1*s* absorption spectrum structure of SWCNTs +F40% were observed only after the fifth annealing carried out temperature close to T_{DF} and T_F of FMWCNTs ($T_A = 350°C$) [Fig. 8(a) and 8(b), curves 2]. The changes manifested themselves in small reduction of bands $B_1{}^*$ and D^* relative intensities and growth of bands *A* and *B* intensities [Fig. 8(a), curve 3]. Bands $B_1{}^*$ and D^* in the C1*s* absorption spectrum correspond to the transitions of C 1*s* electrons to the free hybridized states formed from C 2*p* and F 2*p* states. Bands *A* and *B* correspond to the spectra of the pristine regions of hexagonal graphene mesh without any fluorine atoms attached. At the same time, no changes are observed in F 1s spectrum features [Fig. 8(b), curve 3]. Obviously, such changes in C 1s and F 1*s* absorption spectra of F-SWCNTs [Figs. 8(a) and 8(b), curves 4] demonstrate the onset of defluorination process, i.e., partial breakage of C-F bonds, which, apparently, occurs mainly in the outer SWCNTs in SWCNT bundles. After the eighth annealing performed at $T_{DF} = 420$ °C, defluorination becomes even more evident, namely, the F1s spectrum begins changing as intensely as the C 1s spectrum does [Figs. 8(a) and 8(b), curves 5]. Therefore, F-SWCNT and F-MWCNT defluorination temperatures are equal to each other and to the F-MWCNT fluorination temperature ($T_{DF} = T_F = 420$°C). This fact seems to be quite natural. It is generally recognized that the rolling of the graphene sheet induces strain, but for large-diameter CNTs the strain energy per carbon atom E_s is small. It was shown that $E_s \sim D^{-2}$, where D is the tube diameter (Mitmire & White, 1997; Hernandez et al., 1999). That is why the chemistry of wide CNTs should be similar to that of graphite, whereas the narrower tubes are expected to be more reactive due to larger strain of the carbon framework (Bettinger, 2001). Accordingly, fluorination temperature of SWCNTs with 1.4-1.6 nm in diameter ($T_F \sim 200$°C) is much lower than fluorination temperature of MWCNTs with $10 - 30$ nm in diameter. We have demonstrated that the absorption and photoelectron spectra of F-SWCNTs and F-MWCNTs and, hence, SWCNT and MWCNT fluorination results are similar, which manifests itself in covalent attachment of fluorine atoms to carbon atom $\Pi 2p_z$ orbitals. Thus, at least two of the fluorine-carbon phases formed as a result of fluorination are identical in F-SWCNTs and FMWCNTs, and, hence, the carbon-fluorine chemical bonds within each phase are equally strong. Consequently, the same amount of energy is needed to break those bonds in either of the tube types. It is due to this that the defluorination temperatures of F-SWCNTs and FMWCNTs are equal.

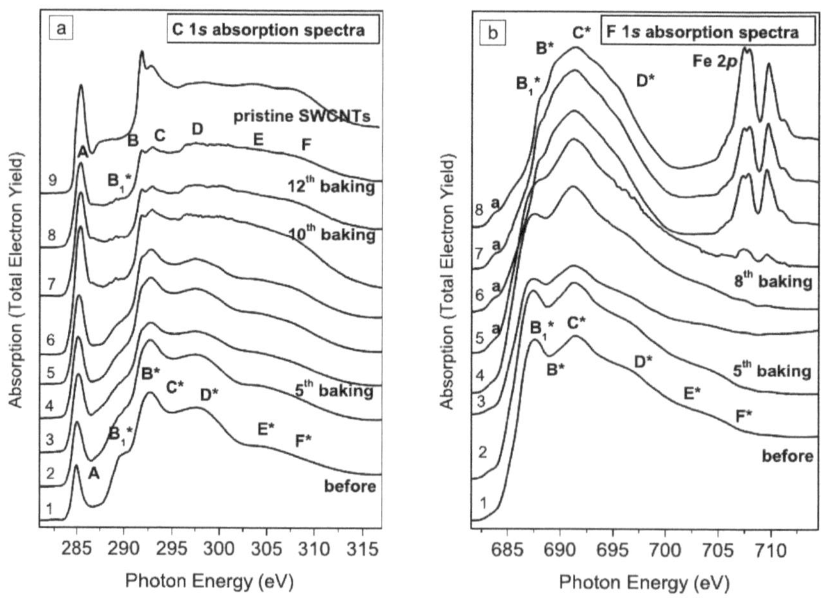

Fig. 8. (a) C 1s and (b) F 1s absorption spectra of SWCNTs +F40% before and during annealing

Notice that even the last 12th annealing did not ensure complete defluorination of FSWCNTs. It appears from the fact that band $B_1{}^*$, though dramatically reduced, is still retained on the C 1*s* absorption spectrum of F-SWCNTs. This fact indirectly confirms the presence of phase 3 with stronger fluorine carbon bonds than in phases 1 and 2 . Contrary to the C 1*s* spectrum, the F 1*s* absorption spectrum [Fig. 8(b)] changes stepwise. The intensities of bands $B_1{}^* - F^*$ drop instantly, the absorption spectrum as a whole becomes less structured. From the above, it is reasonable to believe that at $T_{DF} = T_F$ (MWCNTs) intense breakage of earlier formed C-F bonds takes place. Therefore, we can assume that the hexagonal carbon mesh destruction is insignificant, though more significant than in this case of F-MWCNTs. This is demonstrated by restoration of C 1s spectrum features typical of the initial MWCNTs and HOPG (π resonance A, σ resonance B, bands $D - F$). Such restoration of the $A - F$ bands takes place concurrently with the reduction of bands associated with the transitions of C1*s* electrons to the hybridized C-F states (bands $B_1{}^* - F^*$). Such a behavior of the main spectrum features does not contradict the interpretation of new absorption bands arising in C 1s spectra of F-SWCNTs after fluorination as those reflecting the C 1*s* electrons transition to the hybridized C-F states of the new C-F phase.

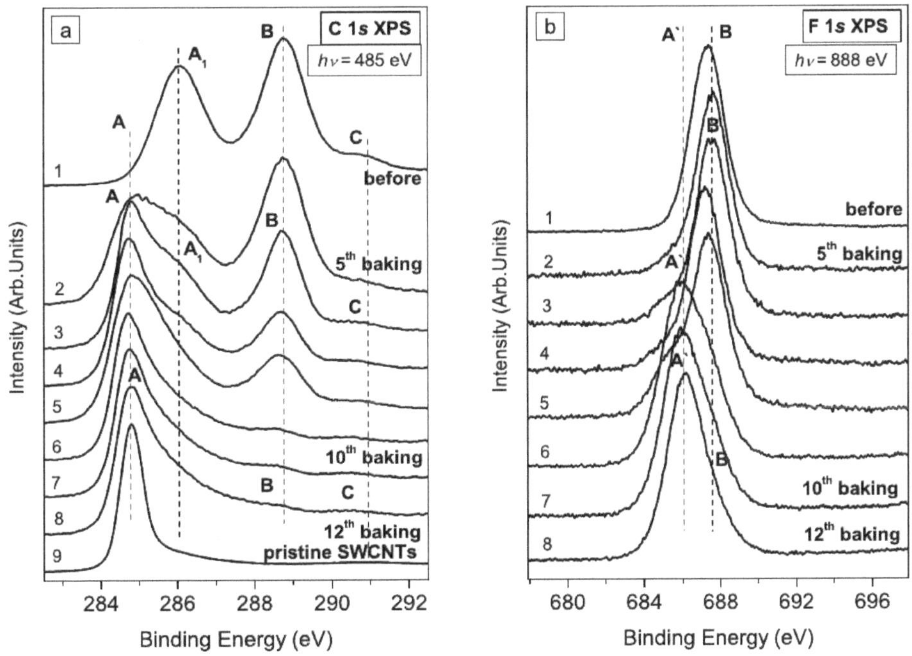

Fig. 9. (a) C 1*s* and (b) F 1*s* x-ray photoelectron spectra of SWCNTs +F40% before and during annealing

Actually, defluorination of F-SWCNTs is accompanied by modification of the $E^* - F^*$ bands typical of sp^3 hybridized electron states of carbon atoms into the E and F bands typical of HOPG with sp^2 hybridization of electron states of carbon atoms. After the further annealing of the sample ($T_{DF} = 440°C$), the C1s spectrum features reduced after fluorination were restored almost fully [Fig. 8(a), curves 6-8], whereas the F 1s spectrum varies only slightly [Fig. 8(b), curves 6-8]. The most important modification of the latter is the emergence of new features at $hv = 704 - 713eV$ after the 8th annealing. Those new features became well pronounced in Fe 2p absorption spectrum [Fig. 8(b), curves 6-8] after the 9th annealing. It should be noticed that, upon the 12th annealing, F1s spectrum does not look like HOPG or CNT spectrum any more. As in the case of the F-MWCNT absorption spectra, we can assume that F 1s absorption spectra of F-SWCNTs become similar to the 3d transition-metal fluorides spectra. It is notable that in the spectrum there is an extra low-energy structure, namely, feature a ($hv = 683.6eV$) that is caused by transitions of the F1s electrons to lowlying empty electronic states with the e_g and t_{2g} type symmetry of the transition-metal $3d$ – fluorine $2p$ hybridized character (Vinogradov et al., 2005). In addition, a broad band (~ 15 eV) at the photon energies of $685 - 700eV$ becomes more and more pronounced in the spectrum structure at the last stages of annealing (after the 9th annealing); this band is caused by core-electron transitions to the unoccupied $4sa_{1g}$ and $4pt_{1u}$ electronic states of the transition-metal atom hybridized with the 2p states of the ligand atoms. Notice that the Fe2p spectrum consists of two bands whose intensities grow with heating. Thus, that it is not only the stainless steel sample holder responsible for the signal detected but that fluorine-iron interaction has happened as a result of F-SWCNT defluorination. The presence of two spectral bands under consideration is, most probably, the result of superposition of the iron spectra from the stainless steel and iron fluorides. The substrate signal is detected not only in the absorption spectra (probing depth of 15 nm) but also in overall photoelectron spectra measured at 1030 eV (probing depth of 1.5 nm). This may be the evidence of micro-cracks formation in the sample (the micro-crack number increasing with the annealing time), since no visible cracks were detected in the sample even after the end of heating.

To our point of view, the scanning transmission microscopy (STM) results (Kelly et al., 1999) support this idea. STM images present the fluorinated zones as well-defined closed rings along the nanotube perimeters. At the same time, complete fluorination of C_2F results in radial distortions of CNTs (van Lier et al., 2005). Therefore, we can assume that thermal defluorination of SWCNTs may result in both destruction of bundles and slight distortion of nanotubes themselves. The first of the processes is more probable and preferable as our FSWCNTs absorption and photoemission spectra show no evidence of any significant amount of amorphous phase in the samples after annealing.

The F 1*s* spectrum shape after the 12th annealing possibly points to formation of iron fluoride FeF_x. These findings (fluorination of metal substrate) testify that separate fluorine atoms are mainly detached during the thermal defluorination of F-SWCNTs, as we found for F-MWCNTs also (Brzhezinskaya et al., 2009). Moreover, formation of FeF_x indirectly proves that in defluorination C-F bonds get broken and fluorine is ejected from the sample not in the form of CF_x clusters but in single atoms as in the case of F-MWCNTs.

At the same time, there is significant difference between F-SWCNT and F-MWCNT defluorination results. For F-MWCNTs, the identity of C 1*s* absorption spectra of defluorinated F-MWCNTs and pristine MWCNTs points to the absence of any remarkable amount of amorphous component in the sample after annealing. As for F-SWCNTs, thermal annealing did not result in complete restoration of resonance *B* associated with C1*s* electron transitions to the unoccupied σ symmetry states in the conduction band, which arise from $\sigma 2p_{x,y}$ carbon atomic states. This may be the evidence of partial distortion of SWCNT graphene mesh in fluorinated zones provided we assume them to be located around the tube circumference (van Lier et al., 2005). On the other hand, defluorination can lead to destruction of the outer (fully fluorinated) tubes of the bundles and to the increase in the C 1*s* absorption spectrum signal from non-fluorinated tubes from the bundles inner parts. The π resonance is restored almost completely. Therefore, we can conclude that SWCNTs are destructed by defluorination only scarcely, the destruction manifesting itself as a small number of point defects in the graphene mesh. At the same time, defluorination still destructed a negligible number of tubes, which resulted in emergence of a small amount of amorphous component in the sample; the evidence of this fact was some blurring in the high-energy region of the C1*s* absorption spectra ($h\nu = 295 - 315\text{eV}$) after the 12th annealing [Fig. 8(a), curve 8]. Such blurring may also be the result of incomplete fluorination of SWCNTs in the bundles or of the existence of two different ways of graphene fluorination: the first one is that fluorine atoms are attached to positions (1,2) of the carbon hexagons, the second one is that they get attached to positions (1,4) (Ewels et al., 2006; Khabashesku et al., 2002). These two types of fluorine atoms attachment to carbon correspond to the two fluorine-carbon phases described above (C_2 F and C_4 F, respectively). As both structures are assumed to be formed on a nanotube wall concurrently (Khabashesku et al., 2002), transition zones with low fluorination degree would appear along the boundaries of the zones with different types of fluorine attachment. On the other hand, defects can emerge in the transition regions in the process of defluorination (Khabashesku et al., 2002).

The photoelectron spectra (Fig. 9) analysis supports our conclusions about the defluorination processes in F-SWCNTs. The first changes in the C 1*s* photoelectron spectrum structure were observed only after the firth annealing carried out at the temperature close to T_{DF} and T_F of F-MWCNTs ($T_{DF} = 350°C$) [Fig. 9(a), curves 2]. The changes manifested themselves in big reduction of peak's B and small reduction of peaks' B and C relative intensities and growth of peak's A intensity [Fig. 9(a), curve 3]. Peak A corresponds to the spectra of the pristine regions of hexagonal graphene mesh. Fig. 9(a) shows that the band A_1 intensity decreases drastically, and peak A increases with respect to band B already at the first stages of annealing. Hence, such a behavior of peak A_1 during annealing is unambiguously consistent with its interpretation as a signal from carbon atoms forming a fluorine-carbon phase with fluorine-carbon chemical bond weaker than the bond in phase 1. At the same time, contrary to F1s absorption spectrum, F1s photoelectron spectrum begins to change after the 5th annealing [Fig. 9(b), curve 3]. Shoulder A' arises on the low-energy side of peak B. These findings become even more evident after the annealing at $T_{DF} = 400°C$. Obviously, such changes in C 1s and F 1s photoelectron spectra of F-SWCNTs [Figs. 9(a) and 9(b), curves 4] demonstrate the onset of defluorination process, i.e., partial breakage of C-F bonds, which, apparently, occurs mainly in the outer SWCNTs in SWCNT bundles. After the annealing at $T_{DF} = 420°C$ defluorination becomes even more evident [Figs. 9(a) and 9(b), curves 5]. Contrary to the absorption spectra, the F 1s photoelectron spectrum changes more dynamically. After the 8th annealing, the new structural feature $A^{'}$ was still looking like a small bulge on the low-energy side of the main peak B; however, immediately after the 9th annealing [Fig. 9(b), curve 6], relative intensities of the two structural features changed dramatically: the $A^{'}$ bulge turned into an intense peak having on the high binding energy side a bulge that had been previously peak B associated with the new C-F phases. After the 10th annealing this bulge becomes even less intense, and the spectrum exhibits a single peak A' with the binding energy of 686 eV and FWHM=2.5 eV [Fig. 9(b), curve 7]; peak A' is asymmetric on side of high binding energies. In effect, annealing caused drastic changes in the F 1s photoelectron spectrum: peak $B(687.6\text{eV}, \text{FWHM} = 2\text{eV})$ has disappeared, and new peak $A'(686\text{eV}, \text{FWHM} = 2.5\text{eV})$ [Fig. 9(b), curve 8] has arisen. Such transformations of the F 1s photoelectron spectrum can be easily explained by our hypothesis that the process of FSWCNT defluorination is accompanied by fluorination of the sample substrate with formation of a new FeF_x phase.

Similarly to the F 1s spectrum, C 1s photoelectron spectrum undergoes principal changes after the 9th annealing: the features associated with the new C-F phases, namely, A_1, B and C, are not clearly seen in the spectrum any more. Notice that, contrary to the A_1 and B features, feature C intensity remains almost the same. This observation supports the idea that band C can be best interpreted as the signal from a fluorine-carbon phase with stronger chemical bonds than those in phases 1 and 2 . The presence of this phase is also confirmed by the asymmetrical shape of F1s photoelectron spectrum and by the fact that, even after the 12th annealing, initial shape of the C1s spectrum is not restored: C1s peak (FWHM $= 1.6$eV, $BE = 284.8$eV) becomes evidently broader, which points to surface deformations of the outer tubes in the bundles.

4. Conclusion

The goal of this work was to study electronic structure of fluorinated single-walled and multi-walled carbon nanotubes. High energy resolution X-ray C 1s and F 1s absorption and photoelectron spectra of F-SWCNTs with different fluorine contents (35 and 40%) and FMWCNTs with fluorine contents ($10-55\%$) were measured. By comparing fine structures of F-SWCNTs and F-MWCNTs absorption spectra with HOPG, WGF and nanodiamond spectra we have found out that fluorination processes in single- and multi-walled carbon nanotubes are partly similar. Fluorine atoms interact with single-walled and multi-walled carbon nanotubes by getting covalently attached to $\pi 2p_z$ orbitals of the carbon atoms similar to WGF. It was found out that SWCNTs and MWCNTs cannot be completely fluorinated at the given fluorination temperatures (190-222 °C and 420°C, respectively). The NEXAFS measurements demonstrated that within the probing depth (~ 15 nm) of F-MWCNTs, the process of fluorination occurs uniformly and does not depend on the fluorine concentration. Our experimental data confirm the previously predicted possibility of existence of two fluorine super-lattices on the F-CNTs lateral surface, which results in developing of fluorinecarbon phases with C_2 F and C_4 F stoichiometries. Phase $1(C_2$ F) content in F-SWCNTs and FMWCNTs does not change with the increase in probing depth. It was assumed that phase 2 (C_4 F) whose electron transfer is smaller than that in phase 1 is formed only on the SWCNTs located in the near-surface layer of the bundles and at the surface of the F-MWCNTs (1-2 monolayers). At the same time, only in the case of F-SWCNTs, phase 3 (CF) exists. It is characterized by greater electron transfer than phase 1. Probably, phase 3 arises due to fluorine atoms attachment to carbon atoms located at the open edges of SWCNTs. We have revealed that at the temperatures of 190°C and 222°C the bundles of single-walled nanotubes are not completely fluorinated within the probing depth of $15-20$ nm. The process of F-SWCNT and F-MWCNT defluorination by thermal annealing has been investigated. Defluorination temperatures of F-SWCNTs and F-MWCNTs have been found to be equal to each other and to the F-MWCNTs fluorination temperature ($T_{DF} = T_F = 420$°C). We have shown that thermal defluorination of F-SWCNTs contrary to F-MWCNTs is impossible without partial deformation of graphene mesh of F-SWCNTs located on the bundle surface.

5. References

An, K.H.; Heo, J.G.; Jeon, K.G.; Bae, D.J.; Jo, C.; Yang, C.W.; Park, C.-Y.; Lee, Y.H.; Lee, Y.S. & Chung, Y.S. (2002). X-ray photoemission spectroscopy study of fluorinated single-walled carbon nanotubes. Appl. Phys. Lett, Vol. 80, No 22 (June 2002) pp. 4235, ISSN 1077-3118

Banerjee, S.; Hemraj-Benny, T.; Balasubramanian, M.; Fisher, D.A.; Misewich, J.A. & Wong, S.S. (2004). Ozonized single-walled carbon nanotubes investigated using NEXAFS spectroscopy. Chem. Commun., No. 7, (2004) pp. 772-773, ISSN 1359-7345

Banerjee, S.; Hemraj-Benny, T.; Sambasivan, S.; Fischer, D.A.; Misewich, J.A. & Wong, S.S. (2005). Near-edge x-ray absorption fine structure investigations of order in carbon nanotube-based systems. J. Phys. Chem. B, Vol. 109, No. 17 (February 2005) pp. 84898495, ISSN 1520-6106

Bettinger, H.F.; Kudin, K.N. & Scuseria, G.E. (2001). Thermochemistry of fluorinated single wall carbon nanotubes. J. Am. Chem. Soc. Vol. 123, No. 51, (December 2001) pp. 12849-12856, ISSN 0002-7863

Brown, F.C.; Bachrach, R.Y. & Bianconi, A. (1978). Fine structure above the carbon K-edge in methane and in the fluoromethanes. Chem. Phys. Lett, Vol. 54, No. 3, (March 1978) pp. 425-429, ISSN 0009-2614

Brzhezinskaya, M.M.; Vinogradov, N.A.; Muradyan, V.E.; Shulga, Yu.M.; Polyakova, N.V. & Vinogradov, A.S. (2008). Characterization of fluorinated multiwalled carbon nanotubes by x-ray absorption spectroscopy. Phys. Solid State, Vol. 50, No. 3 (March 2008) pp. 587-594, ISSN 1063-7834

Brzhezinskaya, M.M.; Vinogradov, N.A.; Muradyan, V.E.; Shul'ga, Yu.M.; Puettner, R.; Vinogradov, A.S. & Gudat, W. (2009, a). Specific features of the electronic structure of fluorinated multiwalled carbon nanotubes in the near-surface region. Phys. Solid State, Vol. 51, No. 9, (September 2009) pp. 1961-1971, ISSN 1063-7834

Brzhezinskaya, M.M.; Muradyan, V.E.; Vinogradov, N.A.; Preobrajenski, A.B.; Gudat, W. & Vinogradov, A.S. Electronic structure of fluorinated multiwalled carbon nanotubes studied using x-ray absorption and photoelectron spectroscopy. Phys. Rev. B, Vol. 79, No. 15, (2009, b) pp. 155439(12), ISSN 1098-0121

Brzhezinskaya, M.M.; Vinogradov, A.S.; Krestinin, A.V.; Zvereva, G.I.; Kharitonov, A.P. & Kulakova, I. I. (2010). Characterization of fluorinated singlewalled carbon nanotubes by X-ray absorption spectroscopy. Phys. Solid State, Vol. 52, No. 4, (April 2010) (in print), ISSN 1063-7834

Burghard, M. (2005). Electronic and vibrational properties of chemically modified singlewall carbon nanotubes. Surf. Sci. Rep., Vol. 58, Nos. 1-4, (August 2005) pp. 1-109, ISSN 0167-5729

Chen, J.G. (1997). NEXAFS investigations of transition metal oxides, nitrides, carbides, sulfides and other interstitial compounds. Surf. Sci. Rep., Vol. 30, Nos. 1-3, (October 1997) pp. 1-152, ISSN 0167-5729

Dresselhaus, M.S.; Dresselhaus, G. (2001). Carbon Nanotubes: Synthesis, Structure, Properties and Applications: Topics in Applied Physics, Springer-Verlag, ISBN 3540410864, Berlin

Ewels, C.P.; Van Lier, G.; Charlier, J.C.; Heggie, M.I. & Briddon, P.R. (2006). Pattern formation on carbon nanotube surfaces. Phys. Rev. Lett. Vol. 96, No. 21, (May 2006) pp. 216103(4), ISSN 0031-9007

Fujita, M.; Wakabayashi, K.; Nakada, K. & Kusakabe, K. (1996). Peculiar localized state at zigzag graphite edge. J. Phys. Soc. Jpn., Vol. 65, No. 7, (July 1996) pp. 1920-1923, ISSN 0031-9015

Gudat W. & Kunz, C. (1972). Close Similarity between Photoelectric yield and photoabsorption spectra in the soft-x-ray range. Phys. Rev. Lett, Vol. 29, No. 3 (July 1972) pp. 169-172, ISSN 0031-9007

Gupta, V.; Nakajima, T.; Ohzawa, Y. & Zemva, B. 2003. A study on the formation mechanism of graphite fluorides by Raman spectroscopy. J. Fluorine Chem., Vol. 120, No. 2, (April 2003) pp. 143-150, ISSN 0022-1139

Hamwi, A.; Alvergnat, H.; Bonnamy, S. & Béguin, F. (1997). Fluorination of carbon nanotubes. Carbon, Vol. 35, No. 6, (May 1997) pp. 723-728, ISSN 0008-6223

Hayashi, T.; Terrones, M.; Scheu, C.; Kim, Y.A.; Rühle, M.; Nakajima, T. & Endo, M. (2002). NanoTeflons: structure and EELS characterization of fluorinated carbon nanotubes and nanofibers. Nano Lett., Vol. 2, No. 5, (May 2002) pp. 491-496, ISSN 1530-6984

Hemraj-Benny, T.; Banerjee, S.; Sambasivan, S.; Fischer, D.A.; Eres, G.; Puretzky, A.A.; Geohegan, D.B.; Lowndes, D.H.; Misewich, J.A. & Wong, S.S. (2006). Imperfect surface order and functionalization in vertical carbon nanotube arrays probed by near edge X-ray absorption fine structure spectroscopy (NEXAFS). Phys. Chem. Chem. Phys., Vol. 8, No. 43, (2006) pp. 5038-5044, ISSN 1463-9076

Hernandez, E.; Goze, C.; Bernier, P. & Rubio, A. (1999). Elastic properties of single-wall nanotubes. Appl. Phys. A, Vol. 68, No. 3, (March 1999) pp. 287-292, ISSN 0947-8396

Huefner, S. (2003). Photoelectron Spectroscopy, Springer, ISBN 978-3-540-41802-3, Berlin.

Kelly, K.F.; Chiang, I.W.; Mickelson, E.T.; Hauge, R.H.; Margrave, J.L.; Wang, X.; Scuseria, G.E.; Radloff, C. & Halas, N.J. 1999. Insight into the mechanism of sidewall functionalization of single-walled nanotubes: an STM study. Chem. Phys. Lett, Vol. 313, Nos 3-4, (November 1999) pp. 445-450, ISSN 0009-2614

Khabashesku, V.N.; Billups, W.E. & Margrave, J.L. (2002). Fluorination of single-wall carbon nanotubes and subsequent derivatization reactions. Acc. Chem. Res., Vol. 35, No. 12, (December 2002) pp. 1087-1095, ISSN 0001-4842

Kiselev, N.A.; Moravsky, A.P.; Ormont, A.B. & Zakharov, D.N. (1999). SEM and HREM study of the internal structure of nanotube rich carbon arc cathodic deposits. Carbon, Vol. 37, No. 7, (June 1999) pp. 1093-1103, ISSN 0008-6223

Klein, D.J. 1994. Graphitic polymer strips with edge states. Chem. Phys. Lett, Vol. 217, No. 3, (January 1994) pp. 261-265, ISSN 0009-2614

Kobayashi, Y.; Fukui, K.; Enoki, T.; Kusakabe, K. & Kaburagi, Y. 2005. Observation of zigzag and armchair edges of graphite using scanning tunneling microscopy and spectroscopy. Phys. Rev. B, Vol. 71, No. 19 (May 2005) pp. 193406(4), ISSN 1098-0121

Krestinin, A.V.; Kiselev, N.A.; Raevskii, A.V.; Ryabenko, A.G.; Zakharov, D.N. & Zvereva, G.I. (2003, a). Perspectives of single-wall carbon nanotube production in the arc discharge process. Eurasian Chem. Tech. J., Vol. 5, No. 1, (January 2003) pp. 7-18, ISSN 1562-3920

Krestinin, A.V.; Raevskii, A.V.; Kiselev, N.A.; Zvereva, G.I.; Zhigalina, O.M. & Kolesova, O.I. (2003, b). Optical activity effect in crystalline structures of purified single-wall carbon nanotubes. Chem. Phys. Lett, Vol. 381, Nos. 5-6, (November 2003) pp. 529534, ISSN 0009-2614

Krestinin, A.V.; Kharitonov, A.P.; Shul'ga, Yu.M.; Gigalina, O.M.; Knerelman, E.I.; Brzhezinskaya, M.M.; Vinogradov, A.S.; Preobrajenski, A.B.; Zvereva, G.I.; Kislov, M.B.; Martinenko, V.M.; Korobov, I.I.; Davidova, G.I.; Gigalina, V.G. & Kiselev, N.A. 2009. The synthesis and characterization of fluorinated single-walled carbon nanotubes. Nanotechnologies in Russia, Vol. 4, No. 1-2, (January 2009) pp. 60-78, ISSN 1995-0780

Kudin, K.N.; Bettinger, H.F. & Scuseria, G.E. (2001). Fluorinated single-wall carbon nanotubes. Phys. Rev. B, Vol. 63, No. 4 (January 2001) pp. 045413(8), ISSN 1098-0121

Kuznetsova, A.; Popova, I.; Yates, J.T.; Bronikowski, M.J.; Huffman, C.B.; Liu, J.; Smalley, R.E.; Hwu, H.H. & Chen, J.G. (2001). Oxygen-containing functional groups on single-wall carbon nanotubes: NEXAFS and vibrational spectroscopic studies. J. Am. Chem. Soc. Vol. 123, No. 43, (October 2001) pp. 10699-10704, ISSN 0002-7863

Lebedev, N.G.; Zaporotskova, I.V. & Chernozatonskii, L.A. (2003). Fluorination of carbon nanotubes within the molecular cluster method. Microelectron. Eng., Vol. 69, Nos. 24, (September 2003) p. 511-518, ISSN 0167-9317

Lee, Y.S.; Cho, T.H.; Lee, B.K.; Rho, J.S.; An, K.H. & Lee, Y.H. J. (2003). Surface properties of fluorinated single-walled carbon nanotubes. J. Fluorine Chem.., Vol. 120, No. 2, (April 2003) p. 99-104, ISSN 0022-1139

Lee, Y.-S. (2007). Syntheses and properties of fluorinated carbon materials. J. Fluorine Chem., Vol. 128, No. 4, (April 2007) pp. 392-403, ISSN 0022-1139

Louie, S.G. (2001). Electronic properties and defects, In: Carbon Nanotubes, Dresselhaus, M.S.; Dresselhaus, G. & Avouris, Ph., (Eds.), pp. 117-152, Springer-Verlag, ISBN 978-3-540-41086-7, Berlin

Martin, C.; Arakawa, E.T.; Callcott, T.A. & Ashley, J.C. (1985). Low energy electron attenuation length studies in thin amorphous carbon films. J. Electron Spectrosc. Relat. Phenom. Vol. 35, No. 2, (March 1985) pp. 307-317, ISSN 0368-2048

Maruyama, M.; Kusakabe, K.; Tsuneyuki, S.; Akagi, K.; Yoshimoto, Y. & Yamauchi, J. 2004. Magnetic properties of nanographite with modified zigzag edges. J. Phys. Chem. Solids, Vol. 65, Nos. 2-3, pp. 119-122 (March 2004), ISSN 0022-3697

Mickelson, E.T.; Huffman, C.B.; Rinzler, A.G.; Smalley, R.E.; Hauge, R.H. & Margrave, J.L. (1998). Fluorination of single-wall carbon nanotubes. Chem. Phys. Lett, Vol. 296, Nos. 1-2, (October 1998) pp. 188-194, ISSN 0009-2614

Mickelson, E.T.; Chiang, I.W.; Zimmerman, J.L.; Boul, P.J.; Lozano, J.; Liu, J.; Smalley, R.F.; Hauge, R.H. & Margrave, J.L. (1999). Solvation of fluorinated single-wall carbon nanotubes in alcohol solvents. J. Phys. Chem. B. Vol. 103, No. 21 (May 1999) pp. 4318-4322, ISSN 1520-6106

Mitmire, J.W. & White, C.T. (1997). In: Carbon nanotubes. Ebbesen, T.W. (Ed.), pp. 191-224, CRC Press, ISBN 0849396026, Boca Raton, FL

Molodtsov, S.L.; Fedoseenko, S.I.; Vyalikh, D.V.; Iossifov, I.E.; Follath, R.; Gorovikov, S.A.; Brzhezinskaya, M.M.; Dedkov, Yu.S.; Puettner, R.; Schmidt, J.-S.; Adamchuk, V.K.; Gudat, W. & Kaindl, G. (2009). High-resolution Russian-German beamline at BESSY. Appl. Phys. A, Vol. 94, No. 3, (March 2009) pp. 501-505, ISSN 0947-8396

Okotrub, A.V.; Yudanov, N.F.; Chuvilin, A.L.; Asanov, I.P.; Shubin, Yu.V.; Bulusheva, L.G.; Gusel'nikov, A.V. & Fryodorov, I.S. (2000). Fluorinated cage multiwall carbon nanoparticles. Chem. Phys. Lett, Vol. 322, Nos. 3-4 (May 2000) pp. 231-236, ISSN 0009-2614

Park, K.A.; Choi, Y.S.; Lee, Y.H. & Kim, C. (2003). Atomic and electronic structures of fluorinated single-walled carbon nanotubes. Phys. Rev. B, Vol. 68, No. 4 (July 2003) pp. 045429(8), ISSN 1098-0121

Rosenberg, R.A.; Love, P.J. & Rehn, V. 1986. Polarization-dependent $C(K)$ near-edge x-rayabsorption fine structure of graphite. Phys. Rev. B, Vol. 33, No. 6 (March 1986) pp. 4034-4037, ISSN 1098-0121

Saito, R.; Yagi, M.; Kimura, T.; Dresselhaus, G. & Dresselhaus, M.S. 1999. Electronic structure of fluorine doped graphite nanoclusters. J. Phys. Chem. Solids, Vol. 60, No. 6, (June 1999) pp. 715-721, ISSN 0022-3697

Sato, Y.; Hagiwara, R. & Ito, Y. 2003. Refluorination of pyrocarbon prepared from fluorineGIC. Solid State Sci., Vol. 5, No. 9, (September 2003) pp. 1285-1290, ISSN 1293-2558

Schiessling, J.; Kjeldgaard, L.; Rohmund, F.; Falk, L.K.L.; Campbell, E.E.B.; Nordgren, J. & Brühwiler, P.A. (2003). Synchrotron radiation study of the electronic structure of multiwalled carbon nanotubes. J. Phys.: Condens. Matter, Vol. 15, No. 38, (October 2003) pp. 6563-6579, ISSN 0953-8984

Seifert, G.; Kohler, T. & Frauenheim, T. 2000. Molecular wires, solenoids, and capacitors by sidewall functionalization of carbon nanotubes. Appl. Phys. Lett, Vol. 77, No 9, (August 2000) pp. 1313(3), ISSN 1077-3118

Stöhr, J. (2003). NEXAFS Spectroscopy, Springer, ISBN-13 978-3540544227, Berlin

Shul'ga, Y.M.; Tien, T.-C.; Huang, C.-C.; Lo, S.-C.; Muradyan, V.E.; Polyakova, N.V.; Ling, Y.-C.; Loufty, R.O. & Moravsky, A.P. (2007). XPS study of fluorinated carbon multiwalled nanotubes. J. Electron Spectrosc. Relat. Phenom., Vol. 160, Nos. 1-3, (August 2007) pp. 22-28, ISSN 0368-2048

Tang, Y.H.; Sham, T.K.; Hu, Y.F.; Lee, C.S. & Lee, S.T. (2002). Near-edge X-ray absorption fine structure study of helicity and defects in carbon nanotubes. Chem. Phys. Lett, Vol. 366, Nos. 5-6, (December 2002) pp. 636-641, ISSN 0009-2614

Touhara, H. & Okino, F. (2000). Property control of carbon materials by fluorination. Carbon, Vol. 38, No. 2, (December 2000) pp. 241-267, ISSN 0008-6223

Van Lier, G. ; Ewels, C.P.; Zuliani, F. ; De Vita, A. & Charlier, J.-C. (2005). Theoretical analysis of fluorine addition to single-walled carbon nanotubes: functionalization routes and addition patterns. J. Phys. Chem. B, Vol. 109, No. 13, (March 2005) pp. 6153-6158, ISSN 1520-6106

Vinogradov, A.S.; Fedoseenko, S.I.; Krasnikov, S.A.; Preobrajenski, A.B.; Sivkov, V.N.; Vyalikh, D.V.; Molodtsov, S.L.; Adamchuk, V.K.; Laubschat, C. & Kaindl G. 2005. Low-lying unoccupied electronic states in 3*d* transition-metal fluorides probed by NEXAFS at the F 1 s threshold. Phys. Rev. B, Vol. 71, No. 4 (January 2005) pp. 045127(11), ISSN 1098-0121

Yudanov, N.F.; Okotrub, A.V.; Shubin, Yu.V.; Yudanova, L.I.; Bulusheva, L.G.; Chuvilin, A.L. & Bonard, J.-M. (2002). Fluorination of arc-produced carbon material containing multiwall nanotubes. Chem. Mater., Vol. 14, No. 4, (April 2002) pp. 14721476, ISSN 0897-4756

Zhou, J.; Zhou, X.; Sun, X.; Li, R.; Murphy, M.; Ding, Z.; Sun, X. & Sham, T.-K. (2007). Interaction between Pt nanoparticles and carbon nanotubes - An X-ray absorption near edge structures (XANES) study. Chem. Phys. Lett, Vol. 437, Nos. 4-6 (April 2007) pp. 229-232, ISSN 0009-2614

Chapter-5

Microwave Dielectric Properties of Carbon Nanotube Composites

1. Introduction

Carbon black (CB) (Donnet et al., 1993) and carbon fiber (CF) (Burchell, 1999) have been used as fillers in radio wave and microwave composites for more than half century. Typical applications include electromagnetic compatibility (EMC) or electromagnetic interference (EMI), microwave absorbing coatings, and anti-electrostatic materials, etc due to their good electric/dielectric properties, stability and chemical resistance as compared to composites with metal powders of nano or micron meter in size. CB composites normally have frequency-dependent permittivity which is useful in the design of broadband absorbing or functional materials at microwave band. The frequency dispersion in CB composites is mainly caused by the effect of large agglomerate of CB spherical particles with diameter below 100 nanometers and usually appears in composites with large concentration of fillers. CF composites can have special microwave properties, such as high permittivity at relatively low concentrations of fibers (Lagarkov et al, 1998). The dispersive property of CF composites at microwave band is determined by the frequency-dependent response of individual fibers or fiber clusters (Liu, et al, 2007a). The dispersion can be observed even at very low concentration of fibers, for example, less than one percent in volume. This feature makes CF composites advantageous for many technical applications, such as lightweight absorbing materials (Neo & Varadan, 2004) and phantom materials for modeling the electromagnetic response of biological issue (Youngs, et al, 2002). However, to obtain strong dielectric dispersion at microwave frequencies, CF must contain fibers which are at least millimeters in length to achieve large aspect ratio considering the diameter of commercial CF which is normally close to 10 microns. The large dimension of the fibrous inclusions restricts the practical application of CF composites. Also, the favorite processing methods, such as spray technique, are not applicable for CF composites. The planar CF composites are anisotropic with degraded dielectric response across the plane. Due to the heterogeneity of the composites, a surface layer of millimeters in thickness appears with properties different from those of the bulk composites (Simovski, et al, 2000 and Matitsine, et al, 2003). This gives rise to difficulty in modeling and designing composites with desired properties.

Carbon nanotubes (CNTs) have been fabricated since the early 1990s (Ijima, 1991). The primary research interests include the synthesis or growth of CNTs, because it was a challenging task to prepare enough amounts of CNTs with desired dimension and purity for measurement purpose during the early stage. With the development of processing method like arc discharge and laser ablation of graphite, as well as more productive chemical vapor deposition (CVD) and plasma enhanced CVD method, high purity CNTs with controllable wall-thickness/length and acceptable price are commercially available now (Meyyappan, 2005). Other research interests include the characterization and theoretical understanding of their basic properties of CNTs. The physical properties of carbon nanotubes were reviewed in literature (Saito, 1998), which have been considered as the classic monograph on CNTs. Due to the unique electronic structure depending on the diameter and chirality, CNTs could be the smallest semiconductor devices likely to be fabricated (Saito, 1998). And the extraordinary mechanical properties of CNTs make them suitable replacement of CF in continuous fiber reinforced composites for defence or aerospace applications. Other potential applications includes single- or multi-walled CNTs sensor for scanning probe microscopy, and biosensor or device applications such as CNTbased diodes, transistors and field emitter, etc (Meyyappan, 2005).

Due to their fibrous shape with extremely large in aspect ratio, CNTs may allow low resistivity and high permittivity and frequency dispersion to be obtained in composites with extremely low concentrations. As compared with CF, one significant difference is that CNTs are nanoscaled particles whose diameter is several orders of magnitude smaller than that of CF (typically about 10μ m). It is well known that the nanosized particles usually display distinct properties from microsized particles of the same composition, which is the primary reason for the great attention currently given to the radio and microwave frequency performance of CNTs composites. A number of novel features have been reported on CNTs in the literature, these results demonstrate the possibility to design CNTs composites with electric/dielectric properties which are more diverse than those obtainable with other carbon fillers. That is why in the last few years, main research interest in CNT was shifted from fabrication methods to the novel properties, such as the conductivity at low concentration and dielectric properties at low frequency (DC to 1 MHz) and high frequency (above 1 MHz , including radio wave, microwave, millimeter wave, Terahertz and Inferred wave frequency). Since the dielectric properties of CNT composites at microwave frequency are the main focus of this chapter, previous research works on the electric properties, dielectric permittivity at high frequency, as well as potential

EMC/EMI and absorbing applications which are most relevant to our research objective, will be reviewed sequentially.

After the observation of a conductivity threshold in polymer/CNTs composites (Coleman, et al, 1998), around 200 publications have reported on the electrical percolation threshold of CNTs in different polymer systems by 2009 (W. Bauhofer & J. Z. Kovacs, 2009). Different CNTs, synthesis methods, dispersion treatment and polymer matrix have been employed in the previous research. Since it is established that the percolation theory or power law could explain the electric conductivity of CNTs composite (Martin et al, 2004). Minimum percolation ratio and maximum conductivity at low frequency becomes the main research focus. Due to the dispersion method and difference in the holding matrix, percolation varies from less than 0.01 percent to a few percent. Theoretically, typical aspect ratio of 1000 produces percolation value about 0.1% (Balberg et al, 1984). Practically, a few order of difference could be found from composite with same CNTs and different polymer matrix (W. Bauhofer & J. Z. Kovacs, 2009). Maximum conductivity of MWCNTs composite could be realized when the weight concentration is larger than 10% (Skakalove et al, 2005). And the maximum conductivity seems insensitive to the types of CNTs and treatments. Nevertheless, it is a challenging task to disperse CNTs uniformly at such high concentration. The mechanism of charge transport in CNTs composites could be explained by hopping and tunneling effect among neighboring tubes (Kim, 2006). However, the morphology of CNTs in holding matrix seems difficult to be represented by the classical percolation theory.

Permittivity ($\varepsilon = \varepsilon' - j\varepsilon''$) instead of AC conductivity is normally used to describe the dielectric properties of CNTs composites as frequency is above 1 MHz . Also, measurement methodologies and potential applications could be different from that at low frequency. Microwave properties are normally measured with impedance or network analyzer through coaxial line or waveguide fixture. A number of special features have been reported on CNT in the literature, such as excitation of localized electronic states resulting in high intrinsic permittivity (Watts et al, 2003). The microwave permittivity is found to exhibit smooth frequency-dependent (Browning, et al, 1998 & Sandler, et al, 2003). These results seem to demonstrate the possibility to obtain CNT composites with microwave properties which are more diverse than those obtainable with other types of carbon fillers, e. g., with various types of dispersive dielectrics, with high dielectric but having relative low dielectric loss (Yang, et al, 2009), and tunable dielectric under small bias voltage (Liu, et al, 2008). In addition, the nanoscaled size of CNT in composites may avoid the drawbacks of CF composites, which are limited by the macroscopic size of the fibers for thin structure and component applications.

Since the low resistivity of CNTs composite with low filling factor as compared with CB composite which are commonly employed as filler of rubber, conductivity polymer and shielding coating, the shielding effectiveness (SE) of CNTs composites are mostly investigated at microwave frequency (Huang et al, 2007). MWCNTs/Ceramic composites were measured at Ku-band with SE up to 25 to 30 dB achieved (Shi & Liang, 2008). Some experimental results showed that absorption is the major shielding mechanism and reflection is the secondary mechanism (M. H. Al-saleh & U. Sundararaj, 2009). SWCNTs thin film demonstrated good shielding of EM wave at terahertz frequency while still maintaining good transparency for visible light (Seo, et al, 2008). Measurement results show that CNT sheets are more effective in providing EMI shielding compared with graphite and carbon black sheets over broad frequency range due to better electron transmission (Wang, et al, 2009). Radar absorbing structures with MWCNTs and polyurethane foam were designed with optimization method before measured with free space setup (Park, et al, 2006). 10 dB reflection loss at X-band could be achieved by MWCNTs layer of thickness about 3.3 mm (Lee et al, 2006). Self-sensing and self-actuation response were found out from CNTs composites which could be explained by electrostriction effect (Lee & Shkel, 2007). The tunability of both real and imaginary parts of the permittivity found from CNTs composites has potential applications for smart materials and structures (Liu, et al, 2008).

The shielding effectiveness and absorption of EM wave are functions of electric and dielectric properties; the objective of this chapter is to investigate the resistivity and dielectric permittivity of CNTs composites for different wall thicknesses. The organization of this chapter is as follow. Theoretical methods and models concerning dielectric properties of CNTs composites are reviewed briefly in Section 2. The materials, fabrication conditions and measurement method of dielectric properties are addressed in Section 3. The results in Section 4 include electric properties, frequency-dependent permittivity, tunable permittivity under bias voltage and potential applications. Last but not least, the conclusion of this research work are summarized in Section 5.

2. Theory

In CB composites, the microwave behavior is attributed mainly to the large agglomerates of carbon particles. In this case, the properties of the composite are consistent with the percolation theory that accounts for the statistical properties of clusters comprising many inclusions in electrical contact within the composite (Bergman & Imry, 1977). The percolation approach (Satuffer & Aharony, 1991) depends on the empirical percolation threshold p_c that cannot be derived from the percolation theory. The percolation threshold is defined as the least concentration at which a composite with conductive inclusions is capable of conducting direct current. For the permittivity ($\varepsilon = \varepsilon' + \mathrm{i}\varepsilon''$) of a composite comprising conductive inclusions and a lossless dielectric host matrix of permittivity ε_{h}, percolation theory predicts a power dependence on the frequency f (Gefen, et al, 1983):

$$\varepsilon'' = Af^{-x}, \varepsilon' - \varepsilon_{\mathrm{h}} = Bf^{-y} \tag{1}$$

In (1), A and B are proportionality factors, x and y are critical exponents. In the classical percolation theory, the range of concentrations p where scaling law (1) is not expected to be observed in a wide frequency range, but typically very narrow, for example $(p - p_{\mathrm{c}})/p_{\mathrm{c}} \leq 0.1$ (Gefen, et al, 1983). However, in many measured data of CB composites, the scaling frequency dispersion is observed in a wide range of both concentrations and frequencies (Brosseau, 2002). If (1) is valid over the entire frequency range, it follows from the KramersKronig relations that (Jonscher, 1999):

$$x = y \tan\,\delta = \frac{A}{B} = \tan\,\frac{\pi y}{2} \tag{2}$$

where tan δ is the dielectric loss tangent, and the second of (2) is valid for $\varepsilon' \ggg_{\mathrm{b}_1}$. Therefore, the critical exponents for ε' and ε'' are the same, while the dielectric loss tangent is frequency-independent. Equation (1) coincides with the "universal relaxation law" established for many types of dielectrics (Jonscher, 1999). In the percolation theory, the critical exponents are considered to be universal, i.e. the same for different composites, independent of the type and properties of inclusions comprising the composite. The physical basis is that the behavior of the composite is determined by the statistical properties of the clusters, but not by the parameters of individual inclusions.

The universal values for critical exponents of frequency dependence are usually derived from the exponents for concentration dependence. For this, two different approaches are known, with exponent value of either 0.27 or 0.42 (Song, et al, 1986). Many numerical and experimental data are obtained within these limits. For example, numerical studies of random RC networks (Straley, 1977) produce 0.28 for 3-D bond problem, but 0.18 for 3-D site problem. For continuum percolation problem, the value of 0.38 is obtained from the studies of water-oil emulsions (Bordi, et al, 1996). Therefore, the critical exponent values are believed to be within the range of 0.2 to 0.4 , with the uncertainty attributed to the difference in geometry between the bond, site, and continuum percolation problems that may result in different properties of the clusters. For CNT composites, the critical exponent of 0.34 , consistent with percolation theory, is obtained (Kim, et al, 2003). In all cases, x is less than 1 . Therefore, the AC conductivity of the composite must increase with frequency.

In actual composites, the percolation dispersion governed by (1) appears within a limited range of frequency. The high frequency limit is determined by the frequency at which the conductivity of the composite is equal to the conductivity of the inclusions, and therefore, no longer increases with frequency. The low frequency limit is as considered (Laibowitz & Gefen, 1984). Below this low frequency limit, the real permittivity or conductivity is independent of frequency at concentrations below p_c or above p_c correspondingly. Due to the finite frequency range for percolation dispersion, (2) is not exact for practical materials, with the discrepancy increases as the frequency approaches the limits of the percolation frequency range. If (1) is valid over restricted frequency range, then the dielectric loss tangent increases relatively with respect to the value given by (2), as the high frequency limit of this frequency range is approached. Near the low frequency limit, the loss tangent decreases for concentrations below p_c and increases for concentrations above p_c.

3. Experiment

3.1 Characterization of CNTs

Table 1. Specifications of three types of CNTs

	SWCNT	DWCNT	MWCNT
Purity	95%	90%	95%
Impurities	MWCNT, Ash, Metal	MWCNT, Ash, Metal	Ash, Metal
Length	~ 30μ m	~ 50μ m	~ 50μ m
Diameters	1 ~ 2 nm	2 ~ 4 nm	8 ~ 15 nm
Aspect ratio	~ 20k	~ 17k	~ 2k

Single-, double- and multi-walled carbon nanotubes (SWCNT, DWCNT, and MWCNT) produced by CVD method were obtained from the Timesnano, Chengdu Organic Chemicals Co. Ltd., the Chinese Academy of Sciences. As produced CNTs were purified by the supplier with Oxidation and HCL wash method. The specifications of these CNTs are given in Table 1. Figure 1 shows the image of these CNTs with high resolution field emission SEM (JEOL JSM-6340F). The diameter of SWCNT and DWCNT seems larger than the values given by the supplier. Possible reasons can be the appearance of bundles or small amount of the CNTs may be multi-walled instead of single-walled. The aspect ratio is calculated based on the length over the average diameter of individual CNTs. It can also been observed from Figure 1 that length of CNTs are larger than the scope observed by SEM which is about a few micron meters.

Fig. 1. High-resolution SEM of CNT (a) SWCNT (b) DWCNT (c) MWCNT

The results of thermal gravimetric analysis (TGA) shown in Figure 2 verify that the carbon content in samples is within the ranges provided by the supplier. The residual weights (which is usually contributed by metallic catalyst particles) of the SWCNT and MWCNT samples are almost zero. It suggests that metallic particles can be efficiently removed by the HCl wash. However, about 6wt% of residues is still remained on the DWCNT sample, shown in Figure 2(b). This may be caused by the different purification method applied by the supplier. Differential Thermal Gravimetry (DTG) results in Figure 2 also demonstrate that most of the carbon materials can be oxidized in a narrow temperature range centered at the peak oxidation temperature (Tg). The narrow oxidation temperature range suggests that carbon species in the samples have similar structures. T_g of the SWCNT sample is at 624°C, which is higher than the oxidation temperature of purified SWCNTs with diameter of 0.7-1.0 nm (Wei, et al, 2008). The increase of T_g may be contributed by the large diameter tubes or certain amount of impurities, such as MWCNTs. The decrease of T_g for DWCNT samples is caused by the metallic residues remained on the sample which may catalyze the oxidation of carbon and result in the lower oxidation temperature at 604°C. The DTG result of MWCNT shows a relative broader oxidation peak, as well as a higher T_g. The relative large range of wall thickness might contribute to the observed difference.

Raman spectroscopy was further applied to analyze the three CNT samples. Samples show three typical bands: radial breathing mode (RBM) between 150 and 350 cm^{-1} characteristic for SWCNT or DWCNT, the D band at approximately 1300 cm^{-1} assigned to defective and disordered carbon species, and the peak complex centered around 1600 cm^{-1}, known as the G band, which is characteristic for ordered carbon species such as CNTs and graphite. The sharp RBM peaks from the SWCNT sample in Figure 3(a) and the DWCNT sample in Figure 3(b) verify the high content of SWCNTs and DWCNTs. The very low D band intensities also suggest that little amorphous carbon exists. Figure 3(c) shows the typical Raman spectrum of MWCNTs. However, it is difficult to quantify the purity of different CNT species. Instead, it is confirmed that there are significant amount of CNTs in the three samples.

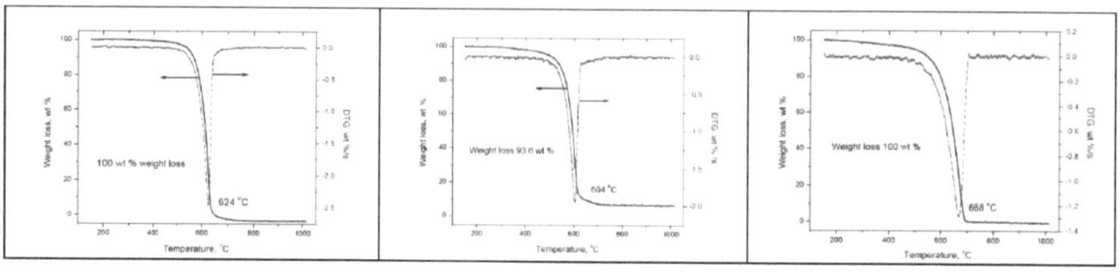

Fig. 2. Purity analysis of (a) SWCNT (b) DWCNT (c) MWCNT with TGA

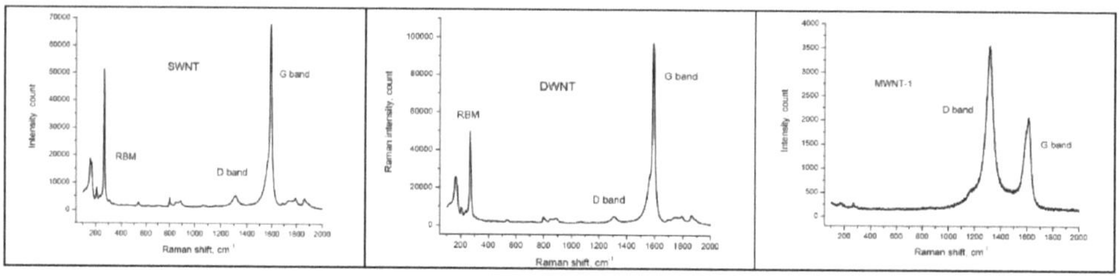

Fig. 3. Raman spectrum of (a) SWCNT (b) DWCNT (c) MWCNT

3.2 Fabrication of CNTs composites

The host polymer to prepare composites was commercial two-component silicone (KE1800T) from Shin Etsu. CNTs and silicone (Component A) were dispersed in Toluene (Methylbenzene) with Ultrasonication processer (Sonics VCX500) for 20 min . The resistance of liquid mixture with 0.01 g of SWCNTs, 1 g of silicone and 25 ml of Toluene was monitored with a parallel-plate sensor connected with insulation multimeter (Megger BMM80). The dependence of resistance on the sonication time plotted in Figure 4 shows that 20 minutes sonication time is probably enough to disperse the mixture to stable stage. A SEM image (shown in Figure 5) of SWCNTs after sonication process shows that big agglomerates in Figure 1(a) were broken and no big bundles (larger than 1μ m) could be found.

Mixture of CNTs, silicone and Toluene in a 25 ml crucible was heated up with a hotplate at 90°C for 8 hours to remove the solvent. After that, the 10% of liquid hardener (Component B) was added and mixed manually before pouring into mould. The mould was put into convection oven to cure at 70°C for 24 hours. As a control, another group of SWCNT, DWCNT and MWCNT was mixed manually with silicone component A and component B for 15 min before cured under the same conditions. And the control group prepared by manual mixing will be used as the benchmark of the sonication process.

Two types of samples were fabricated with moulding method. Disk shape samples have diameter of 16 mm and thickness of 2 mm to measure the resistivity and permittivity from 0.001 GHz to 1 GHz . Toroidal shape samples for permittivity measurement at high frequency (from 0.05 GHz to 5 GHz) have outer diameter of 14 mm , inner diameter of 7 mm and thickness of about 2 mm .

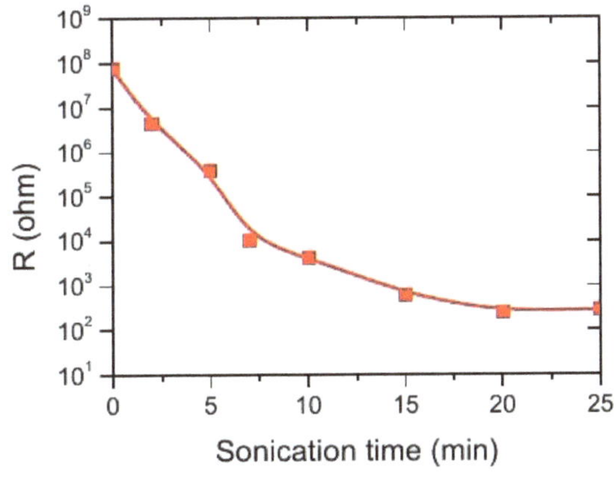

Fig. 4. Resistance vs. sonication time

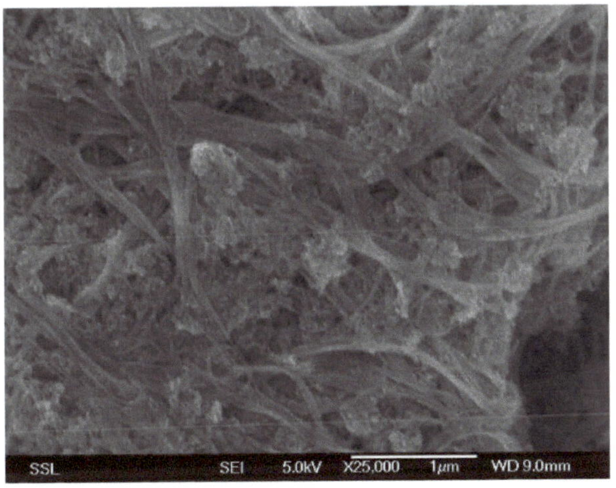

Fig. 5. SWCNT after sonication

3.3 Measurement methods

DC resistance of the samples was measured using a Megger ™ BMM80 ™ insulation multimeter and a parallel plate fixture. Before measurement, the disk-shape samples were cleaned with ethanol to remove the release agent sprayed on the mould prior to casting. This is important since the agent contributed to the measured resistance. Resistivity was calculated from the measured resistance and dimension of the samples. Agilent 4991A RF impedance/materials analyzer was employed for permittivity measurement from 0.001 GHz to 1 GHz . Figure 6 shows the test fixture (Agilent 16453A) for measurement of dielectric permittivity. During measurement, a bias voltage up to 40 volts was applied to the sample in between the upper and lower electrodes. The permittivity measured is in the direction of the bias voltage applied. Special care was taken to keep good electrical contact between the electrodes and the sample by applying certain pressure. Since there is no air gap effect, the impedance method can achieve good accuracy for measurement of conductive samples without correction [Liu, et al, 2007b, 2008].

To validate the measurement results by impedance analyzer, 14 mm coaxial line fixture (as shown in Figure 7) and network analyzer (Agilent N5230A) was used to measure the permittivity of SWCNT composites samples from 0.05 to 5 GHz . The permittivity was extracted from the measured S-parameters of the sample placed within the coaxial line. Since the outer diameter of the sample can be slightly smaller than the inner diameter of the test fixture due to the shrinkage of the sample after curing, the air gap in between is the largest contributor of uncertainty, especially when the concentration is high. Therefore, only 4wt% and below measured by the two different methods have been compared. The real and imaginary permittivity of SWCNTs composites of weight concentration from 0.5% to 4% are plotted in Figure 8. It shows rather good agreement between these two techniques over 0.05 GHz to 1 GHz for concentration up to 3%. Both real and imaginary permittivity measured by 14 mm coaxial line fixture at 4% show lower values than these from impedance method, which could be properly explained through air gap effect (Liu et al, 2007b and reference therein). After correction of air gap effect, the real permittivity (dash line) shows better agreement with impedance measurement. It is proved that impedance measurement has better accuracy even at high concentration as compared with coaxial line method.

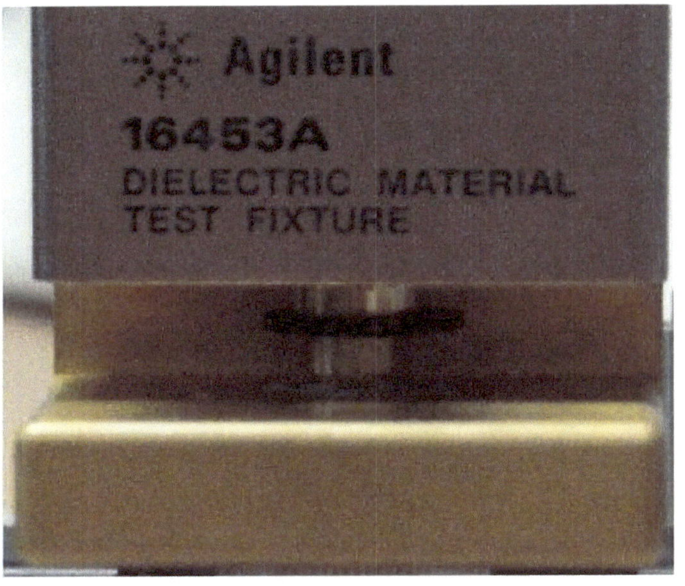

Fig. 6. Test fixture and sample for permittivity from 1 MHz to 1 GHz

Fig. 7. Test fixture and sample for permittivity from 50 MHz to 5 GHz

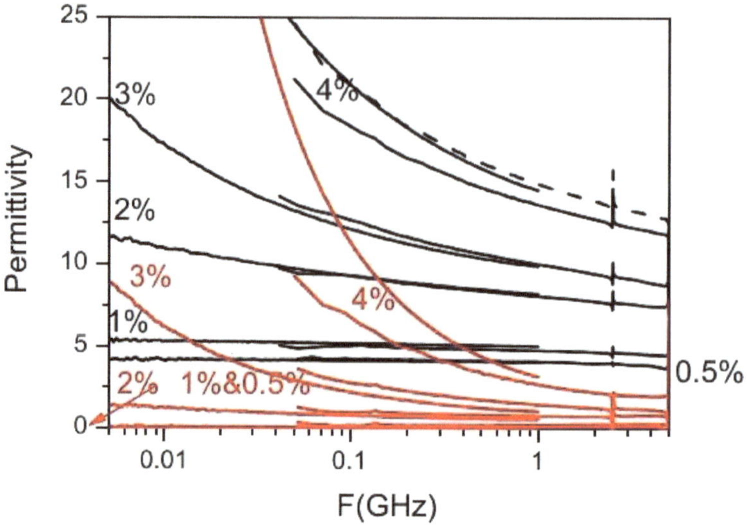

Fig. 8. Permittivity of SWCNT composites measured by Impedance and 14 mm coaxial method

4. Results and Discussion

4.1 Electrical properties

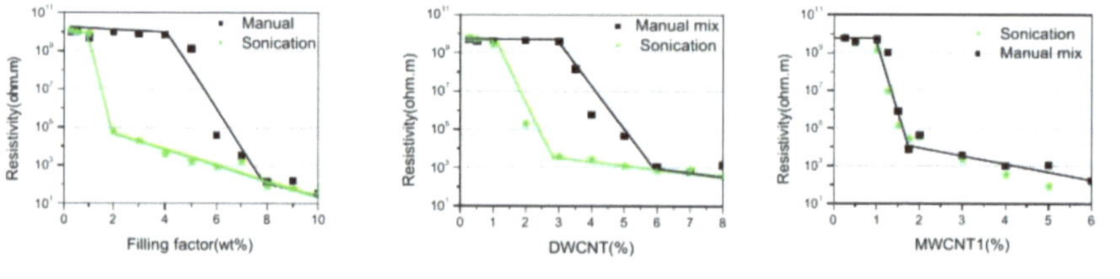

Fig. 9. Resistivity of (a) SWCNT (b)DWCNT (c)MWCNT

The measurement electrical properties of different types of CNTs are plotted in Figure 9. The marks are the measured resistivity of CNTs composites prepared by sonication process (stars) and manual mixing (squares). And the poly-lines are the guide of the dependence of resistivity on the concentration of CNTs instead of fitting curves. Due to the measurement range of the insulation multimeter is up to 200GΩ, the resistibility measured at low concentration might be lower than the real value which could results in larger percolation threshold.

For SWCNT and DWCNT, the percolation threshold is around 1wt% if dispersed by sonication process. The percolation of manually mixed SWCNT and DWCNT composites could be up to 3 ~ 4wt%. For MWCNT composites, it seems the resistivity decreases before

1%. The percolation ratio could be between 0.5% and 1%. No obvious difference could be found from MWCNT composites dispersed by two different processing methods. Individual tubes instead of bundles could be observed from SEM image of MWCNT in Figure 1(c), as compared with SWCNT (Figure 1(a)) and DWCNT (Figure 1(b)). Therefore, manual mixing might be enough to disperse the tubes inside polymer matrix. However, sonication mixing could result in more steep percolation transition for SWCNT and DWCNT composites. And the bundle of SWCNT and DWCNT couldn't be broken by the manual mixing process.

4.2 Frequency-dependent permittivity

Figure 10 to Figure 12 show the measured dielectric properties of SWCNT, DWCNT and MWCNT composites processed with sonication method respectively. The percolation for three types of CNTs is around 1%. It is sufficient to investigate concentration up to 5% from percolation phenomena point of view. In all of three figures, solid lines show the measured real (in black) and imaginary (in red) permittivity and dash lines show fitting curved using (1). The permittivity of silicone host (ε_h) is constant value, 3 , according to impedance measurement. From the measured dependence of the permittivity, the dispersion curves for three types of CNTs appear to be similar and do not show any absorption peak that is typical of the Debye dispersion law and mixing rules. The frequency dispersion is smooth and resembles the dispersion of the percolation type. Therefore, the fitting of date made with scaling dispersion law instead of the Cole-Davidson law is validated (Liu et al, 2007b). The frequency dependence of the real and imaginary permttivity are fitted separatedly, to track deviations from the percolation law given by (2).

The parameters x and y obtained by curve fitting, which depend on the concentration, are shown in Table 2. Both A and B increase steadily with concentration which is understandable due to more conducting channels involved at higher concentrations. Therefore, they are not included in Table 2. Both x and y increase with filling factor. The critical exponent y is smaller than x for most of concentraions for all of three types of CNTs. Exponent x at percolation threshold fall in the the range (0.2 to 0.4) form theoretical calculation, but exponent y is relatively lower (0.1 to 0.2) for 2%. Deviation from the linear dependence (scaling law) can be found from 4% to 5% may explain the higher exponent value than the universal value.

Therefore, the scaling frequency dispersion is observed in the CNTs composites within wide frequency range and concentration around percolation. In the percolation theory, the scaling behavior depends on the geometric spread of the properties of clusters, which is a function of the number of inclusions comprising a cluster. In actual composites, certain reasons lead to an additional spread of the properties of the clusters, such as the dispersion in dimension and in the conductivity of the CNT. Another reason can be the imperfect electrical contacts among the inclusion in the clusters where the conductivity between adjacent inclusions is due to interpaticle tunneling. An additional spread of the dispersion curve may be due to the elongated shape of CNT, because a net of tubes forms clusters of complex shape, which must be described by an assembly of RC circuits rather than by a single circuite. All these phenomana change the distibution of properties of tube clusters and therefore, results different frequency-dispersion from and exponents.

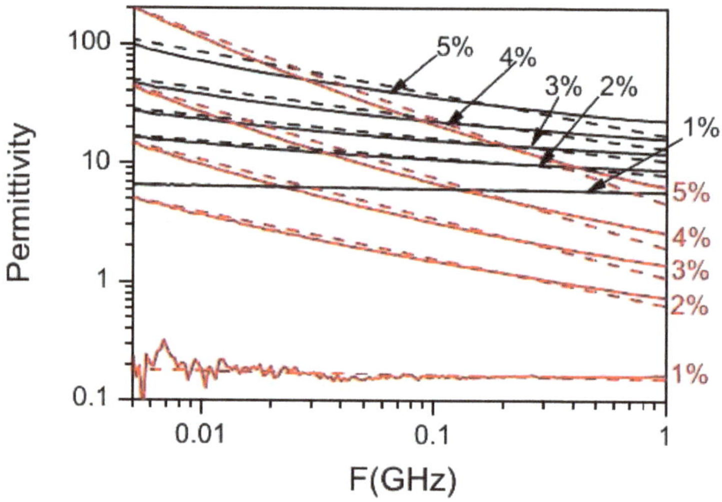

Fig. 10. Permittivity of SWCNT composite

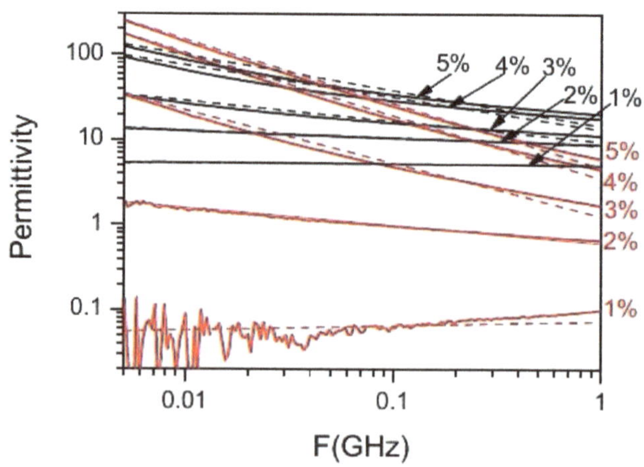

Fig. 11. Permittivity of DWCNT composite

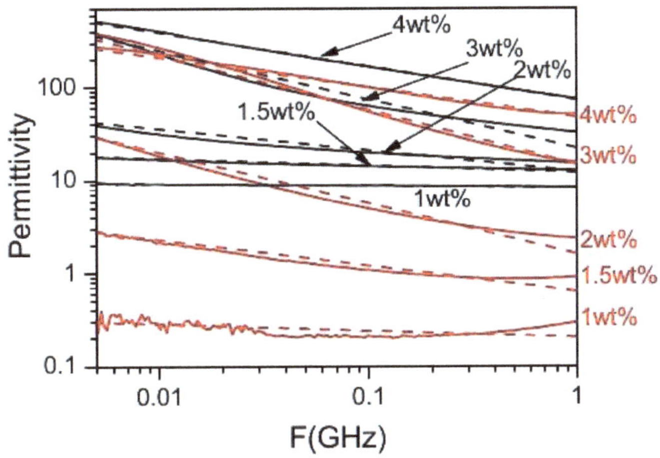

Fig. 12. Permittivity of MWCNT composite

Table 2. Critical exponents as function of concentration obtained by linear fitting of permittivity

SWCNT			DWCNT			MWCNT		
wt%	*y*	*x*	wt%	*y*	*x*	wt%	y	*x*
1	0.04	0.03	1	0.02	-0.05	1	0.03	0.07
2	0.18	0.39	2	0.11	0.21	1.5	0.09	0.28
3	0.217	0.491	3	0.276	0.62	2	0.26	0.55
4	0.275	0.591	4	0.402	0.736	3	0.54	0.59
5	0.368	0.713	5	0.438	0.747	4	0.37	0.32

4.3 Tunable permittivity under bias voltage

Figure 13 shows the dependence of normalized real and imaginary permittivity at 1 MHz , of the SWCNT composites at various concentrations, measured under a biasing voltage. The dependence of permittivity on bias voltage can be found at high frequency till a few hundred MHz. Figure 14 plots the tunability of permittivity of SWCNT composite from 1 MHz to 100 MHz . However, with increasing measurement frequency, the tunability becomes smaller. When the concentration is equal to or smaller than 2%, both real and imaginary permittivities do not change with voltage up to 15 V . It is understandable from the percolation threshold point of view because most of the tubes are out of the conducting network. Therefore, no voltage can be applied to isolate tubes or tube clusters. Once the concentration is higher than 3%, permittivity becomes sensitive to bias voltage. The tunability becomes more pronounced with increasing concentration. Samples with concentration of CNTs higher than 5% were not investigated due to extremely large imaginary permittivity. Obviously, bias voltage can decrease real permittivity but increase the imaginary part as shown in Figure 13. If the whole conducting network is treated as a circuit containing resistors and capacitors, replacing some capacitors with resistors will achieve the effect to reduce the real part and raise the imaginary part of permittivity. This is what was happening when bias voltage was applied to the samples in which junctions are formed by overlapping tubes. Electrons hop among tubes driven by bias voltages. The interaction among tubes becomes more resistive instead of capacitive under bias voltage. The tunability, defined as the change under bias over the original value, is more than 60% for the real permittivity and 80% for the imaginary permittivity for the sample with 5% of tubes. It is important to emphasize that the tunability of the CNT composites was reversible over a few thousand of cycles. Similar effect has also been found out from DWCNT and MWCNT composites when the concentration is large than percolation value within similar frequency range.

The novel tunable permittivity of the CNT composites may have potential applications in electromagnetic smart materials and structures. A smart materials or structures basically possess electrically or magnetically adjustable transmission or reflection coefficient to incident electromagnetic wave. The simplest solution is on-off switch, which was realized by pin diodes. One of the drawbacks of pin diodes is their relatively weak tolerance to high current, which is usually encountered in practical applications. A possible approach is to use a small piece of CNT composite to replace pin diode. Without the application of a bias voltage, the composite is insulating (off), while it becomes conducting under a certain bias voltage (on). Comparing with diodes, CNT composites can withstand higher current, voltage and power. Although the tunability of the SWCNT composite is still not high enough to provide similar on-off effect as a pin diode, we are confident that it can be further improved through proper dispersion or alignment of CNTs within holding matrix.

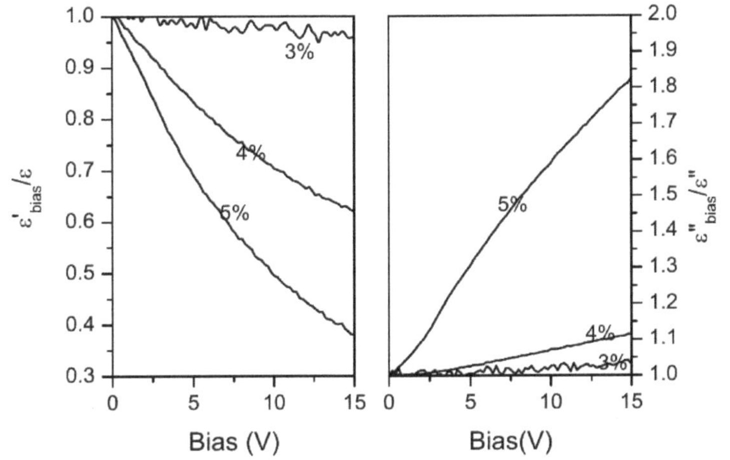

Fig. 13. Tunable permittivity of SWCNT composites (5 MHz)

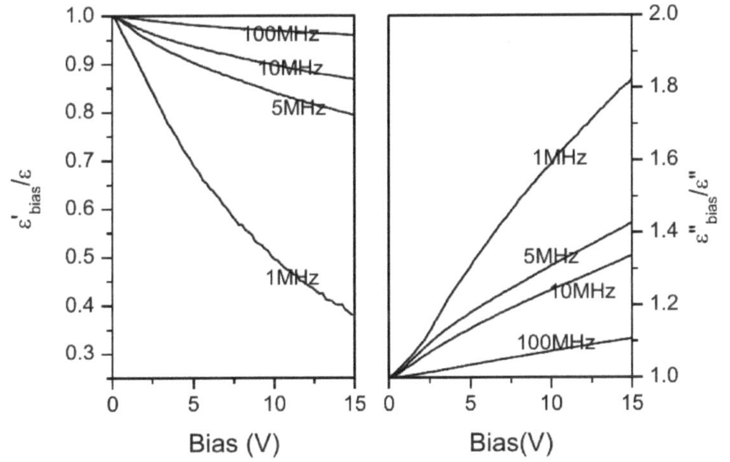

Fig. 14. Tunable permittivity of SWCNT composites (5%)

5. Conclusions

In this chapter, the complex permittivity of SWCNT, DWCNT and MWCNT composites at microwave band is obtained experimentally by the impedance method. The frequency dependence of the permittivity is well described by the scaling law based on the percolation theory. CNT composite is a system comprising clusters and conducting networks, the microwave behavior of which depends on the cluster formation, instead of the properties of individual fibers. Strong frequency-dependent behavior may be observed in measurement due to size effects. The microwave properties of the composites are found to be very sensitive to the preparation method. However, certain deviations are observed in the scaling parameters derived from the percolation theory. Namely the critical exponents are different from conventionally accepted universal values. This may be attributed to the non-perfect electrical contacts within clusters, which typically are not considered in mixing laws.

The CNT composites have large real permittivity at microwaves, adjustable dielectric loss tangent, and noticeable frequency dispersion of permittivity, even at concentrations above the percolation threshold. For these reasons, CNT composites with $p > p_c$ are probably useful as microwave dielectrics and frequency dispersive materials. CNT composites have been found to possess novel tunable effective permittivity under small bias voltages. The tunable properties can be attributed to flowing of electrons in the conducted network formed by tubes and tube clusters. Hence they are only obtainable at concentrations above percolation threshold. The tunability of both real and imaginary permittivity has potential applications in smart material and structure, which is the research subject that is currently on-going in our group.

6. References

A. N. Lagarkov, S. M. Matitsine, K. N. Roznov, and A. K. Sarychev (1998). Dielectric properties of fiber-filled composites, J. Appl. Phys., 84, 3806-14

A.K. Jonscher (1999). Dielectric relaxation in solids, J. Phys. D - Appl. Phys. 32, n 14, pp57-70.

B. K. Kim, J. Lee and I. Yu (2003). Electrical properties of single-wall carbon nanotube and epoxy composites, J. Appl. Phys., 94(10), 6724-8

C. A. Martin, J. K. W. Sandler, M. S. P. Shaffer, M. K. Schwarz, W. Bauhofer, K. Schulte, A. H. Windle (2004), Formation of percolation networks in multi-wall carbon-nanotube-epoxy composites, Composites science and technologies, 64, 2309-16

C. Brosseau (2002), Generalized effective medium theory and dielectric relaxation in particle-filled polymeric resins, J. Appl. Phys. 91(5), 3197-204.

C. P. Neo and V. K. Varadan (2004). Optimization of carbon fiber composite for microwave absorber, IEEE Trans. Electromagn. Compat. 46, 102-6

C. R. Simovsky, S. A. Tretyakov, A. H. Sihvola and M. M. Popov (2000). On the surface effect in thin molecular or composite layers, Eur. Phys. J. Appl. Phys. 9, 195-204

C. Yang, Y. Lin, C. W. Nan, Modified carbon nanotube composites with high dielectric constant, low dielectric loss and large energy density (2009). Carbon, 47, 1096-101

Coleman J. N., Curran S, Dalton AB, Davey AP, Mccarthy B, Blau W, Barklie RC (1998). Percolation-dominated conductivity in a conjugated-polymer-carbon-nanotube composite, Phys Rev B, 58(12), 7492-5

D. Satuffer and A. Aharony (1991), Introduction to Percolation Theory, Taylor & Francis, 2nd edition,

D.J. Bergman and Y. Imry (1977). Critical behaviour of the complex dielectric constant near the percolation threshold of a heterogeneous material, Phys. Rev. Lett., 39, 1222-25

F. Bordi, C. Cametti, J. Rouch, F. Sciortino, and P. Tartaglia (1996). Cluster formation in water-in-oil microemulsions at percolation: evaluation of the electrical properties, J. Phys.: Condens. Matter. 8(25), 19-37.

H. Y. Lee and Y. M. Shkel (2007), Self-sensing and self-actuation response of carbon nanotube composites, Proc. Of SPIE, 6526, 1-10.

I. J. Youngs, A. S. Treen, G. Fixter and S. Holden (2002). Design of solid broadband human tissue simulant materials, IEE Proc. Sci. Meas. Technol. 149, 323-8.

J. B. Donnet, R. C. Bansal, and M. J. Wang (1993). Carbon Black: Science and Technoloy, Dekker, New York

J. K. W. Sandler, J. E. Krik, I. A. Kinloch, M. S. P. Shaffer, and A. H. Windle (2003). Ultra-low electrical percolation threshold in carbon-nanotube-epoxy composites, Polymer, 44, 5893-99

J.P. Straley (1977), Critical exponents for the conductivity of random resistor lattices, Phys, Rev. B, 15, 5733-7.

K. Park, S. Lee, C. Kim, J. Han (2006), Fabrication and electromagnetic characteristics of electromagnetic wave absorbing sandwich structures, Composites science and technology, 66, 576-84

L. Liu, L. B. Kong and S. Matitsine (2008), Tunable effective permittivity of carbon nanotube composites, Appl. Phys. Lett., 93, 113106

L. Liu, S. Matitsine, Y. B. Gan and K. Rozanov (2007a). Cluster effect in frequency selective composites with randomly distributed long conductive fibres, J. Phys. D: Appl. Phys. 40, 7534-9

L. Liu, S. Matitsine, Y. B. Gan, L. F, Chen and L. B. Kong (2007b), Frequency dependence of effective permittivity of carbon nanotube composites, J. Appl. Phys. 101, 094106.

Li-Li Wang, Beng-kang Tay, Kye-yak See, Zhuo Sun, Lin-Kin Tan, Darren Lua (2009), Electromagnetic interference shielding effectiveness of carbon-based materials prepared by screen printing, Carbon, 47, 1905-10

L. Wei, B. Wang, Q. Wang, L. Li, Y. Yang, and Y. Chen (2008). Effect of Centrifugation on the Purity of Single-Walled Carbon Nanotubes from MCM-41 Containing Cobalt, J. Phys. Chem., 112, 17567-75

M. A. Seo, J. H. Yim, Y. H. Ahn, F. Rotermund, D. S. Kim, S. Lee, and H. Lim (2008), Terahertz electromagnetic interference shielding using single-walled carbon nanotube flexible films, Appl. Phys. Lett., 93, 231905

M. Meyyapan (2005). Carbon Nanotubes: Science and Applications, CRC Press, LLC

M. H. Al-Saleh, Uttandaraman, Sundararaj (2009), Electromagnetic interference shielding mechanisms of CNT/polymer composites, Carbon, 47, 1738-46.

P. C. P. Watts, D. R. Ponnampalam, W. K. Hsu, A. Barnes and B. Chambers (2003), The complex permittivity of multi-walled carbon nanotube-polystyrene composite films in X-band, Chem. Phys. Lett., 378, 609-14.

R. B. Laibowitz and Y. Gefen (1984). Dynamic scaling near the percolation threshold in thin Au films, Phys. Rev. Lett. , 53, 380-3

R. Saito, G. Dresselhaus and M. S. Dresselhaus (1998). Physical Properties of Carbon Nanotubes, Imperial College Press, London.

S. Iijima, Nature (1991). Helical microtubules of graphitic carbon, 354 , 56.

S. L. Browning, J. Jodge, R. R. Price, J. Schelleng, P. E. Schoen and D. Zabetakis (1998). Fabrication and radio frequency characterization of high dielectric loss tubulebased composites near percolation, J. Appl. Phys. , 84, 6109-13.

S. Lee, J. Kang and G. Kim(2006), Fabrication and design of multi-layered radar absorbing structures of MWCNT-filled glass/epoxy plain-weave composites, Composite structures, 76, 397-405

S. Shi and J. Liang (2008), the effect of multi-wall carbon nanotubes on electromagnetic interference shielding of ceramic composites, Nanotechnology, 19, 255707

S.M. Matitsine, K.M. Hock, L. Liu, Y.B. Gan, A.N. Lagarkov, K.N. Rozanov (2003), Shift of resonance frequency of long conducting fibers embedded in a composite, J. Appl. Phys. 94, 1146-54.

T. D. Burchell (1999). Carbon Materials for Advanced Technologies, Elsevier Science Ltd, Oxford

Y. Gefen, A. Aharony and S. Alexander (1983), Anomalous diffusion on percolating clusters Phys. Rev. Let., 50(1) 77-80.

Y. Huang, N. Li, Y. Ma, F. Du, F. Li, X. He, X. Lin, H. Gao, Y. Chen (2007), The influence of single-walled carbon nanotube structure on the electromagnetic interference shielding efficiency of its epoxy composites, Carbon, 45, 1614-21

Y. Song, T. W. Noh, S. I. Lee and J. R. Gaines (1983), Experimental study of the threedimensional AC conductivity and dielectric constant of a conductor-insulator composite near the percolation threshold, Phys. Revs. B, 33(2), 904-8

Chapter-6

Environmental Effects on Photoluminescence of Single-walled Carbon Nanotubes

1. Introduction

Spectroscopy is a powerful technique to characterize single-walled carbon nanotubes (SWNTs). In particular, photoluminescence (PL) and excitation spectroscopies have enabled to make great advances in understanding the energy states and transition processes of excitons in SWNTs. Furthermore, PL spectroscopy also has potential for being used to determine the chiral index (n, m) of an SWNT and evaluate the abundance of (n, m) in bulk SWNTs. (Bachilo, Strano et al. 2002; Miyauchi, Chiashi et al. 2004) However, in order to establish PL spectroscopy as a reliable standard technique to characterize SWNTs, it is necessary to understand not only the intrinsic physics of excitons but also the effects of extrinsic factors, namely, environmental effects. (Fantini, Jorio et al. 2004) The optical properties of an SWNT are modified sensitively by its environmental conditions because of the excitons existing on the SWNT surface. In this chapter, we focus on the environmental effects on the optical properties of SWNTs. After considering the fundamental physics of the optical transitions in SWNTs and application of PL spectroscopy to characterize SWNTs, various types of environmental effects are discussed with practical examples.

2. Optical transition and photoluminescence

2.1 Density of states and optical transition

In an SWNT, electrons are confined to its surface, and the wave vectors along the circumference are quantized by a periodic boundary condition; however, the carriers are free to move in the direction of the SWNT axis. Therefore, the SWNT has a one-dimensional electron system, where the density of states is characterized by the sharp peaks originating from a van Hove singularity, as shown in Fig. 1(a). (Saito, Dresselhaus et al. 1998).

An optical transition in an SWNT occurs dominantly between two subbands located at van Hove singularities. However, because of the one-dimensional structure, it depends on the polarization of the incident light. If the incident light is polarized linearly with the polarization vector parallel to the SWNT axis, a strong optical transition occurs. In this case, the optical transitions from a subband of a valence band to a subband of the conduction band, and vice versa, with the same quantum number are allowed by the selection rule. For example, transitions such as those from C1 to V1 and from V2 to C2 are allowed transitions, as shown in Fig. 1(a). On the other hand, if the incident light is polarized perpendicular to the SWNT axis, the optical transition is strongly suppressed by the depolarization effect in which the induced charges cancel the electric field of the light in the SWNT. (Duesberg, Loa et al. 2000).

The energy of an optical transition is given by the difference in the energies of the two subbands, in the case of the single-particle approximation. However, in practice, it is modified by many-body effects, as shown in Fig. 1(b). (Ando 2005) The single-particle bandgap $E_{ii}\mathrm{gr}^{\mathrm{r}}$ is renormalized by the self-energy of an electron due to the repulsive interaction between electrons. In addition, for an optical transition, the exciton effect should be taken into account, that is, the optical transition energy is decreased by the binding energy of the attractive Coulomb interaction between an electron and a hole. In the case of SWNTs, since the exciton binding energy is as large as a few hundred millielectronvolts, the exciton scheme is more suitable than the single-particle scheme. (Wang, Dukovic et al. 2005) Consequently, the optical transition energy E_{ii} between the i-th valence band and the i-th conduction band ($i = 1,2,3,\ldots$) is given by

$$E_{ii} = E_{ii}\mathrm{gr}^{\mathrm{gr}} + E_{ii}{}^{\mathrm{ee}} + E_{ii}{}^{\mathrm{eh}}, \qquad (1)$$

where $E_{ii}{}^{\mathrm{ee}}$ denotes the energy of repulsive interaction between electrons, and $E_{ii}{}^{\mathrm{eh}}$ (< 0) represents the binding energy of an exciton. Note that every term is perturbed by the environment around the SWNT, as described later.

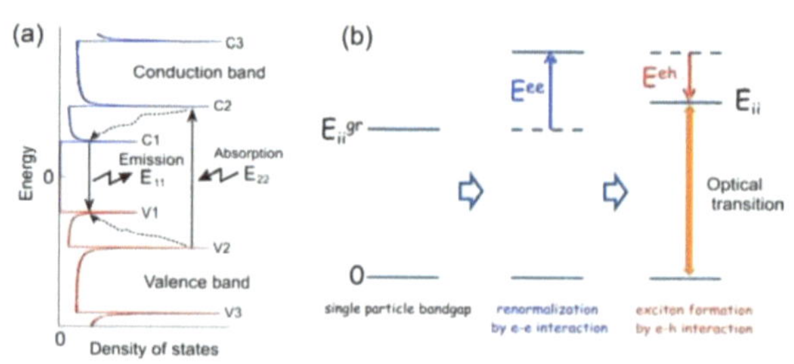

Fig. 1. (a) Density of states of a semiconducting SWNT and optical transition processes in the single particle scheme. (b) Energy diagram of an exciton state, showing optical transition energy in the exciton scheme.

2.2 Photoluminescence spectroscopy

Since the optical transition energy E_{ii} depends on the chiral index (n, m) of an SWNT, (n, m) can be determined by spectroscopic techniques such as Raman scattering spectroscopy (Kataura, Kumazawa et al. 1999) and photoluminescence (PL) spectroscopy. (Bachilo, Strano et al. 2002) In the case of PL spectroscopy of a semiconducting SWNT, the E_{22} state is resonantly excited by incident light so as to generate excitons, as shown in Fig. 1(a). (O'Connell, Bachilo et al. 2002) The excitons relax via phonons to the E_{11} state, where radiative recombination occurs. Figure 2(a) shows the PL spectra of SWNTs measured at various excitation wavelengths. Such measurement is called the PL mapping measurement. Each peak originates from a different (n, m) species. From the relative intensity of each peak, the abundance of each (n, m) species can be evaluated. It should be note that the quantum efficiency of light absorption and emission depends on the diameter and chirality of SWNTs. (Oyama, Saito et al. 2006) Therefore, in order to make accurate evaluation of the (n, m) abundance, a correction of the (n, m) dependence of the quantum efficiency is required. (Okazaki, Saito et al. 2006) Figure 2(b) shows the E_{22} vs. E_{11} plot for various (n, m) species, which are derived from the PL map of Fig. 2(a). The points connected by a solid line belong to the family of SWNTs that have the same value of $(2n + m)$. (Saito, Dresselhaus et al. 1998) It can be observed that the family patterns are separated into two groups, depending on the value of $(2n + m\text{mod}3)$. In this chapter, we define these two groups as Type I if $(2n + m\text{mod}3) = 1$, and Type II if $(2n + m\text{mod}3) = 2$. When a graphite layer is rolled up to form an SWNT, the quantized energies of the subbands primarily depend on the diameter of the SWNT. In addition, they also depend on the direction to be quantized, namely the chiral angle, because the equi-energy contours of graphite are warped to form a triangular shape around the K point, which is called the trigonal warping effect. (Saito, Dresselhaus et al. 2000) This is the main reason for the appearance of the family pattern in the plot of optical transition energies. The family patterns are technically important for the analysis of PL spectra. Even though the optical transition energies of SWNTs vary by more than 100 meV , depending on the methods used to prepare SWNTs and on the environment, the identification of the family pattern allows us to easily determine the chirality index (n, m) of the SWNTs.

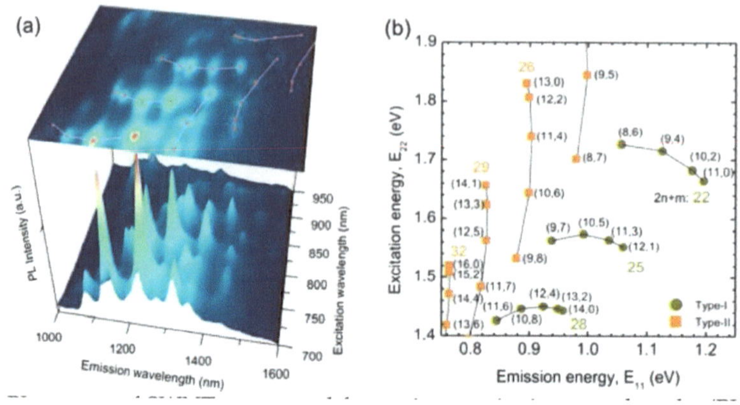

Fig. 2. (a) PL spectra of SWNTs measured for various excitation wavelengths (PL map). (b) E_{22} vs. E_{11} plot. A solid line connects SWNTs belonging to the same family, i.e., SWNTs having the same value of $2n + m$.

2.3 Relaxation process

In PL processes, the excitons excited to the E_{22} state relax to the lowest state E_{11} via phonons in a short time such as $\sim$ 0.1ps because the optical phonon energy is as large as 0.2 eV (Ichida, Hamanaka et al. 2002; Ostojic, Zaric et al. 2004) Then, the excitons relax to the ground state via radiative recombination or nonradiative recombination. The radiative recombination lifetime is expected to be about 1 ns . (Spataru, Ismail-Beigi et al. 2005) In contrast, the experimental results of pump-probe measurements and time-resolved PL measurements have shown a much shorter recombination lifetime, of the order of several tens of picoseconds. (Hagen, Moos et al. 2004; Wang, Dukovic et al. 2004; Hirori, Matsuda et al. 2006; Ohno, Kishimoto et al. 2006) Such a short recombination lifetime suggests that the dominant recombination process is nonradiative. The luminescence quantum efficiency has been estimated experimentally to be several percent. (Lefebvre, Austing et al. 2006) Recombination centers formed by defects and interface states may enhance such nonradiative recombination. It has been suggested that nonradiative recombination rate is enhanced if the exciton is localized. (Perebeinos and Avouris 2008)

2.4 Sample preparation for achieving photoluminescence

The soot of SWNTs generally consists of SWNT bundles formed by van der Waals interaction. In a bundle of SWNTs, the energy of photoexcited excitons is transferred to neighboring metallic SWNTs or SWNTs with a smaller bandgap, and then, these excitons relax via phonons. Therefore, in order to obtain luminescence from SWNTs, it is necessary to debundle and individualize SWNTs from a bundle. The PL from individual SWNTs has been observed for the first time by O'Connell et al. by wrapping SWNTs with a surfactant such as sodium-dodecyl-sulfate (SDS) in order to form micelles encasing individual SWNTs. (O'Connell, Bachilo et al. 2002) After dispersing the soot of SWNTs in an aqueous solution of SDS by performing ultrasonication, they obtained individualized SWNTs in the supernatant liquid of ultracentrifugation, as shown in Fig. 3(a). Another method to obtain individualized SWNTs is to grow SWNTs directly between pillars or on trenches formed on a Si or quartz substrate, as shown in Fig. 3(b). (Lefebvre, Fraser et al. 2004; Ohno, Iwasaki et al. 2006) The free-standing SWNTs bridging such a microstructure provide strong PL, although the PL is suppressed if an SWNT touches the substrate, as will be described later. Further, the PL signal from a free-standing SWNT is sufficiently strong and can hence be observed by a microscopic PL technique. For the purpose of studying the environmental effects on the PL of SWNTs, free-standing SWNTs grown between pillars or on trenches are suitable because the environmental conditions can be intentionally changed by exposing the SWNTs to a gas (Finnie, Homma et al. 2005; Chiashi, Watanabe et al. 2008) or by immersing the SWNTs in a liquid. (Iwasaki, Ohno et al. 2007; Ohno, Iwasaki et al. 2007)

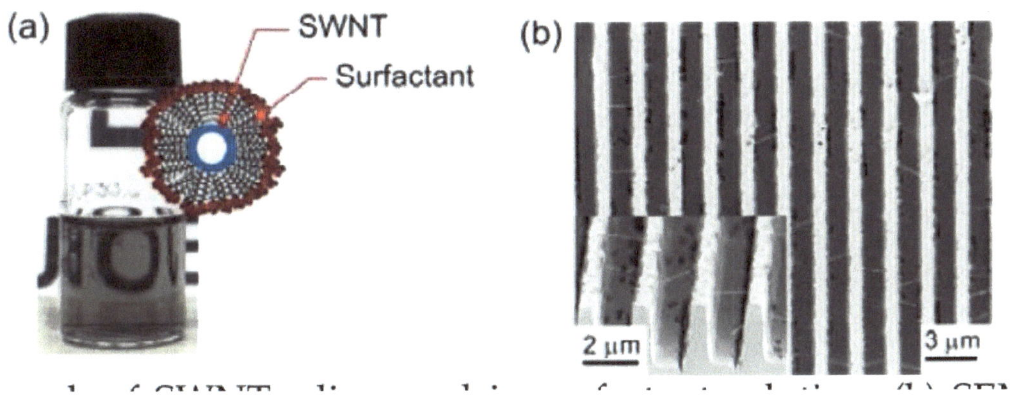

Fig. 3. (a) Photograph of SWNTs dispersed in surfactant solution. (b) SEM image of freestanding SWNTs grown on the trenches formed on a quartz substrate.

3. Environmental effects

The environmental effects on the PL of SWNTs is one of the most important topics as regards the optical properties of SWNTs. (Moore, Strano et al. 2003; Chou, Ribeiro et al. 2004; Fantini, Jorio et al. 2004) It is known that the optical transition energy varies, depending on the type of surfactants used to individualize SWNTs. Lefebvre et al. have also reported that the optical transition energies of SWNTs bridging micropillars show a blueshift as compared to the SDS-wrapped SWNTs. (Lefebvre, Fraser et al. 2004) A later study by Ohno et al. has shown that the energy differences depend on (n, m), particularly on the chiral angle and the type of SWNTs, i.e., whether the SWNTs are of Type I or Type

II. (Ohno, Iwasaki et al. 2006) One of the causes of the variation in the optical transition energy is the dielectric screening of the many-body interactions between carriers. (Perebeinos, Tersoff et al. 2004; Ando 2005) In addition to the dielectric screening effect, various environmental factors such as chemical interactions (Finnie, Homma et al. 2005; Hertel, Hagen et al. 2005) and mechanical interactions (Arnold, Lebedkin et al. 2004) should be considered. Comprehensive understanding of the environmental effects is necessary to understand the optical properties of SWNTs. In this section, we discuss the effects of the following factors on the PL of SWNTs: dielectric screening, wrapping SWNTs with surfactants, exposing SWNTs to air, doping SWNTs with carriers, and allowing SWNTs to make contact with a substrate.

3.1 Dielectric screening effect

The variation in the optical transition energy caused by the environmental conditions is primarily due to the dielectric screening effect, whereby the many-body Coulomb interactions between carriers are screened by the surrounding material. As shown in Fig. 4, the electric force lines contributing to the Coulomb interactions pass through not only the inside of an SWNT but also the outside, where the surrounding material screens the electric force lines depending on its dielectric constant.

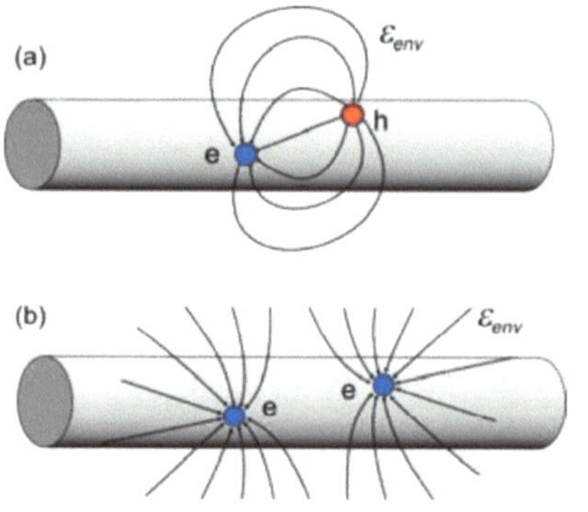

Fig. 4. Schematics of electric force lines contributing to many-body Coulomb interactions: (a) electron-hole attractive interaction, which forms an exciton, and (b) electron-electron repulsive interaction.

Figure 5 shows the typical PL spectra measured in the ambient air, hexane (dielectric constant: 1.9), chloroform (4.8), and acetonitrile (37). Note that although in this chapter, the dielectric constants are used considering a dc field, they depend on the frequency of the field. At the frequency corresponding to the exciton lifetime, a dielectric constant is expected to approach unity. Further, although the exact values of ac dielectric constants of these liquids have not been known, they could be related to the dc dielectric constants.

The peaks exhibit a redshift and broadening with the increasing dielectric constant of liquids. As indicated by the E_{22} vs. E_{11} plots shown in Fig. 6, both E_{11} and E_{22} exhibit a redshift with the increasing dielectric constant of liquids, with a small (n,m) dependence. The amounts of the redshifts are $20 \sim 43$ meV for E_{11} and $15 \sim 33$meV for E_{22}.

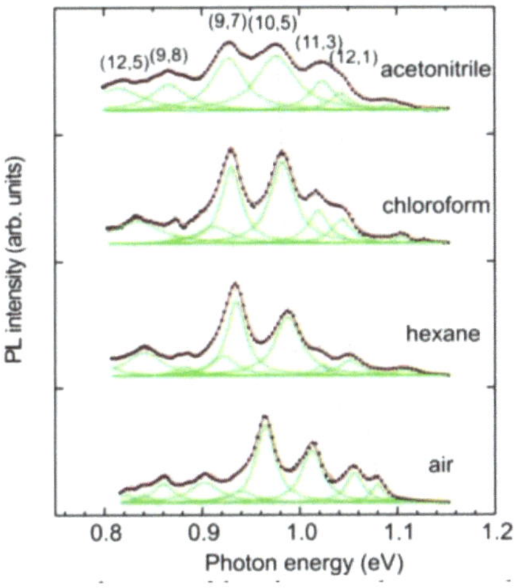

Fig. 5. PL spectra in ambient air, hexane, chloroform, and acetonitrile.

As the dielectric constant increases, the energies of both the electron-electron and electronhole interactions decrease. The reduction in the exciton binding energy causes a blueshift in the optical transition energy because of the negative energy, whereas the reduction in the electron-electron repulsion energy causes a redshift. The occurrence of redshifts suggests that the variation in the electron-electron repulsion energy is larger than that in the exciton binding energy. This is due to the fact that the repulsion energy is larger in magnitude than the exciton binding energy. (Ando 2005; Spataru, Ismail-Beigi et al. 2005) Figure 7 shows E_{11} and E_{22} of (8,6)-SWNTs as a function of the dielectric constant of the surrounding liquid. The energy shift is significant in the regime of a low dielectric constant, and exhibits a tendency to saturate at a dielectric constant of ~ 5. Empirically, the dependence of E_{ii} on the environmental dielectric constant $\varepsilon_{\mathrm{env}}$ is given by

$$E_{ii} = E_{ii}{}^{\infty} + A\varepsilon_{\mathrm{env}}{}^{-\alpha}. \tag{2}$$

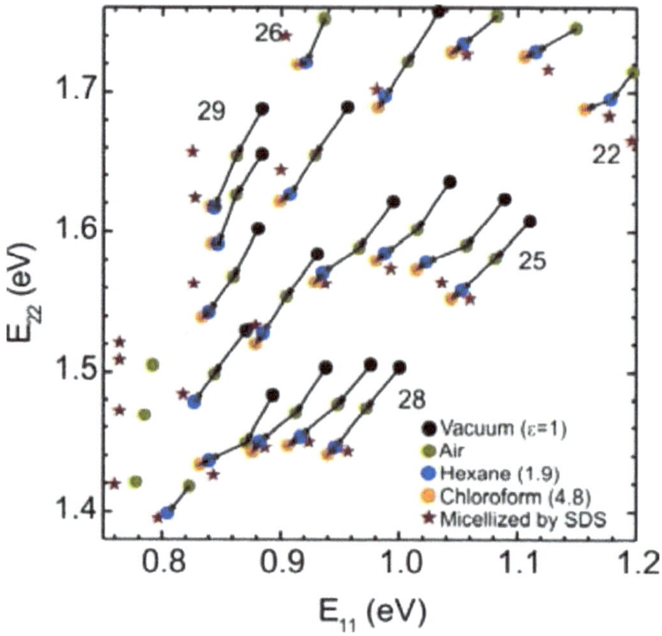

Fig. 6. E_{22} **vs.** E_{11} **plots of SWNTs grown on microtrenches in various environmental conditions: in vacuum (black dots), air (yellow green), hexane (blue), chloroform (orange), and aqueous solution of SDS (red stars).**

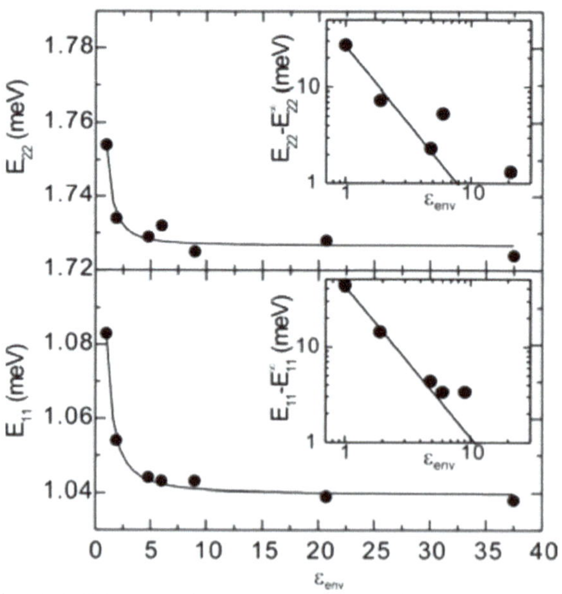

Fig. 7. E_{11} **and** E_{22} **of (8,6)-SWNTs as a function of the dielectric constant of the surrounding liquid. The insets are the log-log plots.**

In Eq. (2), the first term on the right-hand side corresponds to the transition energy when ε_{env} is infinity; it consists of the single-particle bandgap and the energies of the many-body interactions, which are determined by the electric force lines traveling only inside the SWNT. The second term shows the screening of Coulomb interactions, including both electron-hole binding and electron-electron repulsive interactions, which are related to the electric force lines traveling outside the SWNT. In the equation, " A " denotes the maximum energy variation. In Fig. 6, the solid curves are fitted to the data by using Eq. (2). Here, A and α are 43 meV and 1.6 for E_{11}, and 27 meV and 1.8 for E_{22}, respectively. Miyauchi et al. have also obtained a power-law-like dependence of the exciton transition energies on the dielectric constant of the material surrounding SWNTs by the tight-binding calculation.(Miyauchi, Saito et al. 2007) The variation in E_{ii} due to the dielectric screening effect also depends on (n, m), particularly, on the chiral angle, and the type (Type I or Type II). This may be due to the (n, m) dependence of the effective mass of carriers. As descried before, the effective mass depends on the chiral angle because of the trigonal warping effect. (Jorio, Fantini et al. 2005) In addition, the effective mass is one of the scaling parameters for exciton energy. (Perebeinos, Tersoff et al. 2004) Similarly, the repulsive energy of electron-electron interaction could depend on the effective mass.

It should be noted that in addition to exhibiting the redshift, the PL spectrum became broader when the SWNTs were immersed in liquids with a large dielectric constant, as shown in Fig. 5. The full width at half maximum (FWHM) increased from 23 meV in the ambient air to 40 meV in acetonitrile for (9,7)-SWNTs. This linewidth broadening may be attributed to inhomogeneous broadening due to the fluctuation in the local dielectric constant of the environment occurring on the scale of the size of a liquid molecule. It should be mentioned that the dielectric constants used in this chapter are macroscopic and static values. In practice, the size of a molecule of an organic solvent is comparable to that of an exciton; therefore, the local dielectric constant around an exciton could fluctuate depending on the position and orientation of the surrounding organic molecules, which could not follow the dynamic electric field formed by the exciton within the lifetime of the exciton. This results in the dispersion of the energies of many-body interactions.

3.2 Effect of ambient air and intrinsic transition energies

On measuring the PL of free-standing SWNTs in a vacuum, it is found that the transition energies are 20 ~ 40meV higher than those corresponding to the PL measured in the ambient air, as shown by the black dots in Fig. 6. Further, sequential measurements of the PL in the ambient air after the growth of SWNTs exhibit a phenomenon that every E_{ii} shifts to a lower energy with an incubation time ranging from a few hours to several days. The incubation time required for the occurrence of a redshift depends on the diameter of

SWNTs, that is, the redshift occurs earlier in the case of SWNTs with a small diameter than that in the case of SWNTs with a large diameter. This variation in E_{ii} is reversible, namely, E_{ii} returns to its former value when the sample is heated up above 50°C in a vacuum. The shifts in E_{ii} can be attributed to the dielectric screening effect because their behavior is similar to that observed in the liquid immersion experiments.

In air, the ambient molecules such as H_2O and C_xH_y are adsorbed on the surface of SWNTs and cause the screening of many-body interactions. Chiashi et al. have investigated the effect of gas adsorption in detail. (Chiashi, Watanabe et al. 2008) They have shown that a redshift occurs at a gas pressure exceeding the transition pressure, which agrees with Langmuir's adsorption model.

Carrier doping due to oxygen molecules should be discussed as one of the effects of the ambient air because the energies of many-body interactions may be modified depending on the carrier density. The effect of carrier doping performed by the field-effect doping technique has been studied by measuring the PL of a free-standing SWNT placed in a fieldeffect transistor structure. (Ohno, Kishimoto et al. 2006) In addition, the effect of chemical doping with F_4TCNQ (tetrafluorotetracyano-p-quinodimethane), which is a strong *p*-type dopant for SWNTs (Nosho, Ohno et al. 2007), has also been studied. Even though the carrier densities introduced by these doping techniques were as high as that introduced by the ambient oxygen, the optical transition energies were not significantly modified by doping. This may be because the doping density was probably not sufficiently high to dope carriers in the range of the exciton diameter.

Now, by heating the free-standing SWNTs in a vacuum, the adsorbed molecules can be desorbed, and then the intrinsic transition energies of SWNTs can be measured without any environmental effects.

3.3 Effect of wrapping with surfactants

When the free-standing SWNTs grown on trenches were immersed in an aqueous solution of SDS (1wt%), the E_{22} vs. E_{11} plots shifted to the positions indicated by the red stars in Fig. 6. The plots of these SWNTs are almost identical to those obtained for the SWNTs wrapped with SDS by using O'Connell's method, as mentioned in section 2.4, which implies that the free-standing SWNTs were wrapped with SDS by immersing them in the SDS solution.

Most E_{11} and E_{22} of SDS-wrapped SWNTs exhibit redshifts as compared to the free-standing SWNTs in air, except for E_{22} of near-zigzag Type II SWNTs, which exhibit blueshifts. (Ohno, Iwasaki et al. 2006) This behavior is different from the results obtained for the dependence of E_{11} and E_{22} on the dielectric constant of the surrounding material, as described above, in which all E_{11} and E_{22} exhibited redshifts with an increase in the dielectric constant. The positions of the E_{22} vs. E_{11} plots of the SDS-wrapped SWNTs shift so that the $(2n + m)$-family pattern spreads more widely, in addition to the shift due to the dielectric screening effect. This behavior of the energy shifts is similar to stress-induced shift. (Arnold, Lebedkin et al. 2004) One plausible explanation for the effect of SDS wrapping is that the SWNT is compressed by the surfactant molecules in the radial direction, resulting in the uniaxial strain that modifies the bandgap of the SWNT.

3.4 Effect of substrate

Even though a substrate is one of the indispensable parts for device application of SWNTs and supports the SWNTs and electrodes, the contact of SWNTs with the substrate seriously affects the optical properties of SWNTs. For example, in the case of SWNTs grown on a substrate such as SiO_2, PL is strongly suppressed if the body of the SWNT lies on the substrate. (Ohno, Kishimoto et al. 2006) The quenching of PL suggests that the excitons excited in an SWNT recombine by a nonradiative recombination process, probably via interface states between the SWNT and the substrate. The interface states also cause serious degradation of not only the optical properties but also the electrical properties of SWNTs, for example, the capture and emission of carriers at the interface states may cause noise in SWNT FETs.(Lin, Tsang et al. 2007).

Even though PL spectroscopy is not applicable to such SWNTs directly grown on a substrate, photocurrent spectroscopy is useful to study the optical properties of such SWNTs.(Ohno, Kishimoto et al. 2004) Figure 8 shows the photocurrent spectra of two types of transistors; one transistor is a conventional SWNT transistor with a channel lying on a SiO_2/Si substrate, and the other is a transistor with a channel of a free-standing SWNT, as shown in the inset. The peaks in the spectra are attributed to E_{22} of the semiconducting SWNT. In the case of the free-standing SWNT, the peak is quite sharp and has an FWHM of 30 meV , which is comparable to the excitation spectrum of the PL of an SWNT. On the other hand, in the case of the SWNT lying on SiO_2, the peak structure is not clear, and the FWHM is as broad as 60 meV . The broadening of the spectrum suggests that the energy band of this SWNT is modified inhomogeneously, probably depending on the surface structure of amorphous SiO_2.

Recently, Xie et al. have reported that PL can be observed from SWNTs lying on a SiO_2 surface when the SWNTs are transferred to another substrate after the growth. (Xie, Liu et al. 2007) At room temperature, it is difficult to form interface states between an SWNT and a SiO_2 substrate; this is because both the SWNT and SiO_2 are chemically stable. Nevertheless, during the growth of SWNTs directly on a substrate, chemical bonds may be formed at the interface due to high growth temperature. Such chemical bonds may act as nonradiative recombination centers. Moreover, Xie et al. have also reported that the PL was enhanced when the substrate surface was covered with a self-assembled monolayer. It has been explained on the basis of carrier transfer to the traps present in the SiO_2 layer of the substrate.

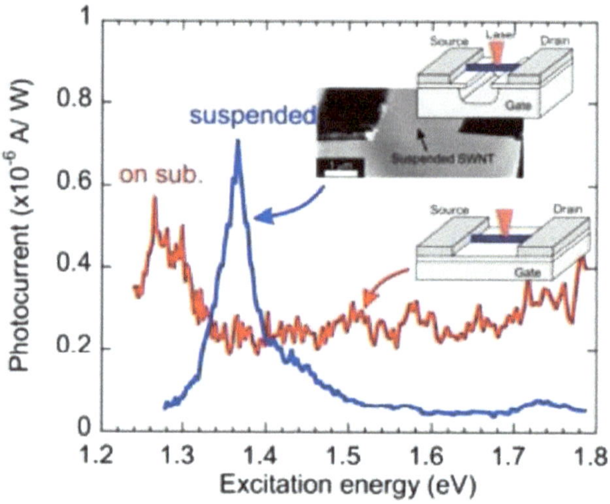

Fig. 8. Photocurrent spectra of CNTFETs with a free-standing SWNT (blue) and a SWNT in contact with a substrate (red).

4. Summary

In this chapter, we discussed environmental effects on the optical properties of SWNTs. Detailed understandings of the environmental effects help us to make a spectroscopic technique a standardized methodology to evaluate the chirality abundance and the quality of SWNTs. It is also interesting to develop sensing tools of nano dimensions for bio and medical applications by utilizing the environmental effects.

5. References

Ando, T. (2005). "Theory of electronic states and transport in carbon nanotubes." Journal of the Physical Society of Japan 74(3): 777-817.

Arnold, K., S. Lebedkin, et al. (2004). "Matrix-imposed stress-induced shifts in the photoluminescence of single-walled carbon nanotubes at low temperatures." Nano Letters 4(12): 2349-2354.

Bachilo, S. M., M. S. Strano, et al. (2002). "Structure-assigned optical spectra of single-walled carbon nanotubes." Science 298(5602): 2361-2366.

Chiashi, S., S. Watanabe, et al. (2008). "Influence of Gas Adsorption on Optical Transition Energies of Single-Walled Carbon Nanotubes." Nano Letters 8(10): 3097-3101.

Chou, S. G., H. B. Ribeiro, et al. (2004). "Optical characterization of DNA-wrapped carbon nanotube hybrids." Chemical Physics Letters 397(4-6): 296-301.

Duesberg, G. S., I. Loa, et al. (2000). "Polarized Raman spectroscopy on isolated single-wall carbon nanotubes." Physical Review Letters 85(25): 5436-5439.

Fantini, C., A. Jorio, et al. (2004). "Optical transition energies for carbon nanotubes from resonant Raman spectroscopy: Environment and temperature effects." Physical Review Letters 93(14): 147406.

Finnie, P., Y. Homma, et al. (2005). "Band-gap shift transition in the photoluminescence of single-walled carbon nanotubes." Physical Review Letters 94(24): 247401.

Hagen, A., G. Moos, et al. (2004). "Electronic structure and dynamics of optically excited single-wall carbon nanotubes." Applied Physics a-Materials Science & Processing 78(8): 1137-1145.

Hertel, T., A. Hagen, et al. (2005). "Spectroscopy of single- and double-wall carbon nanotubes in different environments." Nano Letters 5(3): 511-514.

Hirori, H., K. Matsuda, et al. (2006). "Exciton localization of single-walled carbon nanotubes revealed by femtosecond excitation correlation spectroscopy." Physical Review Letters 97(25): 257401.

Ichida, M., Y. Hamanaka, et al. (2002). "Ultrafast relaxation dynamics of photoexcited states in semiconducting single-walled carbon nanotubes." Physica B-Condensed Matter 323(1-4): 237-238.

Iwasaki, S., Y. Ohno, et al. (2007). "Environmental dielectric screening effect on exciton transition energies in single-walled carbon nanotubes." arXiv:0704.1018v1 [cond-mat.mtrl-sci].

Jorio, A., C. Fantini, et al. (2005). "Resonance Raman spectroscopy (n,m)-dependent effects in small-diameter single-wall carbon nanotubes." Physical Review B 71(7): 075401.

Kataura, H., Y. Kumazawa, et al. (1999). "Optical properties of single-wall carbon nanotubes." Synthetic Metals 103(1-3): 2555-2558.

Lefebvre, J., D. G. Austing, et al. (2006). "Photoluminescence imaging of suspended singlewalled carbon nanotubes." Nano Letters 6(8): 1603-1608.

Lefebvre, J., J. M. Fraser, et al. (2004). "Photoluminescence from single-walled carbon nanotubes: a comparison between suspended and micelle-encapsulated nanotubes." Applied Physics a-Materials Science & Processing 78(8): 1107-1110.

Lin, Y. M., J. C. Tsang, et al. (2007). "Impact of oxide substrate on electrical and optical properties of carbon nanotube devices." Nanotechnology 18(29): 295202.

Miyauchi, Y., R. Saito, et al. (2007). "Dependence of exciton transition energy of singlewalled carbon nanotubes on surrounding dielectric materials." Chemical Physics Letters 442(4-6): 394-399.

Miyauchi, Y. H., S. H. Chiashi, et al. (2004). "Fluorescence spectroscopy of single-walled carbon nanotubes synthesized from alcohol." Chemical Physics Letters 387(1-3): 198-203.

Moore, V. C., M. S. Strano, et al. (2003). "Individually suspended single-walled carbon nanotubes in various surfactants." Nano Letters 3(10): 1379-1382.

Nosho, Y., Y. Ohno, et al. (2007). "The effects of chemical doping with F(4)TCNQ in carbon nanotube field-effect transistors studied by the transmission-line-model technique." Nanotechnology 18(41): 415202.

O'Connell, M. J., S. M. Bachilo, et al. (2002). "Band gap fluorescence from individual singlewalled carbon nanotubes." Science 297(5581): 593-596.

Ohno, Y., S. Iwasaki, et al. (2006). "Chirality-dependent environmental effects in photoluminescence of single-walled carbon nanotubes." Physical Review B 73(23): 235427.

Ohno, Y., S. Iwasaki, et al. (2007). "Excitonic transition energies in single-walled carbon nanotubes: Dependence on environmental dielectric constant." Physica Status Solidi B-Basic Solid State Physics 244(11): 4002-4005.

Ohno, Y., S. Kishimoto, et al. (2006). Carrier transport property in single-walled carbon nanotubes studied by photoluminescence spectroscopy. Journal of Physics: Conference Series. 38: 5-8.

Ohno, Y., S. Kishimoto, et al. (2006). "Photoluminescence of single-walled carbon nanotubes in field-effect transistors." Nanotechnology 17(2): 549-555.

Ohno, Y., S. Kishimoto, et al. (2004). "Chirality assignment of individual single-walled carbon nanotubes in carbon nanotube field-effect transistors by micro-photocurrent spectroscopy." Applied Physics Letters 84(8): 1368-1370.

Okazaki, T., T. Saito, et al. (2006). "Photoluminescence and population analysis of singlewalled carbon nanotubes produced by CVD and pulsed-laser vaporization methods." Chemical Physics Letters 420(4-6): 286-290.

Ostojic, G. N., S. Zaric, et al. (2004). "Interband recombination dynamics in resonantly excited single-walled carbon nanotubes." Physical Review Letters 92(11): 117402.

Oyama, Y., R. Saito, et al. (2006). "Photoluminescence intensity of single-wall carbon nanotubes." Carbon 44(5): 873-879.

Perebeinos, V. and P. Avouris (2008). "Phonon and electronic nonradiative decay mechanisms of excitons in carbon nanotubes." Physical Review Letters 101(5): 057401.

Perebeinos, V., J. Tersoff, et al. (2004). "Scaling of excitons in carbon nanotubes." Physical Review Letters 92(25): 257402.

Saito, R., G. Dresselhaus, et al. (1998). Physical Properties of Carbon Nanotubes. Singapore, World Scientific Publishing Co. Pte. Ltd.

Saito, R., G. Dresselhaus, et al. (2000). "Trigonal warping effect of carbon nanotubes." Physical Review B 61(4): 2981-2990.

Spataru, C. D., S. Ismail-Beigi, et al. (2005). "Theory and ab initio calculation of radiative lifetime of excitons in semiconducting carbon nanotubes." Physical Review Letters 95(24): 247402.

Wang, F., G. Dukovic, et al. (2004). "Time-resolved fluorescence of carbon nanotubes and its implication for radiative lifetimes." Physical Review Letters 92(17): 177401.

Wang, F., G. Dukovic, et al. (2005). "The optical resonances in carbon nanotubes arise from excitons." Science 308(5723): 838-841.

Xie, L. M., C. Liu, et al. (2007). "Photoluminescence recovery from single-walled carbon nanotubes on substrates." Journal of the American Chemical Society 129(41): 12382.

Chapter-7

Charge Transport in Carbon Nanotube Films and Fibers

1. Introduction

Electrical properties and magnetotransport in carbon nanotubes (CNTs) have attracted much attention due to their importance in verification of existing theories of modern condensed matter physics and a number of possible applications (Robertson, 2007; Dai, 2002). Single-wall carbon nanotube (SWCNT) is a graphene sheet rolled up into a hollow cylinder and show metallic or semiconducting properties dependently upon their diameter and chirality. Due to their unique structure SWCNTs allows to study a large variety of different quantum phenomena like single-electron tunneling (Bockrath et al., 1997), Luttinger liquid behaviour (Bockrath et al., 1999), ballistic transport (Krstic et al., 2000), etc. Multi-wall carbon nanotubes (MWCNTs) are more complicated systems. They consist of a several shells of different diameter and chirality. Due to weak coupling between the shells the conductivity in bulk-contacted MWCNTs is defined mostly by the outermost shells. Diffusive transport in majority of experiments for individual MWCNTs was observed. Therefore quantum interference effects inherent for mesoscopic systems (weak localization and universal conductance fluctuation) were reported (Schonenberger et al., 1999). Besides that, MWCNTs with large diameter allows to observe Aahronov-Bohm effect at experimentally available values of magnetic fields (Bachtold et al., 1999; Lassagne et al., 2007). Ballistic transport even at room temperatures was observed by some authors (Frank et al., 1998; Urbina et al., 2003). Possibility to switch between ballistic and diffusive transport regime in the same MWCNTs sample using gate voltage by virtue of an electrostatic change of electron density was reported as well (Strunk et al., 2006; Nanot et al., 2009).

Processing of nanotubes on macroscopic scale and investigation of their synergetic properties is a most important task for realistic application of these materials, especially for fabrication of carbon nanotubes-based gas-, bio- and chemical sensors where signal from the sensor output depends on the conductivity of device (Stetter & Maclay, 2004). Different examples of morphology of the samples of arrays of nanotubes involve definitions of bundles (ropes) (Fischer et al., 1997; Krstic et al., 2000), mats (Fischer et al.,

1997; Kaiser et al., 1998), networks (Kim et al., 2001) and films (Baumgartner et al., 1997). The intertubes barriers and defects, length of the individual nanotubes in the assemblies as well as contact geometry play an essential role in the electrical transport properties of the carbon nanotube arrays. Therefore different charge transport mechanisms can be observed in the arrays of nanotubes: metallic conductivity, variable range hopping (VRH), weak localization (WL), and fluctuation induced tunneling. Combination of various mechanisms is possible as well. Verification of charge transport mechanisms (as well as the ranges of the temperature, electric and magnetic fields etc.) in which these mechanisms play dominating role is very important and non-trivial task for such systems.

In this chapter we show some examples of different CNTs assemblies in which contribution from different charge transport mechanisms can determine electrical properties of the systems. The chapter organized as follows. We briefly describe several methods of preparation of various types of CNTs assemblies (monolayers of MWCNTs and SWCNTs, SWCNTs fibers and SWCNTs coatings of silica fibers) and present experimental results on investigation of their transport properties. Besides traditional experimental technique used for investigations of charge transport (measurements of current-voltage (I-V) characteristics and resistance as a function of temperature and magnetic field), effect of the influence of strong microwave field and terahertz radiation on the conductivity of SWCNTs fibers and SWCNTs coatings of silica fibers was studied as well. Mechanisms responsible for the charge transport in CNTs assemblies were identified. Sensitivity of the electrical properties of carbon nanotubes arrays to the types of pristine carbon nanotubes, contacts geometry and to the preparation methods is considered. Special emphasize on the role of quality of intertube contacts in CNTs assemblies is done as a result of experimental data analysis.

2. Methods of preparation of carbon nanotubes arrays

2.1 Assembly of carbon nanotubes monolayers using Langmuir-Blodgett technique

Thin film geometry of CNT arrays is a most promising for their possible commercial applications like antistatic shielding and transparent conducting layers, if the layers are thin enough (Robertson, 2004), interconnections and heat sinks in integrated circuits, nano- and optoelectronic devices. However, assembly of CNTs into thin films or monolayers is hindered by processing and manipulations difficulties. The Langmuir-Blodgett (LB) technique can be used as a method for depositing of monolayers of molecularly ordered ultra-thin films with controlled thickness. However, due to absence of free bonds on the surface of nanotubes, they possess a high chemical stability and are insoluble in the organic solvents. Therefore, to find a possibility of arrangement of CNTs into arrays by means of LB technique, a chemical modification (functionalization) of their surfaces is necessary. Different approaches for CNTs functionalization were proposed, including covalent sidewall functionalization, noncovalent functionalization with surfactants and polymers, defect group functionalization and plasma modification of nanotubes sidewall (Bahr & Tour, 2002; Hirsch, 2002). One of the main advantage of the organic covalent functionalization of CNTs for their further assembly in thin films using LB technique is a possibility to obtain large surfaces covered by CNTs without free space. In contrast, quite low concentration of CNTs in layers was achieved when noncovalent functionalization was used (Krstic et al., 1998). Besides that, applications for chemical and biosensors may require the presence of different functional groups bounded to CNTs.

We used prior organic functionalization of MWCNTs based on the scheme described in (Georgakilas et al., 2002). The MCWNTs produced by chemical vapor deposition (CVD) method with catalyst 5%Fe, Co/CaCO3 with average diameter of about 30 nm and length less than 1μm were used. The purified MWCNTs were suspended in DMF [N,NDimethylformamide HCON(CH3)2] together with an excess of *p*-Anisaldehide (4 – methoxybenzaldehyde) CH3OC6H4CHO and 3-methylhippuric acid [m-toluric acid, N-(3-methyl-benzoyl)glycine] CH 3 C 6 H 4 CONHCH 2 CO 2 H . The heterogeneous reaction mixture was heated at 130°C for approx. 120 hours. After the reaction was stopped, the organic phase was separated from unreached material by centrifugation and washing five times with chloroform ($CHCl_3$) and vacuum drying. The obtained dark solid phase was easily soluble in $CHCl_3$ up to a few mg/mL without sonication. The deposition of the layers (arrays) of nanotubes on the surface of the devices with the electrodes was done by using the cell, imitating the LB trough. Once a droplet of the solution of functionalized nanotubes in chloroform was spread on the water surface, a droplet of diblock-copolymer PS-PMMA solution was added in order to create a surface pressure of $\sim$ 9mN/m. The functionalized nanotubes were self-assembled in a dense arrays (monolayers) covering without free space the whole surfaces of the devices, when the monolayer was picked up from the water surface. The same procedure was done with grown by CVD method SWCNTs nanotubes. Scanning electron microscopy (SEM) image of dense arrays of SWCNTs is shown in Fig. 1.

Fig. 1. SEM image of carbon nanotube arrays obtained by LB technique

Substrates with finger-shaped contact geometry with 10μ m distance between contacts were used for investigation of the electrical transport both for MWCNTs and SWCNTs arrays.

2.2 Preparation of SWCNTs fibers using wet-spinning process

Due to the high aspect ratio of the nanotubes and intertube coupling in conductivity of the arrays of carbon nanotubes, enhanced electrical transport properties are expected from the more aligned material. Therefore, different techniques for the alignment of nanotubes in the arrays are proposed: orientation of nanotubes in the films by using high magnetic fields (Smith et al., 2000), fabrication of fibers of aligned nanotubes by using an electrophoretic process (Gommans et al., 2000), synthesis of the fibers with the preferential nanotubes orientation by means of wet spinning process (Vigolo et al., 2000; Vigolo et al., 2002). The last technique can be potentially used for the large scale material production due to their particular simplicity and similarity to the industrial process of polymer fibers fabrication. We used SWCNTs fibers produced by polymer-free spinning method. The principle of the spinning method for the polymer/nanotubes composite fibers was reported in (Vigolo et al., 2000). The presence of non-conductive polymer poly vinyl alcohol (PVA) makes problematic the most applications of these fibers. Polymer-free SWCNTs fibers were produced in the process developed at Rice University (Davis et al., 2004; Zhou et al., 2004). The SWCNTs were dispersed in 102% sulfuric acid (2 wt . % excess SO_3) and then wet-spun into either diethyl ether 5% sulfuric acid, or water. The SWCNT fibers with high electrical and thermal conductivity were obtained using this technique. However, some protonation of material occurred due to long contact of fibers with sulfuric acid. Polymer-free SWCNTs fibers used in our experiments were synthesized without superacids from solutions comprising nanotubes, surfactant and water (Kozlov et al., 2005). SEM image of SWCNTs fiber produced by this method is shown in Fig. 2.

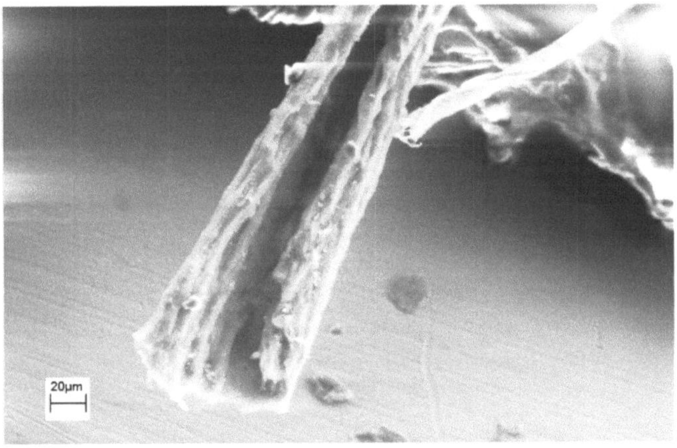

Fig. 2. SEM image of the hollow SWCNTs fiber produced by wet-spinning process

The nanotubes were dispersed using a horn sonicator in an aqueous solution of 1.2wt. −% lithium-dodecyl-sulfate (LDS) surfactant. A narrow jet of this spinning solution was injected into the flocculation bath containing 37% hydrochloric acid, which rotated at 33 rpm . The configuration of the rotating bath and spinneret needle was similar to that described for PVA-based coagulation spinning (Vigolo et al., 2000). Flocculation of nanotubes in the spinning solution to form a gel fiber occurred very close to the point of contact of the spinning solution and the acid in the bath. This gel fiber (containing 90wt% volatilizable liquid, based on gravimetric measurements) was washed in methanol to remove the hydrochloric acid. The fiber, which has very low elasticity both before and after the washing step, was then pulled from the wash bath, stretched over a frame, and dried under tension. The fibers obtained by this method are of $20 - 100\mu$ m diameter. As was mentioned above, one of the main advantages of this method for fabrication of the SWCNTs macroscopic arrays is the existence of the preferential orientation of individual nanotubes in the fibers due to flow-induced alignment of the nanotubes during processing (Vigolo et al., 2000). This alignment can be seen from the scanning electron microscopy (SEM) image of SWCNTs fiber shown in the Fig. 2.

SWCNTs with different geometry (solid, hollow and ribbon fibers) can be obtained by polymer-free wet spinning process in dependence of injection rate into the rotating bath of solution containing SWCNTs and inner spinneret diameter (Kozlov et al., 2005). HiPco SWCNTs obtained from a carbon monoxide process and SWCNTs prepared by laser ablation purchased from Carbon Nanotechnologies Inc. were used as pristine nanotubes for the

SWCNTs fibers preparation. Electrical contacts for further measurements of charge transport in the fibers were made by Ag paint.

2.3 Fabrication of SWCNTs coatings of Silica fibers

Due to the high mechanical strength carbon nanotubes are widely used for reinforcement of different types composites materials, replacing conventional reinforcing fillers such as carbon or glass fibers (Shaffer & Sandler, 2007). Manufacturing of hierarchical composites with carbon nanotubes combined with conventional microscopic fibers has attracted growing interest years recently. Carbon nanotubes coatings on the surface of ceramic (Ci et al., 2005), carbon (Thostenson et al., 2002), quartz and aliminium silicate (Zhang et al., 2008) fibers were synthesized. Multi-wall carbon nanotubes coatings on the surface of silica fibers were obtained by chemical bonding (Liu et al., 2009) and using injection CVD method (Qian et al., 2009).

Procedure of preparation of SWCNTs coatings of silica fibers used in our investigations included the following stages: i) dissolving of polyimide claddings of silica fibers by soaking them in toluene for 2 hours, ii) cleaning of the residues of claddings by sonication in saturated KOH /propanol solution for 20 minutes with following sonication during 5 minutes in distilled water (this step was repeated several times), iii) silanization of silica fibers (creation of abundant carbon long alkyl chains on the silica fibers surfaces) in octadecyltrimethoxysilane (OTMS) solution in ethanol at a concentration of 2 mL/g for 2 hours. After that, the silanized silica fibers were soaked in water in which pristine SWCNTs were dispersed. When the silica fiber turns black due to a good coverage of SWCNTs as a result of clinging presumably driven by hydrophobic interactions, they were heated in an oven at 250°C to anneal the bound SWCNTs. SEM image of SWCNTs coating on the surface of silica fiber obtained as a result of the above described procedure is shown in Fig. 3.

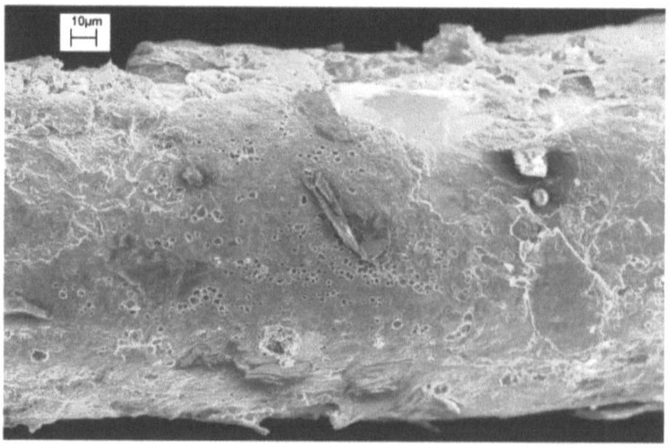

Fig. 3. SEM image of the side view of the silica fiber with good coverage of SWCNTs

For the further electrical and magnetotransport measurements silica fibers with SWCNTs coatings were transferred on solid substrates and electrical contacts were made by Ag paint.

3. Charge transport properties of carbon nanotubes arrays

3.1 Electrical and magnetotransport properties of carbon nanotubes monolayers

Measurements of the electrical and magnetotransport properties is one of the most reliable tool for exploring of different materials and verification of the mechanisms responsible for the charge transport in the system, (for a general revue of electronic and transport properties of carbon nanotubes, see (Charlier et al. 2007)). We used standard four-probe dc- and lock-in technique for the measurements of transport properties of carbon nanotubes assemblies in the temperature range 2 – 300 K. Magnetoresistance (MR) measurements were carried out in pulsed magnetic fields at the Laboratoire National des Champs Magnētiques Intenses de Toulouse (LNCMI) in the temperature range 2 – 300 K and up to 40 T magnitude fields.

We report and compare here the transport properties of different types of carbon nanotubes arrays. The temperature dependence of the resistance of the layers of short MWCNTs shows the negative temperature coefficient of the resistance ($dR/dT < 0$) in the whole investigated temperature range (4.2 – 300 K) (Ksenevich et al., 2008 a) as one can see in the Fig. 4. Assuming possibility of the VRH conduction in our system due to structural defects and impurities, we fitted $R(T)$ dependence by classical law for variable range hopping (VRH) (Shklovskii & Efros, 1984):

$$\mathrm{R} = \mathrm{R_0 exp} \left(\mathrm{T_0/T} \right)^{1/\mathrm{n}} \tag{1}$$

where $T_0 = \beta / k_B \xi^{\xi} g(\mu), \beta$ is a constant, k_B is a Boltzmann constant, ξ denotes the localization length, and $g(\mu)$ is the density of states, $n = d + 1 = 2,3,4$ (d is the dimensionality of the system). In order to find parameter n we used both linearization in the scale log $R - T^{-1/n}$ and method proposed by Zabrodskii (Zabrodskii, 1977). We found that $R(T)$ dependence can be approximated by Eq. (1) in the temperature range $T = 4.2 \sim 60$ K and the best fitting results were obtained at T below 60 K with parameter $n = 3$, indicating 2D VRH.

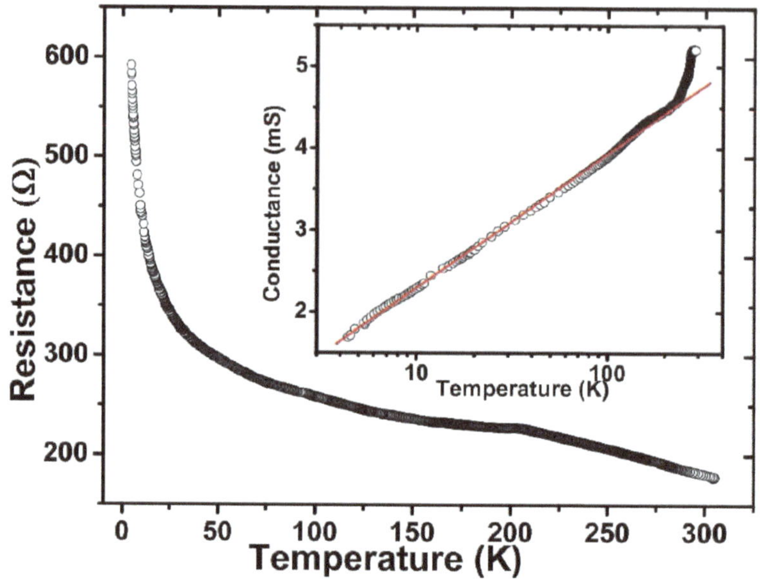

Fig. 4. The temperature dependence of the resistance of the MWCNTs arrays. The inset shows the $R(T)$ dependence in G-LogT scale.

In order to see suitability of the WL theory for our system, we tried to fit $R(T)$ dependence with one-, two- and three-dimensional WL formulas. We found that $R(T)$ dependence follows the 2D weak localization (WL) behavior ($G(T) \sim \ln T$) (Lee & Ramakrishnan, 1985) in the temperature range 4.2 ~ 202 K as one can see from the inset to Fig. 4 In the temperature range 202 – 300 K conductivity increases approximately linearly with temperature.

MR data measured at various temperatures are plotted in Fig. 5. The negative MR in the low magnetic fields range was observed. In the low-temperature range upturn of negative MR was observed. The minimum position of the negative MR shifts to the higher fields as the temperature rises. Both low-field negative MR due to changing of phase between alternate hopping paths enclosing a magnetic flux (Nguen et al., 1985; Sivan et al., 1988) and highfield positive MR due to electronic orbit shrinkage (Shklovskii & Efros, 1984) are predicted for systems with hopping conductivity mechanism. However, monotonic decrease of relative value of negative MR with the temperature rising are usually observed. In our samples the opposite situation is clearly seen: relative value of negative MR increased with the temperature rising in the low temperature range where VRH can be responsible for the charge transport mechanism. From the other side negative MR is inherent for the systems where conductivity can be described in the frame of WL theory (Lee & Ramakrishnan, 1985).

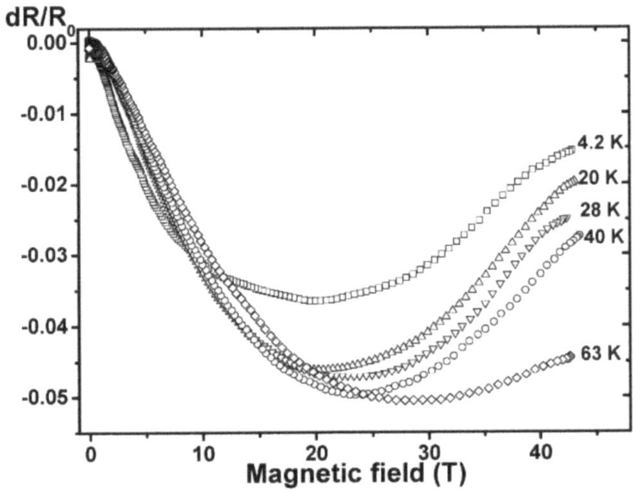

Fig. 5. The dependence of the normalized change in the resistance $dR/R(B=0)$ of the MWCNTs arrays on the magnetic field in the temperature range 4.2-63 K.

Therefore we made assumption that MR data are the sum of the positive and negative contribution due to MR effects in the VRH and WL regimes, respectively. We found that high-field positive part of magnetoristance can be approximated in frame of Kamimura model for spin-dependent VRH conductivity (Kurobe & Kamimura, 1982):

$$\frac{\Delta G}{G} = -A_{KK}\frac{H^2}{H_{KK}^2 - H^2} \tag{2}$$

where $G = 1/R$ is the conductance of the system, H_{KK} is the characteristic field for spin alignment, A_{KK} the saturation value of the magnetoconductance. Using formula (2) and values of parameters H_{KK} and A_{KK} obtained from the approximation data of high field MR data, we calculated positive magnetoresistance for low-field region. Pure negative contribution to MR according our assumption was calculated by subtracting positive MR from the experimental data. These data can be fitted reasonably well by the Eq. (3) for 2D WL:

$$\Delta G = \frac{e^2}{\pi h}\left[\psi\left(\frac{1}{2}+x\right)+\ln(x)\right] \tag{3}$$

where ψ is the digamma function, $x = L_{Th}{}^2 8\pi EH/hc, L_{Th} \sim T^{-p/2}$ is the Thouless length.

The example of the above mentioned procedure for extracting of positive and negative part from the total MR curve for temperature $T = 40$ K is shown in the Fig. 6.

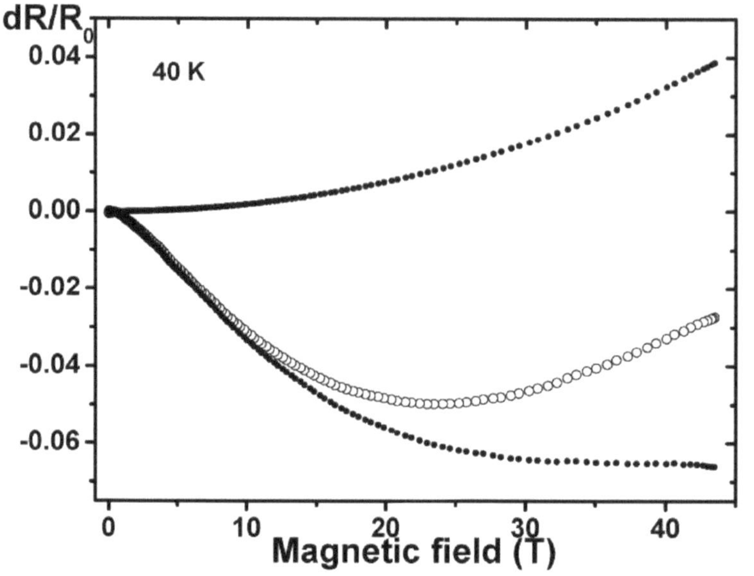

Fig. 6. The dependence of the normalized change in the resistance $dR/R(B = 0)$ of the MWCNTs arrays on the magnetic field measured at 40 K (open circle). Calculated value of positive MR obtained by fitting positive part of MR data to Eq. (2) (upper curve). The negative MR (lower curve) obtained by subtracting the calculated positive MR from the experimental MR data.

With this assumption we can explain positive upturn observed on MR curves by adding curves for positive and negative MR. Positive MR may come from the spin-dependent VRH among the defects on the surface of nanotubes. The negative contribution to MR can originate from the WL effects. It should be noted, that similar magnetotransport properties were observed for SWCNTs entangled networks (Kim et al., 1998). Therefore we can assume that similar MR effects in different type of carbon nanotubes arrays are defined not only by the type of pristine nanotubes, but their length and contact geometry as well. For both types of the samples the distance between contacts was much higher than the length of the separate nanotubes in the arrays, thus enhancing the probability to observe the localization effects.

Investigation of transport in SWCNTs monolayers confirms the importance of intertube contacts between separate nanotubes in their assemblies (Ksenevich et al., 2006). The temperature dependence of the resistance $R(T)$ of the SWCNTs arrays is shown in Fig. 7.

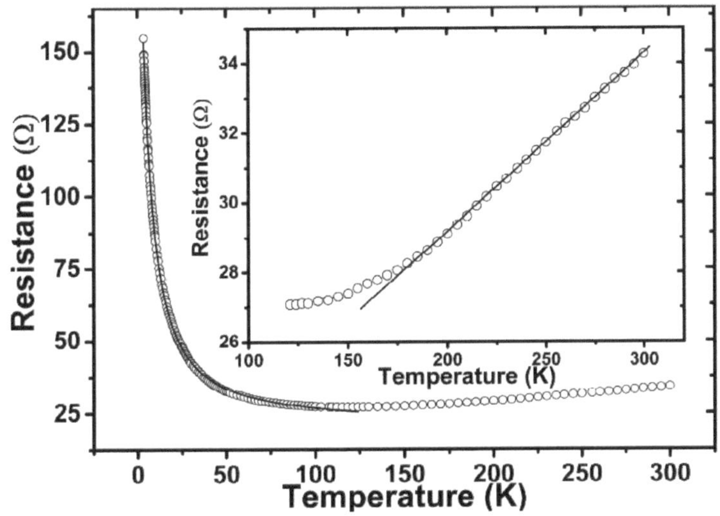

Fig. 7. The temperature dependence of the resistance of the SWCNTs arrays. Solid line is the fitting of the $R(T)$ dependence by means of fluctuation-induced tunneling model (second term of Eq. (4)) in the low temperature range. The inset shows the high temperature metallic term of the $R(T)$ dependence.

The resistivity minimum on the temperature dependence of the resistance of SWCNTs arrays was observed at ~ 125 K. At $T > 125$ K$R(T)$ dependence has a metallic-type behavior ($dR/dT > 0$) while in the low temperature range non-metallic sign for $R(T)$ dependence ($dR/dT < 0$) was observed. In the low-temperature range ($T \sim 4.2 - 100$ K) the temperature dependence of the resistance is consistent with the exponential law proposed for the fluctuation-induced tunneling model (Sheng, 1980). Thus, temperature dependence of the resistance can be written as:

$$R = \alpha T + R_t \exp(T_c/T + T_s), \tag{4}$$

where α is the temperature coefficient arising from metallic conductivity, and the second term represents the conductivity through the barriers between the metallic regions in frame of the fluctuation-induced tunnelling model. Such type of the temperature dependence of the resistance with systematic metallic sign and crossover to the non-metallic sign of the $R(T)$ curves at lower temperatures was firstly observed by Fischer (Fischer et al., 1997) for SWCNTs ropes and mats. Heterogeneous model for conduction was proposed to explain crossover from metallic to non-metallic sign of the temperature dependence of the resistance as the temperature decreases (Kaiser et al., 1998). This model is based on the simple assumption that conduction process at high temperatures is defined by nanotubes with good metallic conductivity. The presence of small electrical barriers (tangled regions, intertube contacts, structural defects insight nanotubes itself) give rise to low temperatures localization effects. Therefore, the conductivity in the low temperature range is determined by tunneling through these electrical barriers. In our samples in the intermediate temperature range ($T \sim 100 - 185$ K) the contribution to the total conductivity of the system from the metallic conductivity and fluctuation-induced tunneling are comparable in the absolute value. This assumption explains the behavior of the $R(T)$ dependence. Linear rise of the resistance with temperature is clearly seen in the temperature range $T \sim 185 - 300$ K as one can see from the inset to Fig. 7. From the other side, due to a weak temperature dependence of the metallic term and its saturation at low temperatures of the temperature dependence the resultant resistance can be well fitted by fluctuation-induced tunneling law (second term of Eq. (4)).

It should be noted that the crossover temperature T between the metallic and nonmetallic type of $R(T)$ curve depends strongly on the type of SWCNTs arrays and quality of intertube contacts between separate nanotubes and varies from 35 K for a single well-ordered rope to 250 K for a rope with tangled regions (Fischer et al., 1997). Moreover, for SWCNTs with tangled regions the nonmetallic term can be described by proposed for variable range hopping conductivity Mott's law (1), indicating localization of charge carriers and hopping transport instead of fluctuation-induced tunneling (Kaiser et al., 1999). We assume that in our samples the localization effects and the resulting hopping transport can be essential only at very low temperatures where the rapid rise of the resistance is observed as the temperature decrease. This conclusion is confirmed by the shape of the MR curves, shown in Fig. 8.Positive magnetoresistance with decreasing amplitude inherent for the variable range hopping conductivity was observed while temperature rising in the low temperature range. At temperatures higher than 7 K the MR effect is negligible. Because localization near defects is a possible cause of the barriers for charge carriers in heterogeneous conductivity model, magnetoresistance can exhibit localization effects at low temperatures, inducing positive MR. However absolute value of the positive magnetoresistance can be lower in comparison with uniformly disordered systems as far as the conductivity in our heterogeneous system is enhanced by the well conductive metallic regions (metallic carbon nanotubes).

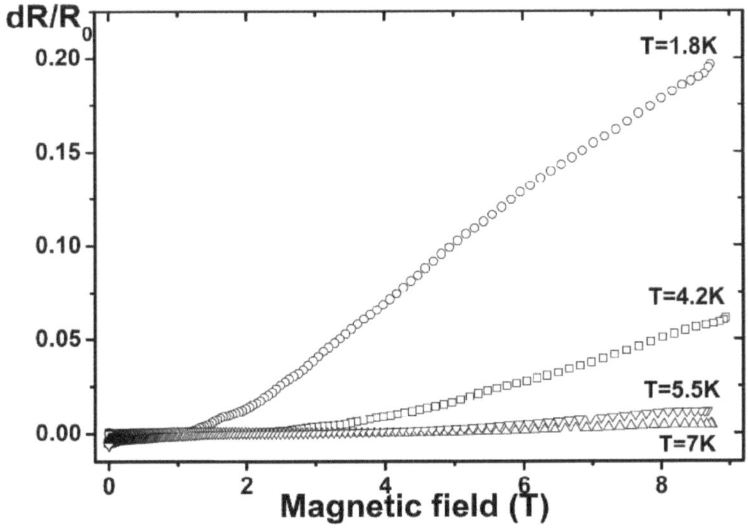

Fig. 8. The dependence of the normalized change in the resistance $dR/R(B=0)$ of the SWCNTs monolayers on the magnetic field in the temperature range $1.8-7$ K.

3.2 Charge transport in SWCNTs fibers

Electrical and thermal transport properties of the SWCNT fibers fabricated by means of wet spinning process in correlation with the orientation degree of nanotubes inside fibers are already reported (Zhou et al., 2004; Badaire et al., 2004). For the optimizing of conducting properties of SWCNT fibers for possible applications deeper understanding of the charge transport mechanism in this material is required.

We applied different experimental techniques in order to study transport properties of SWCNTs fibers and to distinguish mechanisms responsible for the charge carrier's transfer in the system (Ksenevich et al., 2008 b; Ksenevich et al., 2008 c). We have measured resistance as a function of temperature and magnetic field, current-voltage ($I-V$) characteristics, and investigated effect of the influence of the strong microwave field and terahertz radiation on the conductivity of SWCNT fibers. The $R(T)$ dependence of the SWCNT fibers is shown in Fig. 9. In the low-temperature range (4.2-80 K) the resistance as a function of temperature exhibits a classical VRH conduction, following the law (1) with exponent $n=4$, assuming 3D character of hopping conductivity. In the temperature range $T=$ 80-300K K (T) dependence can be approximated by a typical law for fluctuation-induced tunneling conductivity mechanism (Sheng, 1980):

$$R = R_t \exp (T_c/T + T_s), \tag{5}$$

where parameters T_c and T_s are defined by the following equations:

$$T_s = 16\varepsilon_0 \hbar A V_0^{3/2} / (\pi e^2 k_B (2\ m_e)^{1/2} w^2, \tag{6}$$

$$T_c = 8\varepsilon_0 A V_0^2 / (e^2 k_B w), \tag{7}$$

where w is the width of the tunnel junction, A labels its area, and V_0 is the height of the contact potential; k_B stands for the Boltzmann constant, $\hbar$ is a Plank constant, e is the electronic charge, m_e the electron's effective mass, and ε_0 the vacuum permittivity.

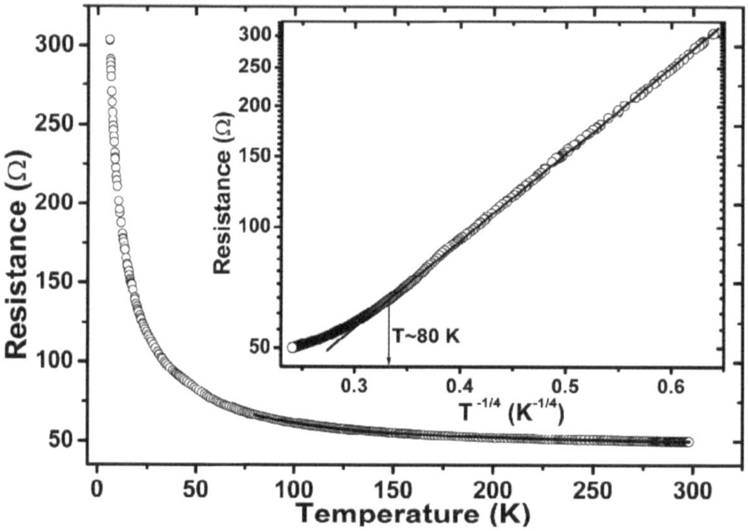

Fig. 9. The temperature dependence of the resistance of the SWCNT fibers. Solid line is the fitting of the $R(T)$ dependence by means of fluctuation-induced tunneling model in the temperature range 80 – 300 K. The inset shows the log R – $T^{-1/4}$ plot. (According to ref. (Ksenevich et al., 2008 b)).

Fluctuation-induced tunneling conductivity model was proposed for disordered heterogeneous systems contrary to the systems with hopping charge carrier transport between localized sites. These systems (for example, conductor-insulator composites, granular metals, disordered semiconductors) consist of large conductive segments (where electrons are delocalized and free to move over very large distances as compared to the atomic dimensions) separated by small insulating gaps. Due to small sizes of tunnel junction, thermal voltage fluctuation resulting from the thermal motion of electrons in the conduction region in the vicinity of tunnel barrier can change drastically the electron tunneling probability through the barrier. Different types of carbon nanotubes arrays with existing electrical barriers can be also considered as heterogeneously disordered systems. Therefore this model was used for describing the temperature dependence of conductivity of SWCNTs fibers (Zhou et al., 2004) and networks (Kim et al., 2001). As was mentioned earlier we observed fluctuation induced tunneling conductivity in the low temperature range for SWCNTs layers obtained by LB technique. As one can see from Fig. 9 for our SWCNTs fibers the fitting results also agree well with experimental data thus indicating the importance of intertube contacts between individual carbon nanotubes. Assuming the area A of the tunnel junction being equal to the cross section of a SWCNT with the diameter of about 1.5 nm and using expressions for parameters T_c and T_s, the effective junction width is estimated to be 4.6 nm and the effective barrier height to be 0.172 eV .

I - V curves of SWCNTs fibers measured in the temperature range 6 – 300 K are plotted in Fig. 10. At room temperature, linear $I-V$ curves were observed. As the temperature decreases, the non-Ohmic behavior becomes pronounced. Nonlinearity of $I-V$ curves is clearly seen at temperatures about 100 K and becomes more significant in the low temperature range. Reproducibility of the $I-V$ curves by number of thermal cycling and different sweeping times indicate that the heating effect is negligible. Nonlinear $I-V$ characteristics for both charge transport mechanisms (VRH and fluctuation induced tunneling) can be observed.

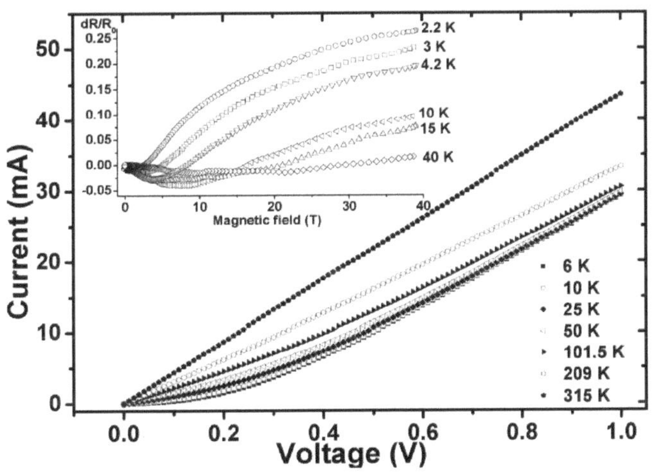

Fig. 10. The current-voltage characteristics of the SWCNT fibers at different temperatures. The inset figure: the dependence of the normalized change in the resistance $dR/R(B=0)$ of the SWCNTs fibers on the magnetic field in the temperature range $2.2-40$ K (According to ref. (Ksenevich et al., 2008 b)).

As far as $R(T)$ dependencies in the low temperature range were approximated by Mott's law (1) for VRH conductivity, we assumed that nonlinearity of the $I-V$ curves can be explained within the framework of classical theoretical models for hopping conductivity (Pollak & Riess, 1976; Grannan et al., 1992). According to these models, in the low electric fields, the conductivity is expected to be nearly independent of the applied electric field. In this regime hopping of the charge carriers between hopping sites is induced by phonons. In the high electric fields, both temperature and electric fields (with energy scales $k_B T$ and eEl, respectively, where E is the electric field and l is the length parameter related to maximum hopping length between the hopping sites) have similar activation effects on the charge carriers. Low fields are defined as fields $E < Ec$, where $Ec = k_B T/el$. In the region of high electric fields $Ec < E < k_B T/e\xi$ (where ξ is a localization length), non-Ohmic VRH conductivity follows the formula (Pollak & Riess, 1976; Grannan et al., 1992):

$$\sigma(\mathrm{E, T}) = \sigma(0, \mathrm{T})\exp(\mathrm{eUl/k_B Td}) \tag{8}$$

where U is the voltage measured between voltage probes and d is the distance between them. However, we found rather large discrepancies between experimental data and fitting curves in the region of very high electric field ($E > k_B T/e\xi$) where theoretical models for VRH predict "activationless" conductivity. In this regime charge carriers acquire the energy necessary for hopping between localized sites not from the phonons but from the electric field, and the conductivity becomes temperature independent: $\sigma(E) \sim \exp(-E_0/E)^m$, where the exponent m is usually equal to exponent in formula (1) for the Ohmic regime of hopping conductivity as was found in most experiments. Therefore in order to check the possibility to explain the origin of nonlinearity of I - V curves in the frame of the VRH models, we estimated the fitting parameter $(k_B d/el)T$ at different temperatures from the slope of $\ln(U)$ dependencies. From the slope $\log(k_B d/el)T$ versus $\log T$ we obtained the temperature dependence of the parameter l (maximum hopping length), $l \sim T^{0.13}$. This means that hopping length slightly increases with the temperature, and this fact contradicts to the classical VRH models. As it was mentioned above, in the low temperature range, the $R(T)$ dependencies of the SWCNT fibers measured in the linear range of I - V curves were fitted well by Mott's law (1). According to the Mott's model for 3D VRH, the maximum hopping length can be described as $l \sim T^{-0.25}$. Therefore, the nonlinearity of $I - V$ curves cannot be explained by the influence of the high electric fields on the hopping conductivity. We assume that at < 80 K, where $R(T)$ dependencies are well fitted by Mott law (1), the VRH conductivity can be considered as the charge transport mechanism in the SWCNT fibers only for low electric fields $E < Ec$. We found that best fitting results of the nonlinear IV curves in the whole range of applied electric fields can be obtained by using the extended Sheng's model (Sheng, 1980) proposed by Kaiser and Park (Kaiser & Park, 2005):

$$\mathrm{G} = \mathrm{I/V} = \mathrm{G_0}\exp(\mathrm{V/V_0})/(1 + \mathrm{h}[\exp(\mathrm{V/V_0} - 1)]) \tag{9}$$

where G_0 is the temperature dependent low-field conductance, parameter V_0 is the voltage scale factor that gives an exponential increase in conductance and depends strongly on the barrier energy, and parameter $h = G_0/G_h$, where G_h reflects the large conductance in the absence of barriers. Eq. (9) was obtained by means of numerical calculations of fluctuation induced tunneling and thermal activation (Kaiser & Park, 2005).

We carried out also measurements of the dependence of the resistance on the magnetic field in order to clarify the charge transport picture in SWCNT fibers. The general features of MR at different temperatures behavior can be seen from the insert to Fig. 10.

At low fields, MR is negative. A positive upturn is observed on the MR curves, of which position is shifted to higher magnetic field as the temperature increases. The upturn field of the MR effect was shifted from 1.5 T at 2 K to value of about 20 T at 40 K . The value of positive MR varies as exp (B^2) which changes to exp $\left(B^{1/3}\right)$ behavior at sufficiently high fields as expected for the VRH transport (Shklovskii & Efros, 1984). The tendency to saturation of the positive MR was observed at low temperatures as one can see from the inset to Fig. 10. However, the negative MR can not be explained in frame of Sivan model (Sivan et al., 1980) proposed for VRH charge transport. According to this model the negative MR should saturate when one flux quantum $\Phi_0 = hc/e$ penetrates an area equal $R_{M^{3/2}}\chi^{1/2}$ (here R_M - the average hopping length and χ - the average distance between the localization centers). The position of the negative MR minimum shifts to the higher fields with increasing temperature due the hopping length decrease and as a consequence, an area bounded by different paths decreases. In addition, the monotonic decrease of the relative value of negative MR with the temperature is usually observed. We found the nonmonotonic behavior of the relative value of the negative MR with the temperature. At $\mathrm{T} = 2$ K the relative value of the negative MR was about 0.4%. The maximum of the relative negative MR value about 4.2% was observed at $\mathrm{T} = 10$ K. At the further temperature increasing the negative MR decreases. Therefore, the contribution of the other mechanism must be considered in order to explain the negative MR behavior. One of the possible explanations of the negative MR can be a contribution of the weak localization effects in total MR (Lee & Ramakrishnan, 1985).

The determined carrier transport features were supported by additional measurements of the change in conductivity in strong 10 GHz microwave fields and measurements of THz radiation induced photocurrent at various lattice temperatures.

Relaxation kinetics of the THz induced photovoltage at different temperatures are shown in Fig. 11. In order to diminish Joule heating of the samples due to absorption of laser radiation pulsed THz radiation with low repetition rate (30 Hz) and duration of pulses of 100μ s (the shortest possible pulse available in this kind of optically-pumped molecular terahertz laser) was used. Two relaxation processes are visible: relatively fast part (with time constant of some hundreds of μ s which corresponds to the duration of THz pulses) and relatively slow part (in the range of about ~ 20 ms) that we associate with the Joule heating. The fast peak we associate with the influence of terahertz radiation on the hopping conductivity. We assume this effect as terahertz - induced hopping conductivity (Ksenevich et al., 2008 b), taking into account the following arguments. In the low temperature range where VRH conduction is observed, the conductivity increases with temperature because of activation of charge carriers by thermal energy k_BT on hopping sites located at higher energy levels. Since the terahertz radiation energy is in the order of magnitude of thermal energy (1THz ~ 4meV ~ 48 K), that terahertz quanta exceeding the thermal energy k_BT should give rise to additional increase of conductivity. On the other hand, within the temperature rise fluctuation induced tunneling mechanism starts to prevail. As it is evident from the experiments, it becomes negligible at T ~ 30 K. Only slow Thz - induced signal due to heating is observed. We assume that contributions from fluctuation induced tunneling and VRH at this temperature are comparable on the absolute value of magnitude.

The dependencies of the variation of the conductivity of SWCNT fibers on applied microwave field power at the temperatures 77 and 300 K measured using technique described in (Pozela, 1985) are shown in the inset to Fig. 11.

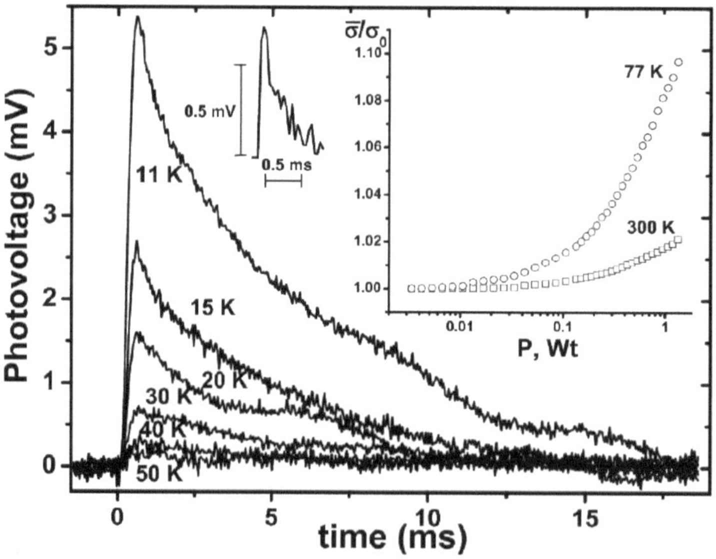

Fig. 11. Relaxation kinetics of THz-induced photovoltage signal at different temperatures. Left inset: terahertz induced photocurrent transient excluding lattice heating. In the right inset the dependence of the conductivity variation on applied microwave power is shown. (According to ref. (Ksenevich et al., 2008 b)).

The increase of the conductivity on microwave power becomes more essential with the temperature decrease as one can see from the right inset to Fig. 11. At the same microwave field power of about 1 W the observed value of relative conductivity increase was about 2% and 10% at the temperatures 300 K and 77 K , respectively. It should be noted that the heating of the charge carriers by strong microwave field can give rise to conductivity both in VRH and fluctuation-induced tunneling regimes due to a possible rising of the probability of charge carriers for the hopping between the localized sites and for the tunneling through the barriers, respectively.

3.3 Conductivity mechanisms in SWCNTs coatings of Silica fibers

According to our knowledge the electrical transport in SWCNTs coatings of silica fibers had not been studied earlier.

We used a standard dc 4-probe technique for measurements of the electrical and magnetotransport properties of the SWCNTs- SiO_2 coatings in a close-cycled cryostat in the temperature range $1.8 - 300$ K and in the magnetic fields up to 8 T (Ksenevich et al., 2009). Charge transport in the system was found to have general features similar to the properties of SWCNTs fibers produced by polymer-free wet spinning process. The $R(T)$ dependence of the SWCNTs coatings of silica fibers can be well fitted by a typical law (5) for fluctuationinduced tunneling model (Sheng, 1980) almost in the whole range of temperatures ($\sim 8 - 300$ K). The best fit to Eq. (5) was obtained with parameters $T_S = 13.3$ K and $T_c = 52.8$ K. Assuming that the area A of the tunnel junction equals to the cross section of a SWCNT with diameter of about 1.5 nm and using equations (6) and (7) for parameters T_s and T_c, the effective junction width was estimated to be 3.2 nm and the effective barrier height of 68 meV . In the low temperature range $(2 \sim 8\ \mathrm{K}) R(T)$ dependence was found to follows the Mott law (1) for 3D variable range hopping (VRH) conduction (Shklovskii & Efros, 1984). It should be noted that both for SWCNTs fibers and SWCNTs coatings of silica fibers nonmetallic sign for $R(T)$ dependence ($dR/dT < 0$) was observed in the whole range of temperatures in contrast to SWCNTs monolayers. We believe that possibility to observe metallic sign for $R(T)$ dependence for SWCNTs fibers and SWCNTs coatings of silica fibers is hindered by large distance between contacts (about ~ 1 mm) in contrast to SWCNTs monolayers, where length of pristine nanotubes exceeded distance between contacts (10μ m) in finger-shaped structures.

Magnetoresistance data measured at different temperatures are shown in Fig. 12.

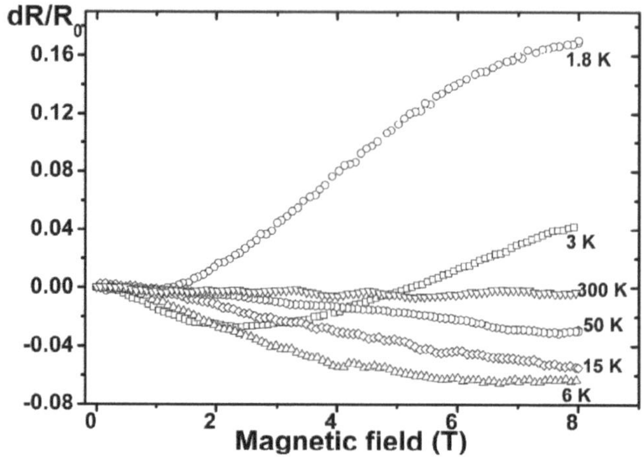

Fig. 12. The dependence of the normalized change in the resistance $dR/R(B=0)$ of the SWCNTs coatings of silica fibers on the magnetic field in the temperature range 1.8-300 K.

Both positive and negative MR was observed in the temperature range 1.8-6 K. The value of positive MR varies as exp (B^2) as expected for the VRH transport and explained by electronic wave function shrinkage in the magnetic field. Tendency to saturation of positive MR was observed at 1.8 K at the value of magnetic field higher than 6 T . Non-monotonic behavior of relative value of negative MR was observed similar to the experimental data for MR of SWCNTs fibers. The relative value of negative MR increases with temperature in the range of $1.8-6$ K. As the temperature exceeds 6 K , only negative MR is observed in the available range of magnetic fields, and the relative value of negative MR then decreases as the temperature rises. One can note, however, that in order to extract the negative contribution to the total magnetoresistance and to clarify the physical origin of this negative MR, a further analysis is needed both for SWCNTs fibers and SWCNTs coatings of SiO_2 fibers.

The experiments in strong microwave field in GHz range and THz radiation-induced photocurrent measurements were carried out as well in order to get an additional view into the electrical transport phenomena in SWCNTs coatings of silica fibers in comparison with transport properties of SWCNTs fibers.

The relaxation kinetics of photocurrent recorded at 1.6 THz frequency measured at different temperatures are shown in Fig. 13. Similar to the THz radiation-induced photocurrent experiments carried out for SWCNTs fibers the relaxation kinetics allows one to distinguish two different processes: relatively fast (with time constant of about of some hundreds of μs, which corresponds to the duration of THz pulses) due to terahertz-induced hopping conductivity, and slow (with time constant in the range of *ms*) due to the Joule heating of the samples. The decreasing of the amplitude of the fast photocurrent signal with the temperature was observed. Quantitative differences between the experimental results on the measurements of the conductivity under THz radiation for SWCNTs coatings of silica fibers and for SWCNT fibers were observed. Firstly, the amplitude of fast photocurrent signal is higher at the same temperatures and at the same frequency of THz radiation for the SWCNTs coatings of silica fibers. Secondly, the time constant for slow part of photocurrent signal is several times lower for SWCNTs- SiO_2 samples than for SWCNT fibers. Thz radiation-induced photocurrent measurements allow us to assume that in the intermediate temperature range both charge transport mechanism (VRH and fluctuation induced tunneling) can coexist. Therefore, the first difference can be explained by the higher degree of alignment of carbon nanotubes in fibers due to specifics of wet spinning method (Kozlov et al., 2005) and as a consequence a higher contribution to the total conductivity of VRH transport in comparison with SWCNTs fibers. A possible reason of lower duration of slow photocurrent signal component in SWCNTs coatings of silica fibers can be the heat exchange between the silica core and SWCNTs coatings.

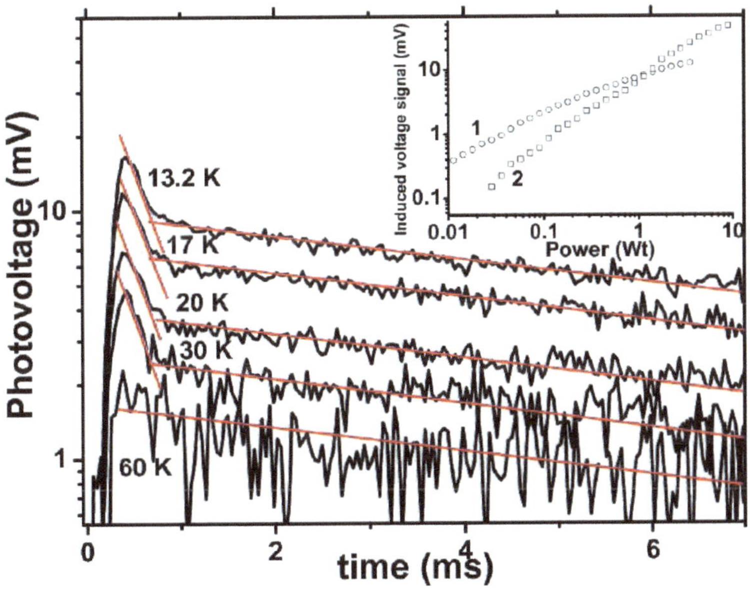

Fig. 13. Terahertz radiation-induced photocurrent at 1.6 THz at different temperatures in SWCNTs coatings of silica fibers. In the inset induced voltage signal proportional to the change of conductivity as function of microwave power at liquid nitrogen temperature for SWCNTs fibers (curve 1) and SWCNTs coatings of silica fibers (curve 2) is shown.

Measurements of the conductivity change of SWCNTs coatings of silica fibers in strong microwave field of 10 GHz at 77 and 300 K have shown similar results with experiments carried out for the SWCNTs fibers produced by wet spinning process. We observed increase in conductivity in microwave fields became more pronounced lowering the temperature. The data measured at 77 K are plotted in the inset to Fig. 13. For comparison, results both for SWCNTs fibers and SWCNTs coatings of silica fibers are given. As one can see, for both types of structures, the dependence is similar, however in SWCNTs- SiO_2 it follows a linear law while for SWCNTs fibers it becomes sublinear with microwave power.

As was mentioned above, heating of charge carriers by strong microwave field can give rise to conductivity both in VRH and fluctuation-induced tunneling regimes. However, according to the $R(T)$ dependencies we can assume that the fluctuation-induced tunneling is a prevailing mechanism of charge carrier transport at 77 K for both types of samples. Hence, the difference in transport properties can be only defined by the barrier height. This can be an argument to explain the more rapid increase of conductivity with microwave fields in SWCNT- SiO_2 samples containing lower effective barrier height in comparison to that of SWCNTs fibers (68 meV vs 172 meV).

4. Conclusion

Electrical and magnetotransport properties of various types of carbon nanotubes arrays were reported. Different fabrication procedures for assembly of separate carbon nanotubes into monolayers, SWCNTs fibers and coatings of silica fibers were used. Langmuir-Blodgett (LB) technique was applied for self-assembling of SWCNTs and MWCNTs monolayers. SWCNTs fibers were synthesized using wet spinning process from solutions comprising nanotubes, surfactant and water. SWCNTs coatings were produced by means of clinging of pristine SWCNTs dispersed in water to previously silanized silica fibers. We experimentally showed that the electrical and magnetotransport properties of investigated carbon nanotubes arrays are defined not only by the types of pristine carbon nanotubes (multi wall or single wall with different chiralities) but also by the quality of intertube coupling between separate nanotubes from which these assemblies consist of.

Magnetotransport properties of MWCNTs arrays can be interpreted in framework of model assuming existing of two charge transport mechanisms in the system: spin-dependent hopping conductivity and 2D weak localization.

Charge transport properties of 2D arrays of SWCNTs can be interpreted in terms of heterogeneous model for conduction where the presence of structural defects and intertube connections give rise to low temperature localization effects. The conductivity of the system at high temperatures is determined by the charge transport trough metallic nanotubes.

Experimental data for SWCNTs fibers and SWCNTs coatings of silica fibers could be consistent both with the charge transport model for homogeneously disordered materials due to low-temperature localization of charge carriers arising from the defects in the tubes themselves and with the model for heterogeneous materials that emphasizes the role of energy barriers between the nanotubes.

5. References

Bachtold, A.; Strunk, C.; Salvetat, J.-P.; Bonard, J.-M.; Forro, L.; Nussbaumer, T. & Schonenberger, C. (1999). Aharonov-Bohm oscillations in carbon nanotubes. Nature, Vol. 397, No. 6721, pp. 673-675

Badaire, S.; Pichot, V.; Zakri, C.; Poulin, P.; Launois, P.; Vavro, J.; Guthy, C.; Chen, M. & Fischer, J. E. (2004). Correlation of properties with preferred orientation in coagulated and stretch-aligned single-wall carbon nanotubes. J. Appl. Phys., Vol. 96, No. 12, pp. 7509-7513

Bahr, J.L. & Tour, J.M. (2002). Covalent chemistry of single-wall carbon nanotubes. J. Mater. Chem., Vol. 12, No. 7, pp. 1952-1958

Baumgartner, G.; Carrard, M.; Zuppiroli, L.; Bacsa, W.; de Heer, W. A. & Forro, L. (1997). Hall effect and magnetoresistance of carbon nanotube films. Phys. Rev. B, Vol. 55, No. 11, pp. 6704-6707

Bockrath, M.; Cobden, D. H.; McEuen, P. L.; Chopra, N. G.; Zettl, A.; Thess, A. & Smalley, R. E. (1997). Single-Electron Transport in Ropes of Carbon Nanotubes. Science, Vol. 275, No. 5308, pp. 1922-1925

Bockrath, M.; Cobden, D. H.; Lu, J.; Rinzler, A. G.; Smalley, R. E.; Balents, L. & McEuen, P. L. (1999). Luttinger-liquid behavior in carbon nanotubes. Nature, Vol. 397, No. 6720, pp. 598-601

Charlier, J.-C.; Blase, X. & Roche, S. (2007). Electronic and transport properties of nanotubes Rev. Mod. Phys., Vol. 79, No. 2, pp. 677-732

Ci, L. J.; Zhao, Z. G. & Bai, J. B. (2005). Direct growth of carbon nanotubes on the surface of ceramic fibers. Carbon, Vol. 43, No. 4, pp. 883-886

Dai, H. (2002). Carbon nanotubes : opportunities and chalenges. Surface Science, Vol. 500, No. 1-3, pp. 218-241

Davis, V. A.; Ericson, L. M.; Parra-Vasques, A. N. G.; Fan, H.; Wang, Y.; Prieto, V.; Longoria, J. A.; Ramesh, S.; Saini, R. K; Kittrell, C; Billups, W. E.; Adams, W. W.; Hauge, R. H; Smalley, R. E. & Pasquali, M. (2004). Phase Behavior and Rheology of SWNTs in Superacids. Macromolecules, Vol. 37, No. 1, pp. 154-160

Fischer, J. E.; Dai, H.; Thess, A.; Lee, R.; Hanjani, N. M.; Dehaas, D. L. & Smalley, R. E. (1997). Metallic resistivity in crystalline ropes of single-wall carbon nanotubes. Phys. Rev. B, Vol. 55, No. 8, pp. R4921-R4924

Frank, S.; Poncharal, P.; Wang, Z. L. & De Heer, W. (1998). Carbon Nanotube Quantum Resistor. Science, Vol. 280, No. 5370, pp. 1744-1746

Georgakilas, V.; Kordatos, K.; Prato, M.; Guldi, D.M.; Holzingeer, M. & A. Hirsch. (2002). Organic Functionalization of Carbon Nanotubes. J. Am. Chem. Soc., Vol. 124, No. 5, pp. 760-761.

Gommans, H. H.; J. Alldredge, W.; Tashiro, H.; Park, J.; Magnuson, J. & Rinzler, A. G. (2000). Fibers of aligned single-walled carbon nanotubes: Polarized Raman spectroscopy. J. Appl. Phys., Vol. 88, No. 5, pp. 2509-2514

Grannan, S. M.; Lange, A. E.; Haller, E. E. & Beeman, J. W. (1992). Non-Ohmic hopping conduction in doped germanium at $T < 1$ K. Phys. Rev. B, Vol. 45, No. 8, pp. 4516-4519.

Hirsch, A. (2002). Functionalization of Single-Walled Carbon Nanotubes. Angew. Chem. Int. Ed., Vol. 41, No. 11, pp. 1853-1859

Kaiser, A. B.; Dusberg, G. & Roth, S. (1998). Heterogeneous model for conduction in carbon nanotubes. Phys. Rev. B, Vol. 57, No. 3, pp. 1418-1421

Kaiser, A. B.; Park, Y. W.; Kim. G. T., Choi, E. S.; Dusberg, G. & Roth, S. (1999). Electronic transport in carbon nanotube ropes and mats. Synth. Met., Vol. 103, No. 1-3, pp. 2547-2550

Kaiser, A. B. & Park, Y. W. (2005). Current-voltage characteristics of conducting polymers and carbon nanotubes. Synth. Met., Vol. 152, No. 1-3, pp. 181-184

Kim, G. T.; Choi, E. S.; Kim, D. C.; Suh, D. S..; Park, Y. W.; Liu, K.; Dusberg, G. & Roth, S. (1998). Magnetoresistance of an entangled single-wall carbon-nanotube networks. Phys. Rev. B, Vol. 58, No. 24, pp. 16064-16069

Kim, G. T.; Jhang, S. H.; Park, J. G.; Park, Y. W. & Roth, S. (2001). Non-ohmic current-voltage characteristics in single-wall carbon nanotube networks. Synth. Met., Vol. 117, No. 1-3, pp. 123-126

Kozlov, M. E.; Capps, R. C.;W. Sampson, M.; Ebron, V. H.; Ferraris, J. P. & Baughman, R. H. (2005). Spinning Solid and Hollow Polymer-Free Carbon Nanotubes Fibers. Adv. Mater., Vol. 17, No. 5, pp. 614-617

Krstic, V.; Duesberg, G.S.; Muster, J.; Burghard, M. & Roth, S. (1998). Langmuir-Blodgett Films of Matrix-Diluted Single-Walled Carbon Nanotubes. Chem. Mater., Vol. 10, No. 9, pp. 2338-2340

Krstic, V.; Roth, S. & Burghard, M. (2000). Phase breaking in three-terminal contacted singlewall carbon nanotube bundles. Phys. Rev. B, Vol. 62, No. 24, pp. R16353-R1635

Ksenevich, V. K.; Galibert, J.; Forro, L. & Samuilov, V. A. (2006). Magnetotransport in 2Darrays of single-wall carbon nanotubes, In: Carbon Nanotubes: From Basic Research to Nanotechnology, NATO ASI Series II: Mathematics, Physics and Chemistry, Vol.222, Popov, V. N. & Lambin, P. (Eds.), pp. 183-184, Springer, ISBN 978-1-4020-4572-1, Dordrecth, The Netherlands

Ksenevich, V. K.; Aboltin, R. S.; Dauzhenka, T. A., Galibert, J. & Samuilov, V. A. (2008, a). Charge transport mechanisms in the arrays of multi-walled carbon nanotubes, Proceedings of the III International conference Materials and structures of modern electronics, pp. 271-273, ISBN 978-985-518-091-4, Minsk, Belarus, September, 2008, BSU publishers, Minsk

Ksenevich, V.K.; Odzaev, V.B.; Martunas, Z.; Seliuta, D.; Valusis, G.; Galibert, J.; Melnikov, A.A.; Wieck, A.D.; Novitski, D.; Kozlov, M.E. & Samuilov, V.A. (2008, b). Localization and nonlinear transport in single walled carbon nanotube fibers. J. Appl. Phys., Vol. 104, No. 7, pp. 073724-1-073724-7

Ksenevich, V.K.; Seliuta, D.; Martunas, Z.; Kasalynas, I.; Valusis, G.; Galibert, J.; Kozlov, M.E. & Samuilov, V.A. (2008, c). Charge Carrier Transport Properties in SingleWalled Carbon Nanotube Fibers. Acta Phys. Pol. A,, Vol. 113, No. 3, pp. 1043-1046

Ksenevich, V.; Dauzhenka, T.; Seliuta, D.; Kasalynas, I.; Kivaras, T.; Valusis, G.; Galibert, J.; Helburn, R.; Lu, Q. & Samuilov, V.A. (2009). Electrical transport in carbon nanotubes coatings of silica fibers. Physica Status Solidi (c), Vol. 1-3, DOI 10.1002/pssc. 200982553 (in press).

Kurobe, A. & Kamimura, H. (1982). Correlations Effects on Variable Range Hopping Conduction and the Magnetoresistance. J. Phys. Soc. Jap., Vol. 51, pp. 1904-1913

Lassagne, B.; Cleuziou, J.-P.; Nanot, S.; Escoffer, W.; Avriller, R.; Rosche, S.; Forro, L.; Raquet, B. & Broto, J.-M. (2007). Aharonov-Bohm Conductance Modulation in Ballistic Carbon Nanotubes. Phys. Rev. Lett., Vol. 98, No. 17, p.p. 176802-1-176802-4

Lee, P. & Ramakrishnan, T.V. (1985). Disordered electronic systems. Rev.Mod.Phys., Vol. 57, No. 2, pp. 287-337.

Liu, H.; Li, J.; Liu, X. & Jiang, S. (2009). A novel multiwalled carbon nanotubes bonded fused-silica fiber for solid phase microextraction-gas chromatographic analysis of phenols in water samples. Talanta, Vol. 78, No. 3, pp. 929-935

Nanot, S.; Escoffier, W.; Lassagne, B.; Broto, J.-M. & Raquet, B. (2009). Exploring the electronic band structure of individual carbon nanotubes under 60 T . C. R. Physique, Vol. 10, No. 4, pp. 268-282

Nguen, V. L., Spivak, B. Z. & Shklovskii, B. I. (1985). Tunnel hopping in a disordered systems. Sov. Phys. JEPT, Vol. 62, No. 5 , pp. 1021-1029

Qian, H.; Bismarck, A.; Greenhalgh, E. S. & Shaffer, M. S. P. (2010). Synthesis and characterisation of carbon nanotubes grown on silica fibres by injection CVD. Carbon, Vol. 48, No. 1, pp. 277-286

Pollak, M. & Riess, I. (1976). A percolation treatment of high-field hopping transport. J. Phys. C, Vol. 9, No. 12, pp. 2339-2352

Pozela, J. K. (1985). Transport parameters from microwave conductivity and noise measurements, In: Topics of Applied Physics, L. Reggiani (Ed.), Vol. 58, pp. 113-147, Springer-Verlag, Berlin

Robertson, J. (2004). Realistic applications of CNTs. Materials Today, Vol. 7, No. 10, pp. 46-52

Robertson, J. (2007). Growth of nanotubes for electronics. Materials Today, Vol. 10, No. 1-2, pp. 36-43

Schonenberger, C.; Bachtold, A.; Strunk, C.; Salvetat, J.-P. & Forro, L. (1999). Interference and Interaction in multi-wall carbon nanotubes. Appl. Phys. A, Vol. 69, No. 3, pp. 283-295

Shaffer, M. & Sandler, J. (2007). Carbon nanotube/nanofibre polymer composites, In: Processing and properties of nanocomposites, Advani, S. G., (Ed.), pp. 1-59, World Scientific, ISBN: 978-981-270-390-3, New Jersey, NJ

Sheng, P. (1980). Fluctuation-induced tunneling conduction in disordered materials. Phys. Rev. B, Vol. 21, No. 6, pp. 2180-2195

Shklovskii, B. I. & Efros, A. L. (1984). Electronic properties of doped semiconductors, Springer, ISBN-10: 0387129952, Berlin, New York

Sivan, U.; Entin-Wohlman, O. & Imry, Y. (1988). Orbital Magnetoconductance in the Variable-Range-Hopping Regime. Phys. Rev. Lett., Vol. 60, No. 15, pp. 1566-1569

Smith, B. W.; Benes, Z.; Luzzi, D. E.; Fischer, J. E. ; Walters, D. A.; Casavant, M. J.; Schmidt, J. & Smalley, R. E. (2000). Structural anisotropy of magnetically aligned single wall carbon nanotube films. Appl. Phys. Lett., Vol. 77, No. 5, pp. 663-665

Stetter, J. R. & Maclay, G. J. (2004). Carbon Nanotubes and Sensors: a Review. In: Advanced Micro and Nanosystems, Vol. 1, Baltes, H.; Brand, O.; Fedder, G.K.; Hierold, C.; Korvink, J.G. & Tabata, O. (Eds.), pp. 357-382, Wiley-VCH Verlag GmbH & Co, ISBN-13: 978-3-527-30746-3, Weinheim, Germany

Strunk, C.; Stojetz, B. & Roche, S. (2006). Quantum interference in multiwall carbon nanotubes. Semicond. Sci. Technol, Vol. 21, No. 11, pp. S38-S45

Thostenson, E. T.; Li, W. Z.; Wang, D. L.: Ren, Z. F. & Chou, T. W. (2002). Carbon nanotube/carbon fiber hybrid multiscale composites. J. Appl. Phys., Vol. 91, No. 9, pp. 6034-6037

Urbina, A.; Echeverria, I.; Perez-Garrido, A.; Diaz-Sanchez, A. & Abellan, J. (2003). Phys. Rev. Lett., Vol. 90, No. 10, pp. 106603-1-106603-4

Vigolo, B.; Penicaud, A.; Coulon, C.; Sauder, C.; Pailler, R.; Journet, C.; Bernier, P. & Poulin, P. (2000). Macroscopic Fibers and Ribbons of Oriented Carbon Nanotubes. Science, Vol. 290, No. 5495, pp. 1331-1334

Vigolo, B.; Poulin, P.; Lucas, M.; Launois, P. & Bernier, P. (2002). Improved structure and properties of single-wall carbon nanotube spun fibers. Appl. Phys. Lett., Vol. 81, No. 7, pp. 1210-1212

Zabrodskii, A. G. (1977). Hopping conduction and density of localized states near the Fermi level. Soviet Phys. Semicond., Vol. 11, pp. 345-348

Zhang, Q.; Qian, W.; Xiang, R.; Yang, Z.; Luo, G.; Wang Y, & Wei, F. (2008). In situ growth of carbon nanotubes on inorganic fibers with different surface properties. Mater. Chem. Phys., Vol. 107, No. 2-3, pp. 317-321

Zhou, W.; Vavro, J.; Guthy, C.; Winey, K. I.; Fischer, J. E.; Ericson, L. M.; Ramesh, S.; Saini, R.; Davis V. A.; Kittrel, C.; Pasqualli, M.; Hauge, R. H. & Smalley, R. E. (2004). Single wall carbon nanotube fibers extruded from super-acid suspensions: Preferred orientation, electrical, and thermal transport. J. Appl. Phys., Vol. 95, No. 2, pp. 649-655

Chapter-8

Doped Carbon Nanotube Properties

1. Introduction

The last twelve years have seen enormous growth in many aspects of doped carbon nanotubes. The idea of doping carbon nanotubes has come from microelectronics and now develops both in a traditional donor-acceptor direction (Miyamoto et al., 1995; Lee et al., 1997; Rao et al., 1997; Grigorian et al., 1998; Esfarjani et al., 1999) and in a number of new aspects of application. Among recent trends of application of a doping of nanotubes it is necessary to term such as forming for them magnetic properties (Esfarjani et al., 2003; Farajian et al., 2002), change of their optical performances (Bondarev, 2007), embodying of hydrogen storages (Zidan et al., 2003), amplification of catalytic properties (Gong et al., 2009) and improving of emission characteristics (Yao et al., 2002).

In the present chapter we discuss influence of impurities on properties of semiconductor carbon nanotubes with chiral vectors (10,0) and (6,5) which are representative of groups of nanotubes of type "zigzag" and the twirled chirality, accordingly. At doping nanotubes usually view an intercalation of impurity inside of a nanotube or in space between separate nanotubes. We view doping nanotubes in traditional sense when impurity ranks one of atoms of a nanotube. It allows to hope, that impurity after such localization will not be lightly to migrate in a nanotube or between separate nanotubes. At such approach to a problem of doping of a periodic nanostructure presence in it of vacancies is supposed. Energy of forming of vacancy in a nanotube makes 5 ... 7eV (Griadun, 2006). It is a little bit more in comparison with energy of forming of vacancy in silicon, but nevertheless leaves probability of substantial existence of vacancies in nanotubes.

Profound knowledges are necessary for an effective utilization carbon nanotubes in nanoelectronic about their local electro-physical performances. In particular extensive data on concentration and parameters of the impurity centers by means of which it is possible to shape fields with the given type and the quantity of an electrical conductivity reproduced by quantities of a lifetime and a diffusion length of non-equilibrium charge carriers are necessary.

Interest to these examinations is called by necessity of development of designs of integrated transistors and logical units for build-up of complete sets of integrated circuits with use semiconductor carbon nanotubes (Martel et al., 1998; Derycke et al., XXXX). The solution of a problem guesses an opportunity to drive equilibrium concentrations of electrons and holes in semiconductor carbon nanotubes by their doping.

2. Doped carbon nanotube with indices (10, 0)

2.1 Method of modeling and base model of the (10, 0) carbon nanotube

Models developed within the limits of a method of molecular mechanics model operation MM + (Berkert & Allinger, 1982) with next parameters:

- a quadratic and a cubic stretch term in the potential; atom types;
- nonbonded electrostatic interactions are calculated using bond dipole, which values come from the MM+ stretch parameters;
- the electrostatic contribution comes from defining a set of bond dipole moments associated with polar bonds;
- cutoffs determine the distance limits for calculating nonbonded interactions of the periodic nanostructure;
- inner cutoff radius is the maximum interatomic distance for full nonbonded interactions - 10Å;
- outer radius is the minimum distance at which nonbonded interactions are set to zero 14 Å.

The base model of the nanotube consisting from two hundred of atoms of carbon, is presented on Fig. 1. In models aromatic bonds, providing interacting of valence electrons of carbon atoms are used.

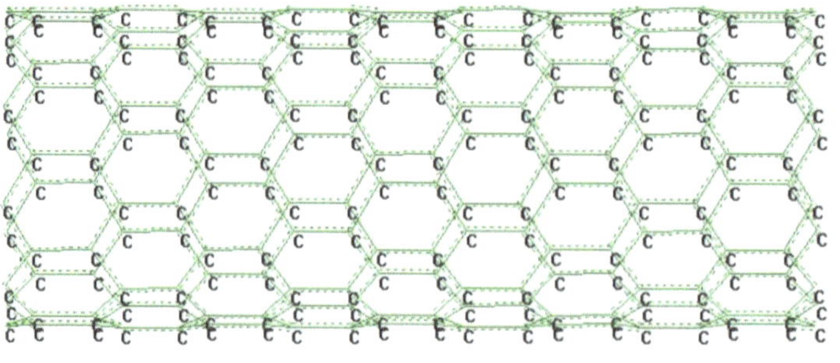

Fig. 1. Basic model of the carbon nanotube with indices (10, 0).

The energy-band structure of a nanotube was calculated by the Extended Hückel method (Hoffmann, 1963). The total energy of 200 -atomic model of a nanotube has made quantity -323079.99kcal/mol. On the diagram of its energy structure of a forbidden gap (Fig. 2) is observed a series of levels with energies:

-10.60226; -10.37579; -10.37578; -10.36441; -10.36440; -10.30684; -10.27539; -10.26634; -10.26634; -10.24621; -10.24621; -10.19875; -10.19875; -9.6362; -9.6362eV - matching orbitals with numbers $401-415$, accordingly.

All from the termed orbitals to a greater or lesser extent match to edges of a nanotube, for example orbital 413 (Fig. 3), and higher orbitals match to a volume part of a nanotube.

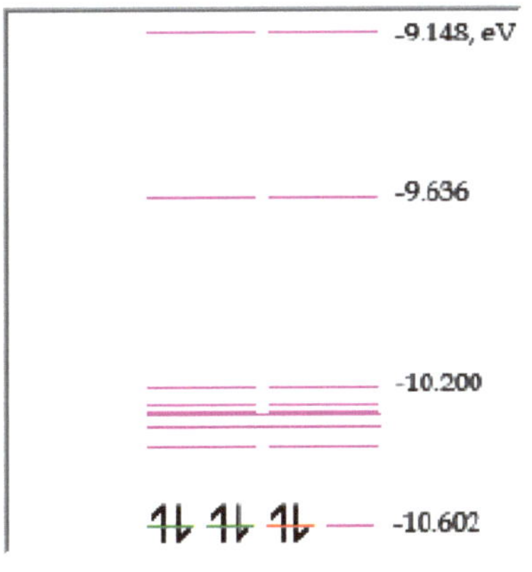

Fig. 2. Structure of a forbidden gap of the nanotube with indices (10,0).

Thus, in a forbidden gap of a nanotube is 30 electronic states matching 15 orbitals. A part from viewed states match to the dangling bonds on edges of a nanotube. For acknowledging this we shall simulate passivation of dangling bonds by atoms of hydrogen.

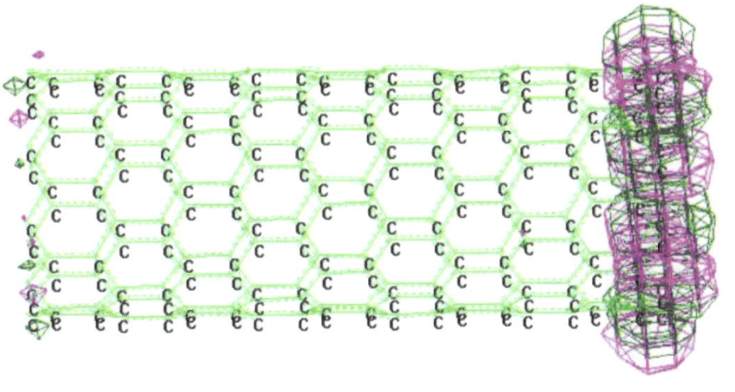

Fig. 3. Orbital 413 of the carbon nanotube with indices $(10, 0)$.

On Fig. 4 the model of a 200-atomic nanotube after hydrogen endings and geometrical optimization of nanostructure within the limits of a method of molecular mechanic modeling MM + is figured. Apparently on the model, atoms of hydrogen have filled 20 dangling carbon bonds - on 10 on each edges of a nanotube. Thus, the remained three orbitals feature six states which parent of origination is unknown. The band of energy diagram in the field of a forbidden gap (Fig. 5) after passivation by hydrogen of the broken off bonds of a nanotube on its edges essentially differs from the diagram of the starting model figured on Fig. 2:

1. The forbidden gap has essentially cleared from energy levels - from 15 remains only 5, that is 10 orbitals that matches to 20 electronic states, were moved in a valence band;
2. Three from five remained energy levels in a forbidden gap should be the acceptor as they are close to a valence band with energies -10.44002 eV (an orbital 411), -10.41608 eV (an orbital 412), and -10.41608 eV (an orbital 413);
3. Passivation by atoms of hydrogen has affected quantity of energy of a ceiling of a valence band (an orbital 410) which has made - 10.44002 eV (before passivation was 10.60226 eV);
4. The magnification of energy in item 3 is caused by electronic states of atoms of hydrogen;
5. Passivation by atoms of hydrogen has affected quantity of energy of a bottom of conduction band (an orbital 416) which has made -9.156353 eV (before passivation was 9.147926 eV);
6. The breadth of a forbidden gap of viewed model has made quantity 1.2837 eV (before passivation was 1.4543 eV);
7. The overestimated value of an energy gap in this case is caused by a trace amount of atoms and, accordingly, small length of a nanotube 19.6 Å.

Fig. 4. Hydrogen endings of the carbon nanotube with indices (10,0).

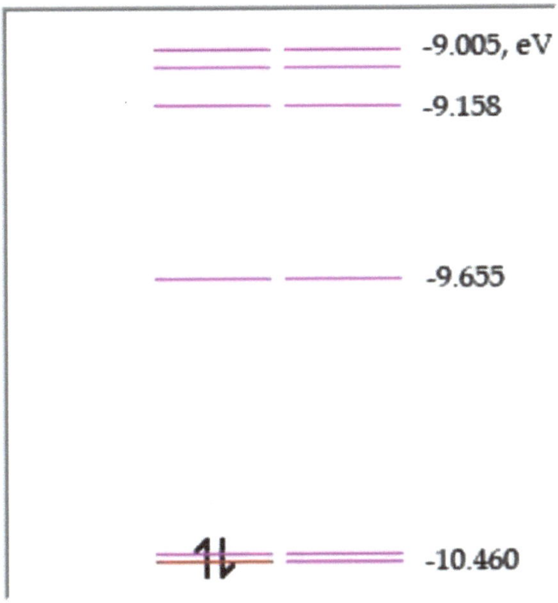

Fig. 5. Structure of a forbidden gap of the (10, 0) nanotube after hydrogen endings.

Analogous calculations are lead also for the nanotubes consisting from three hundred and four hundred of atoms of carbon in length ~ 30.15 and 40.63Å, accordingly. Hydrogen endings of this models also well refines forbidden regions of the energy levels matching broken off bonds. The breadth of a forbidden gap has naturally decreased up to values 1.06 and 0.90 eV , accordingly for 300-atomic and 400-atomic models of nanotubes.

2.2 Doping carbon (10, 0) nanotube by atom of boron

Simulation of doping of the nanotube by atom of boron we shall carry out within the limits of a method of MM +, exchanging in the program of basic model the chosen atom of carbon on atom of boron (Griadun, 2007). Thus we can use aromatic bonds of atom of boron with atoms of carbon or exchange them on single bonds, as boron trivalent. We shall consider also at model operation spin multiplicity of a nanostructure.

The geometry of a 200 -atomic nanotube doped by atom of boron, is presented on Fig. 6. We see, that at modeling of impurity of boron by aromatic bonds with atoms of carbon the nanostructure distorts feebly. In case of single bonds B-C the nanostructure noticeably distorts in the field of localization of atom of boron. It is related by that single bonds B-C are longer than aromatic on 0.08Å.

Extended-Hückel energy of the nanostructure with aromatic bonds of atom of boron is less than energy of the nanostructure with single bonds on 2.22kcal/mol, therefore it is necessary to expect, that in real experiment embodying of the nanostructure with aromatic bonds of atom of boron is more probable.

The energy band diagram of the viewed nanotube doped by atom of boron (Fig. 7a), differs from an energy-band structure of basic model (Fig. 2), however, only quantities of values of energy levels. The expected new acceptor level localized on atom of boron, we do not observe. As well as in basic model three acceptor energy levels, but with other values of energy are observed: -10.60 ; -10.38 and -10.37 eV . It, visible, is related by that bonds B-C the 1S-electron of atom of boron shares. Except for that removal of degeneration of some orbitals is observed, for example orbitals 414 and 415 with energy -9.636 eV borrow levels with energies -9.641 and -9.611 eV .

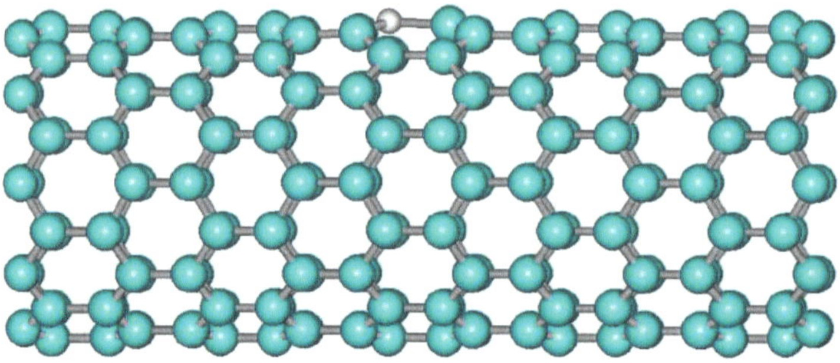

Fig. 6. Geometry optimised model of the carbon B-doped (10,0) nanotube with aromatic bonds B-C.

Hydrogen endings of a nanotube doped by atom of boron practically does not influence geometry of a nanostructure near to atom of boron, however noticeably influences its energy distribution (Fig. 7b). Really, upper occupied level HOMO-0, featured 410-th orbital, has energy -10,46021 eV. Following three orbitals with numbers 411 – 413 are at levels -10.45723, -10.43564 , and -10.42292 eV , accordingly. As a whole, such effect is normal as levels of energy of electrons of hydrogen atoms have placed in a valence band and biased its ceiling on 0.15 eV upwards.

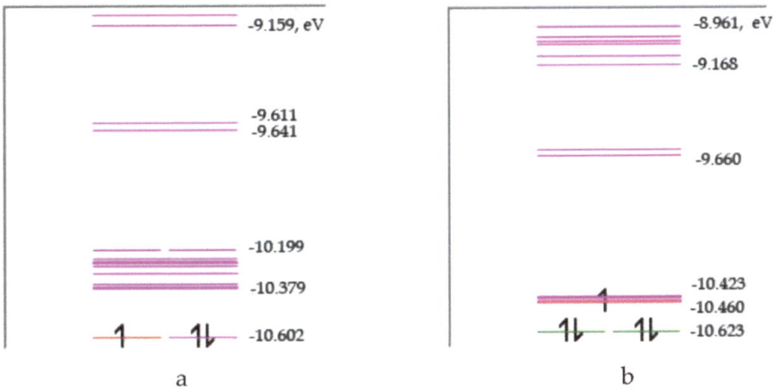

Fig. 7. Energy band structure of the B-doped (10,0) carbon nanotube before (a) and after (b) hydrogen endings.

2.3 Doping of a carbon (10, 0) nanotube by atom of nitrogen

The atom of nitrogen also is well inscribed in geometry of a nanotube, not calling essential strain of its nanostructure (Fig. 8). It is caused by that lengths of bonds N-C and C-C differ a little - 1.28Å and 1.40Å, accordingly.

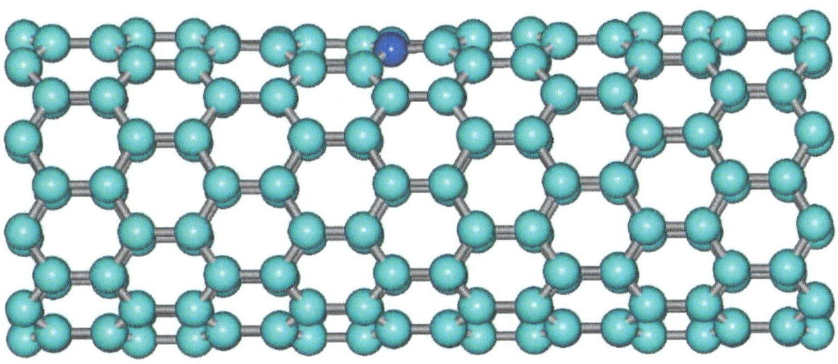

Fig. 8. Model of the nanotube with indices (10, 0) doped by atom of nitrogen.

The energy band distribution of a nanotube without hydrogen endings, doped by atom of nitrogen, is presented on Fig. 9a. Under effect of impurity atom in a spectrum of energy levels of a nanostructure there was an orbital filled by an electron the number 401. This orbital matches to the fifth valence electron of atom of nitrogen and is presented by an energy level equal to quantity -10.6018 eV . The ceiling of a valence band of a model nanostructure is featured by an orbital the number 400 with energy -10.6023 eV . Thus, the extrinsic energy level of atom of nitrogen will not display the donor properties as is near to a valence band, it only will neutralize the inferior acceptor level of a basic nanotube. In case of nitrogen doping a carbon nanotube with hydrogen endings, the fifth valence electron of atom of nitrogen also occupies the inferior acceptor level near to a valence band (Fig. 9b). Quantity of energy of a level which value makes in this case -10.4579 eV varies only.

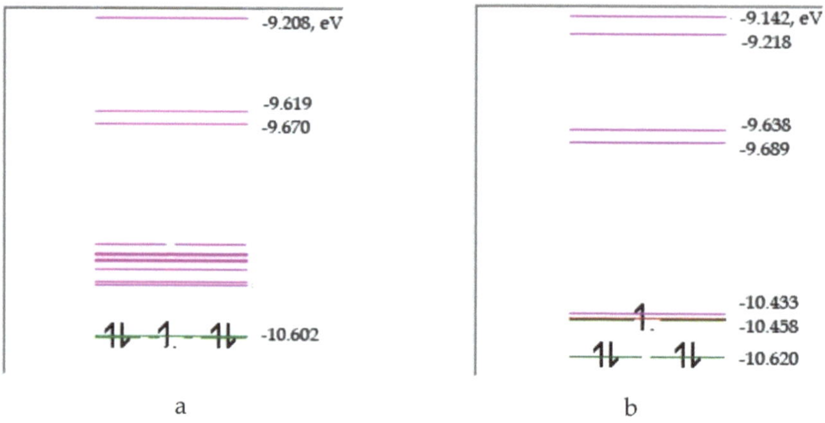

Fig. 9. Effect of impurity atom of nitrogen on energy spectrum of the carbon (10, 0) nanotube before (a) and after (b) hydrogen ending.

Thus, doping by atom of nitrogen of a carbon nanotube with indices (10,0) does not result in to formation of a donor centre, and only results in to blocking of one of acceptor levels of a nanostructure.

2.4 Doping of a carbon (10, 0) nanotube by atom of aluminium

Atom of aluminium as the device of the third group, well approaches for doping carbon nanotubes. As if to its influence on geometrical and other properties of a carbon nanotube with coefficients (10,0) we shall view on model of a nanostructure in which atom №133 of the basic model of nanotube we shall replace with atom of aluminium. After geometrical optimization of a nanostructure in which aromatic bonds Al-C use, the model gets a view figured on Fig. 10.

We see, that the atom of aluminium is not inscribed in wall of a nanotube, but located in immediate proximity from it apart 1Å. Besides atoms of carbon with which bound atom of aluminium, are noticeably drifted for limits of a nanotube. In case of use in model of a nanostructure of single bonds Al-C results in to even greater infringements of geometry of a nanotube. The atom of aluminium settles down thus apart 1.2Å from geometrical wall of a nanotube. The Extended Hückel calculation of energy distribution in area of a forbidden gap of the Al-doped nanotube is presented on Fig. 11.

Fig. 10. Optimised geometry of the Al-doped carbon (10, 0) nanotube.

It is visible, that the type of bonds of atom of aluminium with atoms of carbon very feebly influences quantity of energy levels. In comparison with a spectrum of basic model of the carbon (10,0) nanotube (Fig. 2) has occurred an acceptor level with energy -10.532 eV . Also there were splits of levels -9.636 and -9.148 eV . The parent of this split is spin multiplicity of the nanostructure doped by aluminium which quantity is equal to two.

It is necessary to note, that the nanostructure with aromatic bonds Al-C should be realized with greater probability as the nanostructure with single bonds Al – C is more on 10 kcal/mol.

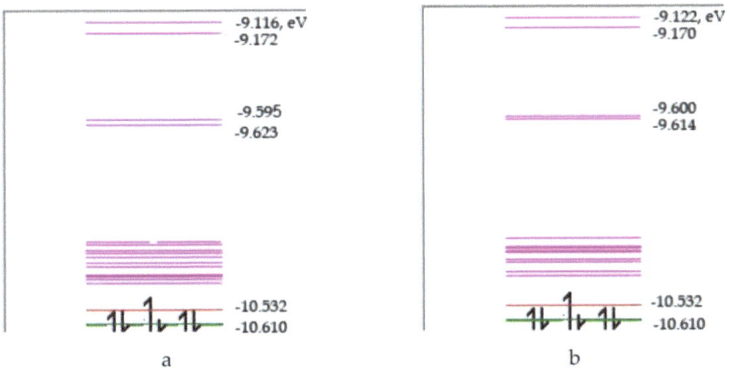

Fig. 11. Energy band structure of a forbidden gap region of the Al-doped carbon (10,0) nanotube: (a) - aromatic bond Al-C; (b) - unary bonds Al-C.

Passivation of the Al-doped carbon (10,0) nanotube by hydrogen endings (Fig. 12a) renders appreciable influence on its energy distribution in the field of a forbidden gap (Fig. 12b). We see, that after passivation of edges of a nanotube by hydrogen in a forbidden region of a nanostructure some level by quantity -10.289 eV which visible is related to the defect created by impurity atom of aluminium was displayed.

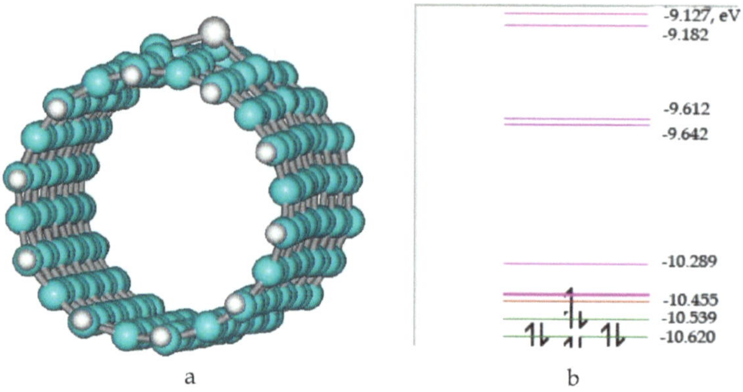

Fig. 12. Al-doped carbon $(10, 0)$ nanotube with hydrogen endings: (a) - geometry of nanostructure; (b) - energy band structure of a forbidden gap.

2.5 Doping of a carbon $(10, 0)$ nanotube by atom of phosphorus

Doping of the carbon (10,0) nanotube by atom of phosphorus results in appreciable strain of geometry of a nanostructure (Fig. 13). It is related by that bonds C-P, having quantities 1.80, 1.80 and 1.79Å, longer in comparison to typical length of bonds C-C, equal 1.41Å. Besides the atom of phosphorus has on 0.3 E greater atomic radius than at atom of carbon that results in to its replacement for limits of a nanotube.

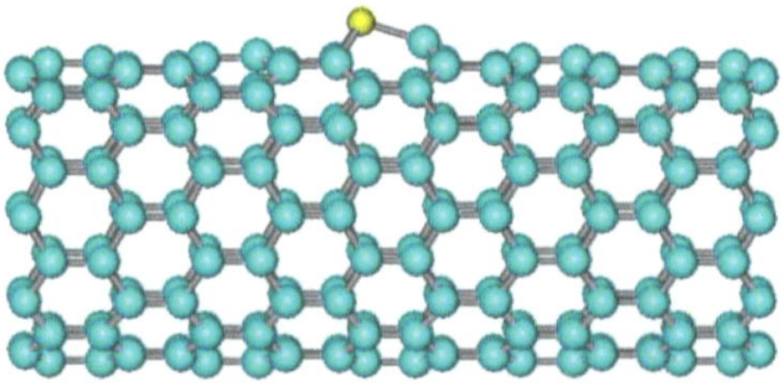

Fig. 13. Molecular mechanics model of the 200-atomic carbon (10.0) nanotube doped by atom of phosphorus.

It is necessary to note, that the atom of phosphorus shapes in a viewed nanostructure an essential lapse rate of mechanical stresses which value equally 0.05kcal/(Åmol). In a basic nanotube geometrical optimization is attained at a root-mean-square lapse rate equal 0.000001kcal/(Åmol).

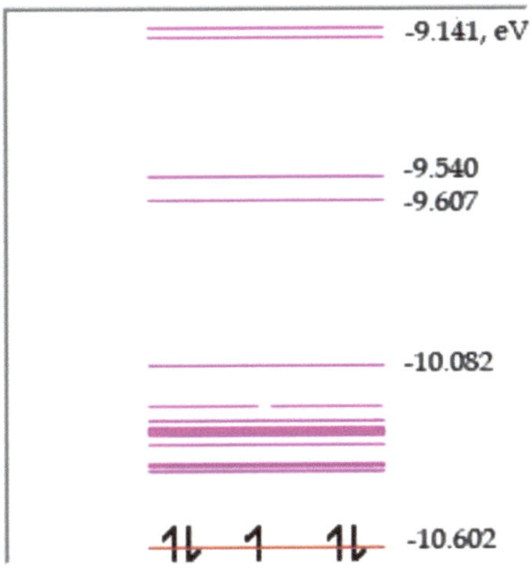

Fig. 14. Energy levels in a forbidden gap of the carbon(10,0) nanotube doped by atom of phosphorus.

The energy diagram of a viewed nanostructure is presented on Fig. 14. We see, that in a forbidden gap of the doped nanotube there was a level (-10.082 eV), which probably matches to the flaw created by atom of phosphorus.

The model of a phosphorus-doped (10,0) nanotube with hydrogen passivation of dangling bonds and its energy band structure are presented on Fig. 15. We see, that the geometry of a viewed nanostructure in the field of localization of atom of phosphorus after hydrogen passivation of dangling bonds on its edges has not changed.

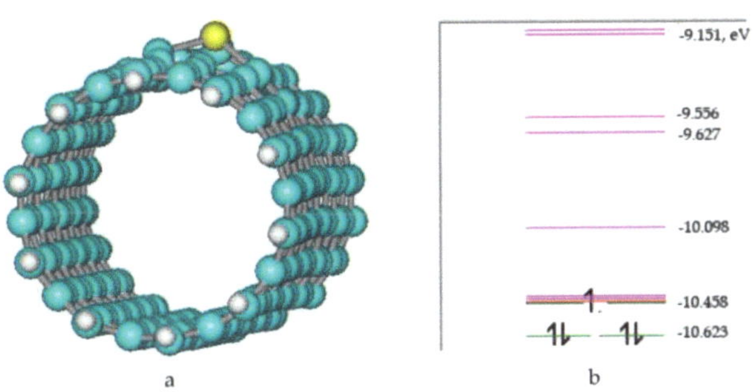

Fig. 15. P-doped carbon (10, 0) nanotube with hydrogen endings: (a) - geometry of the molecular mechanics model; (b) - Extended Hückel calculation of energies of orbitals.

In an energy distribution of a forbidden gap there were changes (Fig. 15b): the upper filled level has risen up to quantity - 10.46 эВ;

degeneration of some energy levels is removed;

quantities of energy levels have changed.

It is necessary to note, that the new level with energy -10.098 eV under effect of hydrogen endings has not changed the quantity.

The magnification of number of atoms of carbon in model of a phosphorus-doped nanotube up to 300 pieces results in to diminution of breadth of a forbidden gap and small bias of a new level up to quantity -10.11 eV .

3. Properties of doped carbon nanotube with indices (6, 5)

3.1 Basic model of a carbon nanotube with indices (6, 5)

There are many methods of model operation of those or other processes and structures. For nanotechnologies the method of the molecular mechanic model operation which basic element is the molecule well approaches. The separate atom too is considered as a molecule. In case of a nanotube the molecule is the nanotube. Atoms of a nanotube are foliated, featured by coordinates and type and also quantity of bonds.

Our model of nanotube consists from two hundred atoms of carbon, which geometrical optimization was led by method MM + with application the block-diagonal of the NewtonRaphson algorithm (Fig 16). The length of a nanotube is equated 22Å, diameter -7.3Å. The distance between atoms is equated 1.41Å. Chirality of the nanotube which it is well visible on figure, defined by its indices (6,5).

At model operation the applied aromatic bonds which well justify itself when atoms of carbon create rings as in this case. There is one more expedient of the plotting of bonds which total for atom of carbon should equate to four.

To calculation of energy properties of a nanotube we applied Extented Hückel Method (Hoffmann, 1963) without taking into account d-orbitals, with unit spin multiplisity and unweighted Hückel constant equal 1.75. Energy has made quantity equal - 323263.1900 kcal/mol.

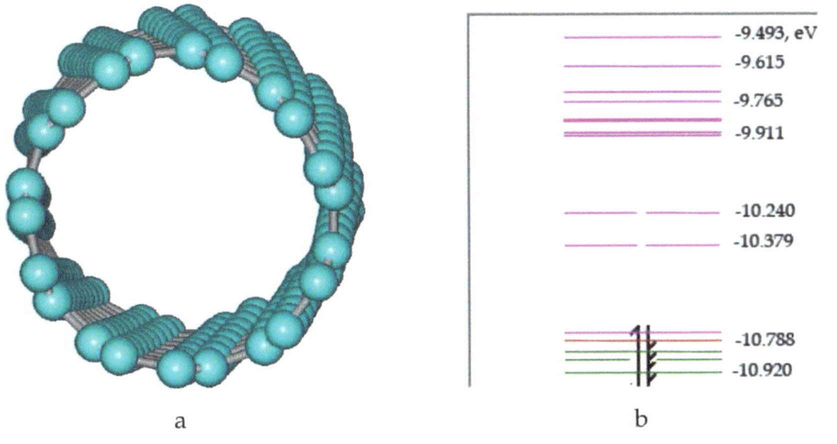

Fig. 16. 200-atoms basic model (a) and energy-band structure (b) of a carbon nanotube with indices (6, 5).

The energy-band structure of a basic nanotube (6,5) is presented on Fig. 16b. The upper filled level (HOMO 0), which is featured by orbital number 400, makes quantity - 10.788 eV

Energy of bottom of a conduction band is equated -9.91 eV . In fact the forbidden region has breadth equal 0.88 eV .

In a forbidden gap there is a series of levels which are caused by dangling bonds on edges of the nanotube. So, it is possible to apply atoms of hydrogen or nitrogen to passivation of the broken off bonds. On Fig. 17 passivation by atoms of hydrogen which are optimum in such cases is shown.

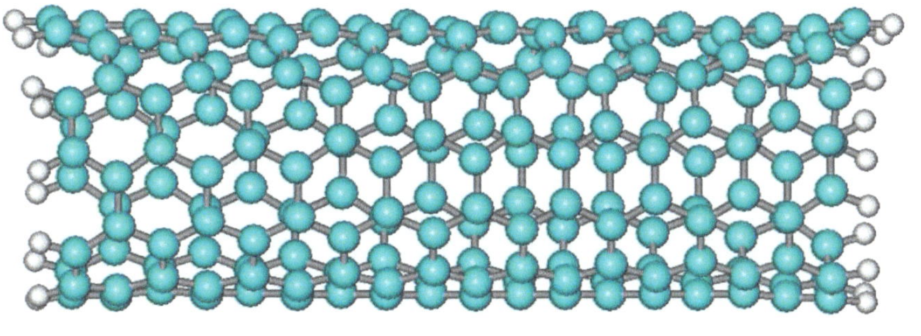

Fig. 17. Atoms of hydrogen fill dangling bonds on edges of the carbon (6, 5) nanotube.

The energy of the nanotube with passivated ends has decreased to value -332889.25 kcal/mol that is caused by both H -atoms and a relaxation of a nanotube on edges where bonds have been broken off. The energy-band structure presented on Fig. 18a, considerably has changed in this case. The forbidden gap became clear from energy levels of the dangling bonds. Apparently from this figure, that the forbidden gap quantity is defined by a difference

$$\Delta E = E_{\text{orb }412} - E_{\text{orb }411} = 1.02\text{eV}.$$

The greater width of the gap is caused by a trace amount of carbon atoms.

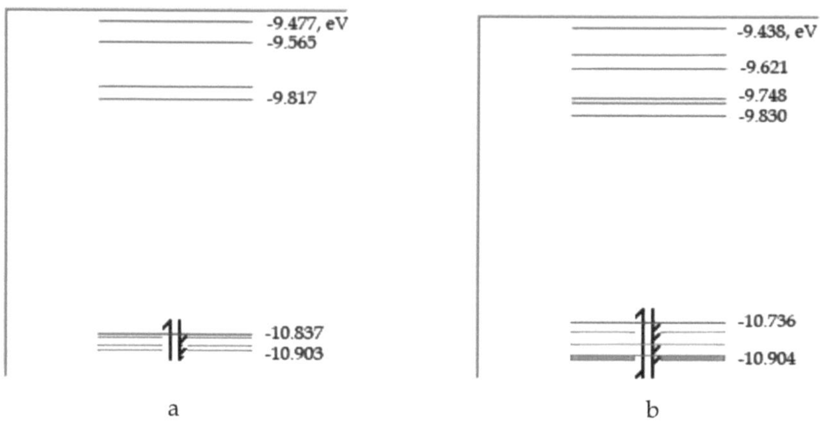

Fig. 18. The energy-band structure of the base nanotube with indices (6,5) after passivation the dangling bonds by atoms of hydrogen: (a) - 200-atomic model; (b) - 400-atomic model.

The magnification of length of a nanotube twice results in to a raising of a ceiling of valence band on 0.1 eV , and the breadth of a forbidden region thus accordingly on as much decreases (Fig. 18b).

3.2. Doping carbon nanotube (6, 5) by atom of boron

Doping of a carbon nanotube with indices (6,5) we shall carry out by replacement in the program of model operation of a nanostructure of the arbitrary atom of carbon by atom of boron. We exchange also aromatic bonds on single bonds, as boron trivalent. The nanotube (Fig. 19a) distorts in the field of localization of the alloyed atom of boron due to longer B-C bonds, equaled 1.65Å, that on 0.24Å it is more than for C-C bonds. Though the nuclear sizes of boron (1.17Å) and carbon (0.91Å) different enough, but covalent radiuses at them close enough, 0.82 and 0.79Å, therefore the geometry of a tube is broken poorly. The energy-band structure of the B-doped nanotube (Fig. 19b) has a little changed in comparison with a spectrum of basic model (Fig. 16b): degeneration of levels -10.240 and -10.379 eV is removed; numerical values of other energy levels have a little changed.

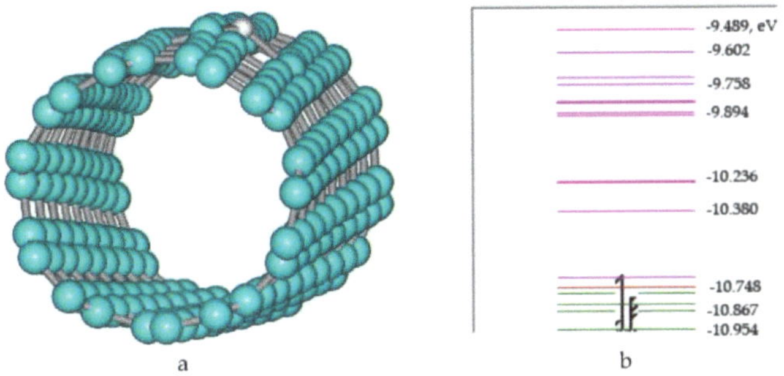

Fig. 19. Geometry of the boron-doped carbon (6.5) nanotube (a) and its energy-band structure (b).

To passivation of dangling bonds on edges of a nanotube we shall apply hydrogen endings (Fig. 20a).

Apparently from Extended-Hückel calculations of the energy band structure diagram (Fig. 20b), that under effect of atom of boron and hydrogen passivation of dangling bonds the structure of the forbidden gap and its width which has become equal 0.898 eV (it was equal 1.03 eV) has changed. The local orbital the number 411 is filled only by one electron, as amount of valence electrons equally to eight hundred twenty one. It is acceptor with energy -10.71eV(0,08eV above valence band).

Thus atoms of carbon with numbers 157, 158 charged by the negative low fidelity equal charges on quantity 0.41 e (e - electron charge), and the atom of carbon number 161 has negative charge 0.37 e . The atom of boron has positive charge equal -0.66 e .

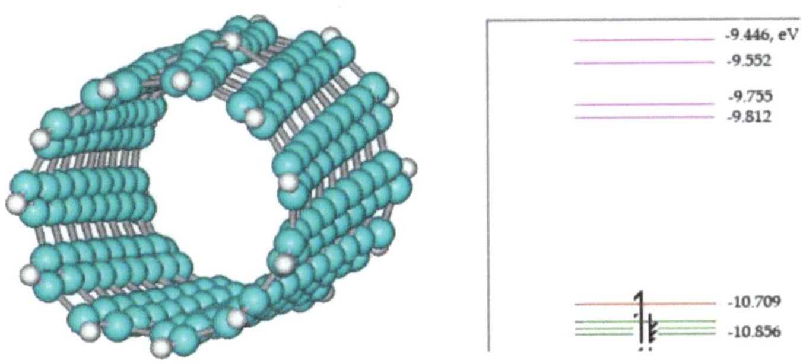

Fig. 20. The B-doped carbon (6,5) nanotube with hydrogen endings: (a) - geometry of the nanostructure; (b) - energy band structure.

Thus around of atom of boron in a carbon nanotube complicated enough electronic pattern which corresponds to an orbital 411 and as which it is possible to term as acceptor local centre is created. It is necessary to note, that acceptor properties of this orbital can be realized at a heat (more than 900 K), as an energy level deep enough.

3.3 Doping carbon (6, 5) nanotube by atom of nitrogen

For forming model of the N-doped carbon nanotube, we shall replace 163-rd atom of carbon with atom of nitrogen. The nanostructure does not distort almost under influence of the alloyed atom of nitrogen (Fig. 21) as lengths of bonds C-C and N-C differ feebly - 1.40Å and 1.28Å, accordingly.

Covalent and nuclear radiuses of atom of nitrogen are equaled 0.75Å, i.e. smaller radiuse of atom of carbon which equal 0.77Å and 0.91Å, accordingly. In this connection, intensity around of atom of nitrogen is not minimal - the RMS gradient of mechanical stress is equal 0.0045kcal/(Åmol).

Total energy of the nanotube doped by nitrogen is calculated by Extended Hückel method and has made quantity −323849.25kcal/mol. It is less than for the base nanotube which energy is equaled −323263.19kcal/mol that speaks in the smaller sizes of radiuses and bonds of atom of nitrogen.

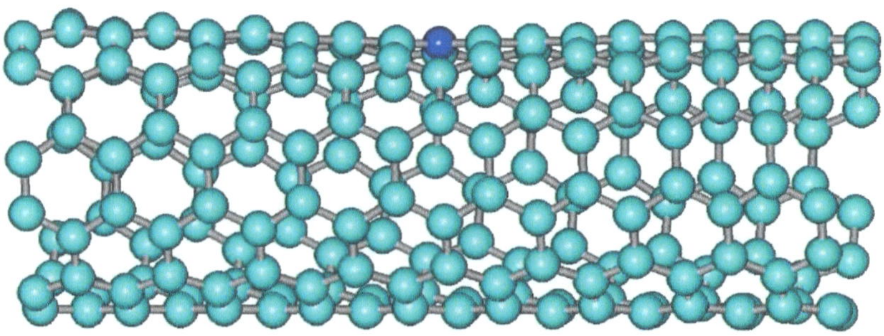

Fig. 21. Cross view of the nanotube with indices (6, 5) doped by atom of nitrogen.

As we see on the diagram of energy spectrum of the nanostructure (Fig. 22), the fifth electron of atom of nitrogen has formed and has filled a level with energy -10.75 eV . On the one hand this level donor as it is filled by an electron and is level HOMO 0 (the highest among the levels filled by electrons), and from the second side - it is near to valence band and should be the acceptor. So it is necessary to consider this level as very deep donor level.The forbidden gap in case of presence of impurity of atom of nitrogen is defined by a difference $\Delta E = -9.92 + 10.79 = 0.87\text{eV}$.

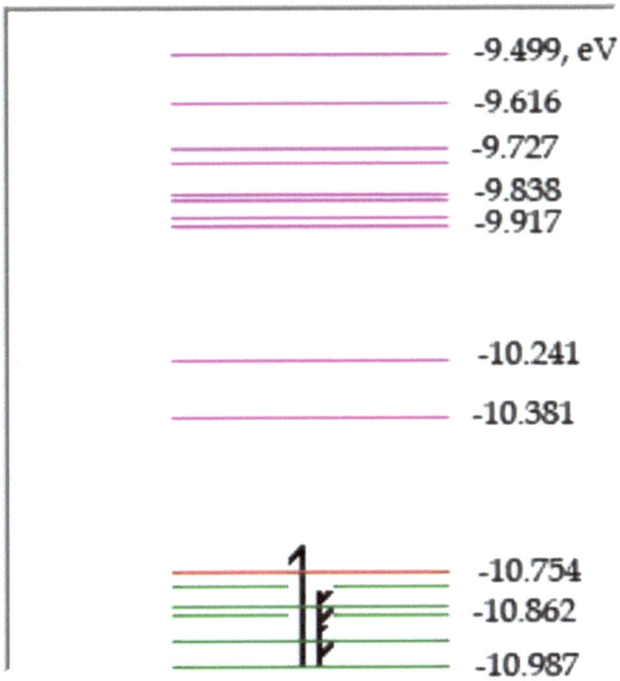

Fig. 22. Energy-band structure of a N-doped carbon nanotube with indices (6,5).

Also we see on Fig. 22 in a forbidden region there are some levels which are caused by the broken off bonds on edges of a nanotube. To neutralization of these levels we shall apply atoms of hydrogen (Fig. 23). Energy band diagram of the nanostructure in this case is shown on Fig. 24.

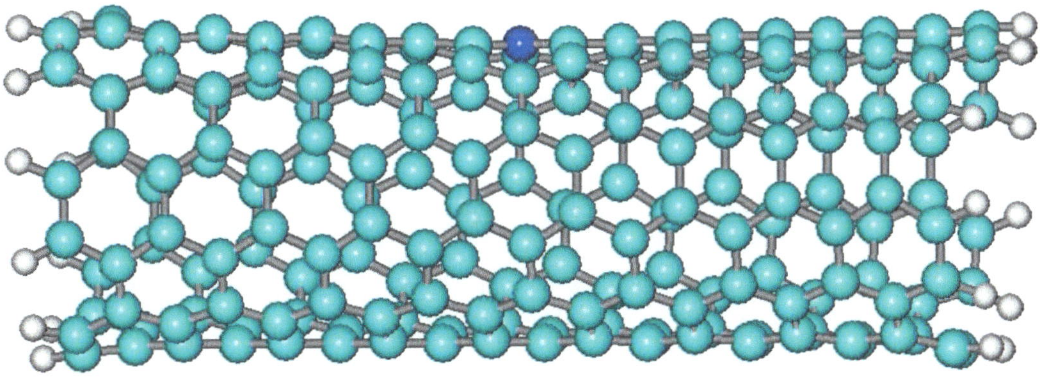

Fig. 23. Passivation of the torn off bonds on edges of the N -doped carbon (6, 5) nanotube.

As we see on Fig. 24, the atom of nitrogen has formed occupied level with energy −9,829eV, which is level HOMO 0 in this case.

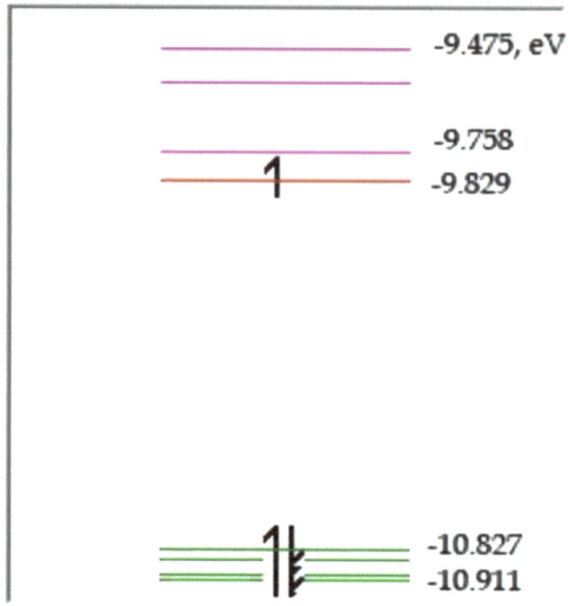

Fig. 24. Energy band diagram of the N-doped carbon (6, 5) nanotube with hydrogen endings.

To term its donor it is impossible, as it deep enough (0.071 eV from level LUMO 0 with energy - 9.758 eV). Presence of such energy level filled by an electron highly above valence band in the tubes alloyed by nitrogen should result in to improvement of their properties of electron emission.

3.4 Properties of the carbon (6, 5) nanotube doped by atom of aluminium

For construction of model of the doped nanotube the 116-th atom of carbon on the atom of aluminium is replaceable in the program of its basic model. We shall replace also aromatic bonds of atom with atoms of carbon on unary. After embodying the given changes in the program and carryings out of geometrical optimization of a nanostructure, we shall receive model of the aluminium-doped carbon (6,5) nanotube (Fig. 25a). We see, that the nanotube essentially distorted under influence of the doped atom. It speaks the following parents:

a) Atom of aluminium has twice greater radius (1.92Å) than radius of atom of carbon (0.91Å));

b) b) Covalent radius of atom of aluminium also greater than at atom of carbon, 1.18Å against 0.77Å;

c) c) Length of bonds C – Al which is equal 1.94Å, is more than length of bonds C – C equal 1.41Å.

d) It results in to that the atom of aluminium is pushed out of limits of the nanotube and its effective diameter increases up to quantity equal 8.51Å.

e) Besides under influence of elastic forces of strain, atoms of carbon which are bound to atom of aluminium also are superseded from the optimum places and take new geometrical positions which are normal for the doped nanotube.

f) Extended Hückel calculation of energy of the aluminium-doped nanostructure yields value 322454.33kcal/mol. It on 808kcal/mol is more than for a basic (6,5) nanotube. Thus for doping a nanotube (6,5) by aluminium the significant energy is required.

g) The zone diagram of the Al-doped nanostructure in this case is shown on Fig. 25b. We see, that the orbital number 400 (HOMO 0) is filled half, hence, there was a vacant level with energy - 10.41 eV . Thus, the atom of aluminium shapes an acceptor level near to valence band.

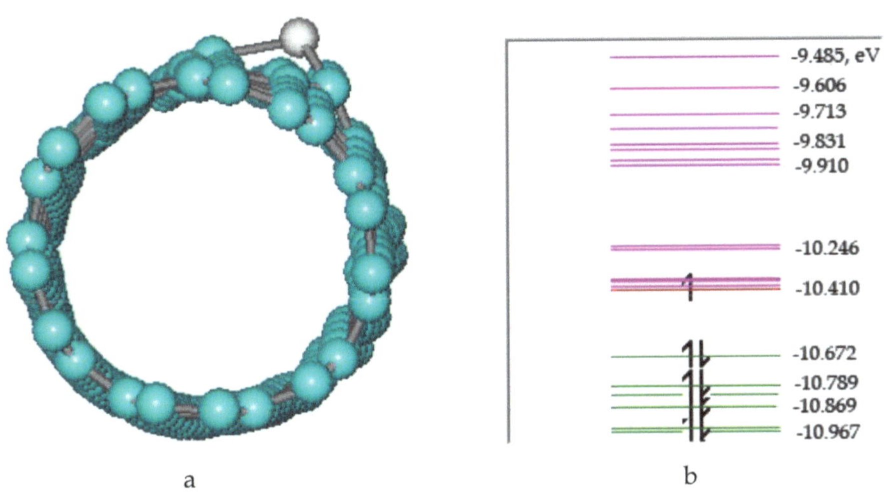

Fig. 25. Al-doped carbon (6,5) nanotube: (a) - a dilatational view; (b) - an energy-band structure.

The nanostructure with hydrogen endings is shown on Fig. 26. Its Extended-Hückel energy is equaled -332121.573kcal/mol. Electrical dipole moment of a nanotube, is equal 297.4 D greater enough in comparison with a boron-doped nanotube (6,5).

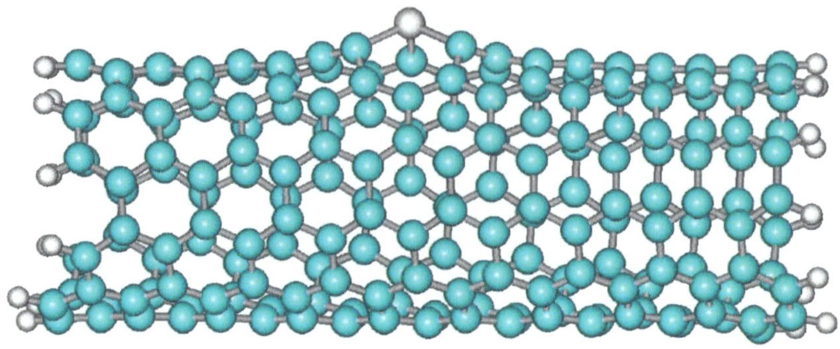

Fig. 26. Doped by atom of aluminium the carbon (6, 5) nanotube with hydrogen endings.

Doping by atom of aluminium very strongly influences an energy band spectrum (Fig. 27) nanotubes that is related to its significant strain after doping, and also spin density of a nanostructure. In this case the impurity atom of aluminium creates an acceptor level (-10.373 eV) with an ionization energy equal 0.015 eV .

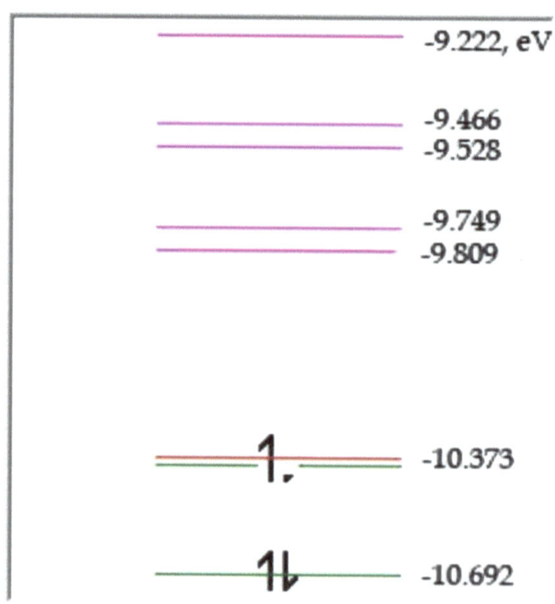

Fig. 27. Energy-band structure of the Al-doped (6,5) nanotube with hydrogen endings.

3.5 Doping by atom of phosphorus of a carbon (6, 5) nanotube

The atom of phosphorus has five valence electrons, therefore it is necessary to expect occurrence in a nanotube of a donor centre near to it, as for aromatic bonds with atoms of carbon enough four electrons. We shall exchange in basic model of a 200 -atomic nanotube the any atom of carbon with atom of phosphorus, having applied aromatic bonds C-P. For carrying out of geometrical optimization of the constructed P-doped nanostructure we shall use the molecular mechanic method MM + with application of the block-diagonal algorithm of Newton-Raphson. We shall gain model of the P -doped carbon (6,5) nanotube presented on Fig. 28. Apparently on pattern, that the atom of phosphorus has taken a place in exterior area near to a nanotube. It is caused by that bond C-P and atomic radius of phosphorus is more than bonds C-C and atomic radius of carbon on 0.4 and 0.3Å, accordingly.

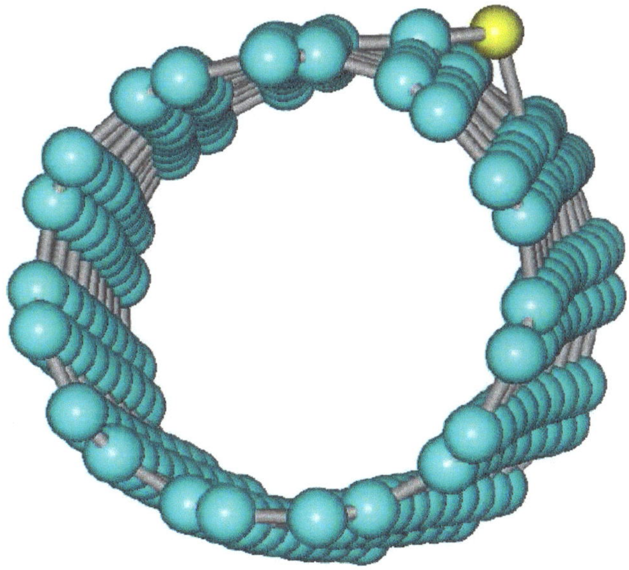

Fig. 28. The molecular mechanics model 200-atomic carbon (6.5) nanotube doped by atom of phosphorus.

In connection with mechanical stresses near to the impurity atom it is necessary to note, that the considered nanostructure is optimized only at RMS gradients of greater 0.015 kcal/(Åmol), and for a basic nanotube analogous magnitude makes 0.000001kcal/(Åmol). The band diagram of a viewed nanostructure is presented on Fig. 29. Comparing the gained diagram with the diagram of a basic nanotube (Fig. 16b), we see, that the fifth electron of atom of phosphorus has occupied the free orbital 401 with an energy -10.758 eV and that in a forbidden region the level with an energy -10.082 eV was generated.

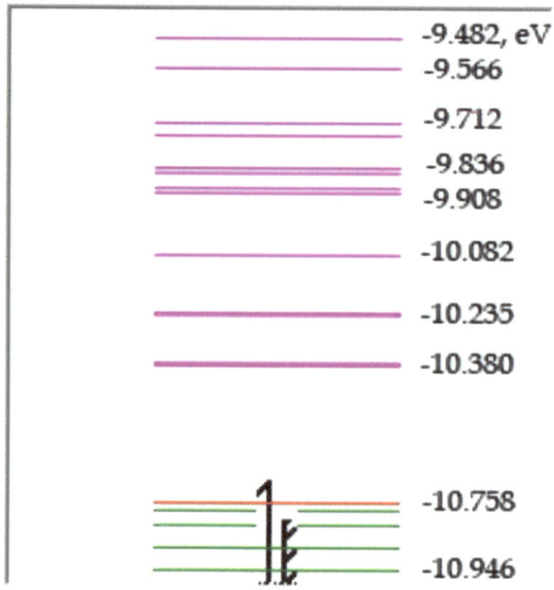

Fig. 29. Energy band structure of the 200 -atomic carbon (6.5) nanotube doped by atom of phosphorus.

The models of a phosphorus-doped (6,5) nanotube with hydrogen passivation of dangling bonds is presented on Fig. 30.

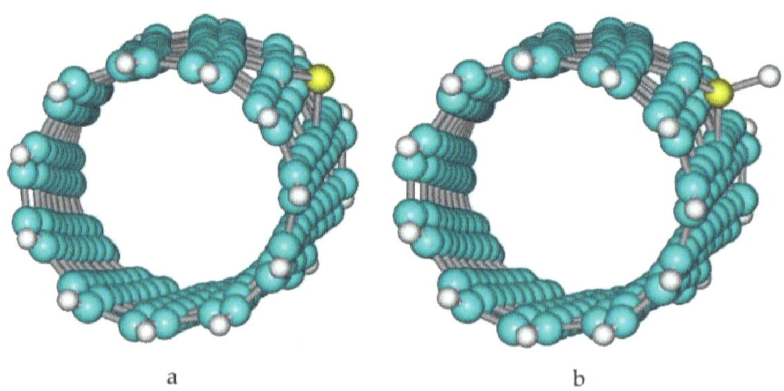

Fig. 30. Molecular mechanics models of the P-doped carbon (6,5) nanotube : (a) - hydrogen endings on edges of a nanostructure; (b) - the complete hydrogen endings.

We see, that the doped nanotube has one broken off bond on atom of phosphorus. If such bond to leave without hydrogen passivation to it any other radical including the same nanotube can be affiliated.

Apparently on Fig. 31, passivation of the dangling bonds has essentially changed an energy distributions of a viewed nanostructure:

the upper filled levels have risen up to magnitudes -10.077 eV (a) and -10.487 eV (b); degeneration of some energy levels is taken off; quantities of energy levels have changed.

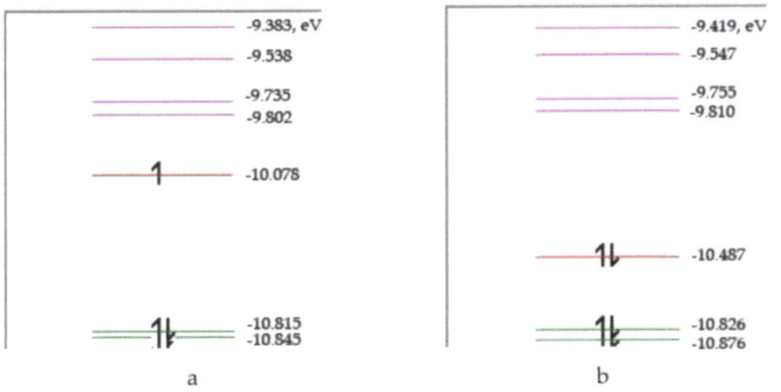

Fig. 31. The phosphorus-doped 200 -atomic carbon (6, 5) nanotube: (a) - hydrogen endings on edges of a nanostructure; (b) - complete hydrogen endings.

The magnification of number of atoms of carbon in model of a phosphorus-doped nanotube up to 410 pieces results in to diminution of breadth of a forbidden gap and bias of a HOMO 0 levels up to quantities -10.466 eV (Fig. 32a) and -9.852 eV (Fig. 32b).

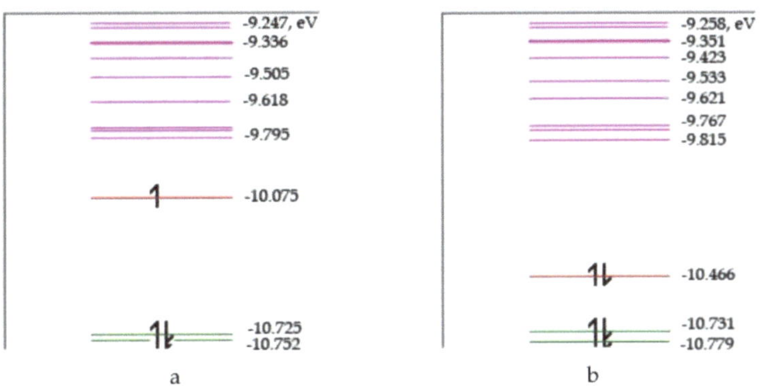

Fig. 32. The phosphorus-doped 410 -atomic carbon (6, 5) nanotube: (a) - hydrogen endings on edges of a nanostructure; (b) - the complete hydrogen endings.

Thus, the impurity atom of phosphorus without hydrogen ending of its dangling bond can create in a nanotube a deep donor level with an ionization energy equal 0.28 eV . In case of when the free bond of atom of phosphorus is neutralized by atom of hydrogen the recombination centre is formed. Besides in any case, the impurity atom of phosphorus will reinforce emissivity of a nanotube as its valence electron occupies higher energy levels.

4. Conclusion

On models of semiconductor carbon nanotubes with chiral vectors (10,0) and (6,5), consisting of 200 atoms, effect of impurities of atoms Al, B, N and P on performances of their energy-band structure is viewed. The magnification of number of atoms of carbon of the doped nanostructure twice feebly influences its energy distribution, but results in to diminution of a forbidden gap of a nanotube approximately on 0.1 eV .

Process of doping of a carbon nanotube was simulated by replacement of one of its atoms of carbon on an impurity atom. Such doping of a nanotube in experiment can be implemented at existence of vacancies in it. Theories about equilibrium concentration of vacancies in nanotubes are not present, therefore it can become the purpose of the further developments. It is necessary to note, that intercalation atoms in nanotubes can be viewed as interstitial atoms in crystals. Then their presence at a nanotube can determine process of doping also.

It is shown, that the energy distribution of levels in the field of a forbidden gap of the doped nanostructure strongly depends on passivation of its torn off bonds. In the present investigation modeling of process of such passivation was carried out by hydrogen endings. Also it is shown, that doping of a carbon nanotube by atom of phosphorus results in to occurrence of dangling bond on its fifth valence electron, as for support of aromatic bonds with atoms of carbon enough four electrons. Passivation of this bond by atom of hydrogen results in to changes in an energy band structure of the doped nanotube.

Impurity atoms of boron and aluminium are representatives of the third group of a periodic system of elements, therefore they create similar acceptor levels in carbon nanotubes with chiral vectors (10,0) and (6,5) :

a) atom of boron creates levels with activation energies 0.17 and 0.08 eV , accordingly;

b) atom of aluminium creates levels with identical activation energies 0.08 eV .

Atoms of nitrogen and phosphorus are representatives of the fifth group of a periodic system of elements, therefore it was expected, that they will create similar donor levels, but it has not proved to be true. The impurity atom of nitrogen creates in a carbon nanotube with a chiral vector (6,5) donor level with an ionization energy 0.07 eV , and in a nanotube (10,0) - very deep donor level near to valence band. The impurity atom of phosphorus creates in a carbon nanotube with a chiral vector (6,5) donor level with an ionization energy 0.28 eV , and in a nanotube (10,0) - very deep donor level near to valence band and a level with energy 10.1 eV .

5. References

Berkert U., Allinger N.L. (1982). Molecular mechanics. Amer. Chem. Soc. Mon. 177, Washington, D.C.

Bondarev, I. V. (2007). Quantum Optics Phenomena in Atomically Doped Nanotubes. Optics and Spectroscopy, Vol. 103, No 3, (2007) p. 366.

Derycke, V.; Martel, R.; Appenzeller J., Avouris, Ph. (XXXX). Carbon nanotube inter- and intramolecular logic gates. Nano Letters, vol. 0, 0, (XXXX) A-D.

Esfarjani, K.; Farajian, A. A.; Hashi, Y.; Kawazoe, Y. (1999). Electronic and transport properties of N-P doped nanotubes. Appl. Phys. Lett., 74, (1999) p. 79.

Esfarjani, K.; Chen, Z. and Kawazoe, Y. (2003). Electronic properties of magnetically doped nanotubes. Bull. Mater. Sci., Vol. 26, No 1, (January 2003) pp. 105-107.

Farajian, A. A.; Mizuseki, H. and Kawazoe, Y. (2002). Spin-polarized transport through doped nanotube junctions in presence of applied magnetic field, Proceeding of 10th Int. Symp. "Nanostructures Physics and Technology", pp. TN.06p, St Petersburg. Russia, June 17-21 (2002), IOFFE Institute, St Petersburg.

Gong, K.; Du, F.; Xia, Z.; Durstock, M.; Da, L. (2009). Nitrogen-doped Carbon Nanotube Arrays with High Electrocatalitic Activity for Oxygen Reduction. Science, Vol. 323, No 5915, (6 February 2009) pp. 760-764.

Griadun V. I. (2006). Vacancies in nanotubes and fullerenes, Proceedings of 16-th International Crimean Conference "Microwave & Telecommunication Technology" (CriMiCo'2006), vol. 2, pp. 669-670, ISBN 966-322-006-6, Sevastopol, September 11-15, 2006, Weber Publ. Co.

Griadun V. I. (2007). Doped carbon nanotube (10,0) characteristics, Proceedings of 17 -th International Crimean Conference "Microwave & Telecommunication Technology" (CriMiCo'2007), vol. 2, pp. 573-574, ISBN 978-966-335-013-4, Sevastopol, September 10-14, 2006, Weber Publ. Co.

Grigorian, L.; Sumanasekera, G. U.; Loper, A. L.; Fang, S.; Allen, J. L. and Eklund, P. C. (1998). Transport properties of alkali-metal-doped single-wall carbon nanotubes. Phys. Rev., B 58, (1998) R4195-R4198.

Hoffmann, R. (1963). An Extended Hückel Theory. J. Chem. Phys., (1963), vol. 39, No 6, (1963) pp. 1397-1412.

Lee, R. S.; Kim, H. J.; Fisher, J. E.; Thess A.; Smalley, R. E. (1997). Conductivity enhancement in single-walled carbon nanotube bundles doped with K and Br. Nature, 388, (17 July 1997) pp. 255-257.

Martel, R.; T. Schmidt, T.; Shea, H. R.; Hartel, T. and Avouris, Ph. (1998). Single- and multiwall carbon nanotube field-effect transistors. Appl. Phys. Lett., vol. 73, No 17, (1998) pp. 2447-2449.

Miyamoto, Y.; Rubio, A.; Blase, X.; Cohen, M. L. and Louie, S. G. (1995). Ionic Cohesion and Electron Doping of Thin Carbon Tubules with Alkali Atoms. Phys. Rev. Lett., 74, (1995) pp. 2993-2996.

Rao, A. M.; Eklund, P. C.; Bandow, S.; Thess, A.; Smalley, R. E. (1997). Evidence for charge transfer in doped carbon nanotube bundles from Raman scattering. Nature, 388, (17 July 1997) pp. 257-259.

Yao, J. B.; Han, J. H.; Choi, S. H.; Lee, T. Y.; Park, C. Y.; Jeong, T. W.; Lee, J. H.; Yu, S.; Park, G.; Yi, W. K.; Baik Y. -J. and Kin, J. M. (2002). Emission Characteristics of boron nitride coated carbon nanotubes. Physica B: Condensed Matter, Volume 323, Issues 1-4, (October 2002) pp. 180-181.

Zidan, R.; Rao, A. M.; Au, M. (2003). Doped Carbon Nanotubes for Hydrogen Storage. FY 2003 Progress Report.

Chapter-9

Fundamental Physical Aspects of Carbon Nanotube Transistors

1. Introduction

The increasing demand for ultra-high speed processors, smaller dimensions and lower power consumption of integrated circuits has made the technology scaling of the electronic components a challenging issue for device designers. In past few decades, miniaturization of transistors has always obeyed the moore's law: the number of transistors that can be placed inexpensively on an integrated circuit has doubled approximately every two years. Nanoscale field effect transistors in the sub- 10 nm regime, suffer from short channel effects such as direct tunneling from source to drain, increase in gate-leakage current and punchthrough effect. These effects have posed severe problems for miniaturized transistors and directed the recent research toward better alternative semiconductors than silicon. Semiconducting carbon nanotubes (CNTs) because of their properties like large mean free path, excellent carrier mobility and improved electrostatics at nanoscales as the result of their non-planar structure, have been known as the best ideal replacement for silicon. In particular, they exhibit ballistic transport over length scales of several hundred nanometers (Heinze, et al., 2006). Absence of the dangling bond states at the surface of CNTs and purely one-dimensional transport properties improve gate control while meeting gate leakage constrains and allows for a wide choice of gate insulators. That's why CNTs suppress the short channel effects in transistor devices. Symmetry of the conduction and valence bands makes CNTs advantageous for complementary applications. CNTs are very attractive for nanoelectronic applications and can be used to achieve high speed ballistic carbon nanotube field effect transistors (CNTFETs). Theoretically, CNTFETs could reach a higher frequency domain (terahertz regime) than conventional semiconductor technologies. CNTFETs, only a few years after the initial discovery of CNTs in 1991 by Sumio Iijima (Iijima, 1991) were first demonstrated in 1998 by Dekker et al. at Delft University (Tans, et al., 1998) and soon after by groups at IBM (Martel, et al., 1998) and Stanford University (Soh, et al., 1999). Intensive research has led to a significant progress in understanding the fundamental properties of CNTs. By using a single-wall CNT as the channel between two electrodes which work as the source and drain contacts of a FET, a coaxial CNTFET can be fabricated. Coaxial devices are of special interest because their geometry allows for better electrostatics because the gate contact wraps all around the channel (CNT) and has a very good control on carrier transport. Type of Metal-CNT contacts plays crucial role in the output characteristics of the transistor. Heavily doped semiconductors because of the ability to form Ohmic contacts can be used as ideal electrodes but they suffer from high parasitic resistance. Existence of potential barrier at the metal-CNT interface, changes the device to a CNTFET resembling to Schottky barrier MOSFETs. However, heavily doped CNT contacts can be used to get to a behavior similar to conventional MOSFETs.

Understanding CNTFETs from electronic point of view requires a deep insight for mesoscopic physics. For modelling a CNTFET, a powerful methodology with the abilities of solving Schrödinger equation under non-equilibrium conditions in the presence of selfconsistent electrostatics and treating coupling of the channel to contacts is needed. The nonequilibrium Green's function (NEGF) formalism provides a sound basis for quantum device simulations. In this chapter we aim to introduce different types of CNTFETs, their electrostatics and output characteristics.

The chapter is organized as follows: section 2 introduces different types of CNTFETs and typical design parameters, section 3 explains different physical issues involved with CNTFETs and ultimately in section 4 we briefly describe the NEGF formalism and WKB treatment for simulating CNTFETs. We use these methods to plot the curves for implying physical aspects through the chapter. While explaining the important issues, new challenges for achieving a well-designed CNTFET will be explained. After reading this chapter, the reader should be able to distinguish how a specific type of CNTFETs behaves (both DC analysis and high frequency response) and why.

2. Common CNTFET Designs

2.1 Types of CNTFETs

There are two main types of CNTFETs that are being currently studied, differing by their current injection methods. CNTFETs can be fabricated with Ohmic or Schottky contacts. The type of the contact determines the dominant mechanism of current transport and device output characteristics. CNTFETs are mainly divided into Schottky barrier CNTFETs (SBCNTFETs) with metallic electrodes which form Schottky contacts and MOSFET-like CNTFETs with doped CNT electrodes which form Ohmic contacts. In SB-CNTFETs, tunneling of electrons and holes from the potential barriers at the source and drain junctions constitutes the current. The barrier width is modulated by the application of gate voltage, and thus, the transconductance of the device is dependent on the gate voltage (Raychowdhury, et al., 2006).

The other type of the CNTFETs takes advantage of the n -doped CNT as the contact. Potassium doped source and drain regions have been demonstrated and the behavior like MOSFETs have been experimentally verified (Javey, et al., 2005). In this type of transistors a potential barrier is formed at the middle of the channel and modulation of the barrier height by the gate voltage controls the current.

2.2 Typical Designs

At this point, we consider two typical designs for two types of the CNTFETs and use them for investigation and comparison of the transistors' behaviors. Let's consider two coaxial CNTFETs with a (13,0) zigzag CNT as the channel which corresponds to a bandgap of about 0.83 eV and a diameter of 1 nm . A 2 nm thick ZrO_2 with a relative permittivity of 25 separates the gate electrode and the CNT. The gate is 10 nm long and wrapped around the channel. The gate thickness is assumed 6 nm . SB-CNTFET employs an intrinsic CNT and 20 nm long metallic contacts as the source and drain. A typical structure and the distribution of the potential energy on the channel are shown in figure 1. The gate voltage makes the barriers near the source/drain thinner and increases the tunneling current.

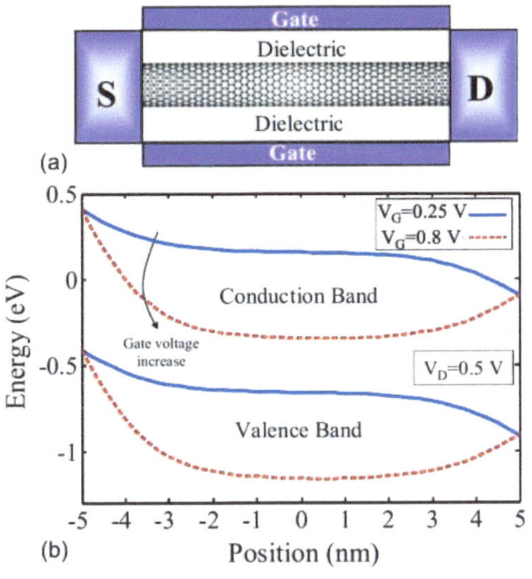

Fig. 1. Schottky Barrier CNTFET, a) 2-D cross section of the coaxial structure with intrinsic CNT as the channel and metal source/drain contacts, b) Energy band diagram obtained from Poisson equation. The metal Fermi level is taken to be at the midgap of the CNT.

On the other hand the MOSFET-like CNTFET benefits from heavily doped ends of the CNT which act as Ohmic source and drain contacts. A typical structure of the MOSFET-like CNTFETs with 20 nm doped sections and its energy band diagram is shown in figure 2. The source/drain doping is $10^{-9}\ \mathrm{m}^{-1}(\sim 0.01$ dopant per atom). In the following sections of this chapter these two transistors will be compared to each other and the results can be generalized to other designs. The curves of this chapter have been obtained using the described methods in section 4.

3. Important Aspects of CNTFETs

3.1 Ambipolarity

One of the important aspects of nanotube transistors is the ambipolarity or unipolarity of their current-voltage characteristics. SB-CNTFETs exhibit strong ambipolar behavior. For high enough gate voltages the tunneling probability of electrons through the source Schottky barrier in the conduction band is high and for the low and negative gate voltages, a Schottky barrier is formed at the drain in valence band and tunneling of holes increases the current.

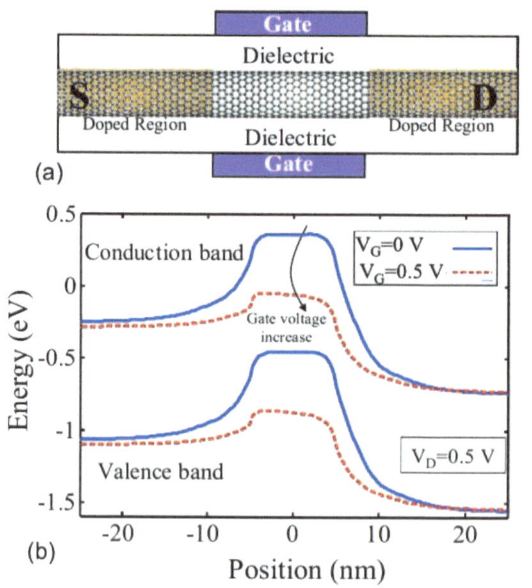

Fig. 2. MOSFET-like CNTFET, a) 2-D cross section of the coaxial structure with intrinsic CNT as the channel and doped CNT sections as source/drain contacts, b) Energy band diagram obtained from Poisson equation.

The energy bands for low and high gate voltages and the Schottky barriers are shown in figure 1(b). However for MOSFET-like CNTFETs only positive gate voltages because of lowering the barrier in the channel increase the current. The energy bands for low and high gate voltages and the potential barrier in the channel are shown in figure 2(b). It is clear from the figures why the characteristics of these two transistors differ. The currentvoltage characteristics of the devices have been compared in figure 3.

Ambipolar behavior of the SB-CNTFETs constraints the use of these transistors in conventional CMOS logic families. We have proposed some methods to reduce or eliminate the ambipolarity in CNTFETs. Using asymmetrical contact types we have introduced Schottky-Ohmic CNTFET which has a Schottky barrier at the CNT-metal interface at the source and an Ohmic contact at the drain at the channel-doped CNT interface (Kordrostami, et al., 2008). However, the device still suffers from band to band tunneling. That is, in low or negative voltages the electrons from the valence band can tunnel to the conduction band and contribute to the total current of the transistor and cause increase of the current in negative voltages. We have compared the potential energy distribution of the channel of the transistors in figure 4 (Kordrostami, 2007).

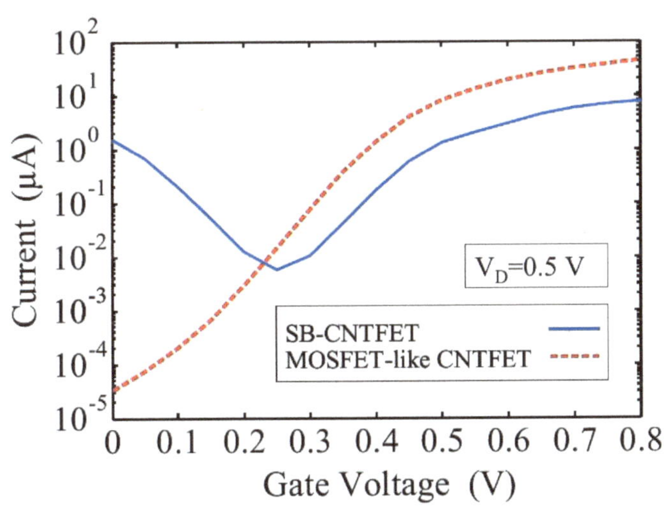

Fig. 3. Comparison between the current-voltage characteristics of the SB and MOSFET-like CNTFETs. Ambipolar behavior of the SB-CNTFET and the unipolar characteristic of the MOSFET-like CNTFET can be seen from the curves.

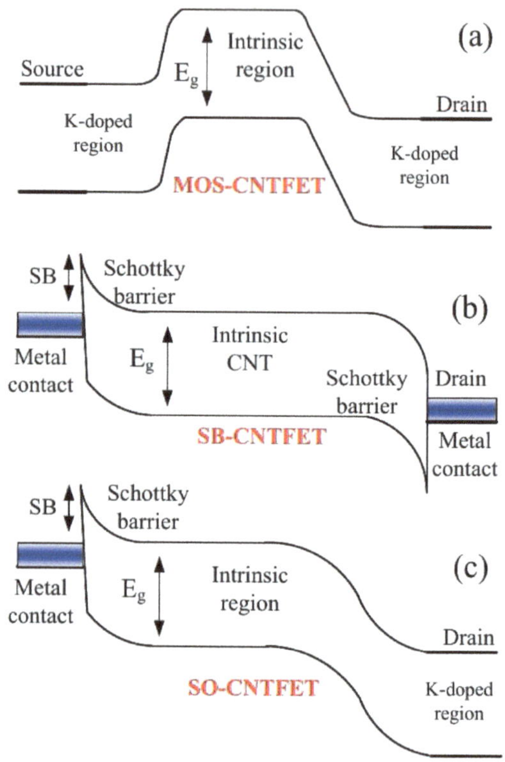

Fig. 4. Potential energy distribution along the carbon nanotube in three different CNTFET structures, a) MOSFET-like, b) Schottky barrier, c) Schottky-Ohmic (Kordrostami, 2007)

Double-gate structures can also reduce the ambipolarity while the CNT is still intrinsic. In the structure shown in figure 5 the first gate controls carrier injection at the source and the second one acts like an electrostatic doping and controls carrier injection at the drain which can be used to suppress the hole tunneling current, for example by applying the same voltage as the drain voltage to the second gate (Pourfath, et al., 2005). In this case, at any drain voltage the band edge profile near the drain contact will be flat like the band diagram in figure 4(c). On the other hand, an advantage of an ambipolar SB-CNTFETs is that they can be used as either an n-type and p-type FET in a CMOS application (Guo and Lundstrom, 2009).

In MOSFET-like CNTFETs with heavily doped source and drain regions, when applying a negative gate voltage, the band to band tunneling may lead to ambipolarity. For suppressing this effect in MOSFET-like CNTFETs we proposed a non-uniform doping profile as shown in figure 6 (Hassaninia, et al., 2008b). This reduces the gradient of the channel potential barrier at each interface between the intrinsic and doped sections of the CNT and suppresses the band to band tunneling and ambipolar conduction.

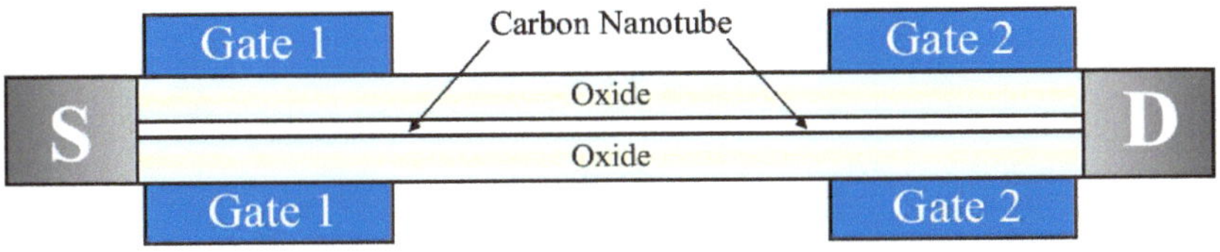

Fig. 5. Double gate CNTFET (Hassaninia, et al., 2008a). The first gate modulates the source tunneling current and the second gate in fact eliminates the Schottky barrier at the drain (path for tunneling of the holes in the valence band).

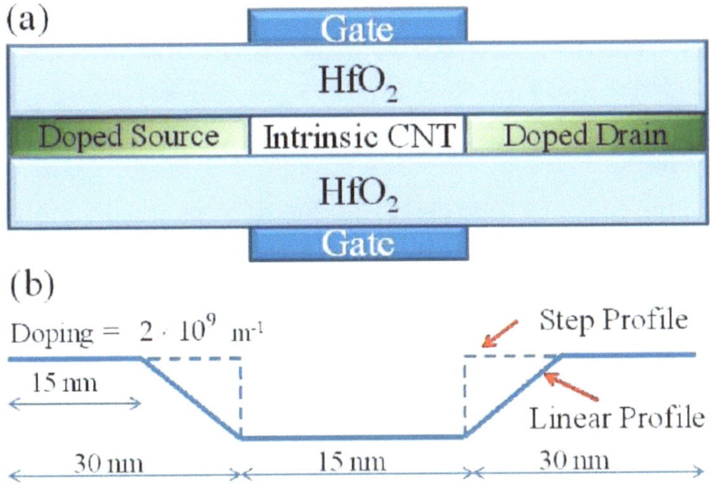

Fig. 6. a) Cross section of the CNTFET. (b) Step and linear doping profiles for the MOSFETlike structure versus position.

3.2 High Permittivity Gate Dielectrics

The relatively low dielectric constant of SiO_2 (at 3.9) limits its use in transistors as gate lengths scale down to tens of nanometers. As the device dimensions approach the 10 nm regime, strong electrostatic coupling of source/drain electrodes arises fundamental challenges especially about the gate control over the channel. Enhancing the gate efficiency requires thinner gate oxides. However, due to excessive direct tunneling leakage currents through the ultra-thin dielectric, the gate dielectric layer is already approaching its limit (~ 1 nm) and the only feasible way is to use high- κ gate dielectrics which afford high gate capacitance without relying on ultra-small film thickness (Javey, et al., 2002). The most common gate oxides for nanotransistors are Zirconium [1](ZrO_2) and Hafnium dioxide (HfO_2) with relative permittivities about 25 and 16 respectively. The current versus the dielectric constant of the oxide in both types of transistors has been shown in figure 7.

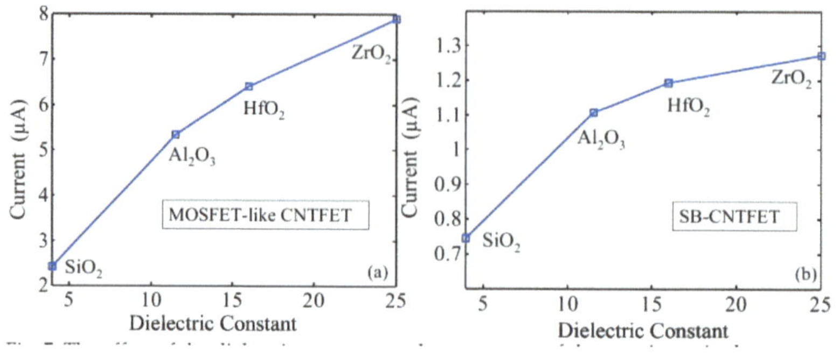

Fig. 7. The effect of the dielectric constant on the on current of the transistors in the on state. ($V_G = V_D = 0.5$ V).

3.3 On/Off Current

Comparison shows that MOSFET-like CNTFETs due to absence of Schottky barriers have a lower off leakage current. On the other hand, in the on state, there is no barrier at the sourcechannel junction and hence, the device demonstrates significantly higher on current. The minimum current of the SB-CNTFETs occurs when the contribution of the electron and hole tunneling currents to the total current becomes equal, that is when $V_G = V_D/2$ where the energy band diagram is symmetric. For each power supply voltage (V_D), the off-current is defined at the minimal leakage point $V_G(\text{off}) = V_D/2$ and the on-current is defined at $V_G(on) = V_G(off) + V_D$. There exists a trade-off: reducing the off-current by lowering the power supply voltage degrades the on-current.

3.4 Transconductance

The transconductance (g_m) of the SB-CNTFETs is severely limited by the Schottky barriers in the on state. The transconductances of the devices as a function of the gate voltage have been compared to each other in figure 8(a). It is seen that transconductance of the MOSFETlike CNTFET is higher than the SB-CNTFET. Figure 8(b) plots CNTFET on state transconductance versus the gate dielectric constant for four types of widely used gate insulators. The high- κ gate insulator improves the CNTFET performance by reducing the selfconsistent potential produced by the charge on the tube (Guo, 2004). For the SB-CNTFETs the transconductance tends to saturate when the gate insulator dielectric constant is large. The reason is that the self-consistent potential produced by the charge on the tube is already small and further improving the gate dielectric constant does not help to significantly reduce the Schottky barrier thickness and the transistor performance (Guo, 2004).

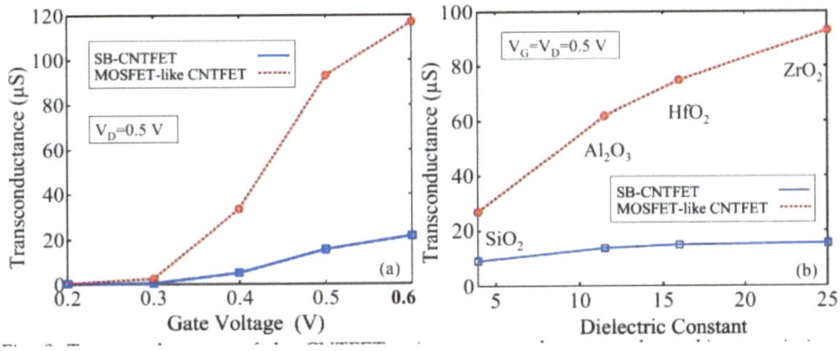

Fig. 8. Transconductance of the CNTFETs, a) g_m versus the gate voltage, b)g_m variations with respect to the dielectric constant.

3.5 Gate Capacitance

For a one dimensional nanotube FET, the total capacitance between the gate and the channel (gate capacitance) depends both on the geometry and the density of states (DOS). If the electrostatic potential on the channel (V_S) is uniform, then the voltage drop over the gate oxide is also uniform and the gate voltage V_G is the summation of the channel potential and the voltage drop over the gate oxide. This can be modeled as the potential division between two capacitances which means the gate capacitance (total capacitance) can be modeled as the series of the electrostatic (geometry dependent) capacitance and the quantum (DOS dependent) capacitance as shown in figure 9. Gate capacitance can be calculated from:

$$C_g = \partial Q/\partial V_G \tag{1}$$

where Q is the total charge on the gate electrode (which has the same magnitude but an opposite sign as the total net charge of the CNT) and V_G is the gate voltage. The gate oxide plays the role of the dielectric between the plates of the oxide (electrostatic) capacitance which can be expressed as:

$$C_{ox} = 2\pi\varepsilon L_g/Ln\left(\frac{t_{ox} + r_{CNT}}{r_{CNT}}\right) \tag{2}$$

where t_{ox} is the gate dielectric thickness, ε is the dielectric constant of the gate insulator, L_g is the gate length and r_{CNT} is the nanotube diameter. Since the charge on the channel (Q) can be derived from CNT density of states and the Fermi level for a given channel potential, the quantum capacitance can be calculated from:

$$C_q = \partial Q/\partial V_s \tag{3}$$

where V_S is the channel potential. For MOSFET-like CNTFETs V_S and Q can be defined as the potential energy and charge density at the top of the channel potential barrier.

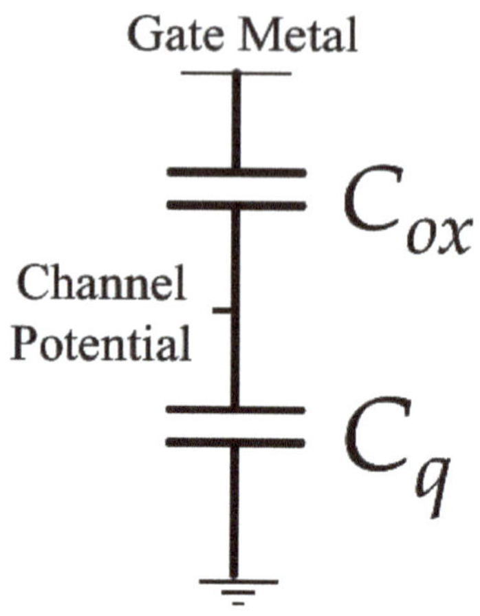

Fig. 9. Series combination of the oxide and quantum capacitances and the voltage drop on the capacitances.

The above derivation is strictly valid only when V_S is position-independent, otherwise the definition of the oxide capacitance breaks down (Guo, et al., 2007). The gate capacitances of the transistors versus the dielectric constant are shown in figure 10.

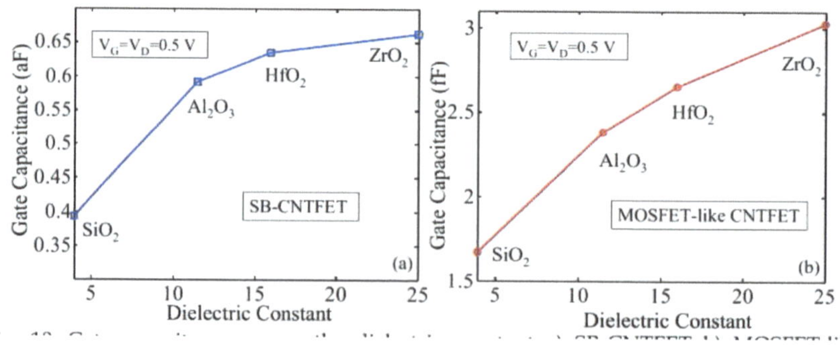

Fig. 10. Gate capacitance versus the dielectric constant a) SB-CNTFET b) MOSFET-like CNTFET.

The energy at the top of the barrier in the channel depends on the gate voltage and on the charge at the top of the barrier. When C_{ox} is small, the gate voltage controls the charge at the top of the barrier (which is independent of the drain voltage) and for large C_{ox}, the gate voltage controls the potential at the top of the barrier, which is independent of the drain voltage (Guo and Lundstrom, 2006). In bulk devices like conventional MOSFETs, the quantum capacitance is larger than the oxide capacitance, however, CNTFETs usually work in the quantum capacitance limit $(C_q \ll C_{ox})$.

We have compared the effect of the gate thickness on the gate capacitances of a completely gated and a partially gated SB-CNTFET. The results are shown in figure 11 and imply that the thicker gate increases C_g in the partially gated and has no effect in a completely gated SB-CNTFET (Kordrostami and Sheikhi, 2009b).

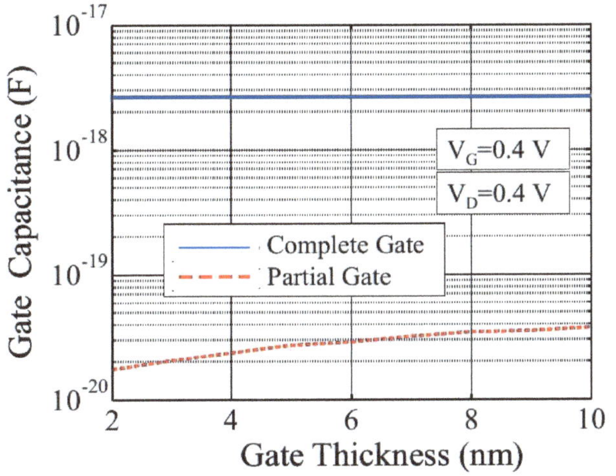

Fig. 11. Gate capacitances for two types of SB-CNTFET. The channel is an intrinsic 50 nm long $(19, 0)$ zigzag CNT, the gate length in partially gated SB-CNTFET is 5 nm and the dielectric is SiO_2 with a permittivity of 3.9.

3.6 Fringing Capacitance

The fringing fields between the gate metal and source and drain contacts result in capacitances which are called fringing or parasitic capacitances. Fringing field becomes more important when the channel length of a CNTFET is reduced. This could be comparable to the intrinsic device capacitances, and hence must be considered. In practice, appropriate electrode geometries are required to minimize the fringing field parasitic capacitances so that the parasitic capacitance due to fringing fields become negligible compared to the gate capacitance required to modulate the conductance, for example, by using 1-D metallic electrodes or nanotubes themselves as the electrodes (Yu, et al., 2006). We have investigated the effect of the contact geometry on the electrostatics of both SB-CNTFETs and MOSFETlike CNTFETs which clearly verifies the contribution of the fringing field to the device electrostatics and frequency response (kordrostami and Sheikhi, 2009a).

3.7 Cutoff Frequency

The unity current gain cutoff frequency (the frequency at which the current gain falls to unity) is usually used to describe high-frequency performance of a transistor. The cutoff frequency of the intrinsic CNTFET is called the intrinsic and the cutoff frequency of the CNTFET with inclusion of the parasitic capacitances is called the extrinsic cutoff frequency. When the parasitic capacitances are small the extrinsic cutoff frequency approaches the intrinsic cutoff frequency. We compute the cutoff frequency (f_T) using the quasi-static approximation. The quasi-static treatment works well when the signal varies slowly compared to the time constant determined by the intrinsic gate capacitance and the channel inductance (Guo, et al., 2005). The small-signal circuit model for a CNTFET based on the quasi-static approximation, which includes the equivalent capacitive and resistive elements, but omits the equivalent inductive elements, is shown in figure 12.

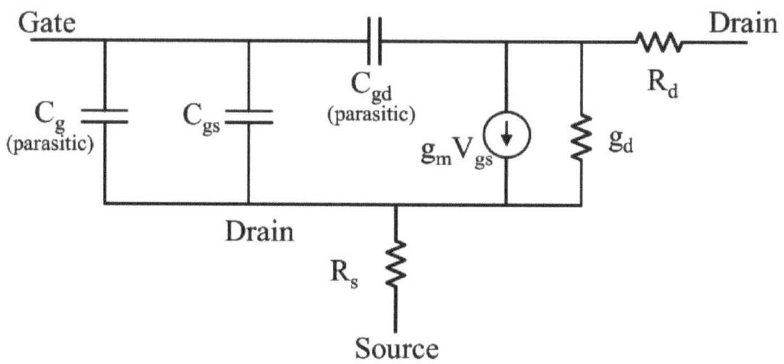

Fig. 12. Small-signal circuit model for a nanotube transistor including intrinsic and extrinsic elements.

By using the circuit model, we can generally write for the cutoff frequency:

$$\frac{1}{2\pi f_T} = (R_S + R_D)C_{gd} + \frac{1}{g_m}\left(C_g + C_{gs} + C_{gd}\right) + \frac{g_d}{g_m}(R_S + R_D)\left(C_g + C_{gs} + C_{gd}\right) \quad (4)$$

Where g_m is the transconductance, R_S and R_D are the parasitic resistances, C_g is the intrinsic gate capacitance and C_{gs} and C_{gd} are the parasitic capacitances (Burke, 2004). Calculation of the parasitic capacitances between the gate and source (drain) electrode requires separate electrostatic computations. If the parasitic resistances are excluded from the calculations we have:

$$f_T = \frac{1}{2\pi}\frac{g_m}{C_g + C_{gs} + C_{gd}} \quad (5)$$

The cutoff frequency of the intrinsic transistor without parasitic capacitances is:

$$f_T = \frac{1}{2\pi}\frac{g_m}{C_g} \tag{6}$$

In CNTFETs the intrinsic cutoff frequency can be determined in terms of the ratio of the change in the current to the change in the channel charge:

$$f_T = \frac{1}{2\pi}\frac{g_m}{C_g}\bigg|_{V_D=V_D(on)} = \frac{1}{2\pi}\frac{g_m}{C_g}\frac{\partial V_G}{\partial V_G} \approx \frac{1}{2\pi}\frac{\partial I_D}{\partial Q_{Ch}}\bigg|_{V_D=V_D(on)} \tag{7}$$

Where, I_D is the source-drain current and Q_{Ch} is the total charge in the CNT channel and is given as (Yoon, et al., 2006):

$$Q_{Ch} = q\int_0^{L_{Ch}} N_e(x)dx \tag{8}$$

where $N_e(x)$ is the electron density as a function of the channel position and L_{Ch} is the channel length. The intrinsic cutoff frequencies of the transistors with respect to the channel length have been compared in figure 13. The shorter the channel the larger the cutoff frequency of the CNTFET.

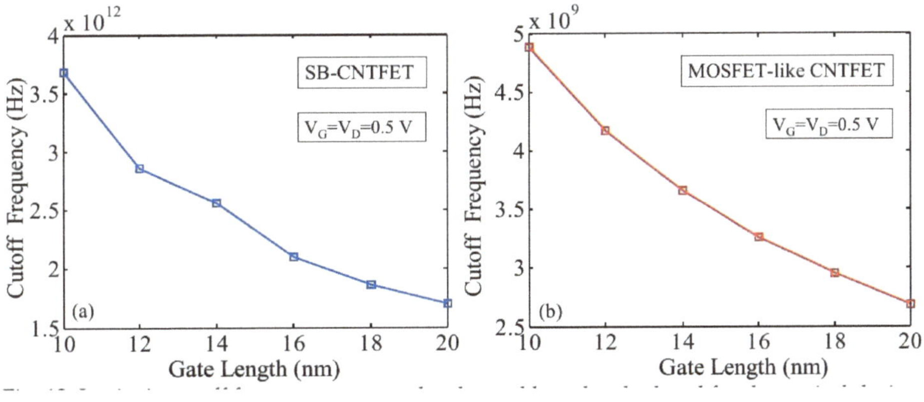

Fig. 13. Intrinsic cutoff frequency versus the channel length calculated for the typical designs in section 2.2, for a) SB-CNTFET, b) MOSFET-like CNTFET.

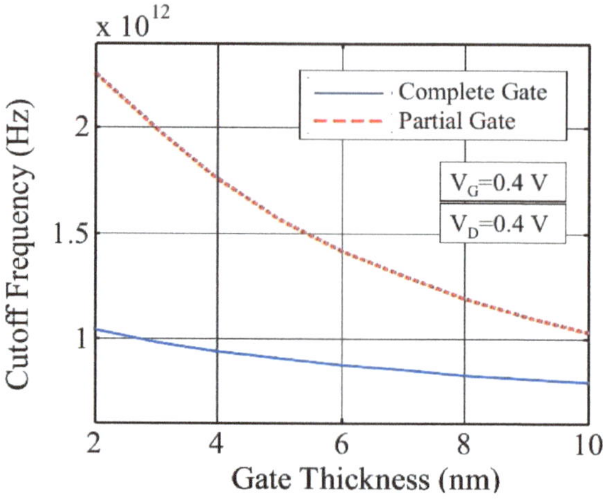

Fig. 14. Cutoff frequency of two types of SB-CNTFETs. The channel is an intrinsic 50 nm long (19, 0) zigzag CNT, the gate length in partially gated SB-CNTFET is 5 nm and the dielectric is SiO_2 with a permittivity of 3.9 .

From the equation 5 it can be concluded that the cutoff frequency is inversely proportional to the fringing capacitances. For example the partially gated SB-CNTFET has a higher cutoff frequency than the complete gate SB-CNTFETs because the fringing capacitances between metal electrodes are lower in partially gated SB-CNTFETs (Kordrostami and Sheikhi, 2009b). figure 14 shows the reduction of the cutoff frequency when the gate metal is thicker because of the increase of the fringing capacitances.

3.8 Intrinsic Delay

The main limitation to a faster intrinsic delay time is the gate capacitance resulting from fringing fields through the high- κ dielectric directly from the gate to source and gate to drain (Khairul and Roger, 2006). The intrinsic delay is one of the most important performance metrics for digital electronic applications and characterizes how fast a transistor switches. The switching delay can be calculated from:

$$\tau = \frac{Q_{on} - Q_{off}}{I_{on}} \tag{9}$$

Where Q_{on} is the total charge of the channel at on state $\left(V_G = V_{D\,(\mathrm{on})}\right)$ and Q_{off} is the total charge of the channel at off state $\left(V_G = 0,\ V_D = V_{D(\,\mathrm{on}\,)}\right)$ and I_{On} is the on current (Yoon, et al., 2006). The calculated intrinsic delays of the typical transistors are shown in figure 15.

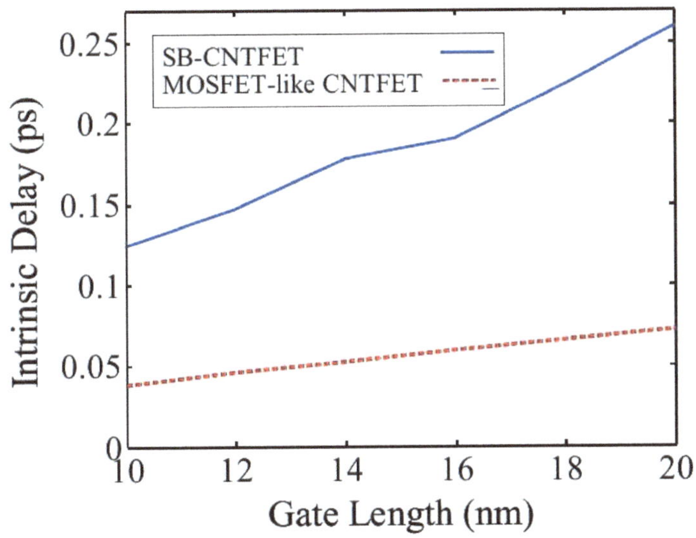

Fig. 15. Intrinsic delays of the SB-CNTFET and the MOSFET-like CNTFET versus the gate length. Longer channel length leads to larger latency. The on state is defined when $V_G = V_D = 0.5$ V and the off state is defined as $V_G = 0$ V and $V_D = 0.5$ V.

4. Physical modelling of the CNTFETs

As electronic devices are being downscaled to nanometer range, the validity of the conventional modeling approaches becomes questionable. The quantitative simulation methods for nanoscale devices should incorporate an understanding of both atomistic structures and quantum mechanical effects.

In the following sections, we describe two common methodologies which can simulate the CNTFETs efficiently: NEGF formalism and WKB method for SB-CNTFETs.

4.1 NEGF Formalism

The non-equilibrium Green's function (NEGF) formalism provides a sound basis for modeling quantum devices, due to the following reasons:

- Atomic-level description of the channel.
- Channel-contact interfaces can be treated by describing open boundary conditions for the Schrödinger equation.
- Dissipative scattering processes and other phenomena like light emission can be modeled.

- Considering quantum mechanical tunneling through Schottky barriers at the metal/nanotube contacts, tunneling and reflection at barriers in the nanotube channel and band to band tunneling.

In this section, we give a brief summary of the NEGF simulation procedure and how to apply the approach to a nanotransistor. The procedure is as follows (Guo and Lundstrom, 2009):

- Identify a suitable basis set and Hamiltonian matrix (H) for an isolated channel.
- Compute the self-energy matrixes (Σ_S and Σ_D).
- Compute the retarded Green's function (G).
- Determine the physical quantities of interest from the Green's function.
- Solve NEGF transport equation iteratively with the Poisson equation until selfconsistency is achieved.
- Calculate the drain current.

The procedure starts with an initial guess for the potential of the channel and then the charge density is calculated from the NEGF equations. For a given charge density, the Poisson equation is solved to obtain the electrostatic potential in the nanotube channel. Next, the computed potential profile is used as the input for the NEGF transport equation, and an improved estimate for the charge density is obtained. The iteration between the Poisson equation and the NEGF transport equation continues until self-consistency is achieved. At this time, all the physical quantities are exact and the current can be calculated. The Green's function can be calculated from:

$$G = [(E + i0^{+})I - H - \Sigma_S - \Sigma_D]^{-1} \tag{10}$$

Where H is the Hamiltonian matrix and $\Sigma_{S,D}$ is the self energy matrixes for the source/drain interfaces. The device and the methodology are shown in figure 16. Table 1 describes the NEGF formalism and how the physical quantities can be calculated. By choosing appropriate self-energy matrixes, the procedure can be implemented to model both types of the CNTFETs. The computationally expensive part of the procedure is calculating the Green's function. Using the real space basis set for calculation of the Hamiltonian matrix for the CNT channel leads to a matrix whose size is the total number of carbon atoms in the nanotube. In CNTFETs the mode space basis which uses the periodic boundary condition around the circumference of the nanotube is the appropriate approach for calculation of the Hamiltonian of the channel because this approach significantly reduces the size of the Hamiltonian matrix.

After calculation of the transmission coefficient (Table 1), the current of the transistor can be computed from Landauer-Buttiker formula:

$$I = \frac{4e}{h}\int T(E)[f_S(E) - f_D(E)]dE \tag{11}$$

Where $f_{S,D}(E)$ is the Fermi function at the source/drain contact.

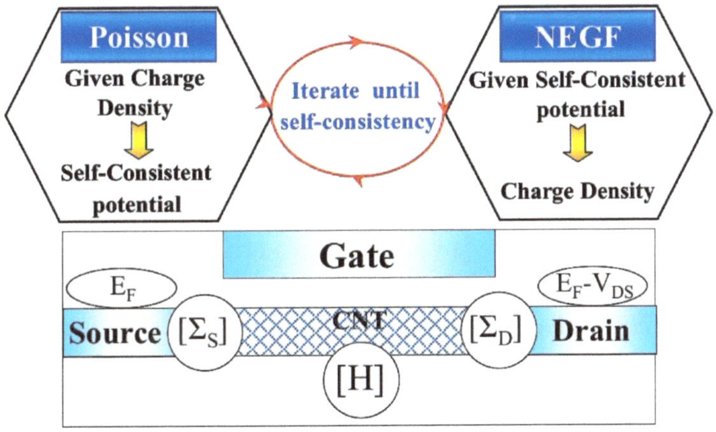

Fig. 16. The iterative method (NEGF) for calculating the potential and the charge density of the channel and the structure of the CNTFET under investigation.

Table 1. NEGF formalism and the description of physical quantities.

Physical Quantity	Matrix Computation	Description
Broadening function	$\Gamma_S = i[\Sigma_S - \Sigma_S^+], \Gamma_D = i[\Sigma_D - \Sigma_D^+]$	Related to the broadened density of states in the channel.
Spectral function	$[A_S(E)] = G\Gamma_S G^+$ $[A_D(E)] = G\Gamma_D G^+$	Diagonal elements give the LDOS at energy E in that representation: $D(E) = \frac{1}{2\pi}\text{Trace}[A(E)]$

Correlation function	$[G^n(E)] = [A_S]f_S + [A_D]f_D$	The matrix version of the electron density per unit energy. From $[G^n(E)]$ density matrix can be calculated.
Density Matrix	$[\rho] = \frac{1}{2\pi}\int [G^n(E)]dE$	Diagonal elements give the electron density: $n(x) = \text{Diag}[\rho]$
Transmission Coefficient	$T(E) = \text{Trace}[\Gamma_S G \Gamma_D G^+]$	Transmission probability from source to drain.

4.2 WKB Treatment of Schottky Barrier CNTFETs

The Wentzel-Kramers-Brillouin (WKB) approximation can be used to solve Schrödinger equation and find the tunneling probability at the source/drain-metal interface. This method is assumed a semi-classical approach. The channel potential distribution can be found by solving Laplace equation. The potential of the channel found from Laplace equation is valid as far as the device works in the quantum capacitance limit. At this limit, the nanotube quantum capacitance is very small and the associated accumulated charge is close to zero (Jiménez, et al., 2006). That's why the Laplace equation is valid under this condition. By using WKB method and solving Schrödinger equation, the transmission coefficient of the channel (probability of tunneling through the Schottky Barriers) is achieved which is a function of the potential energy:

$$T(E) = \exp\left[-2\int_{z_1}^{z_2} k(z)dz\right] \tag{12}$$

Where z_1 and z_2 are classical turning points and $k(z)$ is the wave number which can be calculated from:

$$k(z) = \frac{2}{3aV_0}\sqrt{\left(\frac{E_g}{2}\right)^2 - [E + eV(z)]^2} \tag{13}$$

where $a = 0.144$ nm, $E_g = 0.83$eV and $V_0 = 2.5$eV are the Carbon-Carbon bond length, the CNT band gap and the tight-binding parameter, respectively. $V(z)$ is the electrostatic potential along the CNT and is obtained by solving the Laplace equation. The computed potential profile is used as the input for the transmission coefficient and ultimately the current can be calculated from equation 11 (Kordrostami and Sheikhi, 2009c). The Schottky barrier, potential variations and the classical turning points are shown in figure 17.

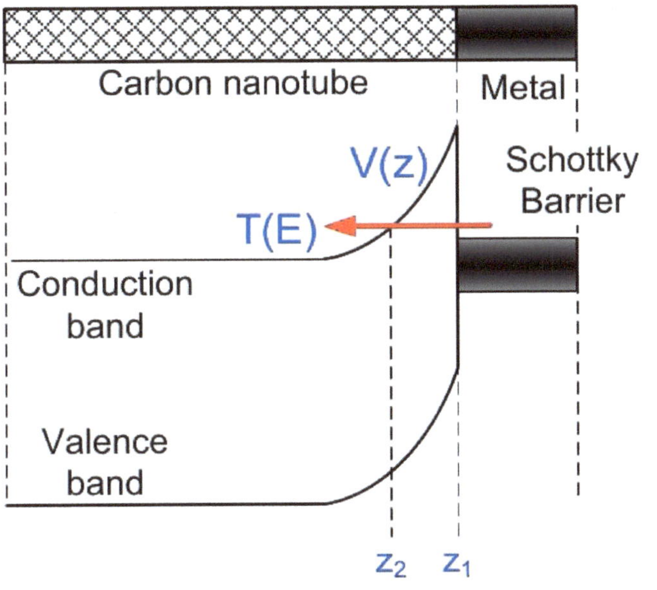

Fig. 17. Schottky barrier, potential variations near the channel-contact interface and the classical turning points which show the tunneling path.

5. Conclusion

The current-voltage characteristic of the SB-CNTFETs is ambipolar and the MOSFET-like CNTFETs exhibit unipolar characteristics. Asymmetric contact types and double gate structures are two ways for suppressing the ambipolarity of the SB-CNTFETs. Linear doping profile in MOSFET-like CNTFETs can reduce the band to band tunneling leakage current in negative voltages. High permittivity dielectrics are needed to ensure the proper control of the gate on the channel at the oxide thickness limit. The cutoff frequency of the CNTFETs severely depends on the gate and fringing capacitances. 1-D contacts can make the fringing capacitances as small as possible. Partially gated SB-CNTFET has smaller fringing capacitances and thus higher cutoff frequency in comparison with completely gated SBCNTFET. Two simulation methodologies are reliable for modelling CNTFETs: NEGF transport equation self-consistently with Poisson equation for both types of the transistors and semi-classical WKB method for calculating the tunneling current through Schottky barriers in SB-CNTFETs. By using the simulation methods we discussed some trade-offs between different parameters of a particular CNTFET design. In order to achieve a welldesigned nanotransistor, a compromise is always needed between different parameters.

6. References

Burke, P. J. (2004). AC performance of nanoelectronics: towards a ballistic THz nanotube transistor. Solid-State Electronics, 48, 10-11, 1981-1986, 0038-1101.

Guo, J. (2004). Carbon Nanotube Electronics: Modeling, Physics, and Applications. Purdue University.

Guo, J., Hasan, S., Javey, A., Bosman, G. and Lundstrom, M. (2005). Assessment of high frequency performance potential of carbon nanotube transistors. IEEE transactions on Nanotechnology, 4, 6, 715-721, 1536-125X.

Guo, J. and Lundstrom, m. (2006). Nanoscale Transistors: Device Physics, Modeling and Simulation, Springer, 978-0-387-28002-8, United States of America.

Guo, J. and Lundstrom, M. (2009). Device Simulation of SWNT-FETs, In: Carbon Nanotube Electronic, Javey, A. and Kong J., (Ed.), Springer, 978-0-387-36833-7, New York, USA.

Guo, J., Yoon, Y. and Ouyang, Y. (2007). Gate Electrostatics and Quantum Capacitance of Graphene Nanoribbons. Nano Letters, 7, 7, 1935-1940, 1530-6984.

Hassaninia, I., Kordrostami, Z., Sheikhi, M. H. and Ghayour, R. (2008a). Non-Equilibrium Green's Function Calculations for Double Gate Coaxial Carbon Nanotube FETs. Proceedings of Nanostructures. March 2008, Kish Island, Iran.

Hassaninia, I., Sheikhi, M. H. and Kordrostami, Z. (2008b). Simulation of carbon nanotube FETs with linear doping profile near the source and drain contacts. Solid-State Electronics, 52, 6, 980-985, 0038-1101.

Heinze, S., Tersoff, J. and Avouris, P. (2005). Carbon Nanotube Electronics and Optoelectronics, In: Introducing Molecular Electronics, Cuniberti, G. and Richter, K., (Ed.), 381-409, Springer, 978-3-540-27994-5, Netherlands.

Iijima, S. (1991). Helical microtubules of graphitic carbon. Nature, 354, 6348, 56-58.

Javey, A., Kim, H., Brink, M., Wang, Q., Ural, A., Guo, J., McIntyre, P., McEuen, P., Lundstrom, M. and Dai, H. (2002). High-[kappa] dielectrics for advanced carbonnanotube transistors and logic gates. Nat Mater, 1, 4, 241-246, 1476-1122.

Javey, A., Tu, R., Farmer, D. B., Guo, J., Gordon, R. G. and Dai, H. (2005). High Performance n-Type Carbon Nanotube Field-Effect Transistors with Chemically Doped Contacts. Nano Letters, 5, 2, 345-348, 1530-6984.

Jiménez, D., Cartoixà, X., Miranda, E., Suñé, J., Chaves, A. F. and Roche, S. (2006). A drain current model for Schottky-barrier CNT-FETs Computational Electronics, 5, 4, 361-364.

Khairul, A. and Roger, K. L. (2006). Dielectric scaling of a zero-Schottky-barrier, 5 nm gate, carbon nanotube transistor with source/drain underlaps. Journal of Applied Physics, 100, 2, 024317.

Kordrostami, Z. (2007). Circuit Modeling of Carbon Nanotube Transistors and Interconnects for Integrated Circuit Applications. Electrical and Electronic Engineering. Shiraz University.

Kordrostami, Z., Hassaninia, I. and Sheikhi, M. H. (2008). Unipolar Schottky-Ohmic carbon nanotube field effect transistor. Proceedings of 3rd IEEE International Conference on Nano/Micro Engineered and Molecular Systems, 529-531, 978-1-4244-1907-4, Jan. 2008.

Kordrostami, Z. and Sheikhi, M. H. (2009a). Contact Geometry dependent Electrostatics of Carbon Nanotube Transistors, Nanoscience and Technology, September 2009, Beijing, China.

Kordrostami, Z. and Sheikhi, M. H. (2009b). High Speed Switching Performance of Carbon Nanotube Field Effect Transistors, Nanotech Europe, September 2009, Berlin, Germany.

Kordrostami, Z., Sheikhi, M. H. (2009c). Schottky Barrier Field Effect Transistors with a Strained Carbon Nanotube Channel. Journal of Computational and Theoretical Nanoscience, 7, 6, 216-221, 1546-1955.

Martel, R., Schmidt, T., Shea, H. R., Hertel, T. and Avouris, P. (1998). Single- and multi-wall carbon nanotube field-effect transistors. Applied Physics Letters, 73, 17, 2447-2449.

Pourfath, M., Ungersboeck, E., Gehring, A., Kosina, H., Selberherr, S., Park, W.-J. and Cheong, B.-H. (2005). Numerical Analysis of Coaxial Double Gate Schottky Barrier Carbon Nanotube Field Effect Transistors. Journal of Computational Electronics, 4, 1, 75-78, 1569-8025.

Raychowdhury, A., Keshavarzi, A., Kurtin, J., De, V. and Roy, K. (2006). Carbon Nanotube Field-Effect Transistors for High-Performance Digital Circuits; DC Analysis and Modeling Toward Optimum Transistor Structure. IEEE Transactions on Electron Devices, 53, 11, 2711-2717, 0018-9383.

Soh, H. T., Quate, C. F., Morpurgo, A. F., Marcus, C. M., Kong, J. and Dai, H. (1999). Integrated nanotube circuits: Controlled growth and ohmic contacting of singlewalled carbon nanotubes. Applied Physics Letters, 75, 5, 627-629, 0003-6951.

Tans, S., Verschueren, A. and Dekker, C. (1998). Room-temperature transistor based on a single carbon nanotube. Nature 393, 49-52, 0028-0836.

Yoon, Y., Yijian, O. and Jing, G. (2006). Effect of phonon scattering on intrinsic delay and cutoff frequency of carbon nanotube FETs. , IEEE Transactions on Electron Devices, 53, 10, 2467-2470, 0018-9383.

Yu, Z., Rutherglen, C. and Burke, P. J. (2006). Microwave nanotube transistor operation at high bias. Applied Physics Letters, 88, 23, 233115, 0003-6951.

Chapter-10

Gate-Controlled Particle–Wave Duality in SWCNT Hole Transistors

1. Introduction

Various electrode materials for single-walled carbon nanotube (SWNT) transistors were investigated. Pd electrodes have been used for Ohmic contacts (Javey et al., 2003). Ti electrodes have been used for Schottky contacts for hole conduction (Heinze et al., 2002; Martel et al., 2001; Kamimura & Matsumoto, 2005), and Mg and Ca electrodes for electron conduction (Nosho et al., 2006). Moreover, in the case of sub- μm order channel length at low temperature, SWNT transistors with Ohmic contacts have shown resonant tunneling transistor (RTT) characteristics (Liang et al., 2001), which are also called as Fabry-Perot characteristics, and SWNT transistors with Schottky contacts have shown single-hole transistor (SHT) characteristics (Suzuki et al., 2001), in which the Schottky barriers act as tunneling barriers. Therefore, electrodes materials should be chosen to obtain the desired characteristics.

In this study, we succeeded in fabricating a multifunctional quantum transistor using the particle nature and wave nature of holes in SWNT. This transistor can operate as an RTT and also as an SHT. An RTT is a device that uses the wave nature of hole and an SHT uses the particle nature of hole in the SWNT. Both devices need tunneling barriers at both sides of the quantum island. The RTT needs strong coupling while the SHT needs weak coupling between the quantum island and the electrodes. Usually, these tunneling barriers are made from thin oxide layers, etc. Therefore, the thickness of the tunneling barriers and the coupling strength cannot normally be controlled in a given device. In the present device, however, the Schottky barriers act as the tunneling barriers between the SWNT quantum island and electrodes. Therefore, the thickness of the tunneling barriers and the coupling strength between the SWNT and electrodes can be controlled by the applied gate voltage V_G.

Moreover, a SWNT is a cylindrical material with a diameter of several nanometers. The small diameter makes it possible to detect an electrical field from even a single-charge. Moreover, by observing the relative energy difference between the conducting carrier and the single-charges to be measured, it is possible to define the potential energy of the singlecharges to be measured. However, as described in many reports (Kim et al., 2003; Radosavljevic et al., 2002), SWNT electron devices show hysteresis characteristics in gate voltage-drain current characteristics. The hysteresis characteristics are caused by gate-voltage-dependent charge fluctuation, e.g., adsorption of water molecules around a SWNT (Kim et al., 2003), charging into insulator layer around a SWNT (Radosavljevic et al., 2002), and charging into amorphous carbon around a SWNT (Martel et al., 2001) By eliminating these origins of the hysteresis characteristics, the number of fluctuating charges becomes small and a single-charge can be detected by a SWNT multi-functional quantum transistor.

2. Materials and methods

2.1 Method

We have eliminated the three origins of the hysteresis characteristics of a SWNT field effect transistor mainly pointed out in current reports (Martel et al., 2001; Kim et al., 2003; Radosavljevic et al. 2002; Kamimura & Matsumoto, 2004). To burn out amorphous carbon, we annealed a SWNT at low temperature in oxidizable atmosphere (Kamimura & Matsumoto, et al., 2004). To reduce the number of adsorbed atmosphere molecules, we covered the channel with a silicon dioxide layer. To reduce the number of trap sites in the insulator, we reduced channel length to 73 nm . The SWNT multi-functional quantum transistor fabricated by the process mentioned above shows almost no hysteresis characteristics in the gate voltage rage from -40 to 40 V . Moreover, an abrupt discrete switching of the source-drain current is observed in the electrical measurements of the SWNT multi-functional quantum transistor at 7.3 K . These random telegraph signals (RTS) are attributed to charge fluctuating charge traps near the SWNT multi-functional quantum transistor conduction channel. The current-switching behavior associated with the occupation of individual electron traps is demonstrated and analyzed statistically.

2.2 Sample preparations

A schematic of the sample structure is shown in Fig. 1. SWNT was prepared as follows. An $n^+ - Si$ wafer with a thermally grown 300 nm thick oxide was used as a substrate. Layered Fe/Mo/Si(2 nm/20 nm/40 nm) catalysts were evaporated using an electron-beam evaporator under a vacuum of 10^{-6} Pa. These layered catalysts were patterned on the substrate using the conventional photo-lithography process. SWNT was grown by thermal chemical vapor deposition (CVD) using the mixed gases of hydrogen and argon-bubbled ethanol. After the growth of the SWNT, it was purified by burning out the amorphous carbon around the SWNT in an air atmosphere at a temperature of several hundred degrees Celsius (Kamimura and Matsumoto, 2004). Ti (30 nm) electrodes were deposited on the patterned catalysts as the source and drain, and on the back side of the n^+-Si substrate for the gate, using the electron-beam evaporator under a vacuum of 10^{-6} Pa. The distance (L) between the source and drain was 73 nm . Thus, a back gate type multi-functional quantum transistor with an SWNT channel was fabricated that had the functions of an RTT and an SHT. The single-charge measurement was carried out with the structure that silicon dioxide layer is on the SWNT channel.

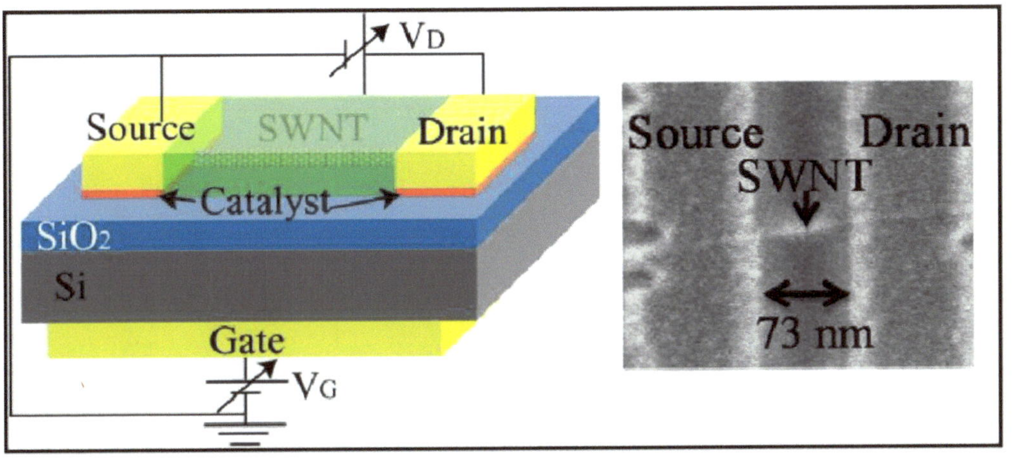

Fig. 1. Schematic structure of SWNT multi-functional quantum transistor covered by silicon dioxide layer. The channel length is 73 nm . The inset shows a SEM image around the channel before silicon dioxide deposition. A few charge storages are fabricated in the SiO_2 layer.

3. Results and discussions

3.1 Particle-Wave duality

Fig. 2(a) shows the differential conductance dI_D/dV_D characteristic as a function of the V_G at 7.3 K , where the drain voltage was set at 8 mV .

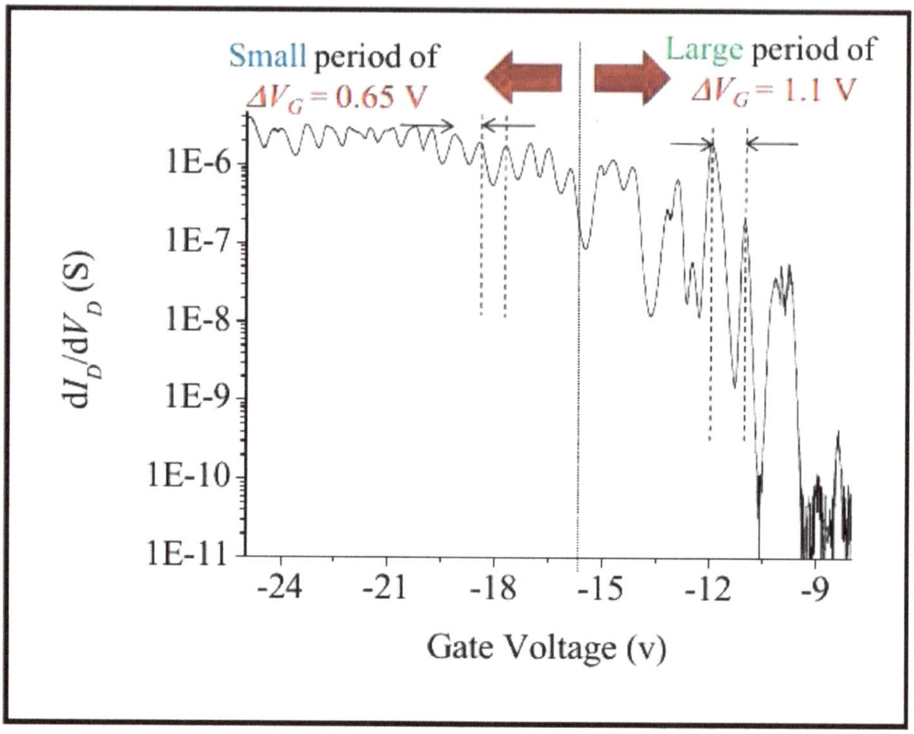

Fig. 2. (a) Differential conductance dI_D/dV_D characteristic as a function of V_G at 7.3 K . Two frequencies of oscillation were observed, which were $\Delta V_G = 1.1$ V and 0.65 V .

An oscillation characteristic with two oscillation periods is also observed in Fig. 2(a). A large oscillation period of $\Delta V_G = 1.1$ V was at $V_G \geq -16$ V and a small oscillation period of $\Delta V_G = 0.65$ V was at $V_G \leq -16$ V. Fig. 2(b) shows the dI_D/dV_D peak on a linear scale at $V_G \geq -16$ V and $V_G \leq -16$ V. A clear difference in the oscillation period is observed as follows. The peaks of dI_D/dV_D at $V_G \geq -16$ V are well fitted by the Gaussian, i.e.

$$G \approx \frac{2e^2}{h}\frac{T_L T_R}{\Gamma}\frac{1}{2kT}\exp\left(\frac{\varepsilon_F - \varepsilon_0}{2kT}\right), \tag{1}$$

and at $V_G \leq -16$ V by the Lorentzian, i.e.

$$G = \frac{2e^2}{h}\frac{4T_L T_R}{(\varepsilon_F - \varepsilon_0)^2 + \Gamma^2} \tag{2}$$

as shown in Fig. 2(c) and (d), respectively, where e is the elementary charge, h is the plank constant, T_L and T_R are the tunneling probabilities at the left and right tunneling barriers,

Γ is the full width at half maximum, k is the Boltzmann constant, T is the temperature, ε_F is the Fermi level, and ε_0 is the quantum level.

(b)

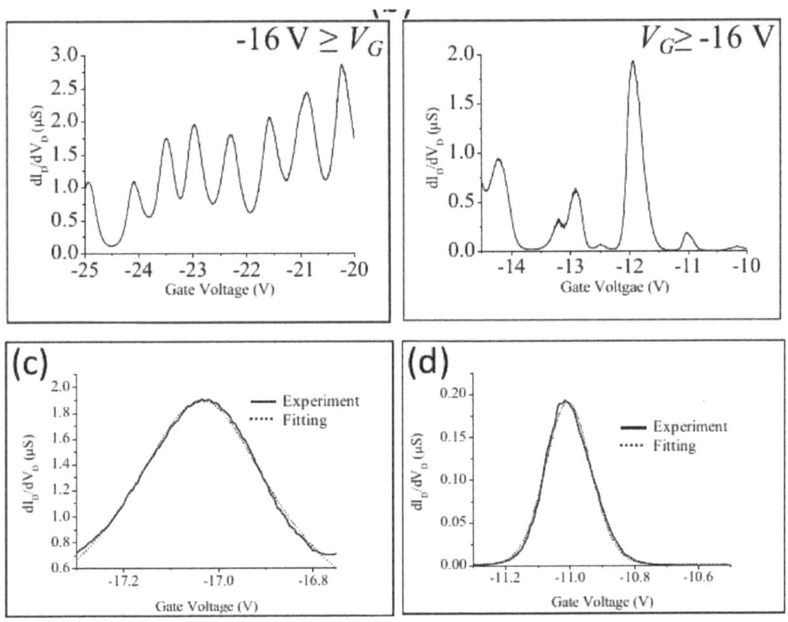

Fig. 2. (b) dI_D/dV_D peak in linear scale at low and high V_G. (c) dI_D/dV_D peak at $V_G = -17$ V (solid) and Lorentzian fitting line (dotted). The peak is well fitted by the Lorentzian function, which means the shape of the peak can be attributed to energy uncertainty broadening. Therefore, the peak must be a resonant tunneling current peak. (d) dI_D/dV_D peak at $V_G = -11$ V (solid) and Gaussian fitting line (dotted). The peak is well fitted by the Gaussian function, which means the shape of the peak can be attributed to thermal broadening. Therefore, the peak must be a Coulomb oscillation peak.

The shape of the Coulomb oscillation peaks must be Gaussian, which is attributed to thermal broadening, while the resonant tunneling current peaks must be Lorentzian, which is attributed to energy uncertainty (Radosavljevic et al., 2002). Therefore, the dI_D/dV_D oscillation at $V_G \geq -16$ V in Fig. 2(a) should be Coulomb oscillation peaks, and at $V_G \leq -16$ V in Fig. 2(a), they should be resonant tunneling current peaks. In other words, the device operates in particle nature mode at $V_G \geq -16$ V, and in wave nature mode at $V_G \leq -16$ V.

Fig. 3 shows the dI_D/dV_D characteristic as a function of the drain voltage V_D at 7.3 K and $V_G = -12.795$ V. The dI_D/dV_D peak spacing of 26 mV is observed in the plot, which corresponds to the quantum energy level separation in the SWNT quantum island. The separation of the quantum energy levels is indicated by

$$\Delta E_Q = \left(\frac{hv_F}{2L}\right)\left[1 + \left(\frac{2L}{3rn}\right)^2\right]^{-\frac{1}{2}}, \tag{3}$$

where r is the radius of the SWNT, L is the length of the SWNT, and v_F is the Fermi velocity (Kamimura & Matsumoto, 2004). When n becomes large, the equation becomes

$$\Delta E_Q = \frac{hv_F}{2L} \tag{4}$$

and shows a constant value independent of n.

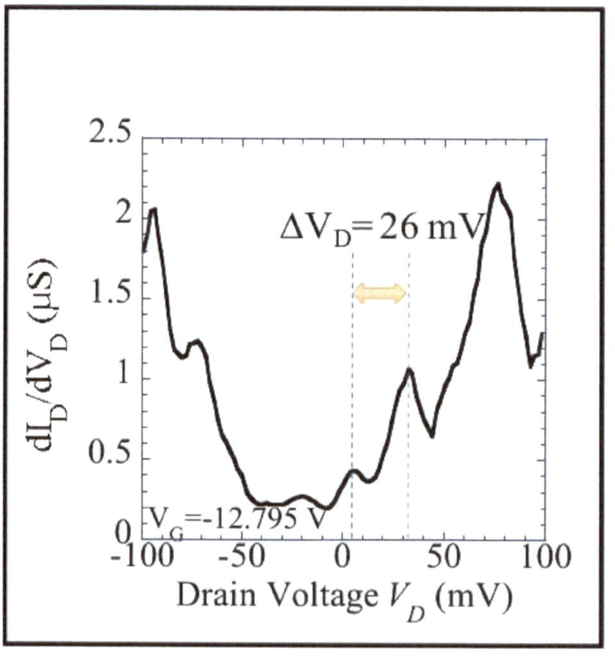

Fig. 3. dI_D/dV_D characteristic as a function of drain voltage V_D at $V_G = -12.795$ V and 7.3 K . The peaks are attributed to quantum energy levels in the SWNT. The estimated energy separation in 73 nm SWNT was 24 meV , which is in good agreement with the peak spacing of 26 mV in the plot.

The energy separation ΔE_Q of the quantum levels for an SWNT length of 73 nm is calculated to be 24 mV from eq. (2). Because this estimated value of energy separation of 24 mV is in good agreement with the dI_D/dV_D peak spacing of 26 mV in Fig. 3, it can be concluded that the entire SWNT channel acts as a single quantum island.

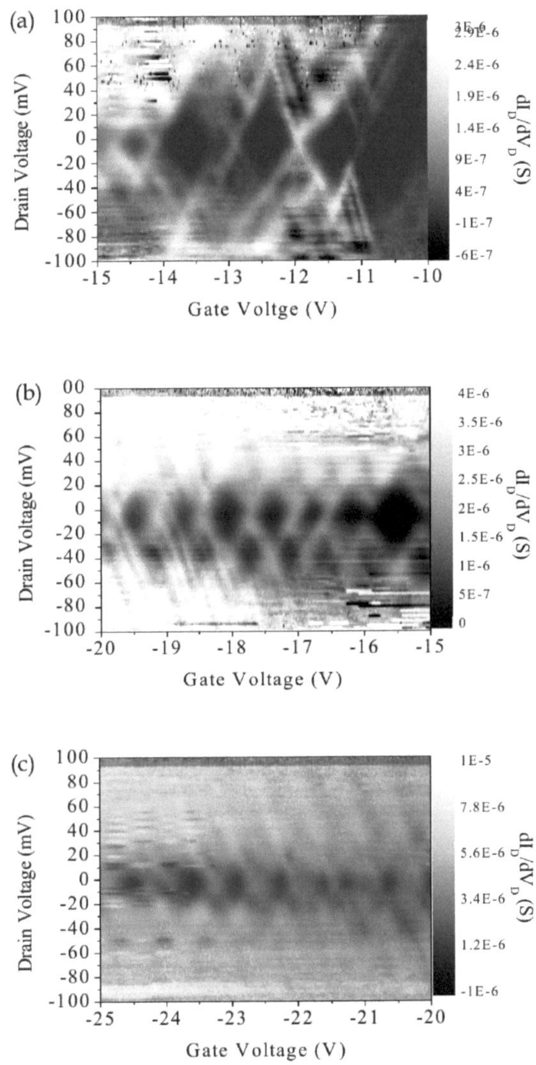

Fig. 4. Contour plot of dI_D/dV_D characteristic as a function of V_G and V_D at 7.3 K . (a) At $V_G \geq -16$ V, Coulomb diamond structures and line shaped quantum levels outside of the Coulomb diamond structures are observed. Moreover, the Coulomb diamonds become smaller with a negatively increasing V_G. (b) At $-16\text{ V} \geq V_G \geq -20$ V, the Coulomb blockade is lifted and the Coulomb diamonds disappear. And the quantum levels become blurred with the negatively increasing V_G. The so-

called Fabry-Perrot quantum interference pattern is observed. (c) At -20 V $\geq V_G$, the blurred quantum levels still remain.

Fig. 4 shows a contour plot of the dI_D/dV_D characteristic as a function of V_G and V_D at 7.3 K . The characteristic can also be divided into two modes, the particle nature mode and wave nature mode. At $V_G \geq -16$ V, as shown in Fig. 4(a) and (b), the plot clearly shows the Coulomb diamond structures, which are getting smaller with negatively increasing V_G. Additionally, line shaped quantum levels are observed outside of these Coulomb diamond structures. Therefore, at $V_G \geq -16$ V, the device operated in the particle nature mode. As shown in Fig. 4(b), at $-16 \geq V_G \geq -20$ V, the Coulomb blockade was lifted and the Coulomb diamond structures disappeared at around $V_G = -16$ V. The quantum levels still remain and the so-called Fabry-Perrot quantum interference pattern is observed at this region. Finally, at $V_G \leq -20$ V, the quantum levels are getting blurred with negatively increasing V_G, as shown in Fig. 4(c). Therefore, at -16 V $\geq V_G$, the device operated in the wave nature mode.

Fig. 5(a) shows the Coulomb charging energy E_C as a function of V_G. E_C is obtained by

$$E_C = \Delta E - E_Q \tag{5}$$

where ΔE is estimated from the size of the Coulomb diamond. E_C drastically decreases with a negatively increasing V_G, almost reaching zero at $V_G \geq -16$ V. Therefore, the Coulomb diamonds disappear at -16 V $\geq V_G$. Tunneling capacitance C_t and gate capacitance C_G as a function of V_G are shown in Fig. 5(b). C_t and C_G are obtained from

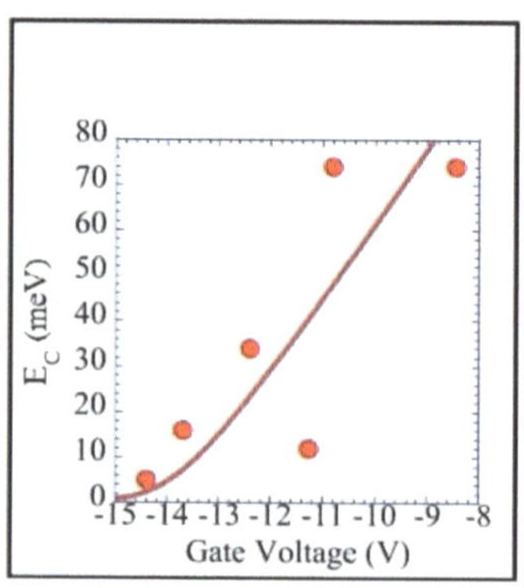

Fig. 5. (a) Coulomb charging energy E_C as a function of V_G. E_C drastically decreases at $V_G \geq -16$ V and reaches zero at around $V_G = -16$ V. Therefore, the Coulomb diamonds disappear at the mid- V_G region.

$$C_t = e/2E_C - C_G \tag{6}$$

and

$$C_C = \mathrm{e}/\Delta V_G \tag{7}$$

respectively. C_t depends on the thickness of the tunneling barrier and C_G depends on the gate structure. C_t drastically increases with a negatively increasing V_G, while C_G is almost constant, independent of the changing V_G. The drastic increase in C_t is attributed to the decrease in the thickness of the Schottky barriers at the contacts between the SWNT quantum island and electrodes (Javey et al., 2003). When the Schottky barriers become thin with negatively increasing V_G, as shown in the inset of Fig.5(b), C_t drastically increases and the coupling strength of the wave function between the outside and the inside of the Schottky barrier becomes stronger. Therefore, the Coulomb blockade is lifted and E_C becomes zero at $-16\text{ V} \geq V_G$.

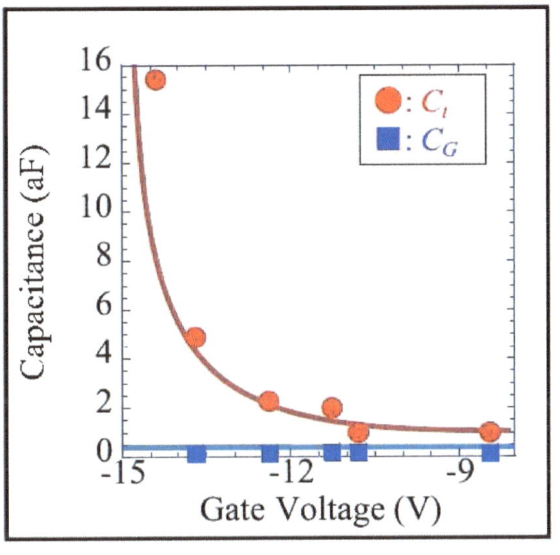

Fig. 5. (b) Tunneling capacitance C_t and gate capacitance C_G characteristics as functions of V_G. C_t drastically increases at $V_G \geq -16$ V. On the other hand, C_G is almost constant, independent of the changing V_G. The drastic increase of C_t is attributed to the change in the thickness of the Schottky barriers at the contact between the SWNT channel and electrodes.

The dependence of the full-width at half maximum (FWHM) of the resonant tunneling current peak characteristic on V_G is shown in Fig. 6(a).

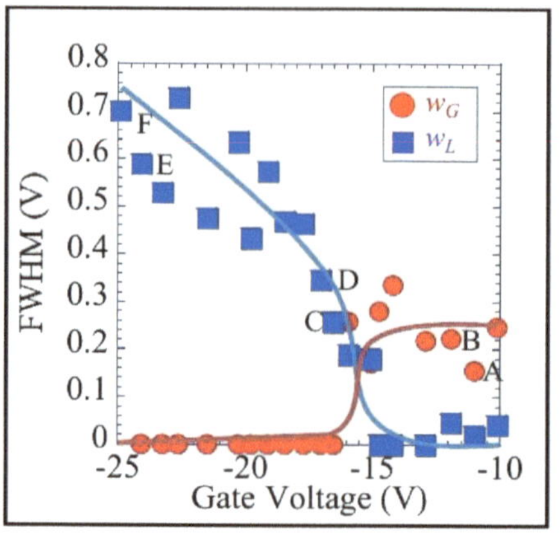

Fig. 6. (a) FWHM of the dI_D/dV_D peak characteristics as a function of V_G. w_G is almost constant, independent of V_G and w_L is negligibly small at $V_G \geq -16$ V. w_G become vanishingly small and w_L increases linearly with a negatively increasing V_G at $-16\ \text{V} \geq V_G$.

The FWHM is estimated from an I_D-V_G plot using the Voigt function, which is a convolution of the Gaussian and Lorentzian functions, and can be used to divide the FWHM into the FWHM of the Gaussian w_G and that of the Lorentzian w_L. Fig. 6(b)-(d) shows several curves fitted by the Voigt function, where the solid line is the experimental data, the dotted lines are the fitting curves, the dashed line is the cumulative fitting curves, and each alphabetic marker for peaks A-F corresponds to a marker in the plot of Fig. 6(a).

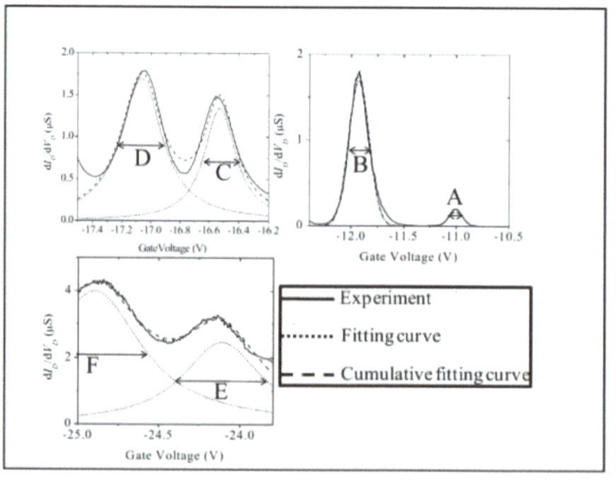

Fig. 6. (b)-(d) FWHMs: the experimental data are indicated by the black line, the fitting curves are indicated by the blue lines, the cumulative fitting curves are indicated by red line, and each alphabetic marker for a peak corresponds to a marker in the plot of Fig. 6(a).

In the particle nature mode of $V_G \geq -16$ V, w_L is negligibly small and the w_G of about 0.2 V is independent of V_G because the FWHM of the Coulomb oscillation mainly depends on the thermal broadening of Fermi dispersion (Foxman et al., 1993). On the other hand, in the wave nature mode of $-16\text{ V} \geq V_G$, w_G becomes negligibly small and w_L increases linearly with the negatively increasing V_G. w_L is proportional to the tunneling probability as follows:

$$\alpha w_L = T\, \mathrm{h} v_F / 2L, \tag{8}$$

where α is the ratio of the modulated energy to applied V_G, and T is the tunneling probability (Kamimura & Matsumoto, 2006). Therefore, the increase in w_L seen in Fig. 6 represents an increase in the tunneling probability, which is attributed to the decrease in the thickness of the Schottky barriers with a negatively increasing V_G. The logarithmic dependence of the drain current I_D on the inverse of the temperature is shown in Fig. 7.

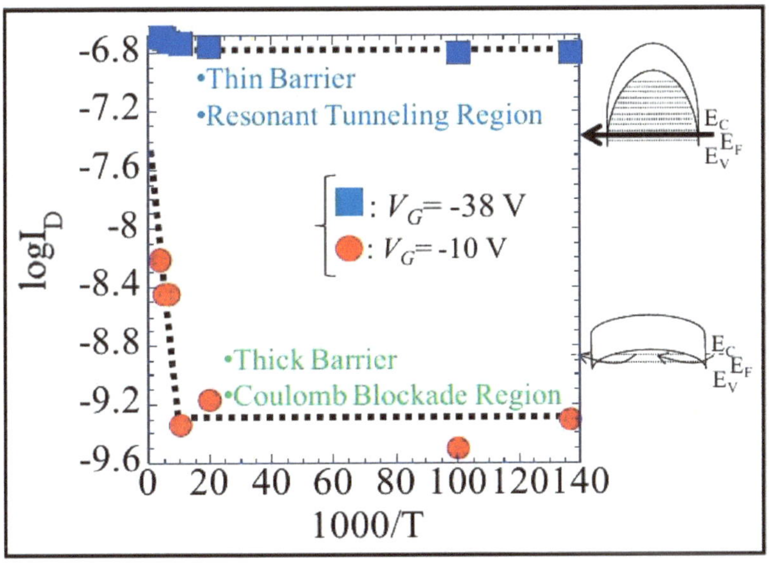

Fig. 7. The logarithmic dependence of the drain current I_D on the inverse of the temperature. At $V_G = -38$ V of the wave nature mode, I_D is almost constant, independent of the temperature. In the wave nature mode, the Schottky barrier is so thin that the tunneling current becomes dominant. In contrast, at $V_G = -10$ V of the particle nature mode, I_D drastically increases at the high temperature region, and the Schottky barrier height estimated from the slope is $\Delta\Phi = 50$mV.

3.2 Hysteresis elimination

Fig. 8 shows the temperature dependence of differential conductance as a function of the gate voltage.

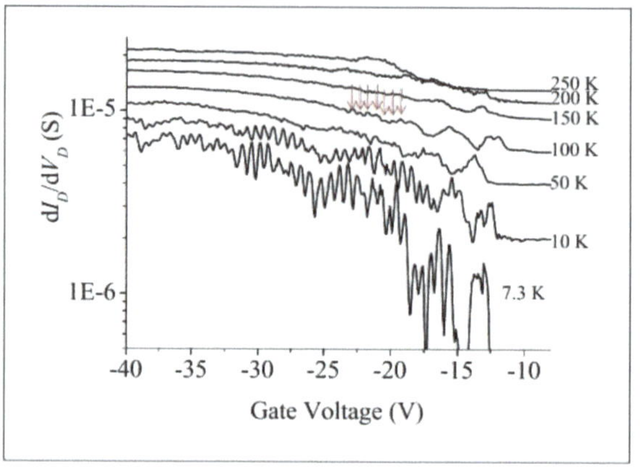

Fig. 8. Temperature dependence of differential conductance on V_G. Each plot is shifted by 3μ S to make it easier to see. The quantum level peaks are becoming blurred with increasing temperature. However, even at = 100 K, the quantum levels remain, as indicated by the arrows.

The quantum levels peaks became blurred with increasing temperature. However, even at $T = 100$ K, the quantum levels still remained, as indicated by arrows.

The static characteristic measurement of the devise with silicon dioxide layer on the SWNT channel shown in Fig. 9 was carried out using an Agilent B 1500.

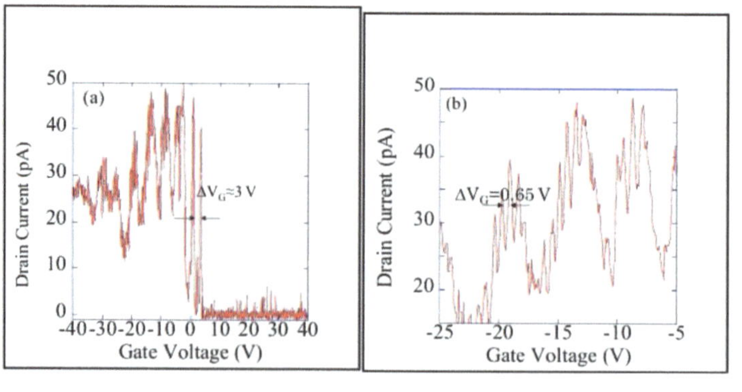

Fig. 9. (a) Coulomb oscillation characteristic at 7.3 K . Drain current was observed only in negative gate voltage regions, which indicates that the measured SWNT is a p-type semiconductor. A large period of ~ 3 V at an applied gate voltage of about

$V_G = 0$ V was observed, which was attributed to Coulomb oscillation characteristic. (b) The small period of $\Delta V_G = 0.65$ V of drain current peaks at 7.3 K , which was attributed to the quantum interference property, namely, Fabry-Perot interference of the hole, the coherent length of which was more than 73 nm . Coherent transport in the entire channel was achieved.

In the measurement, the data were integrated for a few seconds to eliminate the effect of noise, which was set by the equipment automatically. After that, the data were recorded. In the drain current I_D-gate voltage V_G characteristics of the SWNT multi-functional quantum transistor under a drain voltage of 11 mV at 7.3 K , the drain current showed a periodic peakand valley structure with two periods. As shown in Fig. 9(a), I_D was observed only in the negative V_G region, which indicates that the measured SWNT was a p-type semiconductor. Moreover, a large period of ~ 3 V at around $V_G = 0$ V was observed, which was attributed to the Coulomb oscillation characteristic. A part of Fig. 9(a) is expanded in the horizontal scale and shown in Fig. 9(b), where small period of $\Delta V_G = 0.65$ V was observed at a higher V_G, which was attributed to the Fabry-Perot interference property (Liang et al., 2001) of hole, the coherent length of which was more than 73 nm , and coherent transport in the entire channel was achieved. Fig. 9(c) shows a contour plot of the second-order differential conductance as a function of gate and drain voltages, in which a clear Fabry-Perot interference pattern can be seen.

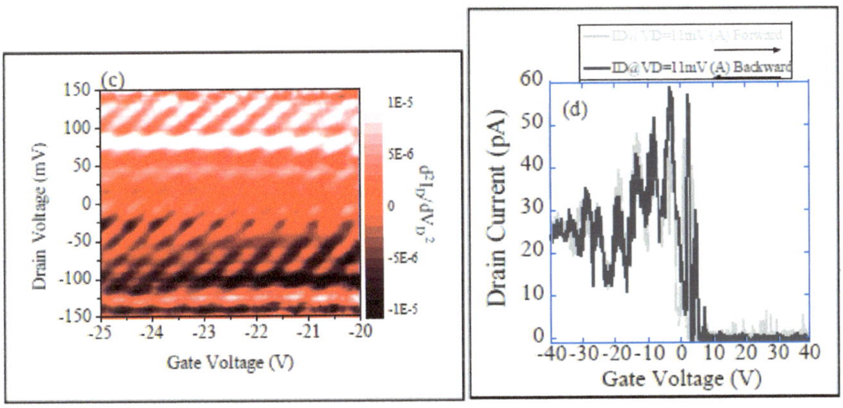

Fig. 9. (c) Contour plot of second-order differential conductance as a function of gate and drain voltages at 7.3 K , which shows a clear Fabry-Perot interference pattern. (d) Drain current-gate voltage characteristics with round trip applied gate voltage sweeping from $V_G = 40$ to -40 V after sweeping from $V_G = -40$ to 40 V at 7.3 K . The two drain current characteristics almost overlapped.

From the small period of drain current oscillation of $\Delta V_G = 0.65$ V and the equation (Kamimura & Matsumoto, 2006) of

$$\Delta E_Q = \alpha V_G = \mathrm{h} v_F / 2L\mathrm{e}, \tag{9}$$

where ΔE_Q is the quantum energy separation, a is the gate modulation coefficient, h is Plank's constant, v_F is the Fermi velocity in graphene, L is the length of the cavity of the hole, and e is the elementary charge, we estimated the width of the quantum well and obtained the length of the cavity of the hole L to be 55 nm . The estimated value is in agreement with the channel length 73 nm of the device which was obtained by scannig electron microscope (SEM) (inset of Fig. 1). The slight difference between the estimated L and the channel length must be attributed to the width of band bending in the SWNT e.g., Schottky barriers.

In Fig. 9(d), the gray (black) line shows drain current vs. applied gate voltage from -40 (40) to $40(-40)$V in a forward (backward) sweep of gate voltage. The SWNT multi-functional quantum transistor showed almost no hysteresis characteristics. The SWNT multi-functional quantum transistor was covered with a silicon dioxide layer, which prevents the SWNT from absorbing and releasing molecules in ambient, because fluctuation of molecules on the SWNT usually induces the hysteresis characteristics in the drain current-gate voltage characteristics. Moreover, the SWNT multi-functional quantum transistor has a small channel length of 73 nm , which reduced the number of trap sites near the SWNT and, at the same time, the number of trapped carriers causing hysteresis characteristics. We believe that the purification was effective to reduce the hysteresis characteristics. Thus, the hysteresis characteristic in the drain current-gate voltage characteristic was eliminated.

3.2 Single charge sensitivity of SWNT multi-functional quantum transistor

Figure 10(a) shows the time dependence of the conductance of the SWNT multi-functional quantum transistor at 7.3 K with a gate bias of $V_G = -25.36$ V. A short sampling time of 10 ms was set in the dynamic characteristic measurements shown in Fig. 10.

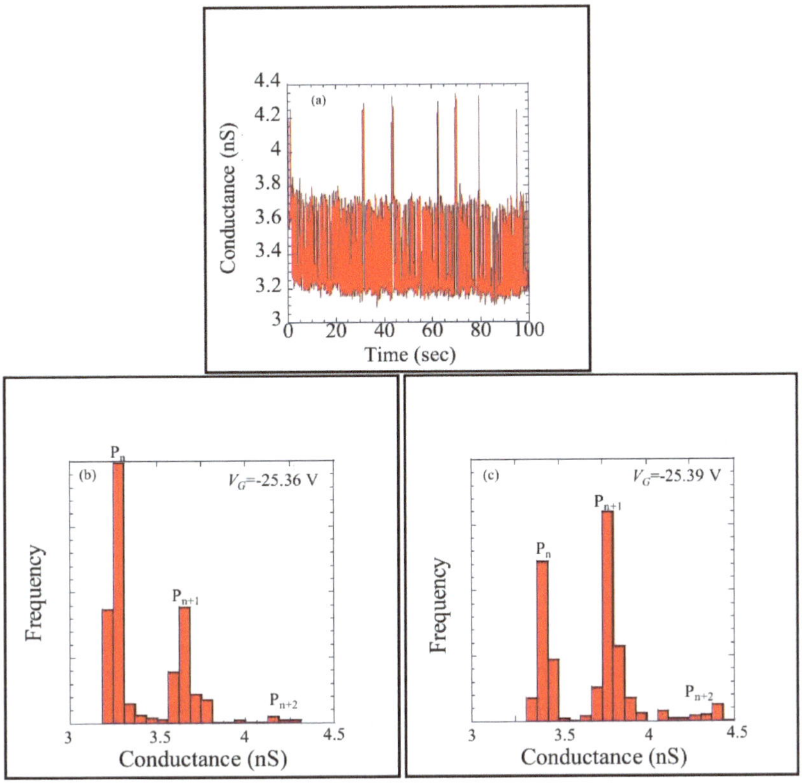

Fig. 10. (a) Time dependence of drain current with RTS at a gate voltage of $V_G = -25.36$ V. The time dependence of drain current was sampled for 100 s and the sampling time was 10 ms ; (b) and (c) show histograms of the conductance levels of RTS at $V_G = -25.36$ and -25.39 V , respectively. P_n decreases and P_{n+1} and P_{n+2} increase with slightly increasing applied gate voltage from $V_G = -25.36$ to -25.39 V .

The applied gate voltage was under the Fabry-Perot interference region. The SWNT multifunctional quantum transistor shows RTS, as shown in Fig. 10(a), The RTS showed three levels, n, n + 1 and n + 2, of the conductance shown in Fig. 10(a). At a lower applied gate voltage, current levels higher than n + 2 such as n + 3 and n + 4 appeared. The multiple levels of RTS are attributed to charge fluctuating charge storages near the conduction channels of the SWNT multi-functional quantum transistor. Moreover, because there was a single-charge storage including multiple energy levels or were some charge storages being at almost the same distances from the conductance channel of the SWNT multi-functional quantum transistor, the RTS appeared. Figures 10(b) and 3(c) show histograms of the conductance levels of RTS at $V_G = -25.36$ V and $V_G = -25.39$ V, respectively. The three peaks of conductance of RTS expressed as P_n, P_{n+1}, and P_{n+2} are shown in Figs. 10(b) and 10(c). The conductance levels of the three peaks directly correspond to the conductance levels of RTS. On the other hand, the relative heights of the peaks correspond to occupation probabilities at each conductance level of the RTS. The heights of the peaks depend on applied gate voltage. P_n decreases and P_{n+1} and P_{n+2} increase with slightly increasing applied gate voltage from $V_G = -25.36$ to -25.39 V , which means that the energy levels in the charge storage are modulated by applied gate voltage.

The gate voltage dependences of the natural log of the ratio between the m_{th} peak and the $(m+1)_{\text{th}}$ peak in the conductance histogram $P_{m+1}/P_m (m = n, n+1, n+2, \cdots, n+4)$ are shown in Fig. 11(a). The natural log of P_{m+1}/P_m linearly depends on applied gate voltage and saturates in each V_G. Each starting point of saturation is marked by an arrow in Fig. 11(a). The charge storage energy levels are floating. Therefore, modulations of energy by V_G may be different at each charge storage. We believe that the reason why the natural log of P_{m+1}/P_m saturates at each V_G may depend on the difference in energy modulation by V_G at each charge storage.

The charge transition is modeled, as shown in Fig. 12(a), in which the energy barrier is between the SWNT and the charge storage. Figures 11(b)-11(e) are enlargement plots of each $P_{m+1}/P_m(\mathrm{m} = \mathrm{n}, \mathrm{n}+1, \mathrm{n}+2, \cdots, \mathrm{n}+4)$. The energy differences ΔE_n between the charge storage energy level E_n and the Fermi level E_f are expressed as $\Delta E_n = E_f - E_n$, which is modulated by the applied gate voltage V_G. According to equilibrium statistical mechanics, P_{m+1}/P_m is given by (Peng et al., 2006)

$$P_{m+1}/P_m = (g_f/g_s)\mathrm{e}^{-\beta(E_f - E_n)} = (g_f/g_s)\mathrm{e}^{-\beta \Delta E_n} \qquad (10)$$

where g_{f} and g_{s} are the degeneracy of the top of valence band and the charge storage, respectively. $\mathrm{g_f}/\mathrm{g_s}$ is assumed to be $1. \beta$ is 1/kT. E_n includes the contributions of e electrostatic potential induced by V_G, intrinsic energy level in the storage, and Coulomb charging energy. The basis of eq. (1) is the Arrhenius equation. In this model, the height of the barrier is the energy difference between E_f and E_n. Assuming a linear dependence of ΔE_n on $V_{\mathrm{G}}, \Delta E_n$ can be written as $\Delta E_n = ae(V_0 - V_{\mathrm{G}})$, where a is the gate modulation coefficient, and is a constant value of 0.062. a is obtained from the periods of Fabry-Perot interference characteristic on V_D and V_G. V_0 is the offset voltage, and is obtained from the intersecting point of the extrapolating line of the fitting lines and the line of $\ln (P_{m+1}/P_m) = 0$. Therefore, eq. (1) is transformed to

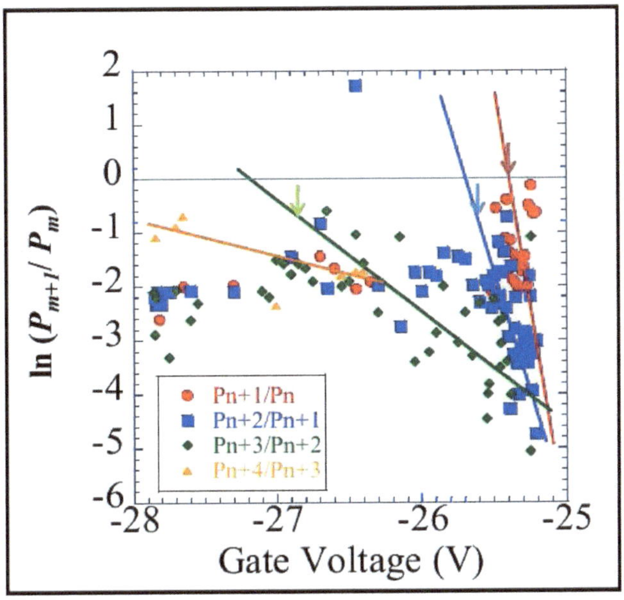

Fig. 11. (a) Gate voltage dependence of the natural log of the ratio between the occupancy probabilities of the m th current levels P_{n+1}/P_n.

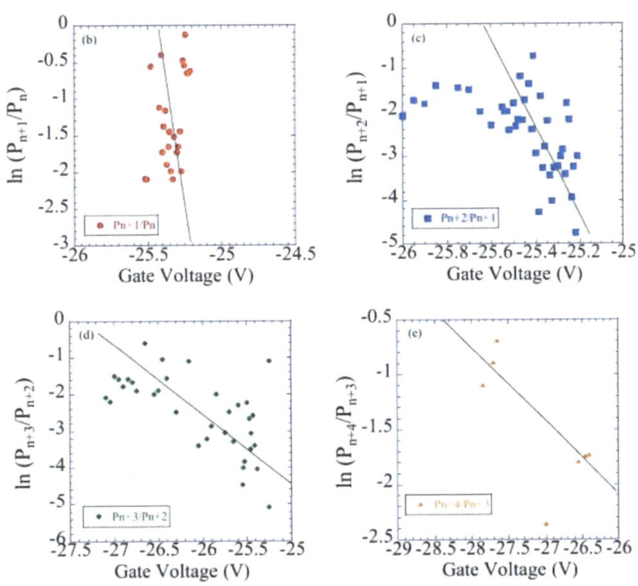

Fig. 11. (b)-(e) Enlargement plots of each $P_{m+1}/P_m(m = \mathrm{n}, \mathrm{n}+1, \mathrm{n}+2, \cdots \mathrm{n}+4)$. The natural log of P_{n+1}/P_n was linearly dependent on gate voltage, the slopes of which are $-10.4, -4.78, -1.47$, and $-0.690\ \mathrm{V}^{-1}$, respectively.

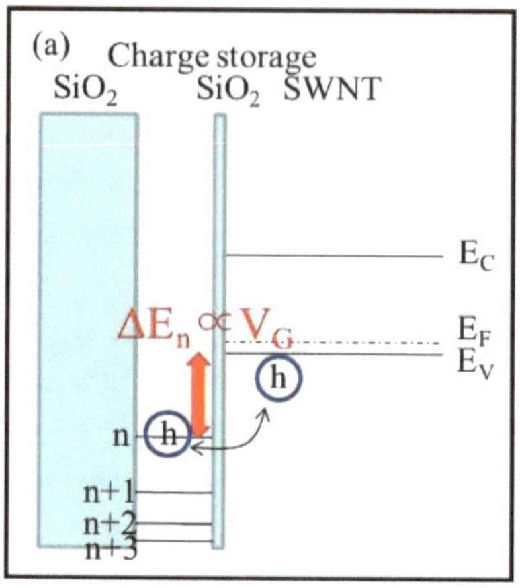

Fig. 12. (a) Schematic model of charge storage. When the charge storage energy level is coincident with the top of the valence band owing to applied gate voltage, the carrier goes and comes between them through the barrier with tunneling. The exiting probabilities of the carrier at the charge storage and the valence band depend on the relative height of their energy levels under equilibrium condition.

$$\ln\left(P_{m+1}/P_m\right) = -\beta e\alpha(V_0 - V_G) \tag{11}$$

Equation (2) is the transformed Arrhenius equation, in which V_G is the parameter. From the dependence of P_{n+1}/P_n on V_G shown in Fig. 11 and eqs. (1) and (2), ΔE_n can be obtained, and is shown in Fig. 12(b). The obtained energy levels are from 1.57 to 1.79 eV .

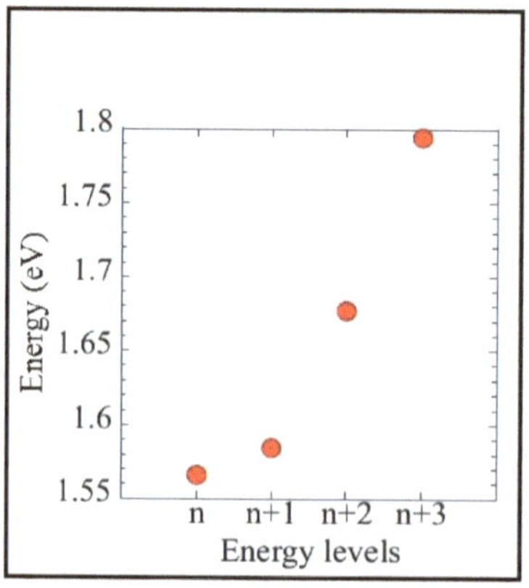

Fig. 12. (b) Estimated charge storage energy levels from the dependence of P_{n+1}/P_n on V_G shown in Figs. 4(b)-4(e) and eq. (1). The energy levels increase from 0.17 to 0.28 eV with increasing number of energy levels in the region of V_G from -25 to -28 V.

4. Conclusion

In summary, we succeeded in fabricating and demonstrating a multi-functional quantum transistor using the particle nature and wave nature of holes in SWNT. This transistor can operate in the wave nature mode as an RTT and in the particle nature mode as an SHT. We were able to reveal that the principle of the characteristic transition from an SHT to an RTT is the modulation of the coupling strength between the SWNT quantum island and the electrodes by the applied V_G.

5. References

Foxman, E. B.; mcEuen, P. L.; Meirav, U.; Wingreen, Ned S.; Meir,Yigal; Belk, Paul A.; & Wind, S. J. (1993). Effects of quantum levels on transport through a Coulomb island. Phys. Rev., Vol. B 47, 10020-10023.

Heinze, S.; Tersoff, J.; Martel, R.; Derycke, V.; Appenzeller, J.; & Avouris, Ph. (2002). Carbon Nanotubes as Schottky Barrier Transistors. Phys. Rev. Lett., Vol. 89, 106801-1-106801-4.

Javey, A.; Guo, J.; Wang, Q.; Lundstrom, M. & Dai, H. (2003). Ballistic carbon nanotube fieldeffect transistors. Nature, Vol. 424, 654-657.

Kamimura, T. & Matsumoto, K. (2004). Reduction of Hysteresis Characteristics in Carbon Nanotube Field-Effect Transistors by Refining Process. IEICE Trans. Electron., Vol. E-87C, 1795-1798.

Kamimura, T. & Matsumoto, K. (2005). Electrical Heating Process for p-Type to n-Type Conversion of Carbon Nanotube Field Effect Transistors. Jpn. J. Appl. Phys., Vol. 44, 1603-1605.

Kamimura, T. & Matsumoto, K. (2006). Coherent Transport of Hole and Coulomb Blockade Phenomenon in Long p-Type Semiconductor Carbon Nanotube. Jpn. J. Appl. Phys., Vol. 45, 338-340.

Kim, W.; Javey, A.; Vermesh, O.; Wang, Q.; Li, Y.; & Dai, H. (2003). Hysteresis Caused by Water Molecules in Carbon Nanotube Field-Effect Transistors. Nano Lett., Vol. 3, 193-198.

Liang, W.; Bockrath, M.; Bozovic, D.; Hafner, J. H.; Tinkham, M. & Park, H. (2001). Fabry Perot interference in a nanotube electron waveguide. Nature, Vol. 411, 665-669.

Martel, R.; Derycke, V.; Lavoie, C.; Appenzeller, J.; Chan, K. K.; Tersoff, J. & Avouris, Ph. (2001). Ambipolar Electrical Transport in Semiconducting Single-Wall Carbon Nanotubes. Phys. Rev. Lett., Vol. 87, 256805-1-256805-4.

Nosho,Y.; Ohno, Y.; Kishimoto, S. & Mizutani, T. (2006). Relation between conduction property and work function of contact metal in carbon nanotube field-effect transistors. Nanotechnology, Vol. 17, 3412-3415.

Peng, H. B.; Hughes, M. E. & Golovchenkoa, J. A. (2006). Room-temperature single charge sensitivity in carbon nanotube field-effect transistors. Appl. Phys. Lett., Vol. 89, 243502-243502-3.

Radosavljevic', M.; Freitag, M.; Thadani,K. V. & Johnson, A. T. (2002). Nonvolatile Molecular Memory Elements Based on Ambipolar Nanotube Field Effect Transistors. Nano Lett., Vol. 2, 761-764.

Suzuki, M.; Ishibashi, K.; Ida, T.; Tsuya, D.; Toratani, K.; & Aoyagi, Y. (2001). Fabrication of single and coupled quantum dots in single-wall carbon nanotubes J. Vac. Sci. Technol., Vol. B 19, 2770-2774.

Chapter-11

Numerical Analysis of CNTFET Current–Voltage Characteristics

1. Introduction

Carbon nanotubes (CNTs) were first discovered in 1991 by Sumio Iijima (Iijima, 1991), and they have been a rapid and successful target of development by many researchers (Baughman et al., 2002). Especially with the end of Moore's law in sight (Wind et al., 2002), CNTs are being considered as possible candidates substituting silicon in the fabrication and design of analog and digital integrated circuits (ICs) (Guo et al., 2002; Wong, 2002).

Carbon nanotubes are basically two dimensional graphene sheets rolled into a one dimensional tubular structure (Martel et al., 1998). Their properties are determined by the chiral vector represented by the indices (n,m) (Tanaka et al., 1999; Wallace, 1947; Wildoer et al., 1998). Depending on the number of layers rolled, carbon nanotubes can be either single walled (one layer), or multi walled (two or more layers) (Dresselhaus et al., 2001). Single walled carbon nanotubes (SWNTs) are used in the fabrication of carbon nanotube field effect transistors (CNT-FETs), the first CNT-FETs were implemented in 1998 (Martel et al., 1998; Tans et al., 1998). The structure of a CNT-FET is similar to the structure of a typical MOSFET where the CNT forms the channel between two electrodes that work as the source and the drain of the transistor. The structure is build on top of an insulating layer and a substrate wafer that works as the back gate, (Nihey et al., 2003) (Wind et al., 2002). The basic structure of a CNT-FET based on a SWNT is shown in Fig. 1. Among the CNT-FETs reported (Martel et al., 1998; Wind et al., 2002), a 40 nm gate length transistor (Lin et al., 2005), a multistage complementary logic (Javey et al., 2002), oscillators (Chen et al., 2006), and an 80 GHz operating field effect transistor (Nougaret et al., 2009) have been achieved.

In addition, carbon nanotubes are one-dimensional conductors (1D), which confines the electrons to only back-scattering effects. This property provides a large electron mean free path in metallic carbon nanotubes of usually a few micrometers (White & Todorov, 1998). Carbon nanotubes also exhibit large current capabilities of $\sim 10^{-9}$ A/cm^2, (Wei et al., 2001; Yao et al., 2000) and it has also been reported that doping can be avoided in the CNT fabrication process, yet still achieving complementary CNT-FETs (Zhang et al., 2007). With carbon nanotubes interconnects being also a major target of research (Koo et al., 2007; Xu et al., 2008) and a completely carbon nanotube based integrated circuit (IC) already reported (Liang et al., 2009), fabrication of fully integrated carbon nanotubes in custom designed ICs and systems on a chip is no longer a future circuit design promise but an attainable grasp.

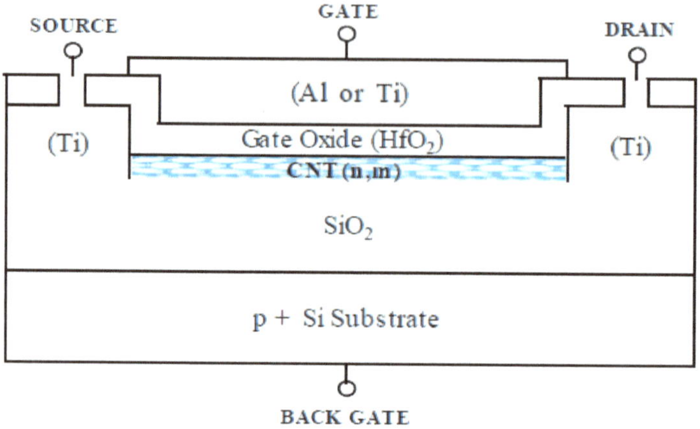

Fig. 1. Cross sectional view of a CNT-FET. Note: HfO_2 (Halfnium dioxide) is a high *k* dielectric (~17-20) (Javey et al., 2004; Wang et al., 2003).

In this paper, we have used our modeling of the potential and charge balance of CNT-FETs (Marulanda, 2009; Marulanda et al., 2008; Srivastava et al., 2009) to develop a numerical model of the current-voltage relation for a CNT-FET of n-type behaviour. This model uses numerical approximation methods to solve this current and voltage equations and present the I-V characteristics of different CNT-FETs.

2. Current Equation

The current per band in a semiconductor (Datta, 1999; Xia et al., 2004) can be expressed as:

$$I_{ds_i} = \frac{q}{\pi \text{h}} \int_{E_{c_i}}^{E_{\max}} T_i(E)[f_s(E) - f_d(E)]dE \tag{1}$$

where E_{ci} and $T_i(E)$ are respectively the conduction band energy minimum and transmission coefficient for the i_{th} band, h is Planck's constant divided by 2π, and $f_s(E)$ and $f_d(E)$ are the Fermi distributions at the source and drain, respectively.

Although the total I_{ds} will be the sum of all I_{dsi} over all bands, the summation in Eq. (1) can be dropped, as only the first band is sufficient. The reason is because the remaining energy bands will have negligible contribution to the current due to the vanishing exponential behaviour of the Fermi distribution for high-energy conduction band values.

By letting the limit of $E_{\max}$ go to infinity and the transmission coefficient be a constant equal to unity for the energy range of integration, Eq. (1) becomes (Marulanda et al., 2008):

$$I_{ds} = \frac{qKT}{\pi h}\left[\ln\left(1 + e^{\frac{E_{Fs}-E_C}{KT}}\right) - \ln\left(1 + e^{\frac{E_{Fd}-E_C}{KT}}\right)\right] \tag{2}$$

where E_{Fs} and E_{Fd} are the Fermi level energies at the source and drain, respectively. This equation for the current can be used in the linear and saturation regions.

3. Gate and Carbon Nanotube Potential Relation

Figure 2(a) shows the basic cross section of a CNT-FET including the charge distributions, Fig. 2(b) shows the corresponding potential distributions between the gate and substrate. In Fig. 2(a), charge distributions are explained as follows: the charge on the gate is Q_g, charges in oxide layers are Q_{01} and Q_{02}, charge inside the CNT is Q_{cht} and the charge in the substrate is $Q_{\mathrm{subs.}}$. In Fig. 2(b), six different potential distributions are shown, which are also described as follows. The voltage between the gate and the substrate (back gate) is V_{gb}, the potential drop across the oxides ψ_{ox1} and ψ_{ox2}, the surface potential in the substrate with respect to the back gate is ψ_{subs}, the potential across the CNT is ψ_{cnt}, and the work function difference between the gate and the substrate materials is ϕ_{ms}. The work function, ϕ_{ms} can be divided in two parts and is expressed as follows,

$$\phi_{ms} = \phi_{mc} + \phi_{cs}, \tag{3}$$

where ϕ_{mc} and ϕ_{cs} are the work function differences between the metal gate and carbon nanotube materials and the carbon nanotube and substrate materials, respectively. We can also combine potentials $\psi_{cnt}, \psi_{ox2}, \phi_{cs}$, and ψ_{subs} into a single potential, $\psi_{cnt,s}$, which describes the potential at the interface of the gate oxide and carbon nanotube with respect to the back gate, we can call $\psi_{cnt,s}$ the carbon nanotube surface potential.

Using Kirchoff's voltage law, the potential balance, and the charge neutrality condition, we can write for Fig. 2,

$$V_{gb} = \phi_{ms} + \psi_{ox1} + \psi_{cnt,s'} \tag{4}$$

$$Q'_g + Q'_{01} + Q'_{cnt} + Q'_{02} + Q'_{subs} = 0. \tag{5}$$

The prime in Eq. (4) denotes the charge per unit area. From basic laws of electrostatics (Cheston, 1964; Langmuir, 1961; Shen & Kong, 1995; Thomas, 1972), we can utilize Eqs. (4)

and (5) and the charge modeling described in (Marulanda, 2009; Marulanda et al., 2008; Srivastava et al., 2009), to express the gate voltage in a CNT-FET as,

$$V_{gb} = \psi_{cnt,s} - \frac{Q_{cnt}}{C_{ox1}} + V_{fb}, \tag{6}$$

where V_{fb} is the flat band voltage, which is the voltage needed at the gate with respect to the back gate to compensate for the band bending at the gate oxide and carbon nanotube interface and it is given by,

$$V_{fb} = \phi_{mc} - \frac{Q_{01}}{C_{ox1}}. \tag{7}$$

In addition, C_{ox1} and C_{ox2} are the oxide capacitances for the respective oxide regions, T_{ox1} and T_{ox2} and are described for a carbon nanotube of length, L and radius, r as follows (Cheston, 1964; Thomas, 1972),

$$C_{ox} = \frac{2\pi\varepsilon_{ox}L}{\ln\left(\frac{T_{ox} + r + \sqrt{T_{ox}^2 + 2T_{ox}r}}{r}\right)}. \tag{8}$$

In Eqs. (6), (7), and (8), Q_{cnt}, Q_{01} and C_{ox1} are the total charges and capacitance, respectively, which are obtained by multiplying Q'_{01}, Q'_{cnt}, and C'_{ox} with their respective areas.

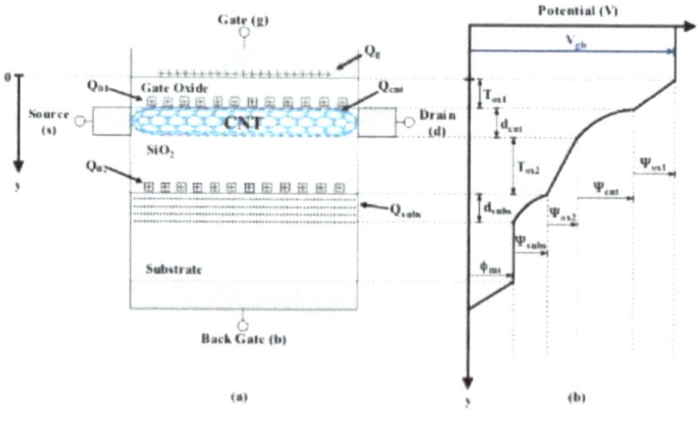

Fig. 2. (a) Plot of the charges from the gate to the substrate and (b) plot of the potential distribution from the gate to the substrate in a CNT-FET.

Using the carrier concentration modeling in (Marulanda & Srivastava, 2008) and charge modeling described in (Marulanda, 2009; Marulanda et al., 2008; Srivastava et al., 2009), we can express Eq. (6) as:

$$V_{gb} = \psi_{cnt,s} + \delta f(\psi_{cnt,s}, V_{cb}) + V_{fb}, \tag{9}$$

where

$$f(\psi_{cnt,s}, V_{cb}) = \begin{cases} Ie^{\frac{\Delta E_F + q(\psi_{cnt,s} - V_{cb} - \phi_0) - E_c}{kT}}; & for\ \psi_{cnt,s} \le V_{cb} + \phi_0 - \frac{\Delta E_F}{q} + \frac{E_c}{q} - \frac{kT}{q} \\ \frac{\sqrt{(\Delta E_F + q\psi_{cnt,s} - qV_{cb} - q\phi_0)^2 - E_c^2}}{kT}; & for\ \psi_{cnt,s} \ge V_{cb} + \phi_0 - \frac{\Delta E_F}{q} + \frac{E_c}{q} + \frac{kT}{q} \end{cases} \tag{10}$$

and

$$\delta = \frac{qLN_c}{C_{ox1}} \tag{11}$$

In Eq. (9), V_{cb} is the induced potential between the carbon nanotube and the substrate due to the drain and source terminal voltages. The potential, V_{cb} varies from V_{sb} (source to back gate potential) to V_{db} (drain to back gate potential). Furthermore, in Eq. (10), k is Boltzmann constant, T is temperature, ϕ_0 is the carbon nanotube surface potential, $q\psi_{cnt,s}$ when $V_{gb} = V_{fb}$ and $\psi_{cnt} = 0, \Delta E_F$ is the shift in the Fermi level of the carbon nanotube depending upon doping, , ϕ_0 and ΔE_F are given by,

$$\phi_0 = \psi_{ox2} + \psi_{subs} + \phi_{cs} \tag{12}$$

$$\Delta E_F = \pm kT \ln\left(1 + \frac{N}{n_{cnt,i}}\right) \tag{13}$$

where $n_{cnt,i}$ is the intrinsic concentration of a CNT (n, m) and N is the impurity doping concentration. ΔE_F is positive for an n-type carbon nanotube (donors impurity concentration, $N = N_D$) and negative for a p-type carbon nanotubes (acceptors impurity concentration, $N = N_A$).

In Eq. (10), I is an indefinite integral involved in finding the carrier concentration of a carbon nanotube, we have found an analytical solution in (Marulanda, 2009; Marulanda & Srivastava, 2008); however, the general form of the integral, I is given by,

$$I = \frac{1}{\sqrt{kT}} \int_0^{\frac{6E_c}{kT}} \frac{(kTx + E_c)}{x^{1/2}(kTx + 2E_c)^{1/2}} e^{-x} dx \tag{14}$$

The term N_c introduced by Eq. (11) is found from our carrier concentration model for carbon nanotubes (Marulanda & Srivastava, 2007; Marulanda & Srivastava, 2008) and is given by,

$$N_c = \frac{8kT}{\pi V_{pp\pi} a\sqrt{3}}, \tag{11}$$

where $V_{pp\pi}$ is the carbon-carbon (C-C) tight binding overlap energy, a is the lattice constant with a value of $\sqrt{3}a_{c-c}$, and a_{c-c} is the nearest neighbor distance between C-C bonds (0.144 nm) (Dresselhaus et al.).

4. I-V Characteristic Simulation

Equation (9) cannot be solved explicitly for the carbon nanotube surface potential, $\psi_{cnt,s}$ in terms of the terminal voltages, hence, a numerical solution technique must be taken in order to find the exact channel potential given a gate voltage input.

In a CNT-FET, when the terminal voltages V_{db}, V_{sb}, and V_{gb} are applied, the two terms ($E_{Fs} - E_c$) and ($E_{Fd} - E_c$), referred in Eq. (2) can be replaced by $\Delta E_F + q(\psi_{cnts} - V_{sb} - \phi_0) - E_c$ and $\Delta E_F + q(\psi_{cnt,s} - V_{db} - \phi_0) - E_c$, respectively (Marulanda, 2009; Raychowdhury et al., 2004; Srivastava et al., 2009; Tsividis, 1999). V_{sb} and V_{db} are the source to bulk and drain to bulk voltages, respectively; furthermore, $\psi_{cnt,s}$ is calculated using Eq. (9) employing a numerical solution approach. Thus, our Eq. (2) becomes:

$$I_{ds} = \frac{qkT}{\pi \text{h}}\left[\ln\left(1 + e^{\frac{\Delta E_F + q(\psi_{crt,s}(0) - V_{sb} - \phi_0) - E_c}{kT}}\right) - \ln\left(1 + e^{\frac{\Delta E_F + q(\psi_{crt,s}(L) - V_{db} - \phi_0) - E_c}{kT}}\right)\right]. \tag{15}$$

The complete process to find the current I_{ds} is as follows:

1. Define the voltages V_{db}, V_{sb}, and V_{gb}.
2. Use a numerical method to find $\psi_{cnt,s}$ from Eq. (9).
3. Compute the current, I_{ds}, using Eq. (15).
4. Repeat steps 1 - 3 for each of the input voltages, V_{db}, V_{sb}, and V_{gb} combination.

Figure 4 shows the I-V characteristics for a CNT-FET with a carbon nanotube of chiral vector (5,3) and diameter of 0.2759 nm , under different overdrive gate voltages. These plots were derived following the four steps just mentioned.

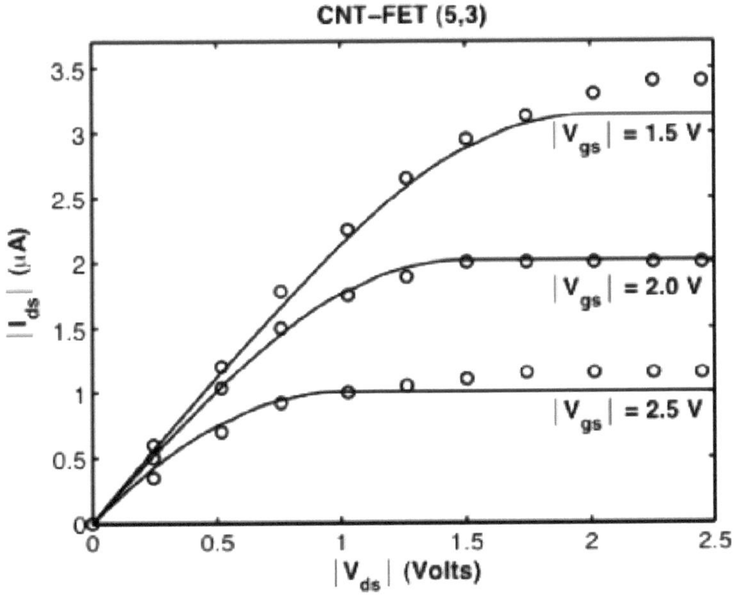

Fig. 3. **I-V characteristics for a CNT-FET (5, 3) with $V_{fb} = 0, \phi_0 = 0, \Delta E_F = 0$, and device dimensions of $L = 3\mu m, T_{ox1} = 500nm$, and $T_{ox2} = 1\mu m$.**

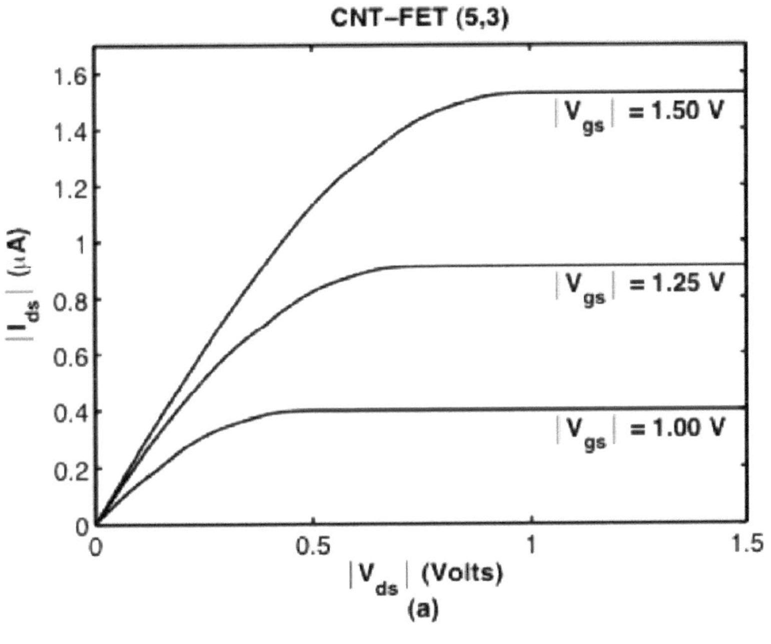

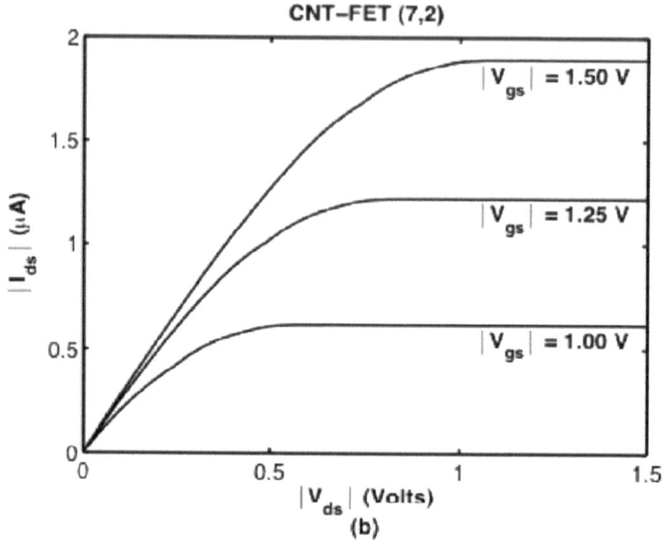

Fig. 4. I-V characteristics for (a) CNT-FET $(5,3)$ and (b) CNT-FET $(7,2)$ with $V_{fb} = 0, \phi_0 = 0, \Delta E_F = 0$, and device dimensions of $L = 25nm, T_{ox1} = 40nm$, and $T_{ox2} = 400nm$.

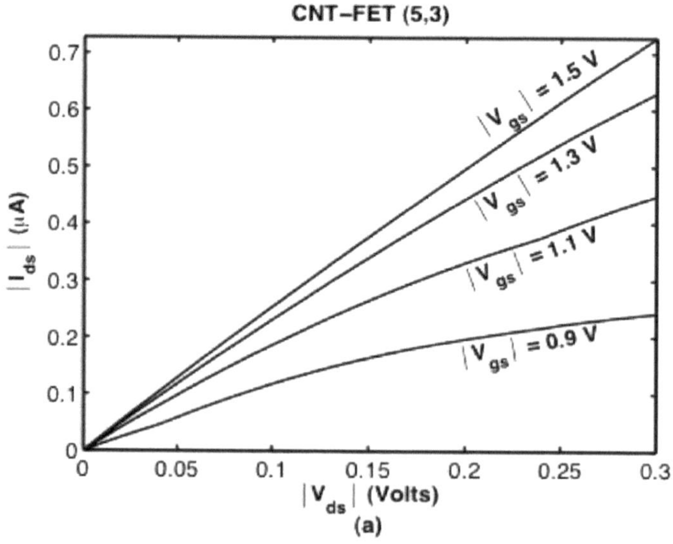

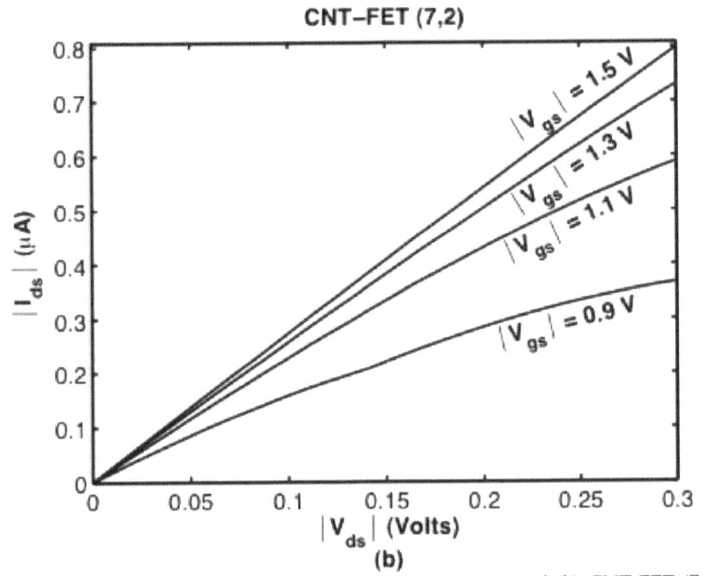

Fig. 5. I-V characteristics under low V_{ds} for (a) CNT-FET $(5,3)$ and (b) CNT-FET $(7,2)$ with $V_{fb} = 0, \phi_0 = 0, \Delta E_F = 0$, and device dimensions of $L = 25nm, T_{ox1} = 40nm$, and $T_{ox2} = 400nm$.

We have included in Fig. 3 experimentally measured data (Zhou et al., 2000) to validate the accuracy of our model. As we can see from Fig. 3, the model equation (Eq. 15) follows closely the experimentally measured data for the linear region. However, as the transistor enters the saturation region (in this region the current I_{ds} begins to saturate and becomes almost independent of V_{ds}), experimental values deviate from the model equation since our model equation does not include a correction term for the saturation region similar to the channel length modulation parameter in a MOSFET (Antognetti & Massobrio, 1988; Tsividis, 1999).

Using our numerical model approach we have also simulated the I-V characteristics for two more CNT-FETs. Figure 4 shows the I-V characteristics for two CNT-FETs with chiral vectors (5,3) and (7,2) and diameters 0.2759 nm and 0.3227 nm , respectively. Using the same procedure described for Figs. 3 and 4, it is also possible to compute the I-V characteristics for small V_{ds}. Figure 5 shows the simulated I-V characteristics for small V_{ds} for the CNT-FET (5,3) and CNT-FET (7,2). As we can see in Fig. 5, for small V_{ds} the current shows approximately a linearly proportional relationship to the voltage of the drain with respect to the source, V_{ds}.

5. Conclusion

Previous works in carbon nanotubes describing a relationship between the gate voltage and the carbon nanotube potential have made possible the implementation of a current equation, which can be solved numerically using any mathematical software, in order to find the current voltage characteristic for any given CNT-FET. The results presented provide designers with useful mathematical relations describing the properties of conductivity of carbon nanotubes and their response in circuit applications. Even though the charge transport in the carbon nanotube is not yet fully understood, and the solutions for the equations presented are strictly based in numerical methods, the results presented in this work set a starting point for device modeling simulation of carbon nanotube field effect transistors (CNT-FETs) and their applications on integrated circuits, including a SPICE compatible model.

6. References

Antognetti, P. & Massobrio, G. (1988). Semiconductor Device Modeling with SPICE, McGrawHill, Singapore.

Baughman, R. H.; Zakhidov, A. A. & Heer, W. A. d. (2002). Carbon nanotubes - the route toward applications. Science, Vol. 297, No. 5582, (August 2002), 787-792.

Chen, Z.; Appenzeller, J.; Lin, Y.-M.; Sippel-Oakley, J.; Rinzler, A. G.; Tang, J.; Wind, S. J.; Solomon, P. M. & Avouris, P. (2006). An Integrated Logic Circuit Assembled on a Single Carbon Nanotube. Science, Vol. 311, No. 5768, (March 2006), 1735-1736.

Cheston, W. B. (1964). Elementary Theory of Electric and Magnetic Fields, John Wiley & Sons Inc., New York.

Datta, S. (1999). Electronic Transport in Mesoscopic Systems, Cambridge University Press, New York.

Dresselhaus, M. S.; Dresselhaus, G. & Avouris, P. (2001). Carbon Nanotube: Synthesis, Properties, Structure, and Applications, Springer Verlag, New York.

Guo, J.; Lundstrom, M. & Datta, S. (2002). Performance projections for ballistic carbon nanotube field-effect transistors. Applied Physics Letters, Vol. 80, No. 17, (April 2002), 3192-3194.

Iijima, S. (1991). Helical microtubules of graphitic carbon. Nature, Vol. 354, No. 6348, (November 1991), 56-58.

Javey, A.; Guo, J.; Farmer, D. B.; Wang, Q.; Wang, D.; Gordon, R. G.; Lundstrom, M. & Dai, H. (2004). Carbon nanotube field-effect transistors with integrated ohmic contacts and high-k gate dielectrics. Nano Letters, Vol. 4, No. 3, (February 2004), 447-450.

Javey, A.; Wang, Q.; Ural, A.; Li, Y. & Dai, H. (2002). Carbon nanotube transistors arrays for multistage complementary logic and ring oscillators. Nano Letters, Vol. 2, No. 9, (July 2002), 929-932.

Koo, K.-H.; Cho, H.; Kapur, P. & Saraswat, K. C. (2007). Performance Comparisons Between Carbon Nanotubes, Optical, and Cu for Future High-Performance On-Chip Interconnect Applications. IEEE Transactions on Electron Devices, Vol. 54, No. 12, (December 2007), 3206-3215.

Langmuir, R. V. (1961). Electromagnetic Field and Waves, McGraw-Hill Book Company, New York.

Liang, X.; Wang, S.; Wei, X.; Ding, L.; Zhu, Y.; Zhang, Z.; Chen, Q.; Li, Y.; Zhang, J. & Peng, L.-M. (2009). Towards Entire-Carbon-Nanotube Circuits: The Fabrication of Single-Walled-Carbon-Nanotube Field-Effect Transistors with Local Multiwalled-CarbonNanotube Interconnects. Advanced Materials, Vol. 21, No. 13, (April 2009), 13391343.

Lin, Y.-M.; Appenzeller, J.; Chen, Z.; Chen, Z.-G.; Cheng, H.-M. & Avouris, P. (2005). HighPerformance Dual-Gate Carbon Nanotube FETs with 40 – nm Gate Length. IEEE Transactions on Electron Devices, Vol. 26, No. 11, (November 2005), 823-825.

Martel, R.; Schmidt, T.; Shea, H. R.; Hertel, T. & Avouris, P. (1998). Single and multi wall carbon nanotube field effect transistors. Applied Physics Letters, Vol. 73, No. 17, (October 1998), 2447-2449.

Marulanda, J. M. (2009). Current Transport Modeling of Carbon Nanotubes: Concepts, Analysis, and Design, VDM Verlag, Saarbrücken, Germany.

Marulanda, J. M. & Srivastava, A. (2007). Carrier density and effective mass calculations for carbon nanotubes, Proccedings of the International Conference on Integrated Circuit Design & Technology (ICICDT), pp. 234-237, Austin, TX, May 30 - Jun 1.

Marulanda, J. M. & Srivastava, A. (2008). Carrier density and effective mass calculations in carbon nanotubes. Physica Status Solidi (b), Vol. 245, No. 11, (November 2008), 25582562.

Marulanda, J. M.; Srivastava, A. & Sharma, A. K. (2008). Threshold and saturation voltages modeling for carbon nanotube field effect transistors (CNT-FETs). NANO, Vol. 3, No. 3, (June 2008), 195-201.

Marulanda, J. M.; Srivastava, A. & Yellampalli, S. (2008). Numerical Modeling of the I-V Characteristic of Carbon Nanotube Field Effect Transistors (CNT-FETs), Proccedings of the IEEE 40th Southeastern Symposium on System Theory (SSST 2008), pp. 235-238, New Orleans, LA, March 16-18.

Nihey, F.; Hongo, H.; Ochiai, Y.; Yudasaka, M. & Iijima, S. (2003). Carbon-nanotube field effect transistors with very high intrinsic transconductance. Japanese Journal of Applied Physics, Vol. 42, No. 10B, (October 2003), L1288-L1291.

Nougaret, L.; Happy, H.; Dambrine, G.; Derycke, V.; Bourgoin, J. P.; Green, A. A. & Hersam, M. C. (2009). 80 GHz field-effect transistors produced using high purity semiconducting single-walled carbon nanotubes. Applied Physics Letters, Vol. 94, No. 24, (June 2009), 243505-243503.

Shen, L. C. & Kong, J. A. (1995). Applied Electromagnetism, PWS Foundations in Engineering Series, Boston.

Srivastava, A.; Marulanda, J. M.; Xu, Y. & Sharma, A. K. (2009). Current transport modeling of carbon nanotube field effect transistors. Physica Status Solidi (a), Vol. 206, No. 7, (May 2009), 1569-1578.

Tanaka, K.; Yamabe, T. & Fukui, K. (1999). The Science and Technology of Carbon Nanotubes, Elsevier, Amsterdam, The Netherlands.

Tans, S. J.; Vershueren, A. R. M. & Dekker, C. (1998). Room-temperature transistor based on a single carbon nanotube. Nature, Vol. 393, (May 1998), 49-52.

Thomas, D. (1972). Engineering Electromagnetics, Pergamon Press Inc., New York.

Tsividis, Y. (1999). Operation and Modeling of the MOS transistor, McGraw-hill, Singapore.

Wallace, P. R. (1947). The band theory of graphite. Physical Review Letters, Vol. 71, (May 1947), 622-634.

Wang, D.; Wang, Q.; Javey, A.; Tu, R.; Dai, H.; Kim, H.; McIntyre, P. C.; Krishnamohan, T. & Saraswat, K. C. (2003). Germanium nanowire field-effect transistors with SiO_2 and high-kappa HfO_2 gate dielectrics. Applied Physics Letters, Vol. 83, No. 12, 2003), 2432-2434.

Wei, B. Q.; Vajtai, R. & Ajayan, P. M. (2001). Reliability and current carrying capacity of carbon nanotubes. Applied Physics Letters, Vol. 79, No. 8, (August 2001), 1172-1174.

White, C. T. & Todorov, T. N. (1998). Carbon nanotubes as long ballistic conductors. Nature, Vol. 393, No. 6682, (May 1998), 240-242.

Wildoer, J.; Venema, L.; Rinzler, A.; Smalley, R. & Dekker, C. (1998). Electronic structure of atomically resolved carbon nanotubes. Nature, Vol. 391, No. 6662, (January 1998), 59-62.

Wind, S. J.; Appenzeller, J.; Martel, R.; Derycke, V. & Avouris, P. (2002). Fabrication and electrical characterization of top gate single-wall CNFETs. Journal of Vacuum Science and Technology B, Vol. 20, No. 6, (November 2002), 2798-2801.

Wind, S. J.; Appenzeller, J.; Martel, R.; Derycke, V. & Avouris, P. (2002). Vertical scaling of carbon nanotube field-effect transistors using top gate electrodes. Applied Physics Letters, Vol. 80, No. 20, (May 2002), 3817-3819.

Wong, H. S. P. (2002). Field effect transistors - from silicon MOSFETs to carbon nanotube FETs, Proccedings of the 23 th International Conference on Microelectronics, (MIEL), pp. 103-107.

Xia, T.-S.; Register, L. F. & Banerjee, S. K. (2004). Quantum transport in carbon nanotube transistors: complex band structure effects. Journal of Applied Physics, Vol. 95, No. 3, (February 2004), 1597-1599.

Xu, Y.; Srivastava, A. & Marulanda, J. M. (2008). Emerging carbon nanotube electronic circuits, modeling and performance, Proccedings of the 51st IEEE International Midwest Symposium on Circuits and Systems (MWSCAS 2008), pp. 566-569, August 10-13.

Yao, Z.; Kane, C. L. & Dekker, C. (2000). High-field electrical transport in single-wall carbon nanotubes. Physical Review Letters, Vol. 84, No. 13, (March 2000), 2941-2944.

Zhang, Z.; Liang, X.; Wang, S.; Yao, K.; Hu, Y.; Zhu, Y.; Chen, Q.; Zhou, W.; Li, Y.; Yao, Y.; Zhang, J. & Peng, L.-M. (2007). Doping-Free Fabrication of Carbon Nanotube Based Ballistic CMOS Devices and Circuits. Nano Letters, Vol. 7, No. 12, (November 2007), 3603-3607, 1530-6984.

Zhou, C.; Kong, J. & Dai, H. (2000). Electrical measurements of individual semiconducting single walled carbon nanotubes of various diameters. Applied Physics Letters, Vol. 76, No. 12, (March 2000), 1597-1599.

Chapter-12

Compact Modeling of Carbon Nanotube Transistor and Interconnects

Silicon based devices have dominated mainstream computing for the last four decades. Achieving sustainable scaling of physical dimensions and device performance (Moore, 1965) has been key to their success. However, due to limitations in fundamental physics, materials, and manufacturing limits, this scaling trend has slowed down. Examples of major bottlenecks for continual scaling include short channel effects, high leakage currents(Wann et al., October 1996), excessive process variations(Bowman et al., 2002) and reliability issues(Chen et al., February 1985). These pitfalls are posing dramatic challenges to fabrication of circuits with scaled silicon devices. As we approach these fundamental limits in planar CMOS process, it becomes imperative to search for alternative materials, structures, devices as well as design paradigm to replace silicon transistor as the building block of future nanoelectronics. Novel structures like FinFETs(Hisamoto et al., 2000) and Trigate devices(Doyle et al., 2003), strained channel to enhance carrier mobility(Welser et al., 1994) and high-K/metal gate to reduce gate leakage current(Chau et al., 2004) have been proposed. These innovations have limited potential and will extend the scaling by a generation or two. Amongst more radical search for new devices and materials, carbon nanotube electronics has attracted significant attention owing to their high intrinsic carrier mobility.

For the sake of simplicity, carbon nanotubes can be defined as hollow cylinders made up of one (single-walled) or more (multi-walled) concentric layer of carbon atoms arranged in a hexagonal lattice structure, which is similar to a rolled-up sheet of graphene. With diameters of 1-4 nm and the length extending to several micrometers, carbon nanotube is essentially a one dimensional object possessing unique properties attributed to low dimensional structures, such as 1-D density of state for carriers (McEuen et al., 2002). This enables reduced phase space for scattering and near ballistic transport of carriers when the device dimension is less than the mean-free path for scattering. Depending on the direction in which the graphene sheet is rolled up, single-walled carbon nanotubes is either metallic or semiconducting. Hence CNT transistor and interconnect can be made out of semiconducting and metallic nanotubes, respectively. Functional field effect transistors with semiconducting carbon nanotube channel(Lin et al., 2005; Zhang et al.,

2006) and metallic nanotubes as interconnects(Close & Wong, 2007) have been demonstrated. Theoretically, it is possible to get current densities much higher than that of silicon devices with a similar dimension(Raychowdhury et al., 2006) using multiple CNTs in parallel.

To speed up the evolution of this novel alternative technology, parallel efforts in circuit design are essential. For this purpose, the development of compact model is a vitally important step that enables circuit simulation and exploration.

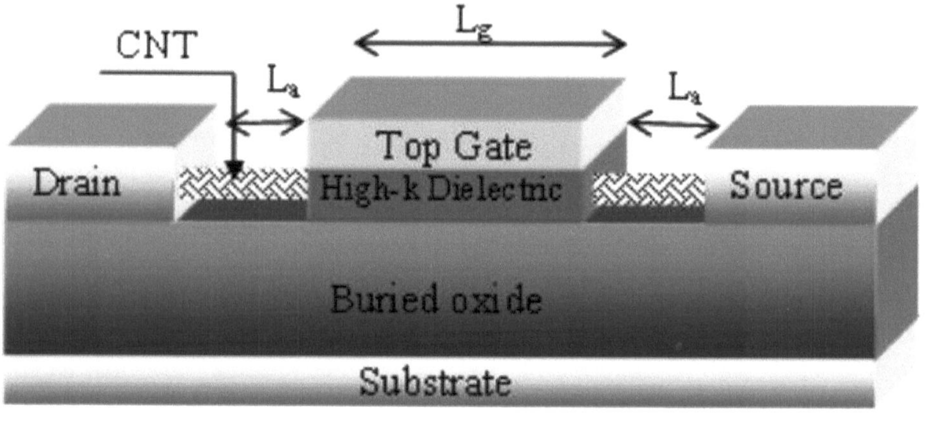

Fig. 1. Cross-section of a generic SW-CNT structure with top gated region as the intrinsic transistor of length L_g and highly doped undated access region of length L_a as the extrinsic part

Early work on carbon nanotube transistor modeling assumed doped source-drain junctions resulting in optimistic performance projections (Guo et al., 2002). The CNT-FET model developed in (Castro et al., 2002) accurately calculated the degradation of current due to presence of Schottky barriers at the contacts. However, it requires self-consistent numerical iterations to calculate the final current and tunneling probability. Currently most of the models developed for carbon nanotube transistors and interconnects employ some kind of numerical approach (Guo et al., 2004; Wong et al., Nov. 2006) to obtain the I-V and C-V characteristics. Though highly physical and accurate, such numerical approaches reduce the computation efficiency and are not suitable for large-scale circuit simulations. Some other modeling approaches include threshold voltage based models(Raychowdhury et al., 2004) and models that resort to SPICE simulator to solve iterative differential equations and compute the surface potential (Deng & Wong, 2007a).

In this chapter, we discuss the development of an integrated compact model for carbon nanotube transistors and interconnects that is non-iterative and SPICE compatible. Initial models concentrated on modeling only the ballistic transport model of the transistor channel. However, the effect of the Schottky barrier at the metal source-drain contacts cannot be decoupled from the channel region. The developed model accounts for the presence of these barriers accurately. The implemented model has been systematically verified with TCAD simulations and measured data. Using this model, we benchmark digital and analog performance metrics and compare them with 22 nm CMOS process to explore design potentials with CNTs.

1. Model Development

The cross-sectional view of a typical carbon nanotube transistor is shown in Fig. 1. The basic structure is similar to a conventional FET with the channel replaced by a semiconducting carbon nanotube. The similarity to the structure of CMOS device improves the compatibility with today's process and design infrastructure.

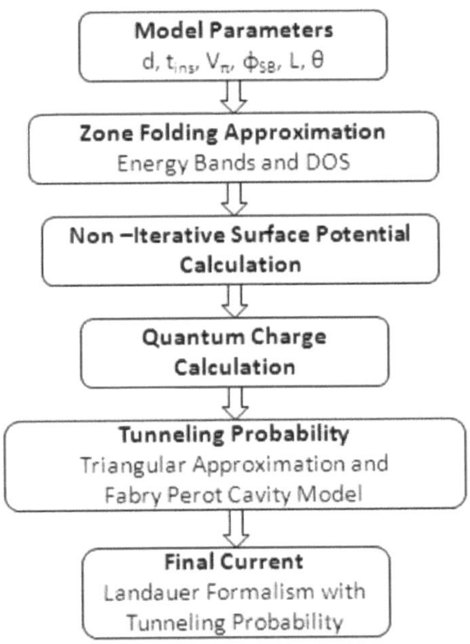

Fig. 2. Flowchart describing the model development

In the ideal case for ballistic transport, the source and drain electrodes would behave as reservoirs that supply and sink mobile carriers without any reflection at the source and drain. This is true only when there are ideal source and drain contacts, i.e., no significant energy barrier between the channel and the contact. There has been extensive work on finding the appropriate contact material for the CNT-FET (Chen et al., 2005) but they all have a finite energy gap when contacting the carbon nanotube forming a Schottky barrier.

The device performance is primarily limited by the Schottky contact, depending on the properties of the contact material and the nanotube. The energy gap is sensitive to the work function of the contact, the diameter of the nanotube, as well as the chirality.

In this section, we will first discuss the intrinsic channel and relate the physical parameters to the electrical equivalents by the Zone-folding approximation. Using linear approximations, we derive an expression for surface potential. Then the impact of the Schottky barrier on the drain current is studied and integrated into the model. Based on similar modeling principles of the channel and the contact, a compact model of CNT interconnect and CNT-FET are developed. Fig. 2 shows a flowchart describing the CNT-FET compact model. Details of the physical mechanisms and model derivations of the CNT interconnect model are discussed in the next section.

1.1 Zone-folding approximation

We begin with characterizing the structure of single-walled carbon nanotubes (SWCNT) and defining its basic electronic properties such as band-gap, density of states etc. A SWCNT device is essentially a one-dimensional nanowire formed by rolling a two-dimensional graphene sheet. The $2s$, $2p_x$ and $2p_y$ orbitals form σ bonds in graphene. Since the σ bonds are weakly coupled to the $2p_z$ orbitals, they form π bonds, which give rise to the electronics properties of graphene. The E-k values for graphene can be obtained from the tight-binding model given by (1),

$$E_{g2D}(k_x, k_y) = +t\left\{1 + 4\cos\left(\frac{\sqrt{3}k_x a}{2}\right)\cos\left(\frac{k_y a}{2}\right) +4\cos^2\left(\frac{k_y a}{2}\right)\right\}^{1/2} \quad (1)$$

To get the bandstructure of carbon nanotubes, we begin with the bandstructure of graphene given in (1), apply periodic boundary conditions along the circumference of the nanotube. The rolling-up of the honeycomb lattice of the graphene sheet along a specific direction, known as the chiral vector (shown in Fig. 3), causes the quantization of the wave-vector space along its direction.

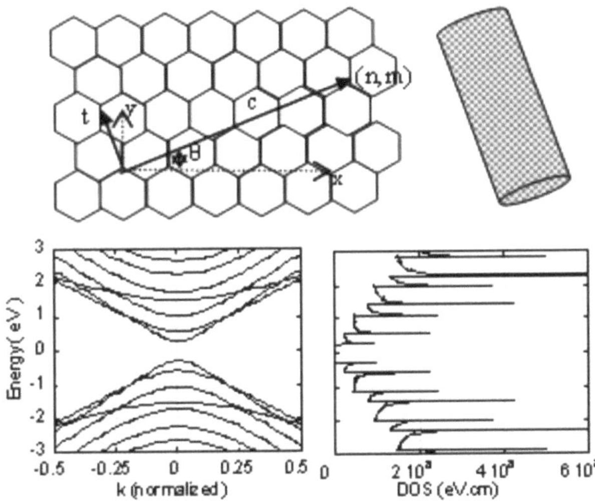

Fig. 3. Honeycomb lattice graphene sheet showing the chiral vectors (n, m) and the corresponding E-k diagram calculated using (1) and DOS using (2)

A chiral vector can be denoted by the coordinates (n, m). If ($n - m$) is a multiple of 3 , the carbon nanotube is metallic, else it is semi-conducting. To calculate the current, the electron density of states (DOS) near the Fermi level is required. Classical tight-binding models are used to accurately compute the DOS. At low bias, the DOS D(E) at energy E can be approximated as expressed in (2) (Guo & Lundstrom, 2006),

$$D(E) = \frac{D_0|E|}{\sqrt{E^2 - E_n^2}} \text{ where } D_0 = \frac{8}{3\pi V_\pi a} \tag{2}$$

All variables used in the above equations are defined in Table 1. For a more detailed discussion on the band-structure of carbon nanotubes, zone-folding approximation and complete derivation the interested reader is referred to (Guo & Lundstrom, 2006).

1.2 Surface-potential based modeling

On applying a gate voltage V_G is applied, the surface potential (ϕ_s) is modulated. The expressions for surface potential and the total charge are as follows :

$$\phi_s = V_{GS} - \frac{|Q_{CNT}|}{C_{ins}} \tag{3}$$

$$Q_{CNT} = N_0 \sum_n \int_{E_n} F\left(\sqrt{E^2 - E_n^2}, \mu_s\right) \tag{3}$$

$$+F\left(\sqrt{E^2 - E_n^2}, \mu_s - V_{DS}\right) dE \tag{4}$$

where

$$F(E,\mu) = \frac{1}{1 + e^{(E-\mu)}} \text{(Fermi-Dirac Integral)} \tag{5}$$

and n is the number of sub-bands under conduction.

The conventional method to compute the ϕ_s (using the conduction-band minima and DOS calculated from Table 1 and (2), respectively) involves numerically solving the 1-D poisson equation and the total charge equation self-consistently. In spite of being accurate, this method is computationally intensive and inappropriate for compact modeling and circuit simulations. Additionally, SPICE solvers may encounter convergence errors when loaded with a task of solving complicated numerical functions. Hence, a linearized equation for ϕ_s has been derived. By eliminating numerical iterations, the simulation speed is considerably improved making the model suitable for large scaling circuit simulation. We derive a closed form linear approximation for surface potential from the fundamental equations of channel charge and bias conditions. From (3) and (4)

$$\phi_s = V_{gs} - \frac{N_0}{C_{ins}} \sum_n \int_{E_n} \frac{1}{1 + \exp\left(\frac{E - V_s - E_n - \phi_s}{KT}\right)} + \frac{1}{1 + \exp\left(\frac{E - V_d - E_n - \phi_s}{KT}\right)} dE$$

$$= V_{gs} - \gamma \sum_n \left[\ln\left(1 + e^{\left(\frac{E_f - V_s - E_n - \phi_s}{KT}\right)}\right) + \ln\left(1 + e^{\frac{(E_f - V_d - E_n - \phi_s)}{KT}}\right)\right] \tag{6}$$

For low bias voltages, the first order approximation (6) is linear. Hence, we may write the equation as

$$\phi_s = V_{gs} - \sum_n \gamma \left[\frac{(E_f - V_s - E_n - \phi_s)}{KT} + \frac{(E_f - V_d - E_n - \phi_s)}{KT} \right] \quad (7)$$

Moving ϕ_s from RHS on to the LHS and re-arranging the terms, we get the non-iterative closed form expression for surface potential as follows

$$\phi_s = \sum_n \left(\frac{V_t \gamma (\xi_s |\xi_s| + \xi_d |\xi_d|)}{2(1 + 2\gamma)} \right) - V_{gs} \quad (8)$$

where $\gamma = N_0 / C_{\text{ins}}$, and

$$\xi_{s,d} = \frac{(E_f - V_{s,d} - E_{o,p} + V_{gs})}{V_t}; \; |\xi_{s,d}| = \begin{cases} 1, & \text{if } \xi_{s,d} > 0 \\ 0, & \text{if } \xi_{s,d} < 0 \end{cases}$$

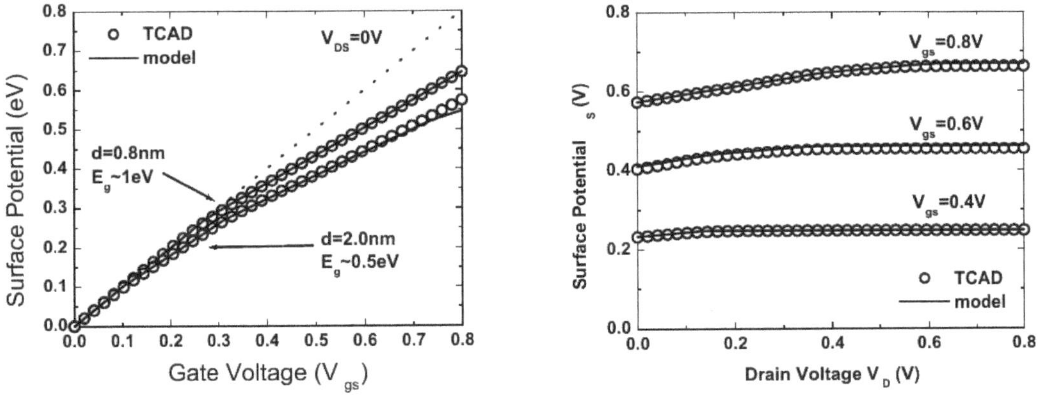

(a) (b)

Fig. 4. ϕ_s as a function of V_{gs} and V_{ds} for d = 0.8 nm and 2 nm The voltage range chosen is the region where there is good gate control and FET-like behavior

This expression forms the basis of the compact model. All existing models (Castro et al., 2002; Deng & Wong, 2007a; Guo et al., 2004; 2002; Raychowdhury et al., 2004; Wong et al., Nov. 2006 use some kind of self-consistent numerical methods to solve for ϕ_s. Fig. 5(b) shows the variation of surface potential as a function of V_{GS} and V_{DS}, for different diameters. At low voltages, the model is in good agreement with the numerical simulations and no regional approximations are required in the expression. The surface potential is a function of the diameter, temperature and gate dielectrics to the first order. At higher voltages, higher sub-bands are filled and therefore the slope of the line in Fig. 5(b) changes and this behavior is modeled by (8).

1.3 Capacitance Model

Surface potential is calculated using (3), a function of Q_{CNT}. However, Q_{CNT} itself is a function of ϕ_s and most other models employ self-consistent iterations to solve for charge and surface potential. As explained in the previous subsection, we approximate the charge in each sub-band to be linear to get a closed form solution for surface potential. With ϕ_s known, the quantum charge Q_{CNT} can be calculated in closed form as well. Fig. 5 shows a plot of quantum charge calculated with respect to varying gate voltage for $V_{DS} = 0.2$ and 0.8 V . Rate of change of quantum charge with respect to gate voltage is given by

$$\frac{\partial Q_{CNT}}{\partial V_G} = \frac{\partial Q_{CNT}}{\partial \phi_s}\frac{\partial \phi_s}{\partial V_G} \tag{9}$$

where the term $\partial Q_{CNT}/\partial \phi_s$ is known as quantum capacitance, C_Q. Since the model shows excellent agreement with TCAD simulations is demonstrated for Q_{CNT} vs. V_G (Fig. 5) and ϕ_s vs. V_G (Fig. 4), the effect of quantum capacitance is implicit and need not be calculated separately. A capacitance branch model similar to (Deng, 2007), shown in Fig. 5(b), is used to calculate the intrinsic voltages that are effective inside the channel.

Extrinsic capacitance such as parasitic fringe capacitance, inter-electrode capacitance and coupling capacitance between adjacent gates in multiple gate/multiple nanotube device geometries can dominate over the intrinsic capacitance and impact the performance of a CNT-FET. A provision for including non-ideal extrinsic capacitance has been incorporated in our compact model in the form of parasitic capacitance, C_p, which is a fitting parameter.

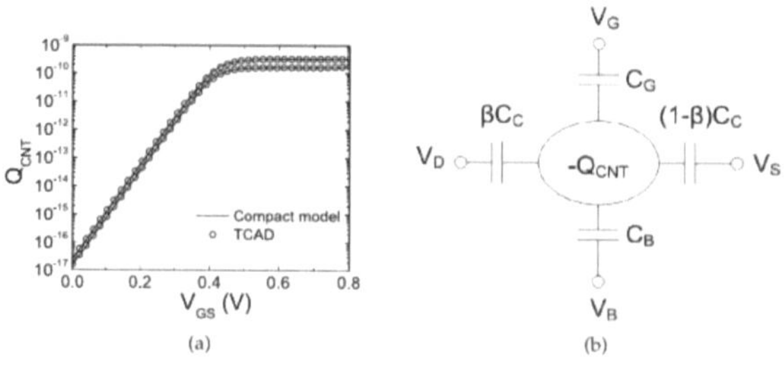

Fig. 5. (a) Q_{CNT} as a function of V_{GS} for $V_{DS} = 0.2$ V and 0.8 V and CNT diameter 1nm. (b) The capacitance divider network used in the model. C_C and β are fitting parameters.

The exact value of C_p is device-geometry dependent and can be estimated from the expressions in (Deng & Wong, 2007b; John & Pulfrey, 2006; Paul et al., 2006).

1.4 Schottky Barrier modeling

The Schottky barrier height ϕ_{SB} depends on the work function difference and the barrier width depends on the insulator thickness. The total current at the junction is the sum of thermoionic emission and the tunneling current through the barrier. Hence it is important to accurately model carrier conduction in a CNT FET through the Schottky barrier. Transmission coefficients for a single barrier are calculated using the WKB approximation (Nakanishi et al., 2002) as follows

$$T(E) = \exp\left[-\int_{z_i}^{z_f} k(z)dz\right] \tag{10}$$

The WKB approximation does not take into account reflection between the source and drain junctions. The conductance of the actual device is lower than calculated (Heinze et al., 2002). The Fabry-Perot Cavity model is used to include the effect of reflection given as

$$T(E) = \frac{T_s(E)T_d(E)}{T_s(E) - T_s(E)T_d(E) + T_d(E)} \tag{11}$$

The WKB approximation (D.Jimenez, Jan 2007) has two exponential terms in the E-k space which does not have a closed form solution. Approximating the barrier profile as a triangle allows us to get a closed form solution and results in negligible loss in accuracy. Extending on the derivation in (D.Jimenez, Jan 2007), we have the following

$$E - E(z) = \pm\frac{3a}{2}V_\pi\sqrt{k_n^2 + k_z^2}$$

Where

$E(z)$ = energy profile of conduction/valence band

k_n = momentum component of nth sub-band

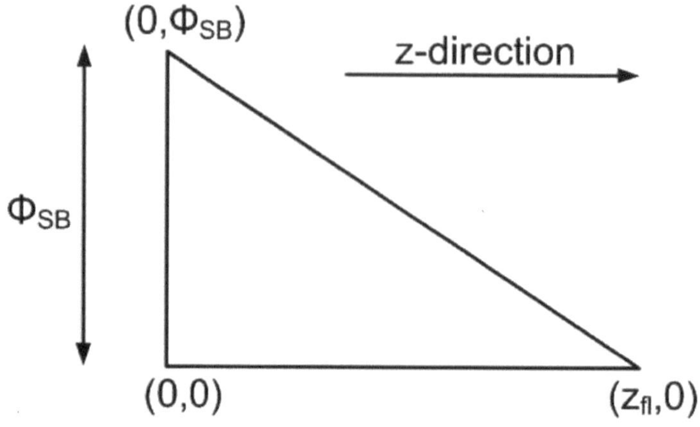

Fig. 6. Triangular Schottky Barrier Height model used in deriving the closed form expression for Tunneling Probability.

k_z = momentum component along direction of electron transport $a = 0.142nm$, the carbon-carbon bond distance

Hence

$$k_z = \sqrt{\frac{4[E - E(z)]}{(3aV_\pi)^2} - k_n^2}$$

Using the value of k_z in (10),

$$T(E) = \exp\left[-\int_{z_i}^{z_f} \sqrt{k_n{}^2 - \frac{16[E - E(z)]^2}{(3aV_\pi)^2}}\right] \tag{12}$$

In order to get an expression for $E - E(z)$, a triangular barrier profile as shown in Fig. 6 is used, which gives us

$$E(z) = -\frac{\phi_{sb}}{z_{fl}}(z) + \phi_{sb}$$

$$E - E(z) = E - \phi_{sb} + \frac{\phi_{sb}}{z_{fl}}(z)(13)$$

From (12) and (13), we get a closed form expression for tunneling probability in (14). The limits of the integration are $z_i = 0$ and $z_f = z_{fl}$, where $z_{fl} = [-t_{\text{ins}}/2]\ln(E/\phi_{sb})$.

$$T(E) = \exp\left[\frac{-t_{ins}k_n}{\phi_{sb'}}\left(E'\sqrt{1-K'^2} + (E-\phi_{sb'})E_t\right)\right] \tag{14}$$

where

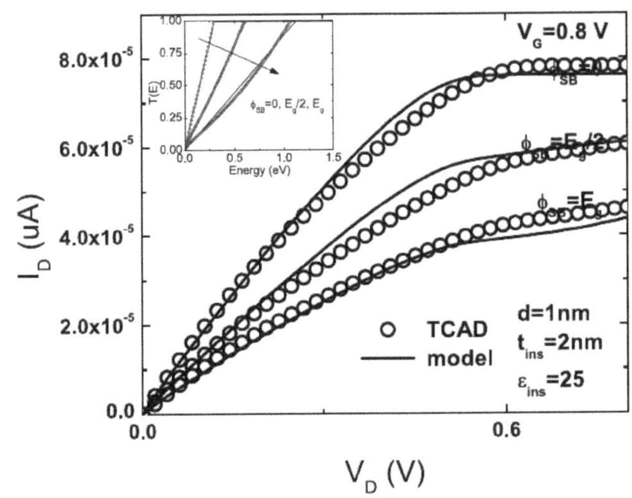

Fig. 7. I_{ds} vs. V_{ds} at $V_{gs} = 0.8$ V for three different barrier heights. The inset shows the variation in the tunneling probability (numerical vs. triangular approximation) for the different contacts.

$$\begin{aligned}
E_t =& \quad \sqrt{\left(1-K(E-\phi_{sb'})\right)^2} - \sin^{-1}\left(-E'\sqrt{1-KE'^2}\right) \\
& \quad +\sin^{-1}(\phi_{sb'} - E) \\
K =& \quad \left(\frac{q\pi}{4k_n N_0}\right) \\
E' =& \quad (E-\phi_{sb'}) + \phi_{sb'}\ln\left(\frac{E}{\phi_{sb'}}\right)
\end{aligned}$$

$$\phi_{sb'} = \mu_{s,d} + \phi_{sb} \tag{15}$$

Fig. 7 demonstrates the excellent agreement between the triangular approximation model and the numerical model for the contact part. The tunneling probability equation given by (14) is solved at the source and drain junctions and (16) is used to compute the final current.

$$\begin{aligned}
I = \frac{4q}{h}\sum_n \int_{E_n} & sgn(E)T(E)\Big[F(sgn(E)(E,\mu_s)) \\
& + F(sgn(E),(E,\mu_s - V_{ds}))\Big]dE
\end{aligned} \tag{16}$$

where $\text{sgn}(E) = 1$ or -1 for conduction and valence band respectively and $F(\mu, E)$ is as defined in (5).

Using the equations and results discussed above, (summarized in Table. 1), a physics based compact model of CNT FET is completed and implemented in VerilogA. It is computationally efficient and supports transient simulations. The I-V characteristics are presented in Fig. 8(a). These results prove that the model is scalable to different diameters and bias conditions. Since we use the surface potential approach, scattering effects that may further affect the I-V characteristics can be easily incorporated in the future.

Table 1. Constants and Parameters used in the Model

Physical constants

V_π	**C-C Bonding Energy**	**2.97 eV**
a	C-C Bonding Length	0.142 nm
q	Charge	$1.6e^{-19}$C
V_t	Thermal Voltage	26 mV

Model Parameters

d	**Diameter (m)**	θ	**Chiral Angle (deg)**
L	Nanotube length (m)	t_{ins}	Insulator thickness (m)
ϕ_{sb}	Barrier height (eV)	ϵ_{ins}	Insulator Dielectric Constant

Derived Parameters

Energy Gap (eV)	$E_g = 2V_{pi}a/d$
Sub-band Energy Levels (eV)	$E_n = (E_g/8)(6n - 3 - (-1)^n)$
Intrinsic Carrier Conc.	$N_0 = 4q/(3\pi V_\pi a)$
Insulator Capacitance	$C_{\text{ins}} = \dfrac{2\pi\epsilon_r\epsilon_0}{\log\left(((_{\text{ins}} + d/2)/(d/2)\right)}$

2. Transistor Model Validation and Extraction

The parameters enlisted in Table 2 comprise the SPICE based circuit model for CNT FET. Running simulations by varying each parameter enables us to gain detailed insight on its impact on the design potential of the CNT FET.

2.1 Extraction procedure

Our compact model can be used to comprehend measurement data in order to gain processrelated insight such as parasitics, variations etc. This is achieved by properly tuning the model parameters enlisted in Table 2. A capacitor divider network similar to (Deng, 2007) is assumed in this model, as shown in Fig. 5(b). C_C, the coupling capacitance and β are the two primary fitting paramters from this network. Scattering effects are not directly incorporated in the model. The fitting parameter mob is used to capture the effect of scattering in the nanotube and is multiplied to the final current computed in (16) in the model. mob value is dependent on the length of the nanotube and lies between 0 and 1 . Appropriate values for this parameter can be obtained from published results (Deng, 2007). The main fitting steps are

1. Define instance parameters; calculate physical parasitics (C_C is set to a very small value, which is about 1/10 of the insulator capacitance).
2. Csubfit: tuned to fit I_{DS} vs. V_{GS} at low V_{DS}(0.1 V) and V_{BS} fixed. This is to match the flat bland voltage.

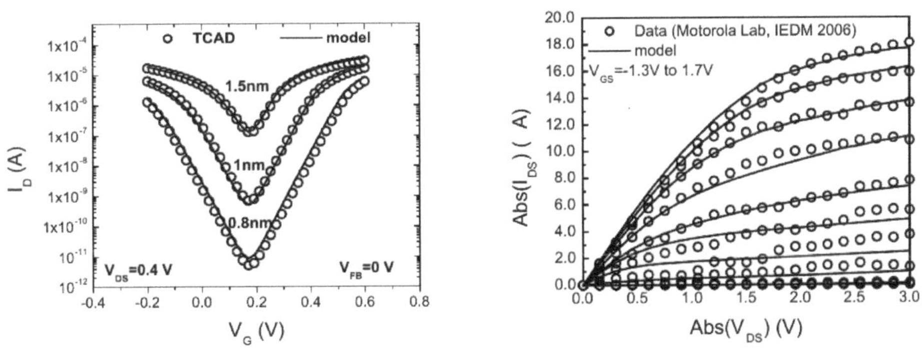

(a) (b)

Fig. 8. (a) I_{ds} as a function of V_{gs} for d = 0.8 nm, 1 nm and 1.5 nm. $V_{FB} = 0V t_{ins} = 2nm\epsilon_r = 25$ and $L = 10nm$. (b) Model validation with experimental data. Both scales show absolute values of drain voltages and currents (Amlani et al., 2006).

3. β : tuned to fit I_{DS} vs. V_{DS} at a high V_{GS} to match the saturation region (basically the shape of the I_{DS} vs. V_{DS} curve).

4. C_p : tuned to match I_{DS} vs. V_{GS} in the subthreshold region, at high V_{DS}; ϕ_{SB} also needs to be tuned to match I_{DS} vs. V_{GS} in the saturation region.

5. $R_{D,S}$: tuned to primarily match I_{DS} vs. V_{DS} in the linear region. $R_{D,S}$ also affects the saturation region. Hence, iterations may be required to get the correct fit.

6. mob: used to match the saturated drain current.

2.2 Model Validation

Using the extraction procedure described in the previous section, the model has been validated with published measurement data (Amlani et al., 2006) as shown in Fig. 8(b). An interesting feature of the fitting is the exact replication of the gap in the I-V plot, which is due to the multiple band conduction in carbon nanotubes.

The I-V characteristics in general have the following trends:

- The off current varies exponentially with diameter and barrier height.
- The on current degrades with barrier height and increases linearly with diameter.

These conclusions have been confirmed in previous models. But the new model helps us run SPICE simulations fast enough to benchmark circuits performance. All the results in the next section are generated using the Verilog-A model that supports DC and transient analysis for a single inverter several times faster than numerical simulations in MATLAB.

3. Interconnect modeling

Metallic CNT possess properties of high mechanical and thermal stability, thermal conductivity and high current carrying capabilities (Naeemi et al., Feb. 2005)(Srivastava & Banerjee, 2005 making them candidates for interconnects. Ideally, metallic SWCNT's have a fermi velocity of about 8×10^5 m/s (M.S. Dresselhaus & Eklund, 1996) but in reality the ballistic motion is degraded by several scattering mechanisms. The scattering mechanisms has been been discussed in several previous publications (Park et al., 2004; Raychowdhury & Roy, 2006). The resistance is normally modeled by a set of piece-wise linear equations. In this work we present a continuous expression for the resistance of the interconnect and the resistance of the contact making them suitable for SPICE simulators. For the sake of simplicity, the effect of temperature on resistance has not been included in this model. For a detailed discussion on the effect of temperature on resistance of metallic CNTs, the reader is referred to (Pop et al., 2007). The circuit model for the interconnect is shown in Fig. 9(b). At high frequencies, the inductance and the capacitance determine the total impedance of the interconnect. The following subsections present the DC and small-signal parameters of the CNT interconnect.

Table 2. Spice Model File Parameters

Parameter	Description	Default (units)
Instance Parameters		
d	Diameter	2 nm
$\boldsymbol{\theta}$	Chiral angle ($0 \leq \theta < 30°$)	0
tins	Insulator thickness	10 nm
eins	Dielectric constant of insulator	9
tback	Backgate insulator thickness	130 nm
eback	Dielectric constant of substrate	3.9(SiO_2)
L	Gate length	100 nm
type	n-type = 1 p-type = −1	1
Model Parameters		
phisb	Schottky barrier height	0 eV
mob	Mobility parameter	1
R_s	Parasitic Source access resistance	0 ohm
R_d	Parasitic Drain access resistance	0 ohm
β	Coupling Coefficient	1
C_C	Coupling Capacitance	7aF
C_p	Parasitic Capacitance	120aF

Due to the nature of the band structure, in an ideal ballistic motion regime, the resistance is constant as expressed in (17)

$$R_{\text{ballistic}} = \frac{h}{4e^2} = \frac{1}{G_0} \tag{17}$$

However, when the length of the interconnect is much longer that the mean free path (MFP), several scattering mechanisms dominate. At low bias, the predominant mechanism is the acoustic phonon scattering with a MFP of $1\mu m - 1.6\mu m$ Park et al. (2004). As the bias voltage increases, the electrons can scatter from band to band and also within the same band. This leads to optical phonon scattering and zone-boundary scattering. These scattering mechanisms are well-known and have been modeled in the past. In this compact model, we derived a single equation to continuously model all these scattering effects across multiple regions as shown in (18),

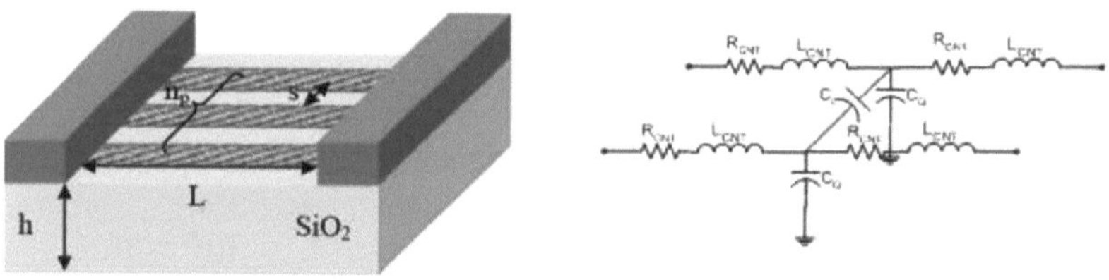

(a) (b)

Fig. 9. (a) Cross-section of a generic interconnect using carbon nanotubes. (b) Circuit model for CNT interconnect

$$G(V,L) = G_{op\_zo} + \frac{V_{eff}\left[G_{acc} - G_{op\_zo}\right]}{V} \tag{18}$$

where

$$V_{eff} = V_{cr} - \frac{1}{2}\left[(V_{cr} - V - \delta) + \sqrt{(V_{cr} - V - \delta)^2 + 4V_{cr}\delta}\right]$$

3.1 Resistance

Here a piece-wise linear model for resistance (Raychowdhury & Roy, 2006) is modified by including V_{eff} which ensures that mobility is a continuous function of the bias voltage and length facilitating convergence in circuit simulators . The dependence of length and bias voltage on the resistance of a CNT interconnect is shown in Fig. 10(a).

3.2 Capacitance and Inductance

As shown in Fig. 9(a) carbon nanotube interconnects are usually formed by arranging arrays of nanotubes aligned next to each other with the terminals at the ends of the two tubes. The coupling capacitance between two adjacent nanotubes C_C, and the quantum capacitance within the nanotube C_q have considerable effect on its conduction properties. The coupling capacitance has the form

$$C_c = \frac{\pi \epsilon L}{\log\left(\frac{d}{S} + \sqrt{\left(\frac{d}{S}\right)^2 + 1}\right)} \tag{19}$$

And the quantum capacitance is given by

$$C_q = \frac{4e^2 L}{\pi h v_f} \tag{20}$$

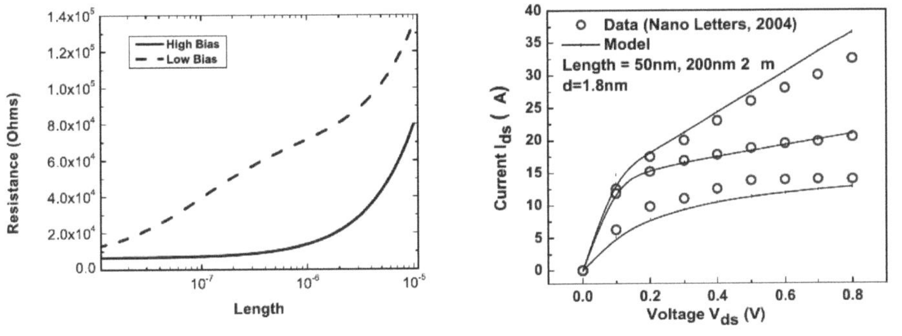

(a) (b)

Fig. 10. (a) Resistance of a carbon nanotube interconnect with varying length for high and low bias across the terminals. (b) Interconnect model validation with measured data for varying length. Diameter of the metallic nanotubes are 1.8 nm .

Carbon nanotubes posses two kinds of inductances, the magnetic or mutual inductance and the kinetic or self-inductance. As discussed in detail in (Raychowdhury & Roy, 2006), kinetic inductance dominates mutual inductance in a one-dimensional structure like carbon nanotubes and hence we only consider kinetic inductance in the model. This is given by

$$L_e = \frac{h}{2e^2 v_f} \tag{21}$$

Carbon nanotubes have two modes of propagation, with two electrons in each mode (spin up and spin down). This results in a total of four modes of propagation giving one-fourth of the total inductance calculated in (21) and four times the quantum capacitance given in (20).

3.3 Interconnect model extraction and verification

Three model parameters, V_{crit}, l_{acc} and l_{zb} are used to model the optical phonon scattering, acoustic phonon scattering and zone boundry phonon scattering respectively. The SPICE circuit parameters for the interconnect model are enlisted in Table 3.

The rest of the parameters are geometry dependent.

1. Coupling capacitance is either calculated by external 2D or 3D solvers such as Raphael (Raphael Interconnect Analysis Program Reference Manual, n.d.) and entered as an instance parameter, or can be calculated internally by (19).
2. If CNT length ranges between 10 nm and 1μ m, V_{crit} is tuned in the range of 0.08 to 0.16 to decrease the resistance. If CNT length is greater than 1μ m, acoustic phonon scattering dominates and therefore l_{acc} changes the slope of the curve.
3. If the contacts are short and ohmic then R_n and R_p can be ignored.
4. At high current values the ϕ_{SB} value is extracted.

The model is validated against measured data in Fig. 10(b).

Table 3. Interconnect Model File Parameters

Parameter	Description	Default value
Instance		
d	Diameter	1 nm
np	Number of CNTs in parallel	1
s	Spacing between CNTs	10 nm
e_{ins}	Dielectric constant of insulator	25
Cc	Coupling Capacitance	0
h	Substrate insulator thickness	100 nm
L	Gate length	100 nm
Model		

phisb	Schottky Barrier height	0 eV
V_{crit}	Optical-phonon scattering parameter	0.16 eV
R_p, R_n	Parasitic Access Resistance	0 ohm
l_{acc}	Mfp for acoustic phonon scattering	1.0 um
l_{zb}	Mfp for zone boundary phonon scattering	20 nm

4. Design Insights

The developed compact model for CNT transistors accurately predicts I-V and C-V characteristics. It is scalable to key process and design parameters such as diameter, chirality, gate dielectrics, and bias voltages. Using the model, we explore design possibilities with CNT in order to identify the optimum design space. CNT with $L = 100$ nm is compared with 22 nm CMOS (PTM) for both analog and digital applications. For consistency we have used $V_{FB} = Vdd/2$(NCNT) and $-Vdd/2$(PCNT). The dielectric material used has $\epsilon_r = 25$. Parasitic capacitances have been lumped into a single parameter based on published values (J. Deng, Sept.2006). Since all the characteristics are mainly dependent on the diameter of the nanotube, our analysis is for varying diameters. Above 1.8 nm , the SB-FET has I_{ON}/I_{OFF} less than 50 and is not included in this study.

4.1 Digital Design

Numerous simulations of FO4 inverter comparing CNT FETs with 22 nm CMOS have been undertaken to study the effect of Schottky barrier height (Source/Drain contact material), gate dielectric thickness, leakage power, supply voltage scaling and process variations on digital design (Balijepalli et al., 2007). It is found that for smaller diameters of the range of $1 - 1.5$ nm and optimum contact materials, up to 10 X improvement in speed, power and energy consumption is observed as compared to 22 nm CMOS. The speed contours have been plotted for adequate scaling in dielectric thickness to ensure the same performance. By varying the diameter, Fig. 11 shows that up to 10 X increase in speed can be achieved when compared to 22 nm CMOS.

The reason for diameters of $1 - 1.5$ nm being optimal is depicted by the shaded region in the Fig. 11. Larger diameters have higher leakage and are harder to switch off. Smaller diameters have a 5X decrease in speed compared to larger diameter CNTs.

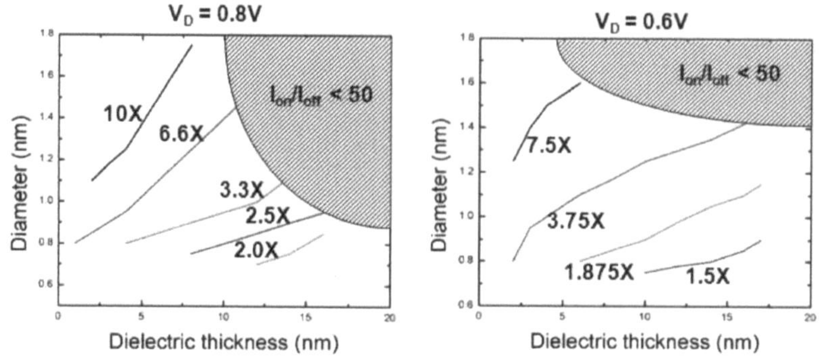

Fig. 11. Speed contours for varying diameters and t_{ins} for $V_{DD} = 0.8$ V and 0.6 V

Thus, there is a trade-off between speed and power in using CNT FET for digital applications, similar as that of CMOS.

4.2 Analog Design

If parasitic capacitance is reduced, CNTs have another advantage in very low quantum capacitance. The device can have very high cut-off frequency given by (Akinwande et al., 2006).

$$f_T = \frac{g_m}{2\pi C_g} \tag{22}$$

Based on the above equation, cut-off frequency for a 2 nm diameter CNT transistor is calculated to be around 120 GHz in our model. An in-depth critique of the high frequency performance of carbon nanotube field effect transistors is given in (Pulfrey & Chen, 2008) which discusses the effect of nanotube chirality (diameter), oxide permittivity and other process parameters on f_T. These effects can be incorporated in our model with minimal effort.

The two major hurdles preventing large scale manufacturing of CNT based devices are lack of measurement technique to characterize analog performance and reducing the parasitic capacitance during fabrication. The AC gain and frequency response are mainly controlled by the transconductance (g_m) and output impedance (R_{out}). Here we plot the variation of output impedance of CNT FET compared to 22 nm CMOS (Fig. 12). For a fair comparison, R_{out} is calculated for the same saturation current of both devices. For CMOS, R_{out} vs. V_{DS} is mainly influenced by the triode region, channel length modulation, drain induced barrier lowering (DIBL) and finally substrate current induced body effect (SCBE) with increasing V_{DS} (Huang et al., 1992). In CNTs, R_{out} is affected by the linear, saturation and ambipolar characteristics of the device. As shown in Fig. 12, due to better saturation characteristics in CNTs, a CNT FET can have up to 25 X higher R_{out} compared to 22 nm CMOS for the same saturation current.

5. Conclusion

In this chapter, a detailed procedure for developing compact models for carbon nanotube transistors and interconnects has been presented. Since the developed model does not use any iteration-based calculations, is scalable with process and design parameters and is highly accurate, it increases the scope of predictive design research. These models have been used for circuit simulation to benchmark the performance of carbon nanotube transistors with 22 nm bulk CMOS transistors providing important perspective on the design trade-offs and potential of CNT based devices. Thus, compact modeling serves as one of the most important bridges between CNT process and design giving key insights into the development of carbon nanotube based electronics.

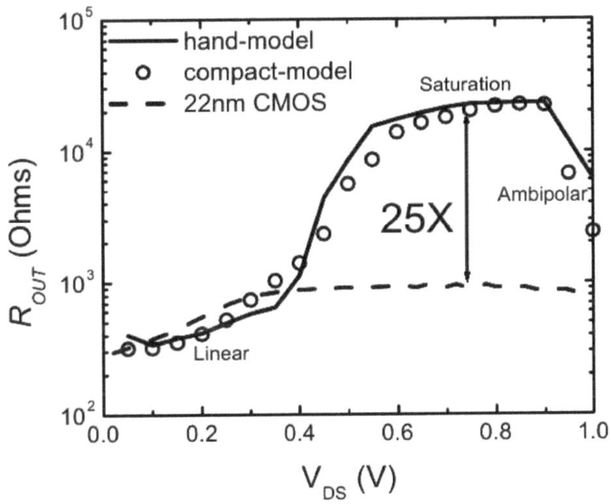

Fig. 12. R_{out} as a function of drain voltage, compared with 22 nm CMOS with same saturation current. Regions affecting R_{out} are annotated.

6. References

Akinwande, D., Close, G. F. & Wong, H.-S. P. (2006). Analysis of the Frequency Response of Carbon Nanotube Transistors, Nanotechnology, IEEE Transactions on 5(5): 599-605.

Amlani, I., Lewis, J., Lee, K., Zhang, R., Deng, J. & Wong, H.-S. P. (2006). First demonstration of ac gain from a single-walled carbon nanotube common-source amplifier, pp. 1-4.

Balijepalli, A., Sinha, S. & Cao, Y. (2007). Compact modeling of carbon nanotube transistor for early stage process-design exploration, ISLPED '07: Proceedings of the 2007 international symposium on Low power electronics and design, ACM, New York, NY, USA, pp. 2 -7.

Bowman, K. A., Duvall, S. G. & Meindl, J. D. (2002). Impact of die-to-die and within-die parameter fluctuations on the maximum clock frequency distribution for gigascale integration, Solid-State Circuits, IEEE Journal of 37(2): 183-190.

Castro, L., John, D. & Pulfrey, D. (2002). Towards a compact model for schottky-barrier nanotube fets, pp. 303-306.

Chau, R., Datta, S., Doczy, M., Doyle, B., Kavalieros, J. & Metz, M. (2004). High- κ/ MetalñGate Stack and Its MOSFET Characteristics, IEEE Electron Device Lett. 25(2004): 408-410.

Chen, I. C., Holland, S. & Hu, C. (February 1985). Electrical Breakdown in Thin Gate and Tunneling Oxides, IEEE Trans. Electron Devices 32: 413-422.

Chen, Z., Appenzeller, J., Knoch, J., Lin, Y.-M. & Avouris, P. (2005). The role of metal-nanotube contact in the performance of carbon nanotube field-effect transistors, Nano Lett. 5: 1497-1502.

Close, G. F. & Wong, H.-S. P. (2007). Fabrication and Characterization of Carbon Nanotube Interconnects, Electron Devices Meeting, 2007. IEDM 2007. IEEE International pp. 203 206.

Deng, J. (2007). A compact spice model for carbon-nanotube field-effect transistors including nonidealities and its applicationópart i: Model of the intrinsic channel region.

Deng, J. & Wong, H.-S. (2007a). A Compact SPICE Model for Carbon-Nanotube Field-Effect Transistors Including Nonidealities and Its ApplicationóPart I: Model of the Intrinsic Channel Region, Electron Devices, IEEE Transactions on 54(12): 3186-3194.

Deng, J. & Wong, H.-S. (2007b). Modeling and analysis of planar-gate electrostatic capacitance of 1-d fet with multiple cylindrical conducting channels, Electron Devices, IEEE Transactions on 54(9): 2377-2385.

D.Jimenez, e. (Jan 2007). A simple drain current model for schottky-barrier carbon nanotube field effect transistors, Nanotechnology 18(2): 025201. ID: 61221.

Doyle, B., Boyanov, B., Datta, S., Doczy, M., Hareland, S., Jin, B., Kavalieros, J., Linton, T., Rios, R. & Chau, R. (2003). Tri-Gate fully-depleted CMOS transistors: Fabrication, design and layout, VLSI Symp. Tech. Dig. pp. 133-134.

Guo, J., Datta, S, Lundstrom & M (2004). A numerical study of scaling issues for Schottkybarrier carbon nanotube transistors, Electron Devices, IEEE Transactions on 51(2): 172 177.

Guo, J. & Lundstrom, M. (2006). Nanoscale Transistors: Device Physics, Modeling and Simulation, Springer.

Guo, J., Lundstrom, M. & Datta, S. (2002). Performance projections for ballistic carbon nanotube field-effect transistors, Applied Physics Letters 80(17): 3192-3194. URL: http://link.aip.org/link/?APL/80/3192/1

Heinze, S., Tersoff, J., Martel, R., Derycke, V., Appenzeller, J. & Avouris, P. (2002). Carbon nanotubes as schottky barrier transistors, Phys. Rev. Lett. 89(10): 106801.

Hisamoto, D., Lee, W.-C., Kedzierski, J., Takeuchi, H., Asano, K., Kuo, C., Anderson, E., King, T.-J., Bokor, J. & Hu, C. (2000). FinFET-a self-aligned double-gate MOSFET scalable to 20 nm, IEEE Trans. Electron Devices 47(12): 2320-2325.

Huang, J. H., Liu, Z. H., Jeng, M. C., Ko, P. K. & Hu, C. (1992). A physical model for MOSFET output resistance, Electron Devices Meeting, 1992. Technical Digest., International pp. 569 -572.

J. Deng, H.-S. W. (Sept.2006). A circuit-compatible SPICE model for enhancement mode carbon nanotube field effect transistors, Synopsys Inc.

John, D. L. & Pulfrey, D. L. (2006). Switching-speed calculations for schottky-barrier carbon nanotube field-effect transistors, 24(3): 708-712.
URL: http://link.aip.org/link/?JVA/24/708/1

Lin, Y.-M., Appenzeller, J., Chen, Z., Chen, Z.-G., Cheng, H.-M. & Avouris, P. (2005). High performance dual-gate carbon nanotube FETs with 40 – nm gate length, IEEE Electron Device Lett. 26: 823-825.

McEuen, P. L., S, F. M. & Park, H. (2002). Single-walled carbon nanotube electronics, Nanotechnology, IEEE Transactions on 1(1): 78-85.

Moore, G. E. (1965). Cramming more components onto integrated circuits, Electronics 38(8).

M.S. Dresselhaus, G. D. & Eklund, P. (1996). Science of Fullerenes and Carbon Nanotubes, Elsevier Inc.

Naeemi, A., Sarvari, R. & Meindl, J. D. (Feb. 2005). Performance comparison between carbon nanotube and copper interconnects for gigascale integration (GSI), IEEE Electron Device Lett. 26(2): 84-86.

Nakanishi, T., Bachtold, A. & Dekker, C. (2002). Transport through the interface between a semiconducting carbon nanotube and a metal electrode, Phys. Rev. B 66(7): 073307.

Park, J.-Y., Rosenblatt, S., Yaish, Y., Sazonova, V., Ustunel, H., Braig, S., Arias, T. A., Brouwer, P. W. & McEuen, P. L. (2004). Electron-Phonon Scattering in Metallic Single-Walled Carbon Nanotubes, Nano Lett. 4(3): 517-520.

Paul, B., Fujita, S., Okajima, M. & Lee, T. (2006). Impact of geometry-dependent parasitic capacitances on the performance of cnfet circuits, Electron Device Letters, IEEE 27(5): 380-382.

Pop, E., Mann, D. A., Goodson, K. E. & Dai, H. (2007). Electrical and thermal transport in metallic single-wall carbon nanotubes on insulating substrates, Journal of Applied Physics 101(9): 093710.
URL: http://link.aip.org/link/?JAP/101/093710/1

Pulfrey, D. L. & Chen, L. (2008). Examination of the high-frequency capability of carbon nanotube fets, Solid-State Electronics 52(9): 1324-1328. Papers Selected from the 37th European Solid-State Device Research Conference - ESSDERC'07. URL: http://www.sciencedirect.com/science/article/B6TY5-ST45Y81/2/35986f4a38a6adca353dfae27fc40cf2

Raphael Interconnect Analysis Program Reference Manual (n.d.). Synopsys Inc.

Raychowdhury, A., Keshavarzi, A., Kurtin, J., De, V. & Roy, K. (2006). Carbon Nanotube FieldEffect Transistors for High-Performance Digital Circuits-DC Analysis and Modeling Toward Optimum Transistor Structure, Electron Devices, IEEE Transactions on 53(11): 2711-2717.

Raychowdhury, A., Mukhopadhyay, S. & Roy, K. (2004). A circuit-compatible model of ballistic carbon nanotube field-effect transistors, Computer-Aided Design of Integrated Circuits and Systems, IEEE Transactions on 23(10): 1411-1420.

Raychowdhury, A. & Roy, K. (2006). Modeling of metallic carbon-nanotube interconnects for circuit simulations and a comparison with Cu interconnects for scaled technologies, Computer-Aided Design of Integrated Circuits and Systems, IEEE Transactions on 25(1): 58 - 65.

Srivastava, N. & Banerjee, K. (2005). Performance analysis of carbon nanotube interconnects for vlsi applications, Computer-Aided Design, 2005. ICCAD-2005. IEEE/ACM International Conference on pp. 383-390.

Wann, C. H., Noda, K., Tanaka, T., Yoshida, M. & Hu, C. (October 1996). A Comparative Study of Advanced MOSFET Concepts, IEEE Transactions on Electron Devices Vol. 43(No. 10): 1742 – 1753.

Welser, J., Hoyt, J. L. & Gibbons, J. F. (1994). Electron mobility enhancement in strained-Si n-type metal-oxideñsemiconductor field-effect transistors, IEEE Electron Device Lett. 15(3): 100-102.

Wong, H.-S., Deng, J., Hazeghi, A., Krishnamohan, T. & Wan, G. (Nov. 2006). Carbon nanotube transistor circuits - models and tools for design and performance optimization, ICCAD pp. 651-654.

Zhang, G., Wang, X., Li, X., Lu, Y., Javey, A. & Dai, H. (2006). Carbon Nanotubes: From Growth, Placement and Assembly Control to 60mV/ decade and Sub- 60mV/ decade Tunnel Transistors, Electron Devices Meeting, 2006. IEDM '06. International pp. 1-4.

Chapter-13

Measurement and Modeling of High-Frequency CNTFET Parameters

1. Introduction

Carbon nanotube field effect transistors (CNTFETs) are high-mobility devices that operate at very high-speeds. Theoretical analyses suggest that the cut-off frequency (f_T) of an ideal CNTFET is between 800 GHz and 1.3 THz when its gate length is 0.1μ m(1; 2). Since this frequency is much higher than that of state-of-the-art Si, GaAs, and InP transistors, CNTFETs are promising candidates for future nanoelectronic devices. Singh et al.(3) measured frequency responses of top-gated CNTFETs up to 100 MHz . Li et al.(4) observed 2.6 – GHz operation of CNTFETs with an LC impedance-matching circuit. However, as Li et al. pointed out (4), measuring high-frequency performance of high-impedance devices, such as CNTFETs, is quite difficult. This is because their output impedances are much higher ($\sim 10^5\Omega$) than the impedance of the measurement system (50Ω) using a network analyzer. To perform accurate high-frequency measurements, especially those to determine f_T values of such devices, we must measure S -parameters with a network analyzer even though large impedance mismatches hinder us from obtaining accurate measurement data. Kim et al.(5) measured S-parameters of multifinger CNTFETs by using a network analyzer and obtained an f_T value of 2.5 GHz . They also concluded a maximum oscillation frequency ($f_{\max}$) of more than 5 GHz was obtained using the maximum stable gain (G_{msg}). Le Louarn et al.(6) obtained intrinsic f_T value of 30 GHz by measuring a CNTFET the channel of which was fabricated using dielectrophoresis to increase the CNT density. They also obtained G_{msg} value of more than 10 dB at 20 GHz .

This chapter will describe a method for accurately measuring and modeling the highfrequency characteristics of CNTFETs, with reference to our experiment and analysis (7). In the experiment, we first decreased the device impedance to be able to measure the S-parameter using network analyzer. This was achieved by developing a high-density multiple-channel CNTFET structure the output impedance of which is much lower than that of the conventional single-channel CNTFETs. Then we used a de-embedding procedure to remove existing errors in measured S-parameters of small-signal devices in order to obtain the current gain and unilateral power gain (U) that can determine accurate f_T and f_{max} values. For accurate RF modeling of CNTFETs, we developed an equivalent circuit RF model that includes parasitic resistances and capacitances of the CNTFET. Then the expression of the $f_T(f_{max})$ was derived as a function of them. Not ignoring the higher order parasitic resistances and capacitances neglected in the cases of current RF transistors, an accurate model was obtained that can fully explain the experimental results.

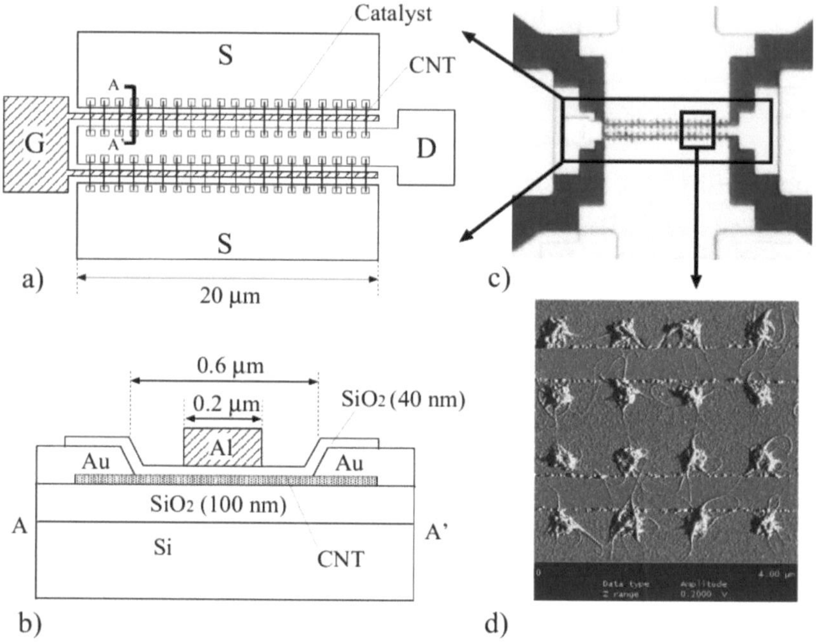

Fig. 1. Multiple-channel CNTFET structure: a) top view, b) cross section, c) optical micrograph, d) atomic force micrograph.

2. Multiple-channel CNTFET Structure

As shown in Figure 1, the evaluated CNTFET was fabricated on a SiO_2 insulator on a highly resistive (10kΩcm) Si substrate. Iron was deposited for a catalyst and was patterned by electron-beam lithography. Single-walled carbon nanotubes (SWCNTs) were grown from the catalyst islands by chemical vapor deposition. The average density of the SWCNTs was 5 per μm, as observed in the AFM analysis (Figure 1-d). The gate oxide was $40 - \text{nm}$ thick SiO_2, which serves as a passivation layer to retain stable characteristics and suppress hysteresis of the CNTFET I-V curve. The top-gated structure was used to reduce parasitic capacitances. The gate consisted of two $20 - \mu\text{m}$ wide fingers. Thus, approximately 200 SWCNT channels were constructed in the total $40 - \mu\text{m}$ gate width. The drain and source electrodes were formed by evaporation of Au , and ohmic contacts were made with CNT channels.

3. De-Embedding Procedure

Using the multiple-channel structure decreases the output impedances of the devices more than those of the single-channel CNTFETs. Therefore, their output signals can be observed directly with the network analyzer. However, output impedances of CNTFETs are still higher than those of conventional RF transistors. This means that the output signal of the device is small and easily disturbed or masked by the parasitic elements. The drain, gate, and source electrodes of the CNTFET were connected with pads for RF probe contacts. The dimensions of the pad were $100 \times 100\mu$ m, and its layout is shown in Figure 2-a. The areas of the pad and the connective wiring region are much larger than the transistor area (shown as DUT in Figure 2-a), and this large area forms parasitic elements and causes large errors in CNTFET S-parameters. Therefore, we applied the de-embedding procedure to effectively eliminate the parasitic error matrix, and only the S -parameters of the transistor were extracted using open-short-through standards on the substrate. This method is basically the same as that described in Vandamme et al.(8) and Temeijer et al.(9). The equivalent circuit of the RF test-structure, including pads and CNTFETs, is shown in Figure 2-b. In the figure, parasitic elements ($Z_1, Z_2, Z_3, G_1, G_2, G_3$) are shown. Z_1, Z_2, Z_3 are parasitic impedances, and G_1, G_2, G_3 are parasitic admittances. To determine the parasitic elements, we made four standard patterns (Open, Short1, Short2, Through) that are the same as the CNTFET measurement patterns but without CNT channels (Figure 3). The equivalent circuits of the four standard patterns are shown in Figure 4. Each standard pattern contains a different combination of the parasitic elements, and so they can be determined by the measured S-parameters of the four standards. Let us transform the measured S-parameters ($s_{ij}: i,j = 1,2$) of the four standards to the Y-parameters and express them as $y_{ijop}, y_{ijsh1}, y_{ijsh2}, y_{ijthr}(i,j = 1,2)$. Here, y_{ijop} is the Y-parameters of the Open standard, y_{ijsh1} is the Y-parameters of the Short1 standard, y_{ijsh2} is the Y-parameters of the Short2 standard, y_{ijthr} is the Y-parameters of the Through standard.

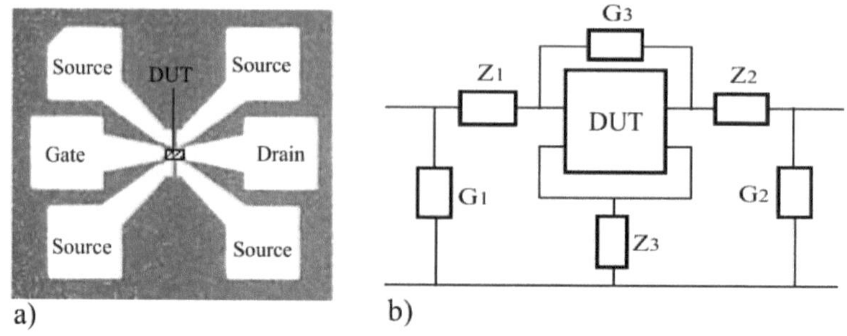

Fig. 2. RF test structure: a) pad layout, b) equivalent circuit.

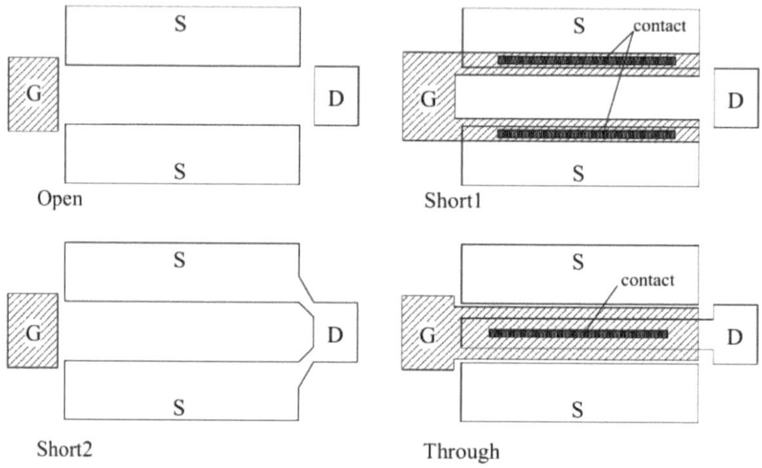

Fig. 3. Standard patterns for de-embedding

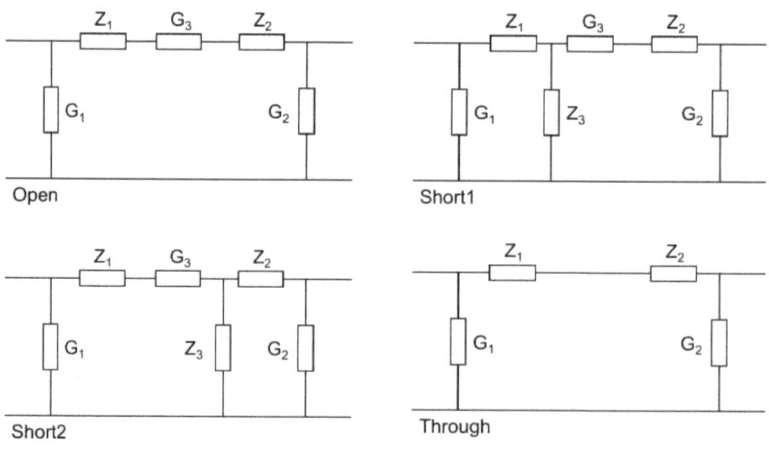

Fig. 4. Equivalent circuit of standard patterns.

Thus the parasitic elements $(Z_1, Z_2, Z_3, G_1, G_2, G_3)$ can be expressed as follows:

$$G_1 = y_{11op} + y_{12op} \tag{1}$$

$$G_2 = y_{22op} + y_{12op} \tag{2}$$

$$G_3 = \left(\frac{-1}{y_{12op}} + \frac{1}{y_{12thr}}\right)^{-1} \tag{3}$$

$$Z_1 = \frac{1}{2}\left(\frac{-1}{y_{12thr}} + \frac{1}{y_{11sh1} - G_1} + \frac{1}{y_{22sh2} - G_2}\right) \tag{4}$$

$$Z_2 = \frac{1}{2}\left(\frac{-1}{y_{12thr}} - \frac{1}{y_{11sh1} - G_1} + \frac{1}{y_{22sh2} - G_2}\right) \tag{5}$$

$$Z_3 = \frac{1}{2}\left(\frac{1}{y_{12thr}} + \frac{1}{y_{11sh1} - G_1} - \frac{1}{y_{22sh2} - G_2}\right) \tag{6}$$

Using the above parasitic elements (parasitic impedance and parasitic admittance), the deembedded matrix can be obtained by the following procedure. Let us transform the measured S-matrix ($\mathbf{S}_{\text{meas}}$) into the Y-matrix and write it as $\mathbf{Y}_{\text{meas}}$. First, we subtract G_1, G_2 from $\mathbf{Y}_{\text{meas}}$ and obtain $\mathbf{Y}_A$ as follows:

$$\mathbf{Y}_A = \mathbf{Y}_{\text{meas}} - \begin{bmatrix} G_1 & 0 \\ 0 & G_2 \end{bmatrix} \tag{7}$$

Transforming the obtained $\mathbf{Y}_A$ to the Z-matrix ($\mathbf{Z}_A$), we next subtract Z_1, Z_2, Z_3 from $\mathbf{Z}_A$ and obtain $\mathbf{Z}_B$ as follows:

$$\mathbf{Z}_B = \mathbf{Z}_A - \begin{bmatrix} Z_1 + Z_3 & Z_3 \\ Z_3 & Z_2 + Z_3 \end{bmatrix} \tag{8}$$

Again, transforming the obtained $\mathbf{Z}_B$ into the Y-matrix ($\mathbf{Y}_B$), we subtract G_3 from $\mathbf{Y}_B$ and obtain $\mathbf{Y}_{\text{DUT}}$ as follows:

$$\mathbf{Y}_{\text{DUT}} = \mathbf{Y}_B - \begin{bmatrix} G_3 & -G_3 \\ -G_3 & G_3 \end{bmatrix} \tag{9}$$

$\mathbf{Y}_{\text{DUT}}$ is the final de-embedded Y-matrix of the DUT part.

4. Measurement Results

4.1 DC Characteristics

The DC characteristics of the multiple-channel CNTFET were measured with a semiconductor parameter analyzer (Agilent 4156C). Figure 5-a shows the drain current (

I_d) versus gate voltage (V_g) curve when the drain voltage (V_d) was -2 V . I_d versus V_d curve is shown in Figure 5-b. These characteristics are like p-type FETs but the drain current is not zero even when the gate voltage is small enough. This is due to the metallic carbon nanotubes. Because the metallic carbon nanotubes do not affect the high-frequency characteristics of the device, we did not perform a special removal process such as a burn out procedure. From the DC curve (Figure 5), transconductance ($g_m = \partial I_d/\partial V_g$) of 226μ S and drain conductance ($g_d = \partial I_d/\partial V_d$) of 1 mS (at $V_g = 5$ V, $V_d = -2$ V) were obtained. The drain current of our multi-channel CNTFET is more than 200 times larger than that of single-channel CNTFETs. We observed hysteresis in the I-V curves; however, the width of the hysteresis is much smaller ($\Delta V_g < 1$ V, $\Delta V_d < 0.1$ V) than that of non-passivated CNTFETs.

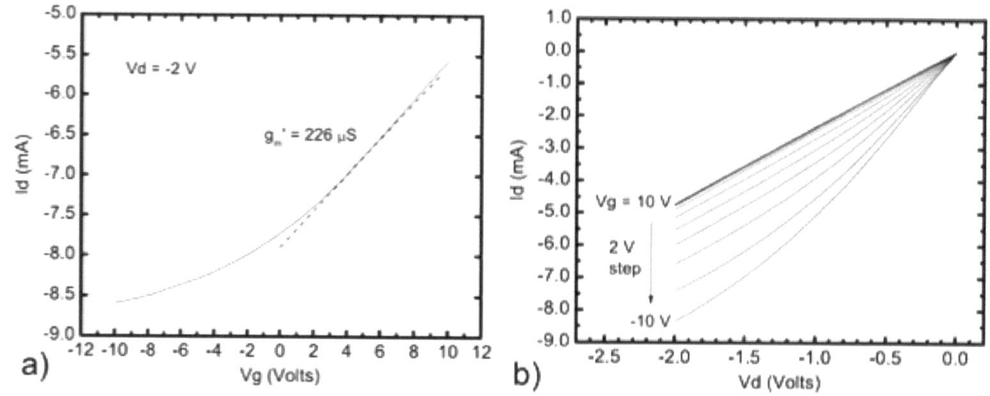

Fig. 5. DC characteristics of multiple-channel CNTFET: a) $I_d - V_g$ curve, b) $I_d - V_d$ curve.

4.2 RF Characteristics

Using RF probes (Cascade Microtech I40-GSG-125), we measured 2-port S-parameters of the device between 100 MHz and 20 GHz with the network analyzer (Agilent PNA N5230A). Standard SOLT-calibration was performed at the probe ends by using the calibration substrate. The measured S-parameters were de-embedded by the previously mentioned error removal procedure, and we obtained current gain ($|h_{21}| = |y_{21}/y_{11}|$) from the de-embedded Y-parameters ($\mathbf{Y}_{\mathrm{DUT}}$) (Figure 6-a). In this figure, the measured and de-embedded data are displayed. The de-embedded data was 15 dB larger than the measured data. The f_T value was determined to be 10.3 GHz by obtaining the frequency when the current gain was unity (0 dB). Also, the unilateral power gain (U) was calculated from the de-embedded Y-parameters ($\mathbf{Y}_{\mathrm{DUT}}$) by the formula below and plotted as a function of frequency in Figure 6-b.

$$U = \frac{|y_{21} - y_{12}|^2}{4[\mathrm{Re}(y_{11})\mathrm{Re}(y_{22}) - \mathrm{Re}(y_{12})\mathrm{Re}(y_{21})]} \tag{10}$$

The $f_{\max}$ value was determined to be 3.5 GHz by obtaining the frequency when the unilateral power gain was unity (0 dB).

5. Equivalent circuit model analysis

Figure 7 shows our proposed equivalent small-signal circuit model for multiple-channel CNTFET. Here, $R_s(R_d)$ is the resistance of the CNT between the source (drain) and gate, and contains the resistance of the CNT extensions and contact resistances. R_s and R_d were extracted from the DC measurements, as described in (10). The extracted values of R_s and R_d were 420 Ω each for our multiple-channel CNTFET. Note that R_s and R_d values of usual RF transistors are negligibly small (a few ohms); however, for CNTFETs, even when using multiple-channel structures, these resistances play a dominant role in the analysis. The g_m and g_d in Figure 7 are intrinsic transconductance and drain conductance and relate to measured transconductance (g_m') and drain conductance (g_d'), like in Chow and Antoniadis (11), as follows:

$$g_m = \frac{g_m'}{1 - g_m' R_s - g_d'(R_s + R_d)} \quad \text{and} \quad g_d = \frac{g_d'}{1 - g_m' R_s - g_d'(R_s + R_d)}. \tag{11}$$

Using these relations and the DC measurement results, we calculated the intrinsic transconductance and drain conductance as $g_m = 3.47\mathrm{mS}$ and $g_d = 15.4\mathrm{mS}$. The intrinsic g_m and g_d values are one order of magnitude larger than g_m' and g_d' because of the large R_s and R_d values. The value $C_{g-\mathrm{cnt}}\left(= C_{g-\mathrm{cnts}} + C_{g-\mathrm{cntd}}\right)$ is the capacitance between the gate electrode and CNTs. We assumed that $C_{g-\mathrm{cnts}} = C_{g-\mathrm{cntd}}$ because of the symmetry of the device. According to Burke (1), $C_{g-\mathrm{cnt}}$ consists of the electrostatic capacitance (C_{ES}) and the quantum capacitance $\left(C_Q\right)$ and is given by $C_{g-cnt}^{-1} = C_{ES}^{-1} + C_Q^{-1} \cdot C_Q$ is about $100\mathrm{aF}/\mu\mathrm{m}$. C_{ES} is calculated from geometry (as shown in Figure 8-a) and is given by $C_{ES} = 2\pi\varepsilon_r\varepsilon_0/\cosh^{-1}(2h/d)$. In our case, $h = 40$ nm and $d = 1$ nm, so C_{ES} was estimated to be $42.5\mathrm{aF}/\mu\mathrm{m}$. Thus, C_{g-cnt} (one CNT) was calculated to be $30\mathrm{aF}/\mu\mathrm{m}$, and taking into account the gate length of 0.2μ m and 200 CNT channels, we calculated $C_{g-cnt} = 1.2\mathrm{fF}\left(C_{g-cnts} = C_{g-cntd} = 0.6\mathrm{fF}\right)$. C_{gs} and C_{gd}, as shown in Figure 7, are parasitic capacitances between the gate and source and the gate and drain. $C_{gs} + C_{gd}$ can be approximated by the electrostatic capacitance between a coplanar stripline and ground planes on a dielectric (Figure 8-b), like in Collin (12), as follows:

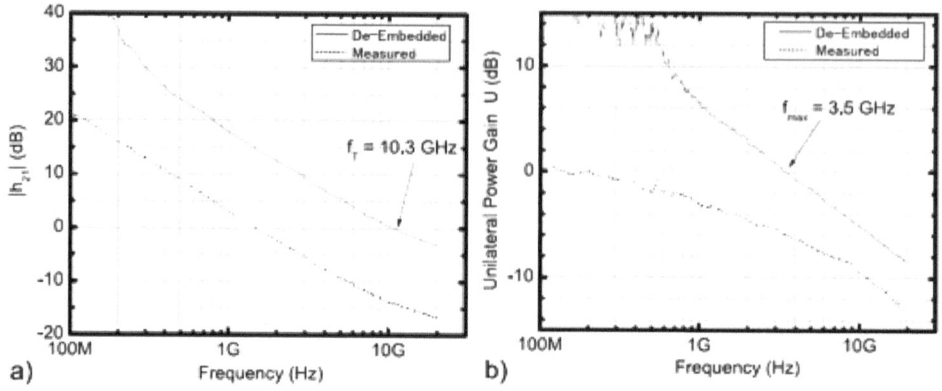

Fig. 6. RF characteristics of multiple-channel CNTFET: a) current gain ($|h_{21}|$), b) power gain (U), bias condition is $V_d = -2$ V, $V_g = 5$ V.

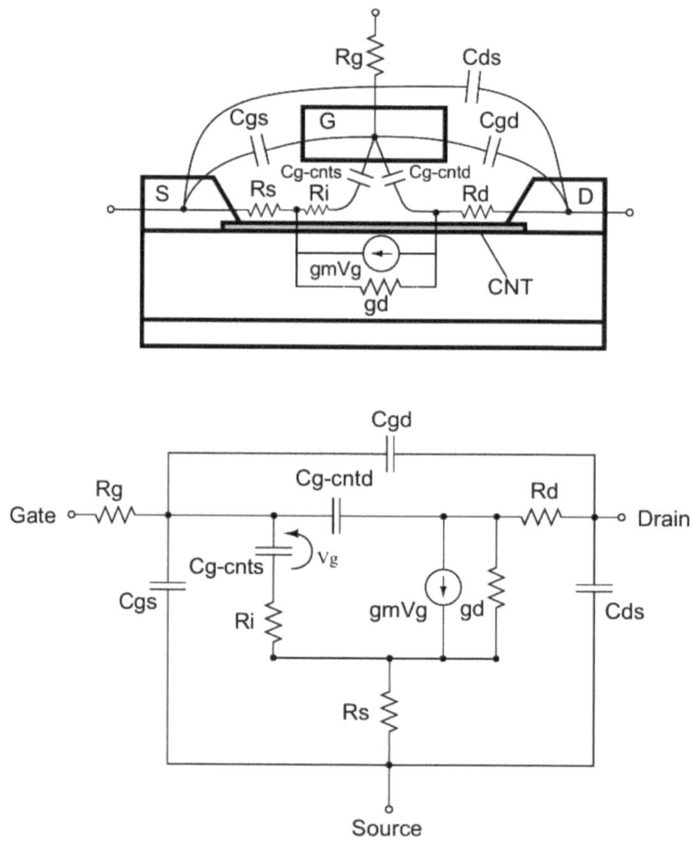

Fig. 7. Equivalent small-signal circuit model for CNTFET.

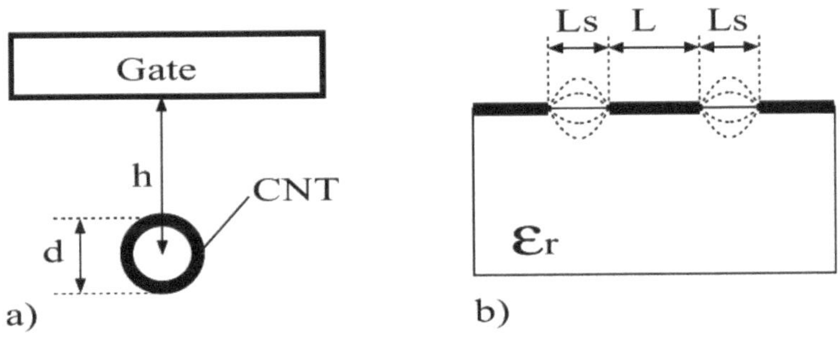

Fig. 8. Capacitance geometry: a) gate-CNT capacitance, b) coplanar stripline on dielectric

$$C_{gs} + C_{gd} = 2(\varepsilon_r + 1)\varepsilon_0 \frac{K(k)}{K\left(\sqrt{1-k^2}\right)}, \quad (12)$$

where $k = L/(L + 2L_S)$, and K is the complete elliptic integral of the first kind. Considering $L = L_s = 0.2\mu$ m and $W = 40\mu$ m, we obtain $C_{gs} + C_{gd} = 2.2$fF. Once the equivalent circuit model is constructed like in Figure 7, we can derive f_T by calculating the H -parameters of the circuit (13). f_T is given as

$$\frac{1}{2\pi f_T} = \frac{1}{g_m} C_g + \frac{g_d}{g_m} C_g (R_s + R_d) + C_{g-cntd}(R_s + R_d) + \left(C_{gs} + C_{gd}\right) R_s, \quad (13)$$

where $C_g = C_{gs} + C_{gd} + C_{g-\text{cnts}} + C_{g-\text{cntd}}$. Substituting the parameters into Equation (13), we obtain f_T (model) $= 10.6$GHz. This value is consistent with the experimental one: f_T(experiment) $= 10.3$GHz.

Similarly, we can derive the expression of f_{max} by calculating the unilateral power gain U (Equation (10)) from the Y-parameter of the equivalent circuit and solving the equation: $U = 1$. The following formula is the calculation result.

$$f_{\text{max}} = \sqrt{\frac{f_T}{2\pi R_g \left[\left(\frac{g_d}{g_m} C_g + C_{gd} + C_{g-\text{cntd}}\right) + \frac{2\pi f_T}{g_m^2 R_g} \Psi\right]}} \quad (14)$$

where C_g and Ψ are defined as follows:

$$C_g = C_{gs} + C_{gd} + C_{g-\text{cnts}} + C_{g-\text{cntd}} \quad (15)$$

$$\Psi = R_s R_d \left(g_d C_{g-\text{cnts}} + g_d C_{g-\text{cntd}} + g_m C_{g-\text{cntd}}\right)^2 \quad (16)$$

$$+R_i g_d C_{g-\text{cnts}} \left(C_{g-\text{cnts}} + g_d R_s C_{g-\text{cnts}} + g_d R_d C_{g-\text{cnts}} + g_m R_d C_{g-\text{cntd}}\right) \quad (16)$$

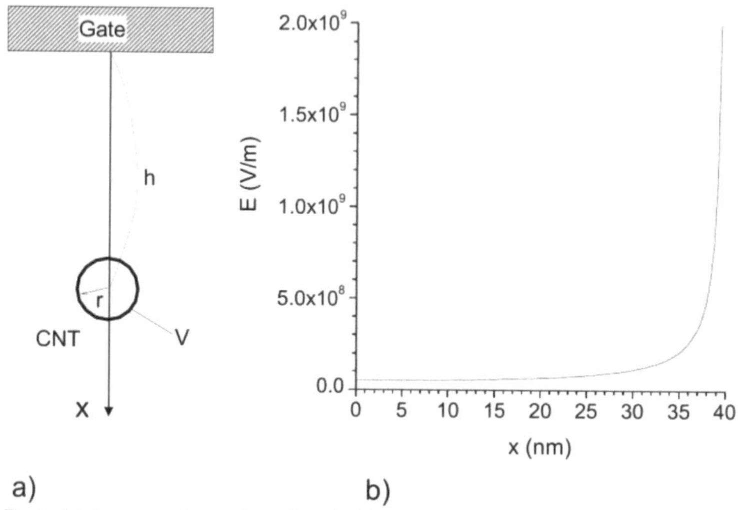

Fig. 9. (a) Geometry of gate electrode and CNT, (b) Electric field at x calculated under the condition of $V = -5$ V, $r = 0.5$ nm, $h = 40$ nm.

The above formula of $f_{\max}$ is more complicated than the expression of f_T (Equation(13)). Note that the expression of $f_{\max}$ contains not only the parameters included in the expression of f_T, but also the parameters R_g and R_i. R_g is the gate resistance and can be estimated from the resistivity of the gate electrode material of aluminum. R_g in our case is about 40 ohms. R_i is equivalent to the channel resistance (R_{ch}) in the case of current transistors and can be expressed as $R_{ch} = \alpha / g_m$. Here, α is a coefficient and has a value less than 0.2 (13). If we assume that $\alpha = 0.2$ and use the g_m of the CNTFET, we obtain $R_i \approx 60$. Using these values of R_g, R_i and the other parameters used for estimation of f_T, we can estimate $f_{\max}$ as $f_{\max}|_{R_i=60} = 19.2\text{GHz}$. This value is much larger than the experimental value: $f_{\max}$ (experiment) $= 3.5\text{GHz}$. To clarify this discrepancy, we should consider the phenomenon peculiar to the CNT channels, and one consistent model was proposed as shown below.

The diameter of the SWCNTs used for the multiple-channel CNTFET is about 1 nm , and Sparameter was measured under the condition that the gate electrode voltage is 5 V . This gate voltage is relatively high enough that the strong electric field exists at the vicinity of the CNTs. To estimate the electric field, we assume that the gate electrode and the CNT are an infinite conductive plane and an infinite length of conductive cylinder that is separate from the plane at the distance h and has a radius of r, as shown in the Figure 9 - a . When the potential of the CNT is V, the potential at x is written as the expression below (14):

$$\phi = \frac{V}{2\cosh^{-1}(h/r)} \ln\left(\frac{x+\sqrt{h^2-r^2}}{x-\sqrt{h^2-r^2}}\right)^2 \quad (17)$$

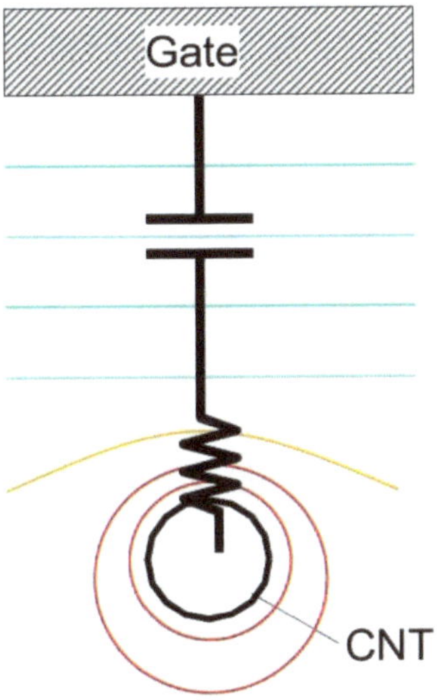

Fig. 10. Equivalent circuit between gate electrode and CNT

Substituting $V = -5$ V, $r = 0.5$ nm, $h = 40$ nm (thickness of the gate oxide) to the above formula, and calculating the electric field $E = -\partial\phi/\partial x$, the result are shown in the Figure 9-b. The result shows that the electric field increases rapidly as the point is getting close to the CNT. The electric field at the CNT surface ($x = 39.5nm$) is 2.0×10^9 V/m. This means that the electric field is extremely strong at the vicinity of the CNT. In this condition, the tunneling current, Fowler-Nordheim(F-N) current, easily flows from the CNT into the gate oxide. This phenomenon is considered also in (15), and they describe that the current density of the F-N current is expressed as below:

$$J_{FN} = aE^2 \exp(-b/E) \quad (18)$$

where a and b are constants and have values $a = 2 \times 10^6 A/MV^2$ and $b = 230MV/cm$. The total F-N current that flows through all CNTs in the transistor can be described as $I = AJ_{\text{FN}}$, where A is the total area through which the F-N current flows. Let us consider the F-N current at the CNT surface ($x = 39.5nm$). In this case, the area is upper half surface of the CNT cylinder under the gate electrode. Multiplied by 200 CNT channels, the total area can be obtained as $A = 200 \cdot L \cdot \pi r$, where L is the gate length.

The variation of the potential ($\delta\phi$) corresponding to the variation of the CNT potential (δV) can be written as $\delta\phi = \frac{\partial\phi}{\partial V}\delta V$. Also the variation of the F-N current (δI) can be written as $\delta I = \frac{\partial I}{\partial V}\delta V$. Thus, the resistance originated from the F-N current can be written as $R_{FN} = \delta\phi/\delta I = \frac{\partial\phi}{\partial V}/\frac{\partial I}{\partial V}$. Calculating R_{FN} using Equations (17) and (18), we obtain $R_{FN} \approx 75k\Omega$. Consequently, we can guess that the equivalent circuit between the gate electrode and CNT consists of the capacitance of the gate oxide and additional serial resistance (R_{FN}), as shown in Figure 10. This adds R_{FN} to R_i in our equivalent circuit model of the CNTFET, so that we can write $R_i = R_{ch} + R_{FN} = 60 + 75k \approx 75k\Omega$. Substituting this into Equation (14), we obtain $f_{\max}$ (model) $= 3.4$GHz. This $f_{\max}$ value is very near to the experimental value: $f_{\max}$ (experiment) $= 3.5$GHz.

From the above discussion, we get a guideline to improve the high-frequency characteristics of the CNTFET. To increase the f_T, the parasitic capacitances $(C_{gs} + C_{gd})$ and the resistances ($R_s + R_d$) need to be reduced. If ($C_{gs} + C_{gd}$) and ($R_s + R_d$) are negligible, then Equation (13) becomes $f_T = g_m/2\pi C_{g-cnt}$, and we obtain $f_T \approx 460$GHz. The parasitic resistance $R_s(R_d)$ is the resistance of CNT extensional part between the source (drain) electrode and the channel. A candidate to decrease this resistance is doping (16). To improve the value of $f_{\max}$, R_g and $R_i(R_{FN})$ should be decreased in addition to reducing the parasitic capacitances and resistances. Because R_{FN} originates from the strong electric field near CNTs, as discussed above, we should develop a CNTFET that operates with low gate voltages.

6. Summary

This chapter described a method for accurately measuring and modeling the high-frequency characteristics of CNTFETs. To directly measure using a network analyzer, we developed a high-density multiple-channel CNTFET structure the output impedance of which is much lower than conventional single-channel CNTFETs. We also focused on accurately measuring S-parameters, e.g., we used a de-embedding procedure that removes existing errors in measured S-parameters of high-impedance devices. Consequently, we obtained a cut-off frequency (f_T) of 10.3 GHz and a maximum oscillation frequency ($f_{\max}$) of 3.5 GHz . We also proposed an equivalent circuit RF model that includes the higher order parasitic resistances and capacitances that are neglected in the case of current RF transistors. The model, therefore, can explain the experimental results very accurately. The analysis also revealed that decreasing the parasitic capacitances of the electrodes and the resistances of the CNT extensions greatly improves the high-frequency performance of CNTFETs.

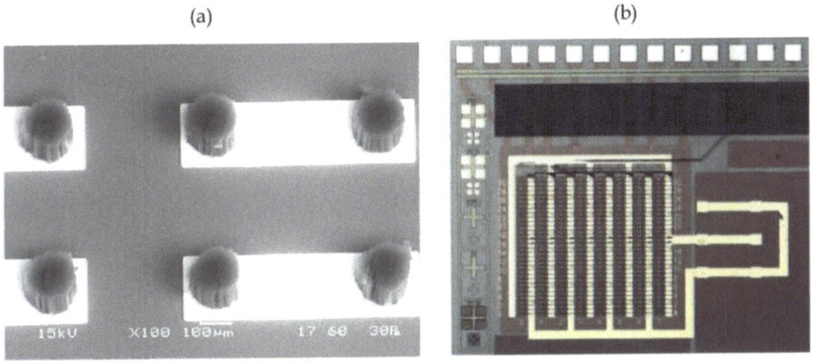

Fig. 2. (a) Vertical CNT bundles as flip-chip bumps (Fujitsu Labs.); (b) carbon nanotubes wiring in an oscillator circuit (Stanford University & Toshiba).

2. Electromagnetic propagation models for CNTs

A carbon nanotube is realized by rolling-up a sheet of a mono-atomic layer of graphite (graphene), see Fig.3a. The rolled graphene may become either metallic or semiconducting, depending on its chirality, i.e. the way in which it is rolled up. A single-wall carbon nanotube (SWCNT) is made by a single sheet of graphene (Fig.3b), whereas a multi-wall carbon nanotube (MWCNT) is made by nested sheets (Fig.3c). In the graphene the conduction phenomenon is related to the motion of the so-called π-electron, i.e. the one out of the four valence electrons which is free to move between the positive ion lattice. Both semiclassical as well as quantum-mechanical approaches have been used to describe the CNT electron dynamics.

We study the electrodynamics of a CNT shell, which may either describe an isolated SWCNT or a shell of a MWCNT, in which the direct coupling of electronic states of adjacent shells may be disregarded. The electrical signal propagation along the CNT shell is analyzed assuming a low frequency regime, where only intraband transitions of π-electrons with unchanged transverse quasimomentum occur. This allows using the semi-classical transport theory expounded, for instance, in (Slepyan et al., 1999; Maksimenko et al., 2000; Miano & Villone, 2006: Maffucci et al., 2008a; Forestiere et al., 2009a and 2009b). Using this approach, the longitudinal dynamic conductivity of a CNT shell of arbitrary chirality may be obtained, expressed in terms of the number of effective conducting channels. This leads to the constitutive equation of this innovative material, which could be coupled to Maxwell equations in order to derive electromagnetic and/or circuital models, following the stream of what done in (Miano & Villone, 2006: Maffucci et al., 2008a; Forestiere et al., 2009).

2.1 Constitutive relation of a CNT shell

The chirality of a CNT is described by two integers n and m, which are the coefficients of he basis vectors $\mathbf{a}_1$ and $\mathbf{a}_2$ depicted in Fig.3a. The CNT is metallic if $|n-m|=3q$, where $q=0,1,2\ldots$ A zig-zag CNT is obtained for $n=0$ or $m=0$, whereas an armchair CNT is given by $n=m$, therefore, armchair CNTs are always metallic, whereas zig-zag ones are metallic only if $m=3q$, with $q=0,1,2\ldots.$ For $m\to\infty$ or $n\to\infty$ any CNT tends to the graphene sheet. The CNT diameter D is given by:

$$D = 0.0782\sqrt{n^2+nm+m^2}\,[\,\mathrm{nm}]. \tag{1}$$

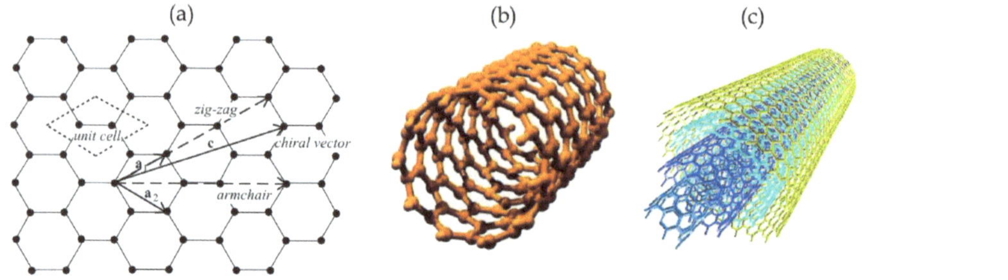

Fig. 3. (a) the graphene lattice; (b) single-wall CNT; (c) multi-wall CNT.

For each segment of the first Brillouin zone of the nanotube we introduce a distribution function for the π electrons in the valence and conduction bands:

$$f_\mu^{(\pm)} = f_\mu^{(\pm)}(z,k,t), \mu = 0,1,\ldots N, -\frac{\pi}{T} < k \le \frac{\pi}{T} \tag{2}$$

where k is the longitudinal wavenumber, t is the time, N is the number of graphene exagons in the primitive cell of the CNT, and T is the translational length. By assuming that the electric field is transversally uniform, $f_\mu^{(\pm)}$satisfy the quasi-classic Boltzmann equation

$$\frac{\partial f_\mu^{(\pm)}}{\partial t} + v_\mu^{(\pm)}\frac{\partial f_\mu^{(\pm)}}{\partial z} + \frac{2\pi e}{h}E_z\frac{\partial f_\mu^{(\pm)}}{\partial k} = -v\left(f_\mu^{(\pm)} - f_{0,\mu}^{(\pm)}\right) \tag{3}$$

where e is the electron charge, h is the Planck constant, $E_z(z,t)$ is the component of the electric field tangent to the CNT surface, v is the frequency of collision of the electrons with the ion lattice, $v_\mu^{(\pm)}$are the longitudinal velocities of the electrons in the conduction/valence subbands and $f_{0,\mu}^{(\pm)}$are the distribution functions at equilibrium. The latter are given by:

$$f_{0,\mu}^{(\pm)}(k) = \frac{1}{\pi^2 D} F\left[E_\mu^{(\pm)}(k)\right], F[E] = \frac{1}{e^{E/k_B T_0} + 1} \tag{4}$$

being $E_\mu^{(\pm)}$the energy of the generic subband, $F[E]$ the Dirac-Fermi function, k_B the Boltzmann constant and T_0 the CNT absolute temperature. Let us assume time-harmonic electric field and surface current density:

$$E_z(z,t) = \mathrm{Re}\{\hat{E}_z e^{i(\omega t - \beta z)}\}, J_z(z,t) = \mathrm{Re}\{\hat{J}_z e^{i(\omega t - \beta z)}\} \tag{5}$$

For small perturbations of functions (2) around the equilibrium values (4) we get

$$\hat{E}_z = \hat{\sigma}_{zz}(\beta,\omega)\hat{J}_z \tag{6}$$

where the spectral domain conductivity $\hat{\sigma}_{zz}(\beta,\omega)$ is given by

$$\hat{\sigma}_{zz}(\beta,\omega) = \frac{2ie^2}{\pi^2 D}\sum_{\mu=0}^{N-1}\int_0^{\pi/T}\left(\frac{1}{\omega - v_\mu^{(+)}\beta - iv} + \frac{1}{\omega + v_\mu^{(+)}\beta - iv}\right) v_\mu^{(+)2}\frac{dF}{dE_\mu^{(+)}}dk \tag{7}$$

For all the subbands that give a meaningful contribution to the conductivity we may assume $v_\mu^{(\pm)} \cong v_F$ in (7), being v_F the Fermi velocity. This leads to

$$\hat{\sigma}_{zz}(\beta,\omega) = \frac{4e^2 v_F}{\pi h D}\frac{1}{i\omega + v}\frac{1}{1 + v_F^2\beta^2/(i\omega + v)^2}M(n,m) \tag{8}$$

where M may be regarded as the equivalent number of conducting channels:

$$M(n,m) = \frac{h}{\pi v_F}\sum_{\mu=0}^{N-1}\int_0^{\pi/T} v_F^2\left(\frac{dF}{dE_\mu^+}\right)dk \tag{9}$$

The constitutive relation of the CNT in the frequency domain comes from (6), (8) and (9):

$$(1 + i\omega/v)J_z + \frac{v_F^2}{v(1 + v/i\omega)}\frac{\partial\rho_s}{\partial z} = \sigma_c E_z \tag{10}$$

where ρ_s is the surface charge density and σ_c is the static limit for the axial conductivity:

$$\sigma_c = \frac{4e^2 v_F}{v\pi h D}M(n,m) \tag{11}$$

In the limit $\omega/v \gg 1$, (11) reduces to the simple fluid model in (Miano & Villone, 2006; Maffucci et al., 2008a) and for $\omega/v \ll 1$ the electrons behave as a cold fluid.

2.2 Approximation of the equivalent number of conducting channels

The number of conducting channels M is a function of the parameters affecting the number of subbands around the CNT Fermi level: CNT chirality, diameter and temperature. Figure 4a shows M for increasing CNT diameter, evaluated at $T = 273$ K. For small diameters it is $M = 2$ for metallic CNTs and $M = 0$ for semiconducting ones, i.e. only metallic CNTs contribute the conduction. For larger diameters the channels of a metallic CNT increase and those associated to semiconducting ones are no longer negligible.

Let us now compute the axial conductivity for the two cases: Fig. 4 b shows these values, normalized to the axial conductivity for a graphene sheet, given by (Slepyan, 1999):

$$\sigma_\infty = \frac{8\pi\ln\,2}{h^2}\frac{e^2 k_B T}{v + i\omega}. \tag{12}$$

Figure 4 b agrees with the condition that for $D \to \infty$ a CNT tends to the graphene.

To obtain a simple analysis tool for studying MWCNTs or bundles of CNTs, it is useful to fit the behavior of M through a simple approximated function of the diameter D and temperature T. A piecewise linear function able to fit the asymptotic behaviors for $D \to \infty$ and $D \to 0$ may be given by the following expression (the parameters are given in Table 1):

$$M \approx \begin{cases} M_0 & \text{for } D < d_0/T \\ a_1 DT + a_2 & \text{for } D \geq d_0/T \end{cases}. \tag{13}$$

Figure 5 shows the fitting results computed at $T = 273$ K and $= 373$ K.

This result is consistent with similar approximations proposed in literature (Naeemi, 2008), where however M is sligthly overestimated, due to the approximations used in evaluating the number of subbands around the Fermi level.

Table 1. Parameters for the fitting formulas (13).

	Metallic	Semiconducting
M_0 [adim.]	2	0
a_1[1/nmK]	$3.3 \cdot 10^{-4}$	$3.3 \cdot 10^{-4}$

a_2[adim].	0.15	-0.20
d_0[nm · K]	$5.6 \cdot 10^3$	$0.6 \cdot 10^3$

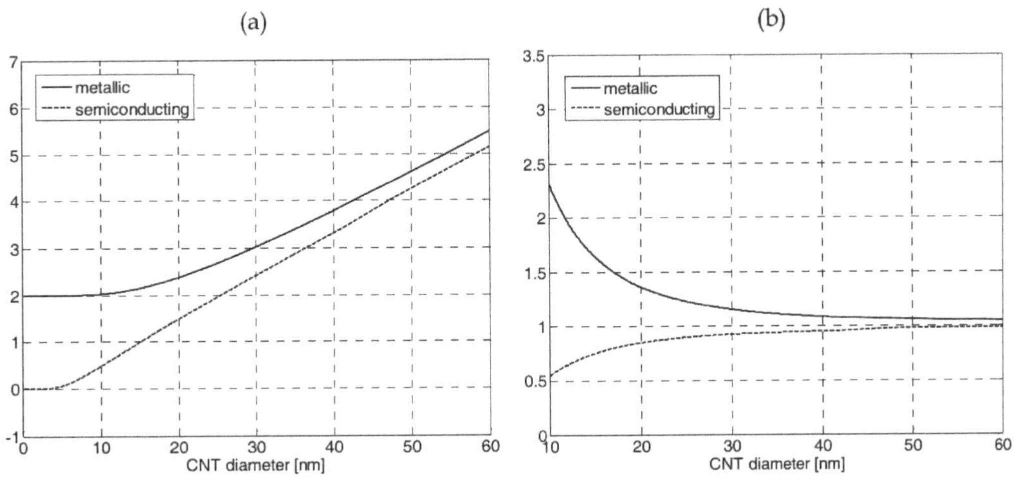

Fig. 4. (a) Number of effective conducting channels versus CNT diameter, at = 273 K; (b) axial conductivity normalized to the graphene value.

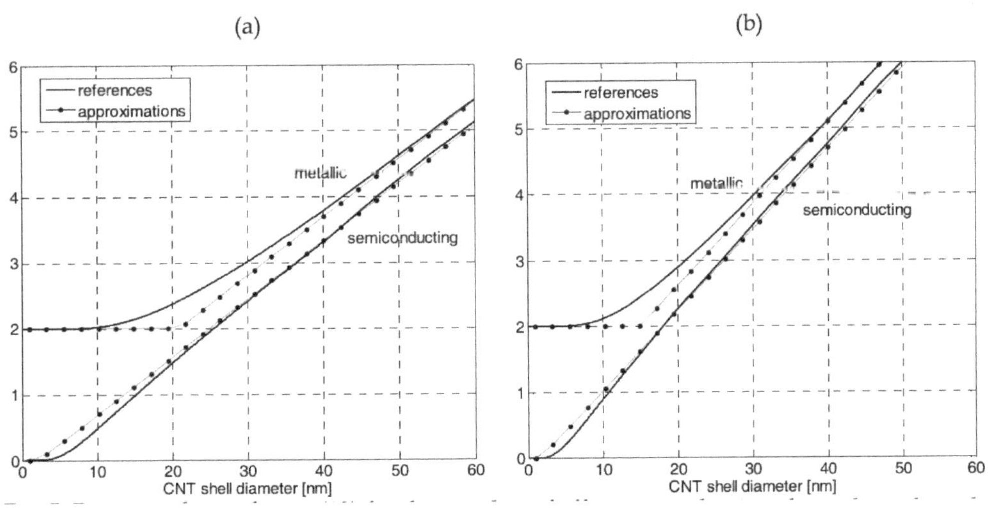

Fig. 5. Piecewise linear fitting (13) for the number of effective conducting channels evaluated at: (a) = 273 K; and (b) T = 373 K.

3. Circuit models for CNT bundle interconnects

3.1 Circuit model for a single CNT above a ground plane

A circuit model for CNT interconnects may be derived in the framework of the transmission line (TL) theory. The problem may be formulated in frequency domain by coupling Maxwell equations to the CNT constitutive relation (10). Then, assuming the propagation to be of quasi-TEM type, a distributed RLC circuit may be derived.

Let us first consider the simple case of a CNT shell of diameter D, located at a distance t above a PEC ground, Fig. 6a. As shown in paragraph 4.3, the distribution of surface currents and charges along the contour of a CNT section may be assumed to be uniform, since the proximity effect is negligible, hence the current and charge at any position along the CNT axis are given by:

$$Q(\omega,z) = \pi D \rho_s(\omega,z), I(\omega,z) = \pi D J_z(\omega,z). \tag{14}$$

Using (14) in (10) we get:

$$(v + i\omega) I(\omega,z) + \frac{v_F^2}{(1 + v/i\omega)} \frac{\partial Q(\omega,z)}{\partial z} = 2 v_F \frac{M}{R_0} E_z(\omega,z), \tag{15}$$

being $R_0 = h/2e^2 = 12.9\text{k}\Omega$ the so-called quantum resistance. The longitudinal component of the electric field E_z may be derived from the scalar electric and vector magnetic potentials:

$$E_z(\omega,z) = -\frac{\partial V(\omega,z)}{\partial z} - i\omega A_z, \tag{16}$$

which may be related to charge and current through the classical expressions:

$$V(\omega,z) = \frac{Q(\omega,z)}{C_e}, A_z(\omega,z) = L_m I(\omega,z) \tag{17}$$

being C_e, L_m the per-unit-length (pul) electrostatic capacitance and magnetic inductance.

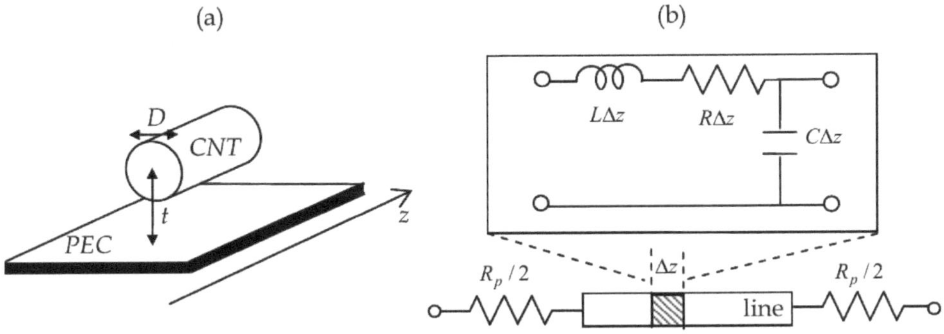

Fig. 6. (a) A CNT shell above a PEC; (b) equivalent circuit: elementary cell (inset) and lumped contact resistances.

Combining (15)-(17) we get the first Telegraphers' equation:

$$-\frac{dV(\omega,z)}{dz} = (R(\omega) + i\omega L(\omega))I(\omega,z) \tag{18}$$

with the pul parameters R and L defined as:

$$L = (L_m + L_k)/a_C, R = \nu L_k/a_C, C = C_e \tag{19}$$

where the parameter α_C is given by:

$$a_C = 1 + \frac{C_e}{C_q}\frac{1}{1+\nu/i\omega} \tag{20}$$

In (19) we have introduced the p.u.l. kinetic inductance L_k and p.u.l. quantum capacitance C_q :

$$L_k = \frac{R_0}{2v_F M} = \frac{h}{4e^2 v_F M}, C_q = \frac{1}{L_k v_F^2} = \frac{2M}{R_0 v_F} = \frac{4e^2 M}{h v_F}. \tag{21}$$

The second Telegraphers' equation is obtained by imposing the charge conservation law:

$$-\frac{dI(\omega,z)}{dz} = i\omega C V(\omega,z) \tag{22}$$

with the pul capacitance C defined as $C = C_e$. Equations (18) and (22) describe a lossy TL where the quantum effects are combined to the classical electrical and magnetic ones in the definition of the p.u.l. parameters R, L, and C. Assuming a frequency operating condition such as $v/\omega \ll 1$, the parameters R and L do not depend on frequency, since (20) becomes:

$$a_C \approx 1 + C_e/C_q \tag{23}$$

Typical orders of magnitude for the collision frequency v are $10^{11} \div 10^{12}$ Hz, hence the above condition is satisfied for frequencies up to the order of hundrends of GHz. Assuming this limit, the propagation along a CNT is then described by a simple lossy TL with constant parameters R, L, and C. The elementary cell is depicted in the inset of Fig.6b. The circuit model obtained here is a generalization of other models proposed in literature: for instance in (Burke, 2003; Salahuddin, 2005; Maffucci et al., 2008a) TL models for metallic SWCNTs are obtained. The elementary cell is however slightly different from those used in (Burke, 2003; Salahuddin, 2005), where the voltage variable is derived from the electrochemical potential rather than from the electrostatic one. A detailed discussion may be found in (Maffucci et al., 2009). The TL models in the quoted references describe metallic CNTs, where the model presented here is more general, since it deals with both metallic and semiconducting CNTs of any diameter. For instance the values of L_k and C_q used in literature may be obtained using $M = 2$ in (21), i.e. assuming metallic CNT with small radius:

$$L_{k0} = \frac{h}{8e^2 v_F} = \frac{R_0}{4v_F}, C_{q0} = \frac{8e^2}{hv_F} = \frac{4}{R_0 v_F} \tag{24}$$

Assuming in (24) $c_s \approx v_F \approx 8 \cdot 10^5$ m/s, we get $L_{k0} \approx 4\text{nH}/\mu\text{m}$ and $C_{q0} \approx 400\text{aF}/\mu\text{m}$. However the actual values deviate from this bulk values as an effect of the increase of the number of channels with the CNT diameter, as shown in Figure 7.

Let us now investigate the influence of the quantistic capacitance and kinetic inductance on the pul parameters (19). Assuming a homogeneous dielectric the classical parameters are:

$$L_m = \frac{\mu}{2\pi}\cosh^{-1}\left(\frac{2t}{D}\right), C_e = 2\pi\varepsilon/\cosh^{-1}\left(\frac{2t}{D}\right) \tag{25}$$

Let us consider the case of a metallic SWCNT with $D = 4$ nm, $\varepsilon_r = 2.2$ and $\mu = \mu_0$. For such a condition $L_k \approx L_{k0}$ and $C_q \approx C_{q0}$. Table 2 shows the ratio between classical and quantistic parameters for different aspect ratios t/D. The kinetic inductance dominates the magnetic one, being $3 - 4$ order of magnitude higher.

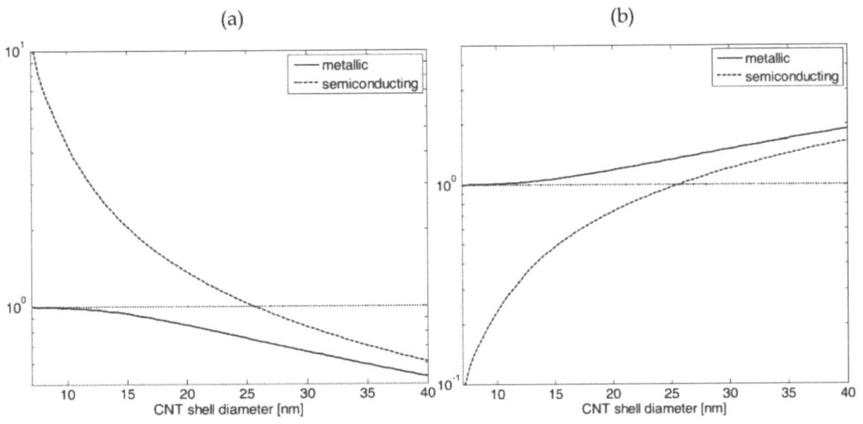

Fig. 7. (a) kinetic inductance, and (b) quantum capacitance for a CNT shell vs diameter, normalized to the bulk values (24).

Table 2. Comparison between classical and quantistic pul TL parameters for a SWCNT.

t/D	L_k/L_m	C_e/C_q	α_C
1.5	$8.1 \cdot 10^3$	0.13	1.13
3	$7.5 \cdot 10^3$	0.14	1.14
5	$5.5 \cdot 10^3$	0.09	1.09
10	$4.6 \cdot 10^3$	0.07	1.07

As for the capacitance, the quantum capacitance is approximately one order of magnitude smaller than the electrostatic one, hence the parameter α_C is close to 1 . A consequence of the huge value of the kinetic inductance is a low propagation velocity and an high characteristic impedance: the values of these two parameters normalized to an ideally scaled PEC interconnect of same dimensions would be

$$\frac{c_{CNT}}{c} \approx \sqrt{\frac{L_m}{L_k}}, \frac{Z_{CNT}}{Z_C} \approx \sqrt{\frac{L_k}{L_m}}. \quad (26)$$

Indeed the lower propagation velocity is not a real limit for the practical application of CNT interconnects. Assuming the same conditions as those used for Table 2, for $t = 3D$ it is $c_{\text{CNT}} \approx 2.5 \cdot 10^6$ m/s, instead of $c \approx 2 \cdot 10^8$ m/s, hence at $f = 10$GHz we have a wavelength $\lambda = 0.25$ mm. This means that at local and semilocal level the line is electrically short and the propagation delay, although higher than that introduced by an ideally scaled conventional interconnect, is negligible compared to the delay introduced by the RC or RLC behavior of the line (e.g., Banerjee, 2008; Maffucci et al., 2008b).

Let us now analyze the p.u.l. resistance, using (19) with $\alpha_C \approx 1$ and assumimg $L_K \approx L_{k0}$:

$$R = \nu L_K = \frac{R_0}{2M}\frac{1}{l_{mfp}}, \tag{27}$$

where the collision frequency ν has been defined as

$$\nu = v_F / l_{mfp}, \tag{28}$$

being l_{mfp} the mean-free-path. The values of l_{mfp} may vary significantly if low or high bias conditions are considered. In the low bias condition l_{mfp} may be assumed constant, whereas for higher bias conditions, l_{mfp} depends on the longitudinal voltage too, hence the parameter R is a non-linear function of the longitudinal voltage, usually described as a piecewise linear function with only two asymptotic values (Raychowdhury & Roy, 2006). A detailed discussion on this point may be found in (Park et al., 2004; Maffucci et al., 2009). In addition, assuming a CNT length greater than the ballistic length, a simple linear dependence of l_{mfp} on the diameter D may be considered (Nieuwoudt & Massoud, (2006).

As a final remark we note that experimental evidences and theoretical limits show a bulk value of $R_0/2 \approx 6.45\mathrm{k}\Omega$ for the CNT resistance, which is a lower bound even when ballistic transport is considered (e.g., Park et al., 2004). Furthermore, if the contacts between the CNT and the metallic electrodes at its terminations are not ideal, we have to consider an additional parasitic lumped resistance, whose value is strongly dependent on the quality of the contacts. For poor contacts this value may easily rise up to $100\mathrm{k}\Omega$, even if recently the fabrication techinques are demonstrated very good quality contacts (Li et al, 2009). We include these phenomena by adding two lumped resistors $R_P/2$ in series to the line (see Fig.6b), taking into account both the bulk resistance and the imperfect contact resistance.

3.2 Circuit model for bundles of CNTs

The huge value of the resistance makes useless an interconnect made by a single CNT shell. For practical application purposes, bundles of SWCNTs or MWCNTs are proposed as material to be used in fabricating interconnect traces (ITRS, 2007): all the CNT shells are fed in parallel, so lowering the total resistance. Following the stream of what done in paragraph 3.1, we can model the propagation along a CNT bundle in the frame of the multiconductor TL theory, starting from the scheme in Fig.8.

Let $\mathbf{V}(\omega,z) = [V_1(\omega,z) \dots . V_N(\omega,z)]^T$ and $\mathbf{I}(\omega,z) = [I_1(\omega,z) \dots . I_N(\omega,z)]^T$ be the vectors of the N voltages and currents at given z , in frequency domain. On the k-th CNT eq. (15) reads:

$$(v_k + i\omega)I_k(\omega,z) + \frac{v_F^2}{(1+v_k/i\omega)}\frac{\partial Q_k(\omega,z)}{\partial z} = 2v_F\frac{M_k}{R_0}E_{zk}(\omega,z). \tag{29}$$

Let us assume that in evaluating the low-energy band structure we can neglect the interactions between the CNT shells, as shown by experimental evidence (Maarouf et al., 2000). This assumption allows using the results obtained for isolated CNT shells in evaluating the parameters M_k and v_k in (29). Using again (16)-(17) to derive $E_{zk}(\omega,z)$ and imposing the charge conservation law, we get the MTL equations:

$$-\frac{d\mathbf{V}(\omega,z)}{dz} = (R + i\omega L)\mathbf{I}(\omega,z), -\frac{d\mathbf{I}(\omega,z)}{dz} = i\omega C\mathbf{V}(\omega,z), \tag{30}$$

with the pul parameter matrices given by:

$$L = \alpha_C^{-1}(L_M + L_k), C = C_E, R = \alpha_C^{-1}L_k v, \alpha_C = I + C_q^{-1}C_e, \tag{31}$$

where I is the identity matrix, and the other matrices are given by:

$$v = \mathrm{diag}(v_k), L_k = \mathrm{diag}(L_{kk}), C_q = \mathrm{diag}(C_{qk}). \tag{32}$$

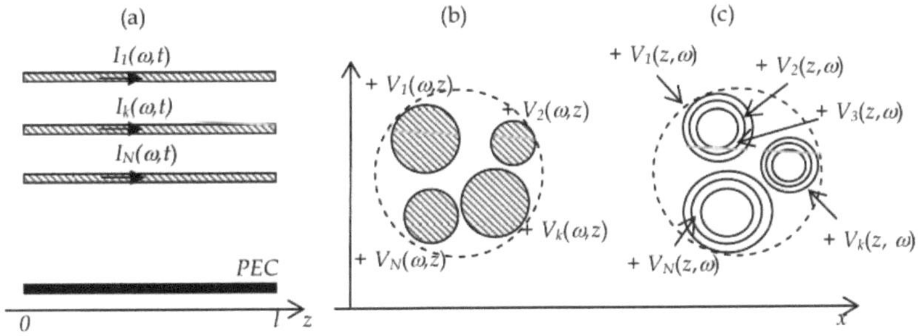

Fig. 8. A bundle of CNTs modeled as a multiconductor interconnect: (a) longitudinal view; transverse section of (b) SWCNTs and (c) MWCNTs bundle.

In practical applications any CNT bundle is used to carry a single signal, hence a CNT bundle above a ground may be described by a two-conductor TL model, which could be rigorously derived from the MTL one (30)-(32) assuming all the CNTs in parallel. Alternatively, we can use approximated expressions for its pul parameters. It is important to stress that statistically a low-cost fabrication process for CNT bundles ends up with 1/3 of metallic CNTs and 2/3 semiconducting ones (Li et al., 2005; Naeemi & Meindl, 2006). In addition we have to recall that typical values of SWCNTs diameters fall in the range $2 \div 10$ nm (Cheung et al., 2002), whereas the external shells of MWCNTs have typically diameters greater than 20 nm (Li, 2005; Naeemi, 2006). Recalling the results of Section 2, we can state that in a bundle of SWCNTs we may assume 1/3 conducting ($M = 2$) and 2/3 non conducting ($M = 0$), whereas all the MWCNTs shells may assumed to be conducting.

Let us first focus on the bundle capacitance. Figure 9 shows two circular bundles made by MWCNTs of external diameter $D_{\text{out}} = 20$ nm, embbedded in a homogeneous dielectric matrix of SiO_2 with $\varepsilon_r = 4.5$. The bundle diameter is $D_b = 146$ nm and the center-center distance is $s = 300$ nm. Figure 9 shows the distribution of the electrostatic potential, assuming a differential mode on the pair, which leads to a computed pul capacitance of 88.7pF/m. The classical pul capacitance of a pair of solid cylindrical conductors would be:

$$C_e = \pi\varepsilon/\cosh^{-1}\left(\frac{2s}{D_b}\right) \tag{33}$$

Using (33) would give a value of 92.8pF/m, with an error of less 5%. In a SWCNTs only 1/3 of the SWCNTs are metallic, hence the distribution of the potential may sensibly vary assuming random distribution of the fraction of metallic CNTs in the bundle. Figure 10 shows two possible cases, referred to a pair made by two bundles of SWCNTs with CNT diameter $D = 2$ nm, bundle diameter $D_b = 18.8$ nm, center-center distance $s = 38.2$ nm, embedded in a SiO_2 matrix ($\varepsilon_r = 4.5$). Table 3 shows the results obtained for the cases of Fig. 10 and assuming all the CNTs metallic (case 3). In this condition approximation (33) would give an error of 20%. This approximation is acceptable or not depending on the values of the load capacitance in practical applications, as will be shown in Section 4.

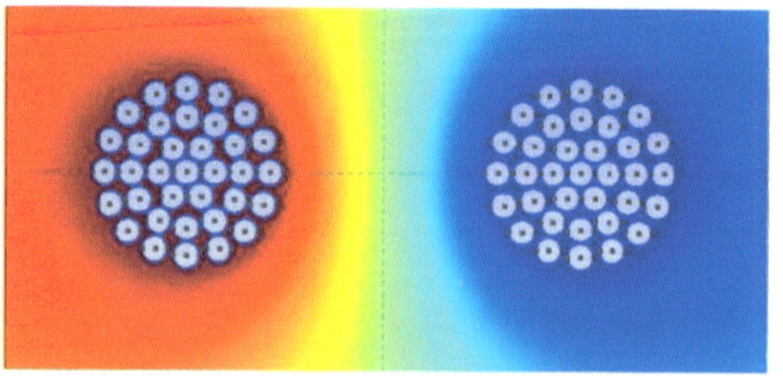

Fig. 9. Distribution of electrostatic potential for a differentially-fed pair of MWCNT bundles.

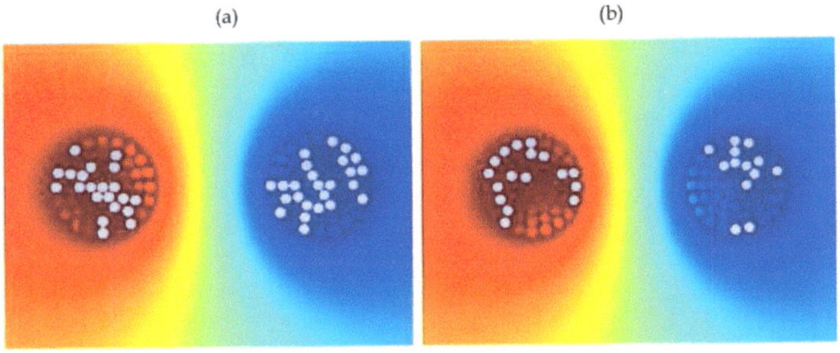

Fig. 10. Distribution of the electrostatic potential for a differentially-fed pair of SWCNT bundles: two random cases obtained assuming that 1/3 of CNTs are metallic (highlighted).

Table 3. Evaluation of the bundle pul capacitance for the SWCNT cases depicted in Fig.10.

	C_b (exact) [pF/m]	C_b (33) [pF/m]	% err
1	76.5	92.8	17.6
2	74.3	92.8	19.9
3	87.9	92.8	5.4

Let us now investigate the behavior of the bundle pul inductance and resistance. The magnetic inductance of the bundle may be computed from the vacuum space electrostatic capacitance, since it is $L_m = \mu_0\varepsilon_0 C_{e0}^{-1}$, where C_{e0} may be computed exactly or may be approximated by (33) if the introduced error is acceptable (which is the case of MWCNT bundles). When considering the kinetic inductance L_{kb} of a bundle of NSWCNTs, we may assume that only 1/3 are metallic, each of them contributing with $L_k = L_{k0}$. In addition, if we consider CNTs with the same diameters, we have a unique value for v, hence:

$$L_{kb} = \frac{L_{k0}}{N/3}, R_b = \frac{vL_{k0}}{N/3} = \frac{3}{4}\frac{R_0}{l_{mfp}N}. \tag{34}$$

Recalling the results in Table 2, for N up to the order of 10^3 we may still assume that $L_{kb} \gg L_m$, and since it is $\alpha_C \approx 1$, we may approximate the total bundle inductance with the first of (34), hence following the simple scaling law obtained in literature (Maffucci et al., 2009). Let us consider the case of a microstrip with effective dielectric constant $\varepsilon_{\text{reff}} = 2.2$, where the signal trace of width $w = 27$ nm, thickness $t = 2$w and distance to ground $2t$ is made by a bundle of NSWCNTs of diameter $D = 2.2$ nm. Figure 11 shows the self admittance of the line computed with the rigorous approach and with the proposed approximation, for $N = 100$ and $N = 200$. Here we have assumed $v_F = 8.7 \cdot 10^5$ m/s and $l_{mfp} = 1\mu$ m.

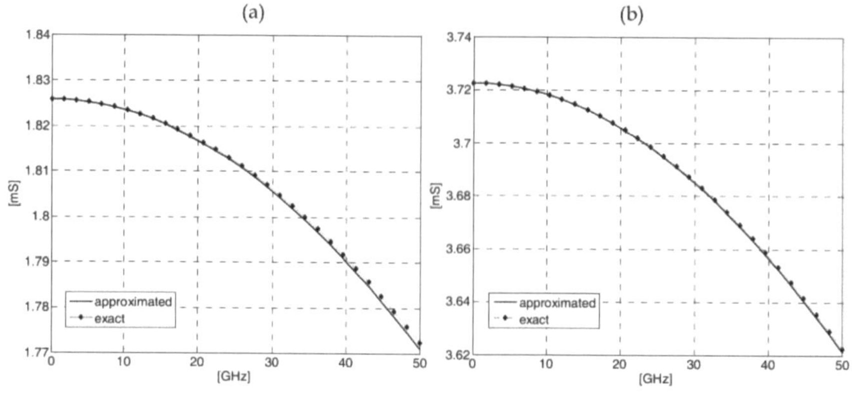

Fig. 11. Self admittance for a bundle of (a) 100 and (b) 200 SWCNTs.

Let us consider MWCNTs bundles. For these structures, the electrostatic capacitance and hence the magnetic inductance may be computed from (33). As for the kinetic inductance, we may introduce a simple approximation based on (13). Let D_{in} and D_{out} be the outer and inner shells diameters, respectively. The inter-shell spacing is the Van-der Waals distance $\delta = 0.34$ nm (Li et al., 2005), hence the number of shells is $N_{\text{shell}} = 1 + (D_{\text{out}} - D_{\text{in}})/2\delta$. Once again we assume 1/3 of the shells to be metallic and 2/3 semiconducting. Assuming the above distribution and using approximation (13) we can compute the equivalent number of channels for a single shell of diameter D_s in a MWCNTs through the simple law:

$$M_{\text{shell}} \approx \begin{cases} 2/3 & \text{for } D < d_s/T \\ a_{1s} D_s T + a_{2s} & \text{for } D \geq d_s/T \end{cases} \tag{35}$$

where $a_{1s} = 3.3 \cdot 10^{-4} 1/\text{nmK}, a_{2s} = -0.08$ and $d_s = 1.9 \cdot 10^3 \text{ nm} \cdot \text{K}$. The kinetic inductance of a bundle of N_{CNT} MWCNTs is then given by:

$$L_{kb} \approx \frac{2L_{k0}}{N_{CNT} N_{\text{shell}} M_{\text{shell}}}. \tag{36}$$

where L_{k0} is the bulk value defined in (24). Using (36) in (21) we can derive the bundle quantum capacitance. Using $\alpha_C \approx 1$, from (19), (21) and (28) we may derive the bundle resistance:

$$R_b \approx vL_{kb} \approx \frac{2vL_{k0}}{N_{CNT} N_{\text{shell}} M_{\text{shell}}} = \frac{1}{2l_{mfp}} \frac{R_0}{N_{CNT} N_{\text{shell}} M_{\text{shell}}}. \tag{37}$$

Note that the mean-free-path l_{mfp} may be a function of the diameter D. A simple linear dependence is proposed in (Nieuwoudt & Massoud, 2006).

4. Performance analysis of carbon nanotube interconnects

This Section is devoted to performance analysis of CNT interconnects, compared to conventional Cu ones. The investigated scenarios refer to typical arrangements foreseen for future 22 nm technology node, expected for year 2016 and beyond (ITRS, 2007). In the following the circuit models for CNT interconnects are those presented in Section 3, whereas the conventional ones are described through classical TL models.

4.1 Signal integrity analysis for CNT on-chip interconnects

Let us consider the on-chip interconnect in Fig. 12 made by two horizontal traces and a vertical via. The electrical and geometrical parameters given in Table 4 are typical values for the intermediate level at 22 nm technology node (ITRS, 2007). The signal traces on the layers and the via barrel may be either constituted by a solid Cu conductor or by a SWCNT bundle. In particular we consider the case of traces made by bundles of SWCNTs of $D = 0.94$ nm, with a fraction of 1/3 metallic. At the ends of the two horizontal tracts, a lumped contact resistance of 50kΩ is considered for each metallic CNT. For the Cu case, the horizontal tracts are described as striplines, whereas the via is modeled through a series impedance R - L and a pad capacitance to ground (e.g., Chiariello et al., 2009):

$$R_{via} = \frac{\rho l_{via}}{\pi (d_{via}/2)^2}, C_{via} \approx \frac{1.41\varepsilon_r D_P T}{D_{AP} - D_P}[pF], L_{via} \approx 5.08 \cdot l_{via} \cdot \left[\ln\left(\frac{4l_{via}}{d_{via}}\right) + 1\right][nH], \quad (38)$$

where d_{via} and l_{via} are the diameter and the length of the via barrel, respectively, T is the dielectric thickness, D_P and D_{AP} are the diameters of the pad and the antipad, respectively.

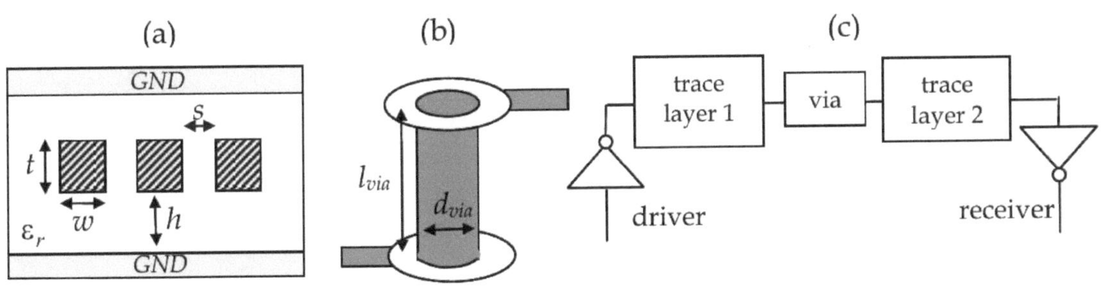

Fig. 12. On-chip interconnect: (a) section of the horizontal tracts; (b) vertical via; (c) circuit.

Table 4. Parameters for the case study in Fig.12.

w [nm]	t [nm]	h [nm]	s [nm]	d_{via} [nm]	l_{via} [μm]	ε_r [SiO_2]	ρ_{Cu} [$\mu\Omega \cdot$ cm]
44	44	39.6	44	60	1.0	4.5	4.235

Table 5. Computed pul and lumped parameters.

	R [Ω/μm]	L [pH/μm]	C [fF/μm]	R_{via} [Ω]	L_{via} [pH]
Cu	22.31	0.30	0.17	10.19	0.03
CNT	2.29	2.62	0.17	1.29	1.47

As for the terminations in Fig.12c, we assume that the driver is modeled as a voltage source of $V_S = 0.8V_{DD}$ and a series resistor of $1.5k\Omega$, whereas the receiver is modeled as a capacitor of 1 fF . These are typical values for the $22 -$ nm node, obtained from the minimum size gate values in the case with a size factor of $20x$ (Li et al., 2009).

Let us investigate the signal integrity performances of these two systems, assuming an high data rate (DR) signal transmission. Figure 13 shows the eye-diagrams obtained for the Cu case (Fig.13a-b) and the CNT case (Fig.13c-d), for values of the data rate DR ranging from 3 to 20Gbit/s. The complete SI analysis is given in Table 6. The jitter is of the same order, whereas the performances of CNTs in terms of eye-opening factor are much better than the Cu ones. For instance at 10Gbit/s the Cu channel could not be used (Fig.13b), whereas the CNT one still shows good performances (Fig.13c). The behavior obtained here is justified from the consideration that the higher inductance of CNT is compensated by its lower resistance and the SI performance of the overall CNT channel is better than Cu.

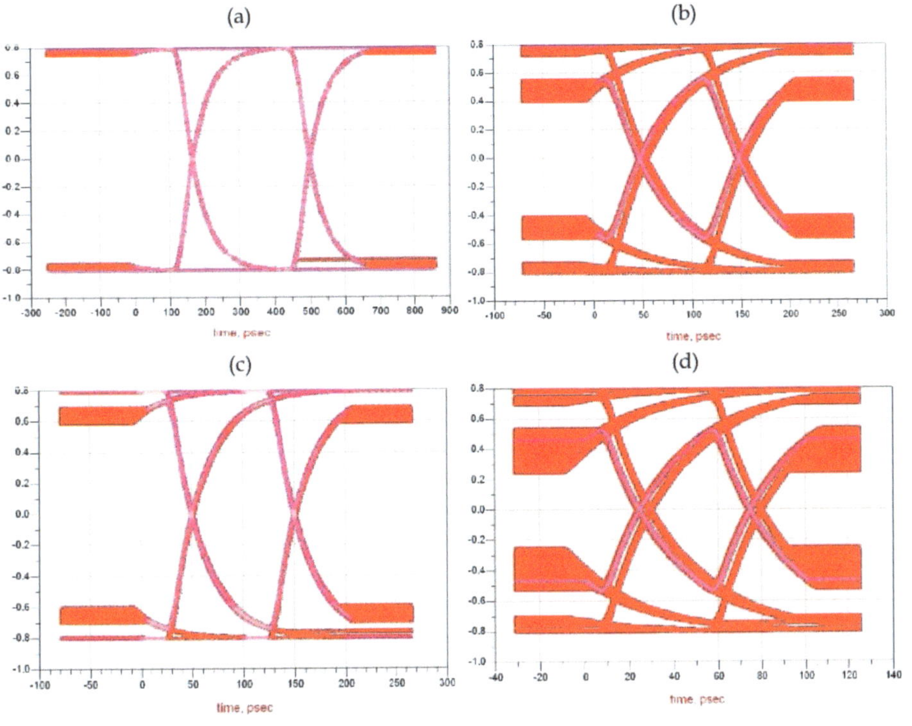

Fig. 13. Eye-diagram analysis for the case in Fig.12: Cu interconnect with (a) $DR =$ 3Gbit/s, $t_r = 33$ps; and (b) $DR = 10$Gbit/s, $t_r = 10$ps; CNT interconnect with (c) $DR = 10$Gbit/s, $t_r = 10$ps; and (d) $DR = 20$Gbit/s, $t_r = 5$ps.

Table 6. Results of the Eye-Diagram analysis for the case-study in Fig.12.

	DR [Gbit/s]	t_r [ps]	Eye jitter [ps - rms]	Eye opening factor
CNT	1	100	2.1	1
Cu	1	100	1.6	1
CNT	3	33	166	0.99
Cu	3	33	167	0.97
CNT	5	20	100	0.98
Cu	5	20	100	0.91
CNT	10	10	50	0.89

Cu	10	10	50	0.76
CNT	20	5	24	0.72
Cu	20	5	22	0.47

4.2 EMC behavior of CNT interconnects

As pointed out in Section 3, the parameters of the circuit model for CNT interconnects are strongly affected by the presence of a kinetic inductance and a quantum capacitance. Within the limits of the model used here, these two parameters are independent on frequency, hence the frequency behavior of CNT interconnects may be completely different from that of conventional ones. Here we investigate two typical problems arising when performing an electromagnetic compatibility (EMC) analysis: the high-frequency effects on the line impedance and the crosstalk noise.

The case-study used for EMC analysis is the stripline configuration in Fig.14a, assuming global level parameters: $w = 0.32\mu\,\mathrm{m}, t = 0.19\mu\,\mathrm{m}, h = 0.15\mu\,\mathrm{m}$ and $\varepsilon_r = 2.2$. Three different realizations of the signal traces are compared: solid copper, a bundle of SWCNTs and a bundle of MWCNTs. At such a scale the copper resistivity is $\rho \approx 2.942\mu\Omega \cdot cm$.

Let us first study the high-frequency impedance. The first realization of the signal traces is made a bundle of SWCNTs, of diameter $D = 1.4$ nm, assuming the area of the trace filled by CNTs with a density of 80%, and considering 1/3 of the total number of CNTs to be metallic. The second realization is made by a bundle of MWCNTs, with the same density, each of them with outer shell diameter of $D_{\mathrm{out}} = 40$ nm, inner shell diameter $D_{\mathrm{in}} = 0.5D_{\mathrm{out}}$ and inter-shell distance $\delta = 0.34$ nm. The temperature is 300 K .

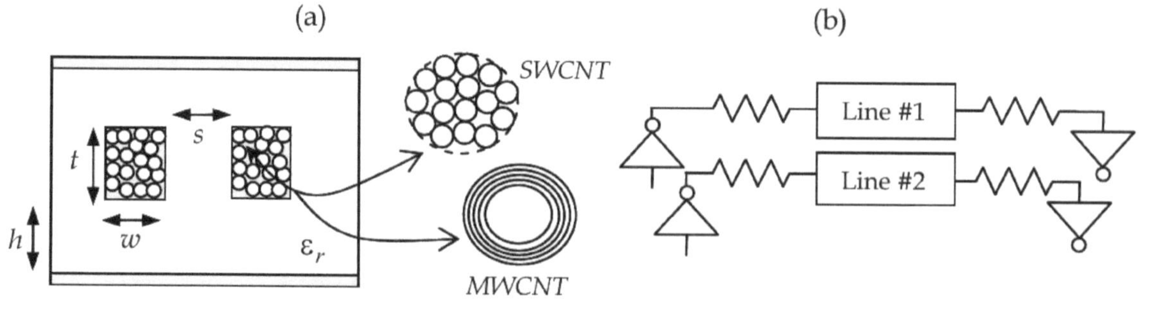

Fig. 14. Interconnect geometry used for EMC analysis: (a) the signal traces are made by either solid Cu or CNT bundles (SWCNT or MWCNTs); (b) circuit for the crosstalk analysis.

The p.u.l. resistances for the three realizations are reported in Fig.15a. For Cu lines this parameter is affected by the skin-effect, whereas for CNTs this effect is not present. For instance, at 200 GHz the penetration depth is 0.19μ m, equal to t and lower than w. The CNT interconnects outperform the Cu one and show a resistance parameter insensitive to the skin effect. In particular, the MWCNTs bundle outperforms the other two realizations in the whole frequency range: this is due to the fact that all the shells contribute to the conduction. As for the p.u.l. inductance (Fig.15b), the Cu line shows a slight decrease of this parameters, due to the decrease of the contribution of the internal inductance with increasing frequency, whereas the CNTs bundles are again insensitive to it. The MWCNT bundle interconnect introduces an inductance comparable to the Cu one, whereas the SWCNT realization shows an higher inductance. Since the kinetic inductance of the bundles decreases as (34) or (36), it may be comparable to the magnetic one, as in this case. As a conclusion we may assume that CNT interconnects exhibit parameters independent on frequency, whereas the conventional ones would suffer from this problem.

For the crosstalk analysis we consider a coupled stripline as in Fig.14a, assuming $s = w$, and typical intermediate level parameters: $w = t = 44$ nm, $h = 39.6$ nm and $\varepsilon_r = 2.2$. The considered circuit is shown in Fig.14b. We assume all the resistors of 1450Ω assuming that line 1 is fed at near end by a time-harmonic voltage source of amplitude $V_{\text{in}} = 1$ (a.u.). The far-end receivers are modeled as capacitors of 1 fF . For the CNT realizations we consider a density of 80%, a temperature of 300 K and a mean free path $l_{mfp} = 0.5\mu$ m. The SWCNTs diameter is $D = 1.4nm$, whereas for the MWCNTs the outer shell diameter is $D_{\text{out}} = 30nm$, the inner shell diameter is $D_{\text{in}} = 0.5D_{\text{out}}$ and the inter-shell distance is $\delta = 0.34$ nm. Because of the large number of CNTs in the bundle, we may disregard the effect of the parasitic contact resistances. Figure 16 shows the computed near-end and far-end crosstalk voltages, normalized to the input voltage V_{in}. The crosstalk noise for the three realizations is of the same order, hence from this point of view we may assume an equivalent EMC performance.

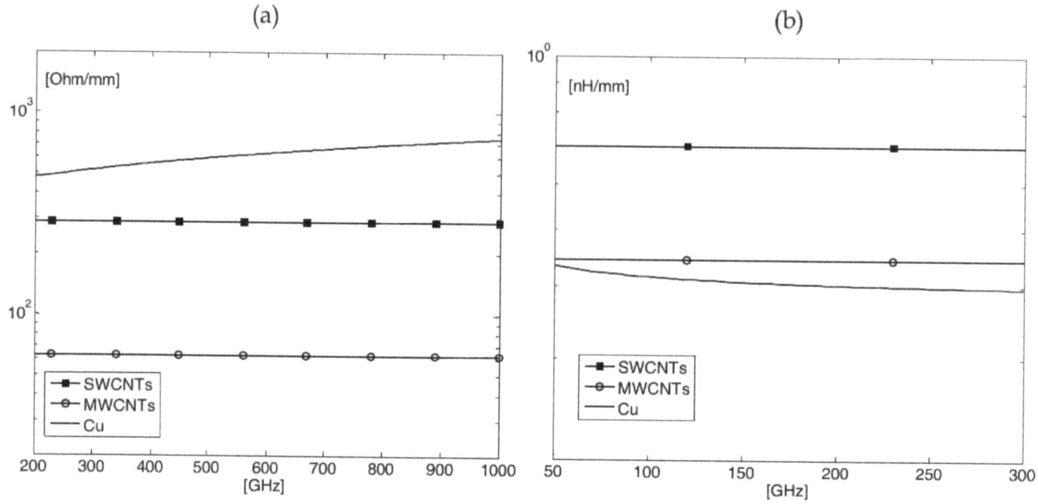

Fig. 15. (a) P.u.l. resistance for a trace made by Cu, a bundle of SWCNTs or MWCNTs; (b) P.u.l. inductance for a trace made by Cu , a bundle of SWCNTs or MWCNTs.

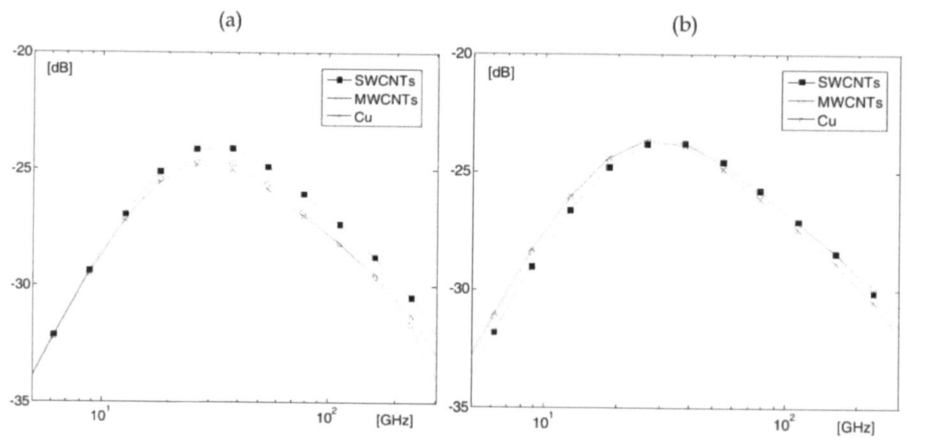

Fig. 16. Crosstalk level for a coupled line made by copper trace, SWCNTs bundle and MWCNTs bundle: (a) near end; (b) far end.

4.3 CNT interconnects as pillars for nanopackaging

Recently the CNTs have been proposed as material for nanopackaging interconnects, since their unique electrical, mechanical and thermal properties allows meeting many of the new requirements for bonding, molding compound, underfill, thermal interface, die attach (Morris, 2008; Maffucci, 2009). One of the main reasons pushing towards the use of CNTs in nanopackaging is the possibility either to achieve good electrical performances and implement new heat removal technologies. The thermal management of future nanoelectronics requires new approaches, since the classical heat removal techniques are inadequate. Using the conventional approach, for instance, the cooler for future Systems in Package would require much greater volume than the semiconductor itself. Recently, CNTs have been proposed as microchannel coolers in thermofluidic cooling approaches and as thermal interface material (TIM). The main limit to the practical use of CNTs is still given by the possibility of integrating a high density of aligned CNTs in a polymer matrix, without degrading thermal conductivity.

In order to investigate the electrical properties of CNT nanopackaging interconnects, let us consider the use of CNTs bundles as chip to package interconnect, following the scheme depicted in Fig.17a: the two vertical pillars are realized using Cu , a bundle of SWCNTs (Fig.17b) or a bundle of MWCNTs (Fig.17c). In terms of electrical properties, the goal of a good packaging interconnect is of course to introduce low parasitics.

As done before, in the following we assume that in the SWCNT bundle only a fraction of 1/3 of the total CNTs is metallic, and similarly in a MWCNT only 1/3 shells are metallic. Using the results in Section 2, since the semiconducting SWCNTs do not give contribution to the conduction, very high density bundles must be fabricated to have low resistance CNT pillars. In MWCNTs, instead, the semiconducting shells give a non-negligible contribution to the conduction and the conduction of the metallic shells themselves is improved, hence the optimal density is a trade-off between the MWCNT outer diameter and its shell number.

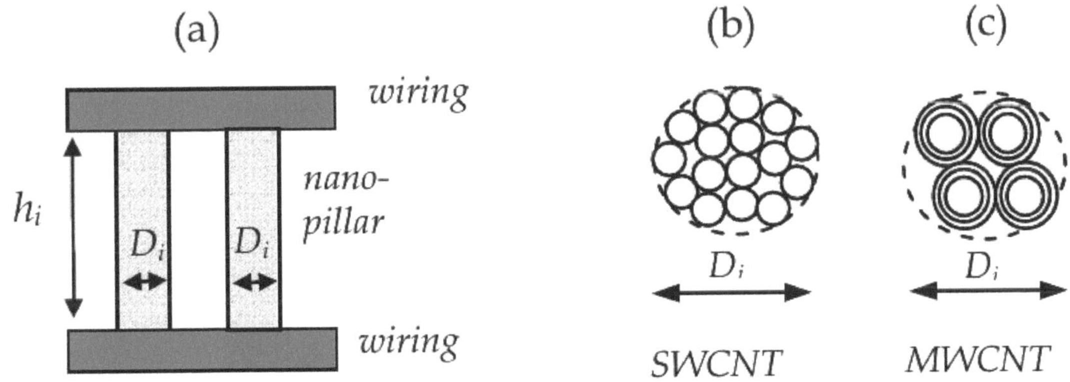

Fig. 17. Chip-to-package interconnect: (a) scheme; (b) SWCNTs and (c) MWCNTs realization

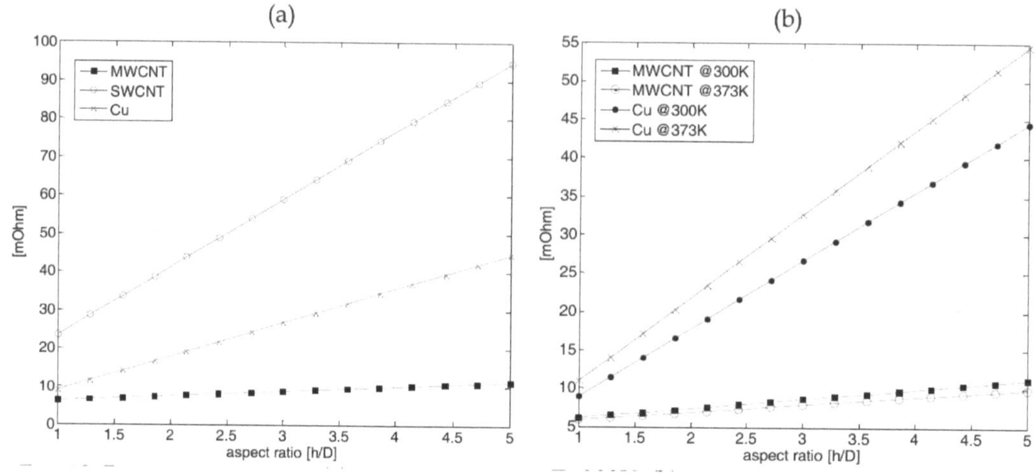

Fig. 18. Parasitic resistance: (a) vs aspect ratio, at T = 300 K; (b) vs temperature.

Let us consider the case of Fig. 17, assuming the pillars to be cylinders of diameter D_i and height h_i, a wire bond pitch of 20μ m and an operating frequency of 10 GHz at a temperature of 300 K . First we consider Cu interconnects with $\sigma_{Cu} = 5.8 \cdot 10^7$ S/m. Next, we fill the interconnect with a bundle of SWCNTs of mean diameter 2.7 nm , assuming a density of 80% of the total area and assuming 1/3 of metallic CNTs. In addition, a parasitic contact resistance of $20k\Omega$ in series to any single CNT is considered. Finally we consider the bundle to be filled by MWCNTs with outer diameter $D_{\text{out}} = 30$ nm, inner diameter $D_{\text{in}} = 0.5D_{\text{out}}$ and inter-shell distance 0.34 nm . We assume again 1/3 of metallic shells and the rest semiconducting, a density of 80% and a contact resistance of $10k\Omega$ per shell. The considered mean free path for all CNTs is $l_{mfp} = 0.5\mu$ m.

For this case the parasitic inductances are well beyond the maximum allowed of $5 - 10$pH, hence we focus on the resistance. Figure 18a compares the resistance values obtained for aspect ratios ranging from 1 to 5. The lowest values are obtained by MWCNT bundle, whereas SWCNT bundle shows higher resistance. The better performance of MWCNTs is due to the contribution to the conduction given by the semiconducting shells.

Let us now investigate the problem of the compatibility of CNT interconnects with the temperatures of CMOS technology. Let us consider a realistic chip operating temperature of T = 373 K. When the temperature increases the kinetic inductance decreases, as shown in Section 2, and the difference is more pronounced for large diameters, hence for MWCNTs rather than for SWCNTs. Let us neglect the variation of the mean-free path with temperature: in the same conditions described above, the parasitic resistance for the MWCNT bundle and the Cu interconnects would be that shown in Fig.18b. Note that for the copper at 373 K we have assumed $\sigma_{Cu} = 4.0 \cdot 10^7$ S/m. The MWCNT interconnect has better performances at higher frequencies, which is opposite to the behaviour of Cu ones.

Finally let us focus on the problem of proximity effect, which is a major problem for packaging interconnect, since it leads to non uniform current crowding which cause local thermal hot-spots. Figure 19 shows the distribution of the current density over two CNT bundles at a distance equal to their width. The coupled line is driven by a differential signal (in the figures the underlying PEC ground is omitted). The simulations have been carried out by means of the full-wave 3D numerical simulator SURFCODE (Miano & Villone, 2005). As shown in Figs.19a-b, the CNT bundles do not suffer from the proximity effect, which is a problem for the conventional interconnects, Figs.18c-d. This is again due to the presence in the CNT electrodynamics of a kinetic term which dominates over the magnetic one. Within the limits of validity of our model, this kinetic term is independent from the distance between the bundles, which, on the contrary, is a crucial parameter in determining the proximity effect when conventional interconnects are analyzed.

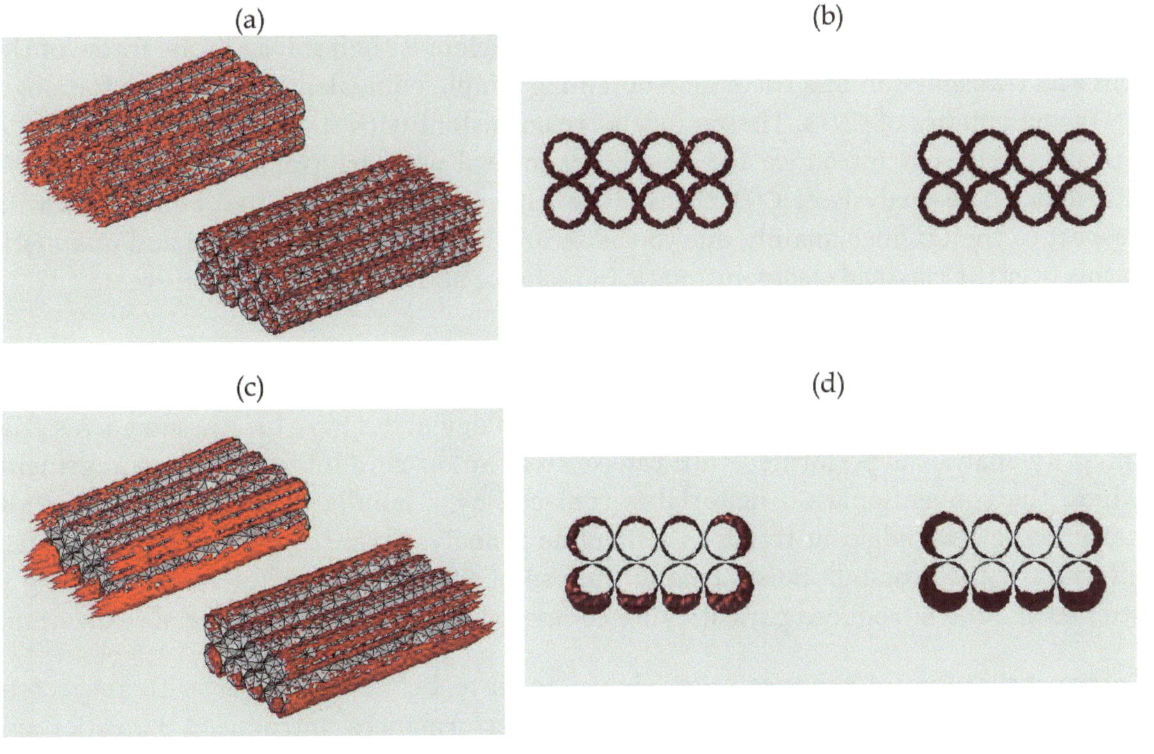

Fig. 19. Current density distributions over two bundles (differential mode): (a) CNT bundles; (b) CNT transverse section; (c) PEC bundles; (d) PEC transverse section.

5. Conclusion

This Chapter has analyzed the use of carbon nanotubes as possible innovative material to fabricate high-speed interconnects at various hierarchical levels of the future nanoelectronics technology.

The electromagnetic model presented here to describe the electrical propagation along CNT interconnects is derived from a semi-classical solution of the transport equation. All the quantistic and kinetic effects affecting the electrodynamics of charge carriers are taken into account through global parameters (kinetic inductance and quantum capacitance) which have been simply related to the number of effective conducting channel per CNT shell. Both the rigorous approach and a simple approximation have been provided to evaluate this number for CNT shells of arbitrary chirality, hence allowing the modelling of either single-wall and multi-wall CNTs.

The electrodynamical model provides a constitutive equation for the CNT interconnect, which is coupled to the Maxwell equation: the problem is then solved in the frame of the classical transmission line theory, so obtaining simple equivalent RLC circuits for single CNTs and bundles of CNTs. These models are consistent with those used in literature for metallic CNTs, but introduce the possibility to deal with arbitrary chiralities and CNT diameters. The analysis of CNT transmission lines highlights unusual behaviours with respect to the Cu lines, mainly due to the strong influence of the kinetic and quantistic terms over the classical electromagnetic inductance and capacitance parameters.

Some case-studies have been carried out referring to a realistic use of CNTs to build interconnects for wiring and for packaging nanotechnology ICs. In the considered examples typical values of the future 22 nm technology nodes have been assumed. A signal integrity analysis is performed with reference to an on-chip interconnect arrangement, where the conventional Cu material is replaced by a bundle of CNTs. Next, the EMC behaviour of two adjacent traces in a stripline is analyzed, checking the high-frequency effects and the crosstalk noise. Finally the use of CNTs as pillars for nanopackaging is studied and the introduced parasitics are compared to conventional material realization.

Carbon nanotube interconnects are shown to have better behaviour with respect to conventional Cu ones in terms of SI and EMC performances, introduced parasitics and response to temperature variations. From a practical point of view, these simulated performances may be achieved provided that good quality bundles are realized, i.e. bundles of high density, good direction control and good terminal contacts.

6. References

Anantram, M.P.; & Léonard, F. (2006). Physics of carbon nanotube electronic devices. Report Progress in Physics, Vol.69, 507.

Avouris, P.; Appenzeller, J.; Marte, R. & Wind, S.J. (2003). Carbon nanotube electronics, Proceedings of IEEE, Vol.91, 1772.

Banerjee, K.; Li, H. & Srivastava, N. (2008). Current status and future perspectives of carbon nanotubes interconnects. Proc. of IEEE Nanotechonology Conference, Arlington, TX, USA, August 18-21, 2008.

Burke, P.J. (2002). Luttinger liquid theory as a model of the gigahertz electrical properties of carbon nanotubes, IEEE Transactions on Nanotechnology, Vol. 1, 129-144.

Burke, P.J. (2003). An RF circuit model for carbon nanotubes. IEEE Transactions on Nanotechnology, Vol.2, 55.

Cheung, C.L. & et al. (2002). Diameter-controlled Synthesis of Carbon nanotubes. Journ. Phys. Chem. B, Vol.106, 2429-2433.

Chiariello, A.G.; Maffucci, A. & Miano, G. (2009). Signal Integrity Analysis of Carbon Nanotube on-chip Interconnects. IEEE Workshop on Signal Propagation on Interconnects SPI 2009, paper no. S4_1, Strasbourg, France, 12-15 may 2009.

Close, G.F.; Yasuda, S.; Paul, B.; Fujita, S. & Philip Wong, H.-S. (2008). A 1 GHz integrated circuit with carbon nanotube interconnects and silicon transistors. Nano Letters, Vol.8, 706-709.

Forestiere, C.; Maffucci, A. & Miano, G. (2009a). A Generalized Model for the Signal Propagation along Single- and Multi-Walled Carbon Nanotubes with Arbitrary Chirality, Proc. of 9th IEEE Nanotechnology Conference, paper #64, Genova (Italy), 2730 Jul 2009.

Forestiere, C.; Maffucci, A.; Miano, G., Maksimenko, S.A. & Slepyan, G.Y. (2009b). Signal propagation in carbon nanotubes of arbitrary chirality. In press on IEEE Trans. On Nanotechnology.

ITRS (2007) International Technology Roadmap for Semiconductors. http://public.itrs.net. Edition 2007.

Iijima, S. (1991). Helical microtubules of graphitic carbon. Nature, Vol.354, 56-58.

Li H.J.; Lu, W. G.; Li, J. J. ; Bai, X. D. & Gu, C. Z. (2005). Multichannel Ballistic Transport in Multiwall Carbon Nanotubes. Physical Review Letters, Vol.95, 086601.

Li, H.; Xu, C; Srivastava, N. & Banerjee, K. (2009). Carbon Nanomaterials for Next Generation Interconnects and Passives: Physics, Status, and Prospects. IEEE Transactions on Electron Devices, Vol. 56, 1799-1821.

Liu, Z. (2009). Fabrication and electrical characterization of densified carbon nanotube micropillars for IC interconnection. IEEE Trans. on Nanotechnology, Vol.8, 196-203.

Maarouf, A.A.; Kane, C.L. & Mele, E.J. (2000). Electronic structure of carbon nanotube ropes. Physical Review B, Vol.61, 11156-11165.

Maksimenko, S.A. & Slepyan, G.Y. (2000). Electrodynamic properties of carbon nanotubes, in Electromagnetic Fields in Unconventional Materials and Structures, O. N. Singh and A. Lakhtakia Eds., pp. 217-255, Wiley, New York.

Maksimenko, S.A.; Khrushchinsky, A.A.; Slepyan, G.Y. & Kibisb, O.V. (2007). Electrodynamics of chiral carbon nanotubes in the helical parametrization scheme. Journal of Nanophotonics, Vol.1, p. 013505.

Maffucci A.; Miano, G. & Villone, F. (2008a). A transmission line model for metallic carbon nanotube interconnects, Intern. Journal of Circuit Theory and Applic., Vol. 36, 31-52.

Maffucci A.; Miano, G. & Villone, F. (2008b). Performance Comparison Between Metallic Carbon Nanotube and Copper Nano-Interconnects, IEEE Transactions on Advanced Packaging, Vol. 31, 692-699.

Maffucci A.; Miano, G. & Villone, F. (2009). A new Circuit Model for Carbon Nanotube Interconnects with Diameter-dependent Parameters. IEEE Transactions on Nanotechnology, Vol.8, 345-354.

Maffucci, A. (2009). Carbon Nanotubes in Nanopackaging applications. IEEE Magazine on Nanotechnology, Vol.3, no.3, 22-25.

Miano, G. & Villone, F. (2005). A surface integral formulation of Maxwell equations for topologically complex conducting domains. IEEE Trans. on Antennas and Propagation, vol. 53, 4001-4014.

Miano, G. & Villone, F. (2006). An Integral Formulation for the Electrodynamics of Metallic Carbon Nanotubes Based on a Fluid Model, IEEE Transactions on Antennas and Propagation, Vol.54, 2713-2724, 2006.

Morris, J.E. (2008). Nanopackaging: Nanotechnologies and electronics packaging. New-York: Springer.

Naeemi, A. & Meindl, J. D. (2006). Compact Physical Models for Multiwall CarbonNanotube Interconnects. IEEE Electron Devices Letters, Vol.27, 338-340.

Naeemi, A. & Meindl, J. D. (2008). Performance Modeling for Single- and Multiwall Carbon Nanotubes as Signal and Power Interconnects in Gigascale Systems, IEEE Trans. on Electron Devices, Vol. 55, n.10, 2574-2582.

Nieuwoudt, A. & Massoud, Y. (2006). Evaluating the impact of resistance in carbon nanotube bundles for VLSI interconnect using diameter-dependent modeling techniques. IEEE Trans. Electron Devices, vol. 53, 2460-2466.

Park, J.Y.; et al., (2004). Electron-phonon Scattering in Metallic Single-Walled Carbon Nanotubes. Nano Letters, Vol.4, 517-520.

Raychowdhury, A.; Roy, K. (2006). Modelling of metallic carbon-nanotube interconnects for circuit simulations and a comparison with Cu interconnects for scaled technologies. IEEE Trans. on Computer-Aided Design for Integrated Circ. and Syst., Vol. 25, 58-65.

Rutherglen, C. & Burke, P.J. (2009). Nanoelectromagnetics: Circuit and Electromagnetic Properties of Carbon Nanotubes. Small, Vol.5, 884-906.

Saito, R.; Dresselhaus, G. & Dresselhaus, M. S. (2004). Physical Properties of Carbon Nanotubes, Imperial College Press, Singapore.

Salahuddin, S.; Lundstrom, M. & Datta, S. (2005). Transport effects on signal propagation in quantum wires. IEEE Transactions on Electron Devices, Vol.52, 1734-1742.

Sarto, M.S.; Tamburrano, A. & D'Amore, M. (2009). New Electron-Waveguide-Based Modeling for Carbon Nanotube Interconnects. IEEE Transactions on Nanotechnology, Vol. 8, 214-225.

Slepyan, G.Y.; Maksimenko, S. A.; Lakhtakia, A.; Yevtushenko, O. & Gusakov, A.V. (1999). Electrodynamics of carbon nanotubes: dynamics conductivity, impedance boundary conditions, and surface wave propagation. Physical Review B, Vol. 60, 17136.

Soga, I.; Kondo, D.; Yamaguchi, Y.; Iwai, T.; Mizukoshi, M.; Awano, Y.; Yube, K. & Fujii T. (2008). Carbon nanotube bumps for LSI interconnect, Proc. Of Electronic Components and Technology Conf., 1390-1394, May 2008.

Wesström, J.J. (1996). Signal propagation in electron waveguides: Transmission-line analogies, Physical Review B, Vol. 54, 11484-11491.

Pu, S-N.; Yin, W.-Y.; Mao, J.-F. & Liu, Q.H. (2009). Crosstalk Prediction of Single- and Double-Walled Carbon-Nanotube (SWCNT/DWCNT) Bundle Interconnects. IEEE Transactions on Electron Devices, Vol. 56, 560-568.

Chapter-15

Interconnect Challenges & Carbon Nanotube as Interconnect in Nano VLSI Circuits

This chapter discusses about the behavior of Carbon Nanotube (CNT) different structures which can be used as interconnect in Very Large Scale (VLSI) circuits in nanoscale regime. Also interconnect challenges in VLSI circuits which lead to use CNT as interconnect instead of Cu , is reviewed. CNTs are classified into three main types including Single-walled Carbon Nanotube (SWCNT), CNT Bundle, and Multi-walled Carbon Nanotube (MWCNT). Because of extremely high quantum resistance of a SWCNT which is about 6.45kΩ, rope or bundle of CNTs are used which consist of parallel CNTs in order to overcome the high delay time due to the high intrinsic (quantum) resistance. Also MWCNTs which consist of parallel shells, present much less delay time with respect to SWCNTs, for the application as interconnects. In this chapter, first a short discussion about interconnect challenges in VLSI circuits is presented. Then the repeater insertion technique for the delay reduction in the global interconnects will be studied. After that, the parameters and circuit model of a CNT will be discussed. Then a brief review about the different structures of CNT interconnects including CNT bundle and MWCNT will be presented. At the continuation, the time domain behavior of a CNT bundle interconnect in a driver-CNT bundle-load configuration will be discussed and analyzed. In this analysis, CNT bundle is modeled as a transmission line circuit model. At the end, a brief study of stability analysis in CNT interconnects will be presented.

1. Interconnect Challenges in VLSI Circuits

As interconnect feature sizes shrink, copper resistivity increases due to surface and grain boundary scatterings and also surface roughness [1]. Furthermore, wires, especially power and ground lines, are becoming more and more vulnerable to electromigration because of rapid increases in current densities [2]. The resistance of copper interconnects, with crosssectional dimensions of the order of the mean free path of electrons (~ 40 nm in Cu at room temperature) in current and imminent technologies [2], is increasing rapidly under the combined effects of enhanced grain boundary scattering, surface scattering and the presence of the highly resistive diffusion barrier layer [3]. The steep rise in parasitic resistance of copper interconnects poses serious challenges for interconnect delay [2] (especially at the global level where wires traverse long distances) and for interconnect reliability [4], hence it has a significant impact on the performance and reliability of VLSI circuits. In order to alleviate such problems, changes in the material used for on-chip interconnections have been sought even in earlier technology generations, for example the transition from aluminum to copper some years back [3].

Carbon nanotubes (CNTs) exhibit a ballistic flow of electrons with electron mean free paths of several micrometers, and are capable of conducting very large current densities [3]. They are therefore proposed as potential candidates for signal and power interconnections [5], [6]. Because of their extremely desirable properties of high mechanical and thermal stability, high thermal conductivity and large current carrying capacity [7], CNTs have aroused a lot of research interest in their applicability as VLSI interconnects of the future. Depending on their chirality (the direction along which the graphene sheets are rolled up), CNTs demonstrate either metallic or semiconducting properties. Fig. 1 shows different structures depending on the chirality.

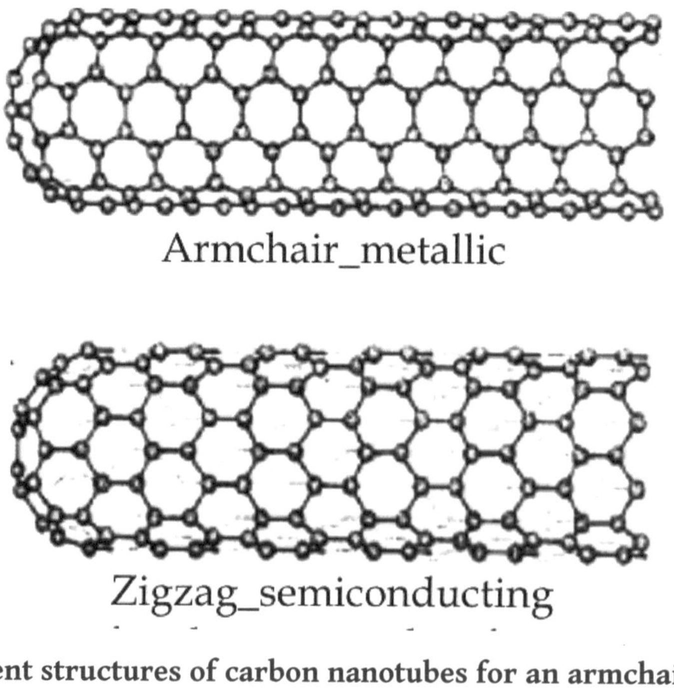

Fig. 1. Two different structures of carbon nanotubes for an armchair-type nanotube and a zigzag-type nanotube [8].

Carbon nanotubes are also classified into single-walled and multi-walled nanotubes.

2. Repeater Insertion as a Technique for the Delay Reduction

With the technology scaling in very deep submicron (VDSM) CMOS circuits, the gate delay decreases rapidly, while the delay of global interconnects tends to increase because of increasing their aspect ratio (thickness to width ratio) with scaling [2], [9], [10]. The repeater (buffer) insertion technique is generally used to reduce the delay of long (semi global) and global interconnects [9], [11]-[16]. An analytical model for obtaining the optimal buffer size and segment length for an equal partitioning network, in which the buffers sizes and segments lengths are constant, has been presented [12], [14], [17]. In [18] we have discussed about the optimization of global interconnects using unequal repeater (buffer) partitioning technique. This method which is discussed and reviewed in this chapter, is based on the segmentation of a long global interconnect into unequal parts, and inserting buffers with unequal sizes between them. The related structure is named as "Unequal buffer partitioning network" against "Equal buffer partitioning network" that was mentioned above. The optimum delay is a function of various parameters such as the buffers sizes, the interconnect segments lengths, the load and so on [9], [11], [12], [15]-[17], [19]-[22]. It is shown that for the optimization of a buffer inserted interconnect behavior, the energy-delay product minimization is better than the delay minimization. Thus in this research, the energy-delay product has been chosen as target function for the minimization. There are different algorithms for minimizing a function, which in this chapter, the genetic algorithm (GA) using MATLAB [23] has been used for minimizing the energy-delay product function.

2.1 Equal Buffer Partitioning Network

Fig. 2 shows a global interconnect with the buffer insertion, in which each segment has equal length and all the buffers have the same size

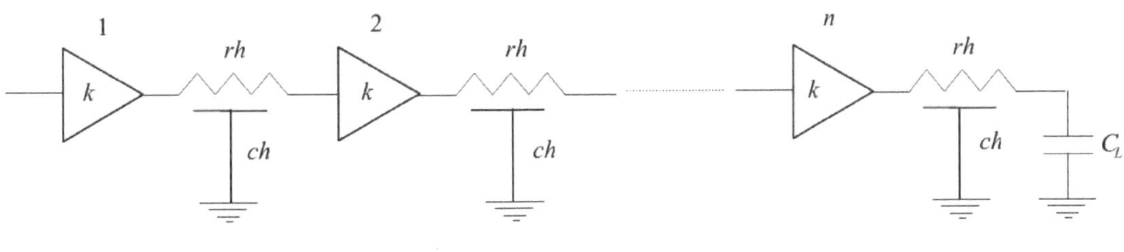

Fig. 2. Equal buffer partitioning network [18]

where r, c, and h are the interconnect resistance per unit length, the interconnect capacitance per unit length, and each segment length of the interconnect, respectively.

Also k, n and C_L are the buffers size, the number of buffers, and the load capacitance, respectively. The total time delay of global line interconnect including the buffers and load, using Elmore relation [12], [21] will be

$$\text{Delay}_1 = n\left[R_{Bo}(C_{Bo} + ch) + \frac{1}{2}rch^2\right] + (n-1)(R_{Bo} + rh)C_{Bi} + (R_{Bo} + rh)C_L. \quad (1)$$

where $R_{BO} = r_0/k, C_{Bi} = kc_0$, and $C_{BO} = kc_p$ are the buffers output resistance, the buffers input capacitance, and the buffers output capacitance, respectively. Also r_0, c_0, and c_p are similar parameters of the minimum sized repeater (buffer), respectively. We can express the total energy as

$$\text{Energy}_1 = n\left[ch + k(c_0 + c_p) + \frac{C_L}{n}\right]V_{dd}^2 \quad (2)$$

where V_{dd} is the power supply voltage. The energy-delay product will be written as

$$EDP_1 = \text{Energy}_1 \times \text{Delay}_1, \#(3)$$

$$EDP_1 = n^2\left[ch + k\left(c_0 + c_p\right) + \frac{C_L}{n}\right] \times \left[\frac{r_0}{k}\left(kc_p + \frac{n-1}{n}kc_0 + \frac{C_L}{n} + ch\right) + rh\left(\frac{n-1}{n}kc_0 + \frac{C_L}{n} + \frac{1}{2}ch\right)\right]V_{dd}^2 \#(4)$$

2.2 Unequal Buffer Partitioning Network

Fig. 3 shows a global interconnect with buffer insertion, in which each segment length is a times of the previous segment length, and each buffer size is f times of the previous buffer size, respectively

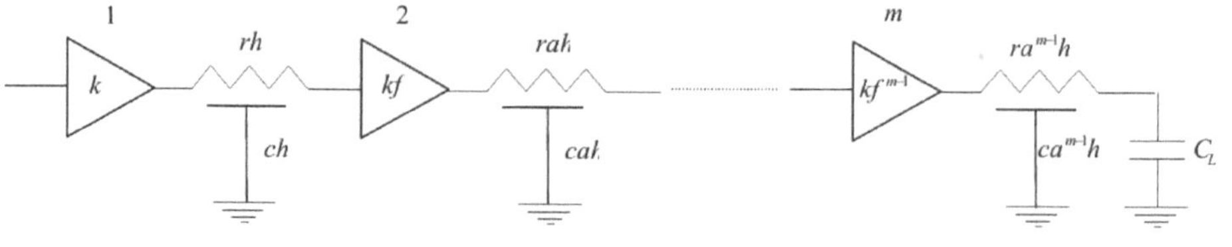

Fig. 3 Unequal buffer partitioning network [18]

where $h = l(1-a)/(1-a^m)$ which l is the total length of line (interconnect), k is the first buffer size, and m is the number of buffers. The other parameters are the same as in Fig.2. The total time delay of global interconnect in this structure, including the buffers and load, using Elmore relation [12], [21] will be

$$Delay_2 = mr_0c_p + (m-1)r_0c_0f + \left(\frac{1-\left(\frac{a}{f}\right)^m}{1-\frac{a}{f}} \right)\frac{r_0}{k}ch + \left(\frac{1-(af)^{m-1}}{1-af}\right)rhkfc_0 + \frac{1}{2}\left(\frac{1-a^{2m}}{1-a}\right)rch^2 + \left(\frac{r_0}{kf^{m-1}} + ra^{m-1}h\right)C_L \quad (5)$$

where r_0, c_0, and c_p are the output resistance, the input capacitance, and the output capacitance of the minimum sized buffer, respectively. We can express the total energy as

$$\text{Energy}_2 = \left[cl + \left(\frac{1-f^m}{1-f}\right)k(c_0 + c_p) + C_L\right]V_{dd}^2. \quad (6)$$

Thus the energy-delay product for "Unequal partitioning network" will be written as

$$EDP_2 = \text{Energy}_2 \times \text{Delay}_2, \quad (7)$$

$$EDP_2 = \left[cl + \left(\frac{1-f^m}{1-f}\right)k(c_0 + c_p) + C_L\right]$$

$$\times \left[mr_0c_p + (m-1)r_0c_0f + \left(\frac{1-\left(\frac{a}{f}\right)^m}{1-\frac{a}{f}} \right)\frac{r_0}{k}ch + \left(\frac{1-(af)^{m-1}}{1-af}\right)rhkfc_0 + \frac{1}{2}\left(\frac{1-a^{2m}}{1-a}\right)rch^2 + \left(\frac{r_0}{kf^{m-1}} + ra^{m-1}h\right)C_L \right]V_{dd}^2. \#(8)$$

2.3 Optimization Procedure

In this section, EDP [1] functions for the two networks "Equal partitioning network" and "Unequal partitioning network", which are defined and obtained in sections II and III, are minimized using the genetic algorithm (GA) of MATLAB [23]. This procedure is performed on the two networks for the three technology nodes 65,90,130 nm, which the specifications of global interconnect and the minimum sized repeater (buffer) in each technology node have been extracted from ITRS [2] [2], [9]. Also the interconnect capacitance per unit length is obtained using the formulations presented in [9]. Moreover, in each step of the optimization (minimization) procedure, the load capacitance has been taken as a parameter which varies from one to hundred times of the minimum sized buffer output capacitance.

In Figs. 4-9, the propagation delay improvement for "Unequal partitioning network" respect to "Equal partitioning network", versus the global interconnect length, and for three technology nodes 65, 90,130 nm, have been plotted whereas the capacitive load varies from one to hundred times of the minimum sized buffer output capacitance (c_p). The genetic algorithm (GA) of MATLAB [23] has been used as a tool for minimizing the energy-delay product (EDP) for the two networks at different technology nodes and various loads. For obtaining the correct results for each step of the minimization procedure, the algorithm has been done 1000 times in each step and the least value has been chosen as the best answer.

It is found from Figs. 4-9 that the improvement of the propagation delay, in unequal partitioning network is more than equal partitioning network. This improvement is obvious for the technology nodes 90,130 nm and goes high with increasing the load capacitance. Also for technology node 65 nm , the delay improvement will be achieved for the high values of the load capacitance, which is cleared from Figs. 8, 9.

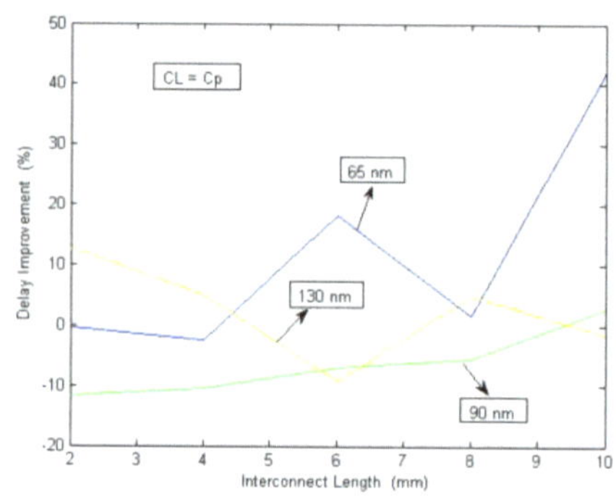

Fig. 4. The delay improvement of unequal partitioning network respect to equal partitioning network, for $C_L = c_p$ [18].

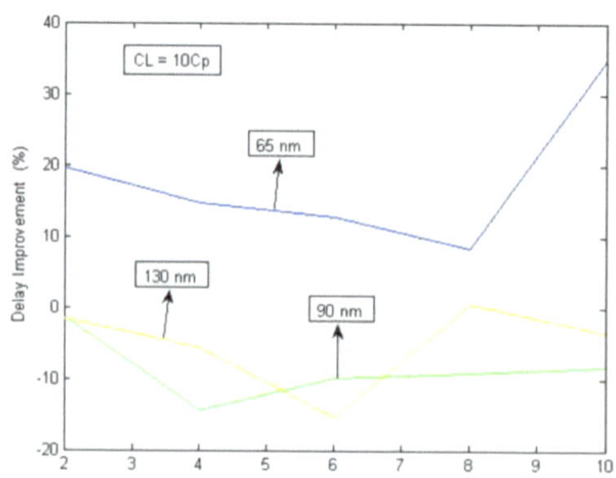

Fig. 5. The delay improvement of unequal partitioning network respect to equal partitioning network, for $C_L = 10c_p$ [18].

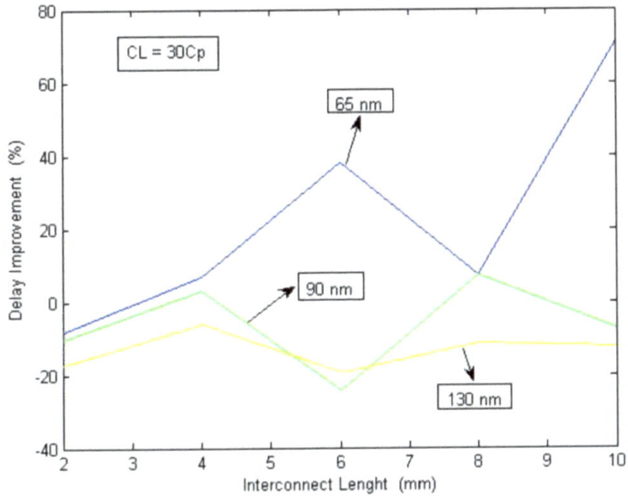

Fig. 6. The delay improvement of unequal partitioning network respect to equal partitioning network, for $C_L = 30c_p$ [18].

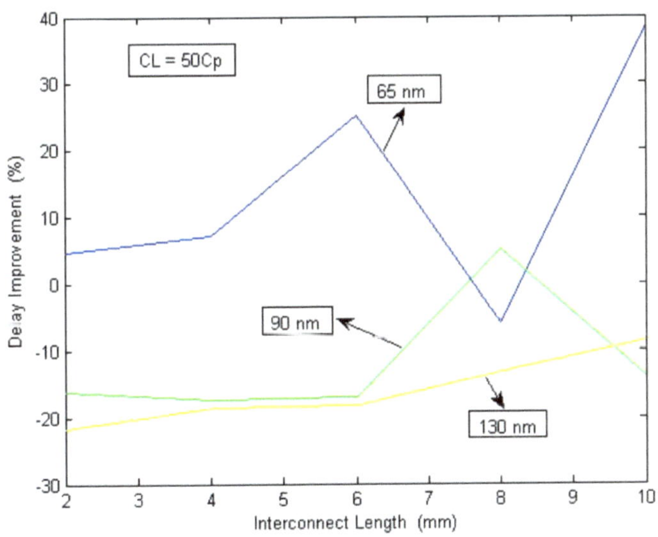

Fig. 7. The delay improvement of unequal partitioning network respect to equal partitioning network, for $C_L = 50c_p$ [18].

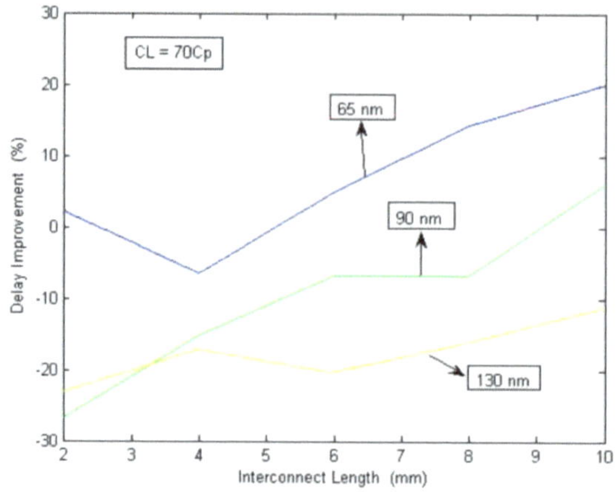

Fig. 8. The delay improvement of unequal partitioning network respect to equal partitioning network, for $C_L = 70c_p$ [18].

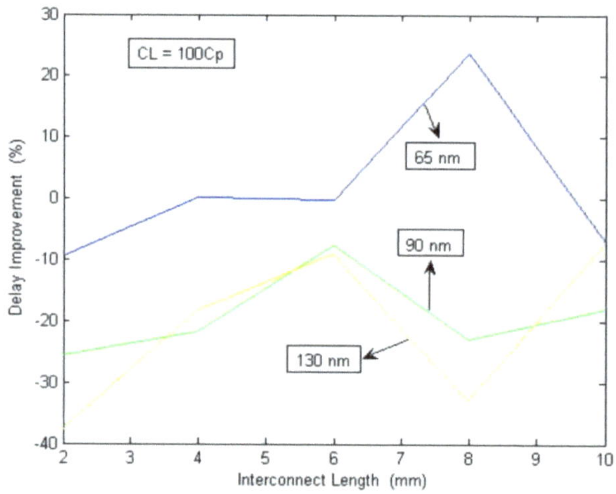

Fig. 9. The delay improvement of unequal partitioning network respect to equal partitioning network, for $C_L = 100c_p$ [18].

3. Circuit Model and Parameters for CNT

Fig. 10 depicts the equivalent circuit for an isolated single-walled carbon nanotube (SWCNT) of length less than the mean free path of electrons in a CNT [24], [25].

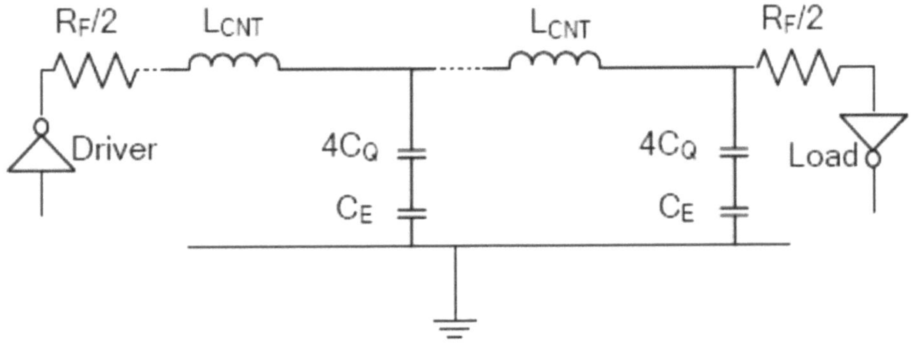

Fig. 10. Equivalent circuit model for an isolated SWCNT of length less than the mean free path of electrons, assuming ideal contacts [25].

In this figure, R_F is the fundamental (quantum) resistance, L_{CNT} is the total inductance, and C_Q and C_E are the quantum and electrostatic capacitances, respectively. In the following subsections, these parameters and their related formulations will be discussed.

3.1. CNT Resistance

Due to spin degeneracy and sub-lattice degeneracy of electrons in graphene, each nanotube has four conducting channels in parallel [3], [26]. Hence, the conductance of an isolated ballistic single-walled CNT (SWCNT) assuming perfect contacts, is $4e^2/h = 155\mu$ S, which yields a resistance of 6.45 KΩ [8], [24]. This is the quantum resistance associated with a SWCNT that cannot be avoided [3], [27]. This fundamental resistance, as shown in Fig. 11, is equally divided between the two contacts on either side of the nanotube and can be expressed as [27]

$$R_F = \frac{h}{4e^2} \tag{9}$$

where h is plank's constant and e is electron charge. The mean free path (MFP) of electrons in a CNT is typically 1μ m [3], [8], [28]. For CNT lengths less than λ_{CNT}, electron transport within the nanotube is essentially ballistic [3]. In this case, the resistance of nanotube with ideal coupling to the two metal contacts at its ends is independent of length and is given by $h/(4e^2) \approx 6.45$ KΩ [26], [27]. However, for lengths greater than the mean free path, scattering leads to an additional ohmic resistance which increases with length as [29]

$$R_{CNT} = R_F \frac{\ell}{\lambda_{CNT}} = \left(\frac{h}{4e^2}\right) \frac{\ell}{\lambda_{CNT}} \tag{10}$$

where ℓ and λ_{CNT} are the length and the mean free path of CNT, respectively. This has also been confirmed by experimental observations [3], [28]. It should be noted that this additional scattering resistance would appear as a distributed resistance per unit length [24] in the equivalent circuit, to account for resistive losses along the CNT length

$$R_{CNT}(\text{ p.u.l }) = \frac{R_F}{\lambda_{CNT}} = \left(\frac{h}{4e^2}\right)\frac{1}{\lambda_{CNT}}. \tag{11}$$

It is necessary to note that there are inconsistent results published in literature, both experimental and theoretical, regarding the dependency of resistance on length [30]. Some of these results indicate an exponential relationship [31], [32]

$$R = R_F \exp\left(\frac{\ell}{2\lambda_{\text{CNT}}}\right) \tag{12}$$

and some show a linear dependency [28], [33]

$$R = R_F\left(1 + \frac{\ell}{\lambda_{CNT}}\right) \tag{13}$$

It can be observed from (10)-(13) that the value of mean free path (MFP) plays an important role in determining the resistance of the carbon nanotube. It has been proven that the MFP of a CNT, both for metallic and semiconductor types, is proportional to the diameter [34], [35]. For the MFP of metallic CNTs, we have [34], [36]

$$\lambda_{\text{CNT}} = \left(\frac{\sqrt{3}\pi\phi^2}{2\sigma_\varepsilon^2 + 9\sigma_\phi^2}\right)\cdot D \tag{14}$$

where D is the diameter of the CNT, φ is the nearest neighbor tight-binding parameter, ε is the on-site energies, and σ_ε and σ_φ are the variances of ε and φ, respectively. For the MFP of semiconducting CNTs, we have [34], [35]

$$\lambda_{CNT} = \left(\frac{v_F}{\alpha T}\right)\cdot D \tag{15}$$

where v_F is the Fermi velocity, a is the coefficient of scattering rate, and T is the temperature. For a typical SWCNT with diameter 1 nm , the value of MFP has been reported about 1μ m based on measurements [37], [38]. Thus irrespective of the nature of SWCNTs (shells in an MWCNT), metallic or semiconducting, we can assume $\lambda_{CNT} \approx 1000D$ [34].

Fig. 11 shows the equivalent distributed circuit model of an individual CNT (shell in a multi-walled CNT)

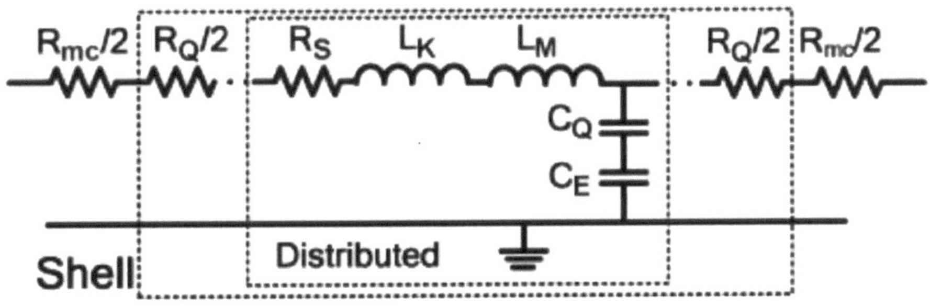

Fig. 11. Equivalent distributed circuit model of an individual CNT [34].

In this figure, R_{mc} is the imperfect contact resistance, R_Q is the quantum resistance (as the fundamental resistance R_F in Fig. 10), R_S is the scattering-induced resistance (as R_{CNT} in (10), (11)), L_K and L_M are the kinetic and magnetic inductances, respectively, and C_Q and C_E are the quantum and electrostatic capacitances, respectively.

The imperfect metal-to-nanotube contacts at each of the two ends of the nanotube, give rise to an additional resistance typically about 100 KΩ in series with the fundamental resistance R_F [24], [39].

3.2. CNT Capacitance

The total capacitance of a CNT arises from two sources: the electrostatic capacitance which is the intrinsic plate capacitance of an isolated CNT, and the quantum capacitance which accounts for the quantum electrostatic energy stored in the nanotube when it carries current [3], [26].

The electrostatic capacitance is calculated by treating the CNT as a thin wire placed away from a ground plane, as shown in Fig. 12, and its value per unit length is given by [3], [26], [30]

$$C_E = \frac{2\pi\varepsilon}{\cosh^{-1}\left(\frac{2y}{d}\right)} \approx \frac{2\pi\varepsilon}{\ln\left(\frac{y}{d}\right)} \tag{16}$$

where $\boldsymbol{\varepsilon}$, $\boldsymbol{d}$ and $\boldsymbol{y}$ are the dielectric permittivity, the CNT diameter, and the distance of CNT from the ground plane, respectively.

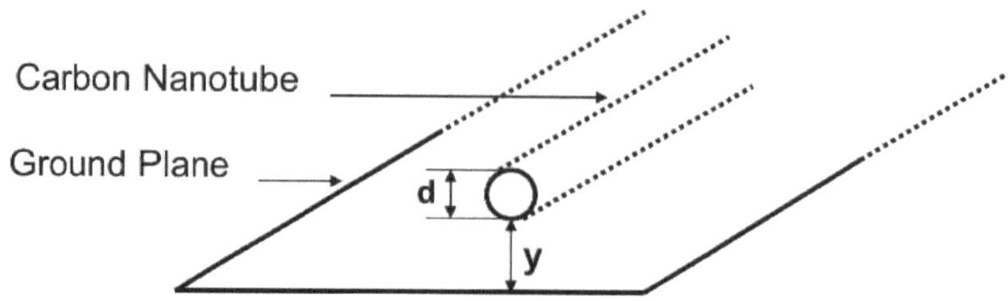

Fig. 12. Structure of an isolated CNT over a ground plane.

The electron cloud in a CNT can be assumed to be a quantum electron gas in one dimension. Hence, this follows Pauli's exclusion principle and it is not possible to add an electron with energy less than the Fermi energy of the system (E_{F}) [40]. The quantum capacitance is used to model the energy needed to add an electron at an available quantum state above the Fermi level [26]. By equating this energy to that of an effective capacitance, the expression for the quantum capacitance per unit length is obtained as [25]

$$C_Q = \frac{2e^2}{hv_F} \tag{17}$$

where v_F is the Fermi velocity in graphite [40] and is approximately 8×10^5 m/s [25]. Also for a CNT, C_Q is approximately $100\mathrm{aF}/\mu\mathrm{m}$ [25], [26]. Since a CNT has four conducting channels as described before, the effective quantum capacitance resulting from four parallel capacitances is given by $4C_Q$.

In [34], the following relations for the quantum capacitance per unit length of a shell in a MWCNT have been expressed, according to the result in [25]

$$C_{Q/\,\text{channel}} = 2 \times \frac{2e^2}{hv_F} \approx 193\mathrm{aF}/\mu\mathrm{m} \tag{18}$$

$$C_{Q/\,\text{shell}} = C_{Q/\,\text{channel}} \times N_{\text{shell}}\,(D) \tag{19}$$

where

$$N_{\text{shell}}\,(D) \approx a.D + b\;D > 3\;\text{nm} \tag{20}$$

is the number of conducting channels (spin degeneracy is already considered) in any shell, D is the diameter of the shell, a $= 0.0612\ \mathrm{nm}^{-1}$, and b $= 0.425$.

On the other hand, in [38], the following relation for the number of conducting channels in any shell has been reported

$$N_{\text{shell}}(D) \approx a \cdot D + b \qquad D > 3 \text{ nm}$$
$$\approx \frac{2}{3} \qquad D < 6 \text{ nm}. (21)$$

It should be noted that the error introduced by (20), (21), due to different chiralities, is within 15% for all values of D [34], [38]. Note that the two regions in (21) have an overlap, and for 3 nm < D < 6 nm, both constant and linear functions can be used without any considerable error [38].

3.3. CNT Inductance

The total inductance of a CNT (L $_{\text{CNT}}$ in Fig. 10) arises from two sources: the magnetic inductance and the kinetic inductance (L_M and L_K in Fig. 11). In the presence of a ground plane, the magnetic inductance per unit length is given by [25], [40]

$$L_M = \frac{\mu}{2\pi}\cosh^{-1}\left(\frac{2y}{d}\right) \approx \frac{\mu}{2\pi}\ln\left(\frac{y}{d}\right). \tag{22}$$

For a typical situation, the nanotube is placed on top of an insulating substrate (typically silicon dioxide), with a conducting medium below. A typical oxide thickness is between 10 nm and 1μ m with a typical nanotube radius of 1 to 2 nm . It can be noted that the magnetic inductance is a relatively weak function of the factor (y/d) and for typical geometries, it can be estimated to be around 1pH/μm [40], [41].

In one-dimensional CNT conductors, apart from the magnetic inductance, another inductive component appears due to the kinetic energy of the electrons. The details of its derivation can be obtained in [40]-[42]. The kinetic inductance per unit length can be expressed as [25], [40], [41]

$$L_K = \frac{h}{2e^2 v_F} \tag{23}$$

It is necessary to note that the four parallel conducting channels in a CNT give rise to an effective kinetic inductance of $L_K/4$. Also as it has been shown from (23), the kinetic inductance per unit length for a one dimensional CNT conductor is around 16nH/μm [26], [40], more than 4 orders of magnitude larger than its magnetic counterpart L_M($\approx$ 1pH/μm) [30], [40], [41], and will essentially play a vital role in high frequency applications [40].

In [34], the following relations for the kinetic inductance per unit length of a shell in a MWCNT have been expressed, according to the result in [25]

$$L_{K/\text{ channel}} = \frac{h}{2e^2 v_F} \times \frac{1}{2} \approx 8\text{nH}/\mu\text{m} \tag{24}$$

$$L_{K/\text{ shell}} = \frac{L_{K/\text{ channel}}}{N_{\text{shell}}(D)} \tag{25}$$

where N_{shell} (D) has been defined in (20), (21).

4. Different Structures of CNT as Interconnect

4.1. CNT Bundle as Interconnect

While SWCNTs have desirable material properties, individual nanotubes suffer from an intrinsic ballistic resistance of approximately 6.5kΩ that is not dependent on the length of the nanotube [43]. As a result, the high resistance associated with an isolated CNT, causes excessive delay for interconnect applications. To alleviate the intrinsic resistance problem, bundles or ropes of CNTs conducting current in parallel, have been proposed and physically demonstrated as a possible interconnect medium for local, intermediate, and global interconnects [3], [43]. Fig. 13 shows a CNT bundle interconnect structure consists of a signal line and two ground return paths.

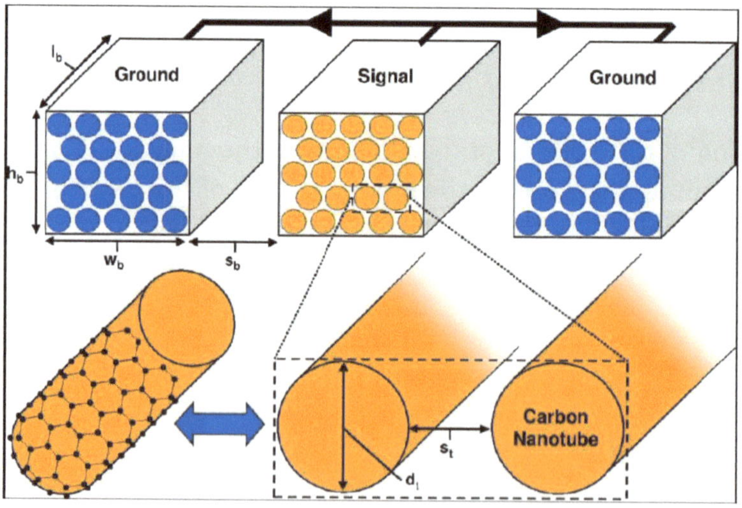

Fig. 13. System of SWCNT interconnect bundles implementing a signal line and two adjacent ground return paths [43].

Due to the lack of control on chirality, any bundle of CNTs consists of metallic as well as semi-conducting nanotubes. The required relations for the parameters of CNT bundles including the magnetic and kinetic inductances, the electrostatic and quantum capacitances, the fundamental and scattering resistances, can be obtained from [5]. In section 5 the time domain behaviour of a CNT bundle as interconnect is discussed based on [44].

4.2. Multi-walled CNT as Interconnect

Fig. 14 shows a geometric structure of a Multi-Walled carbon nanotube (MWCNT) over a ground plane.

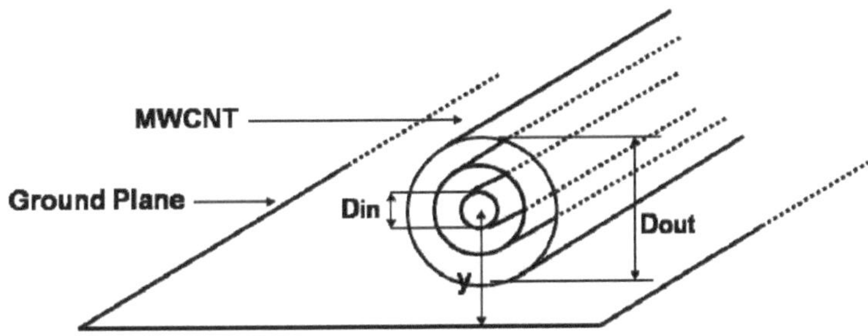

Fig. 14. Structure of a Multi-Walled CNT over a ground plane.

In this figure, D_{in} and D_{out} are the diameter of inner shell and the diameter of outer shell, respectively, and y is the height of inner shell from the ground plane. Recently wide spread studies regarding the benefits of the performance of MWCNTs as interconnect in comparison with CNT bundles and Cu have been performed. In [34] the performance of MWCNT interconnects has been analyzed and their circuit modelling has been discussed. Although MWCNT has an important role in the interconnect applications, the main scope of this chapter which follows in the subsequent section, is dedicated to the analysis of the behaviour of CNT bundle interconnects.

5. Time Domain Response

In [44] we have discussed about the time domain analysis of a CNT bundle interconnect in a driver-interconnect-load configuration and a new relation for the input-output transfer function in the related configuration has been extracted. A review of the discussion presented in [44] is brought in this section. Fig. 15 shows a CNT bundle interconnect with resistance, capacitance and inductance per unit length of R_s, C and L respectively, driven by a repeater of output resistance R_{tr} and output parasitic capacitance C_{out}, and driving an identical repeater with input capacitance equal to the load capacitance C_L.

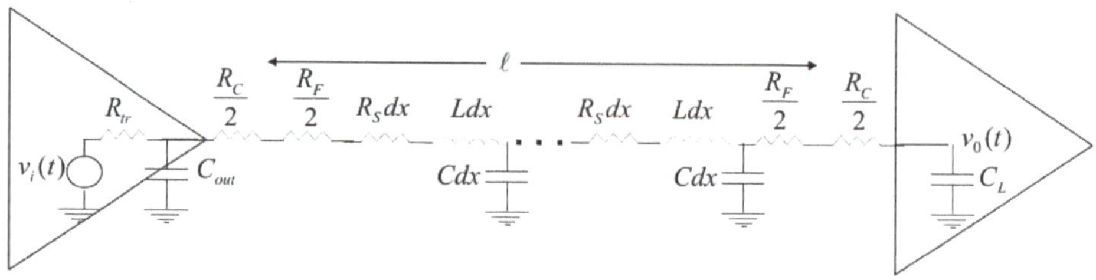

Fig. 15. Equivalent circuit of a driver-CNT bundle-load configuration, based on transmission line modeling [44].

In this figure, the CNT bundle interconnect has been modelled as a transmission line. For calculating the input-output transfer function of the configuration in Fig. 15, we need to derivate the total transmission parameter matrix. Using the ABCD transmission parameter matrix for a uniform RLC transmission line of length ℓ as given in [45]-[47], we can express the total ABCD transmission parameter matrix of the configuration in Fig. 15 as

$$T_{\text{total}} = \begin{bmatrix} 1 & R_{tr} \\ 0 & 1 \end{bmatrix} \begin{bmatrix} 1 & 0 \\ sC_{\text{out}} & 1 \end{bmatrix} \begin{bmatrix} 1 & R_{ex} \\ 0 & 1 \end{bmatrix} \begin{bmatrix} \cosh(\theta^T \ell) & Z_0^T \sinh(\theta^T \ell) \\ \frac{1}{Z_0^T} \sinh(\theta^T \ell) & \cosh(\theta^T \ell) \end{bmatrix} \begin{bmatrix} 1 & R_{ex} \\ 0 & 1 \end{bmatrix} \quad (26)$$

where $R_{ex} = (R_C + R_F)/2, Z_0^T = \sqrt{(R_S + sL)/(sC)}, \theta^T = \sqrt{(R_S + sL)sC}$, and $s = j\omega$ is the complex frequency. The elements of matrix T_{total} can be written, using (26), as

$$A_T = (1 + sR_{tr}C_{out})\cosh(\theta^T \ell) + \frac{(R_{tr} + R_{ex} + sR_{tr}R_{ex}C_{out})}{Z_0^T}\sinh(\theta^T \ell), \quad (27)$$

$$B_T = [R_{tr} + 2R_{ex} + 2sR_{tr}R_{ex}C_{out}]\cosh(\theta^T \ell) + [Z_0(1 + sR_{tr}C_{out}) + \frac{R_{ex}(R_{tr} + R_{ex} + sR_{tr}R_{ex}C_{out})}{Z_0^T}]\sinh(\theta^T \ell), \quad (28)$$

$$C_T = sC_{out}\cosh(\theta^T \ell) + \frac{1 + sR_{ex}C_{out}}{Z_0^T}\sinh(\theta^T \ell), \quad (29)$$

$$D_T = (1 + 2sR_{ex}C_{out})\cosh(\theta^T \ell) + [sC_{out}Z_0^T + \frac{R_{ex}(1 + sR_{ex}C_{out})}{Z_0^T}]\sinh(\theta^T \ell). \quad (30)$$

Therefore the input-output transfer function of the configuration in Fig. 15 can be written as

$$H(s)=\frac{V_0(s)}{V_i(s)}=\frac{1}{A_T+sC_LB_T}=\left(\begin{array}{l}[1+s(R_{tr}C_{out}+R_{tr}C_L+2R_{ex}C_L)+s^2(2R_{tr}R_{ex}C_{out}C_L)]\cosh(\theta^T\ell)\\+\left[\begin{array}{l}\frac{R_{tr}+R_{ex}}{Z_0^T}+s\left(\frac{R_{tr}R_{ex}C_{out}+R_{ex}(R_{tr}+R_{ex})C_L}{Z_0^T}+Z_0^TC_L\right)\\+s^2\left(\frac{R_{tr}R_{ex}^2C_{out}}{Z_0^T}+Z_0^TR_{tr}C_{out}\right)C_L\end{array}\right]\sinh(\theta^T\ell)\end{array}\right)^{-1}. \quad (31)$$

For simulation purposes, we need to extract a parametric linear approximation for (31). For this purpose, we need to calculate the equivalent linear terms for cosh $(\theta^T\ell)$ and sinh $(\theta^T\ell)/Z_0^T$, and put them in (31). Using the relation for the transfer function of a driverCNT interconnect-load configuration, as discussed in [47], and rigorous calculations, we can extract the following linear expressions

$$\cosh(\theta^T\ell)=1+\left(\frac{R_SC\ell^2}{2!}\right)s+\left(\frac{LC\ell^2}{2!}+\frac{R_S^2C^2\ell^4}{4!}\right)s^2$$
$$+\left(\frac{2R_SLC^2\ell^4}{4!}+\frac{R_S^3C^3\ell^3}{6!}\right)s^3+\left(\frac{L^2C^2\ell^4}{4!}+\frac{3R_S^2LC^3\ell^6}{6!}+\frac{R_S^4C^4\ell^8}{8!}\right)s^4, \quad (32)$$

$$\frac{\sinh(\theta^T\ell)}{Z_0^T}=sC\ell\left[1+\frac{R_SC\ell^2}{3!}s+\left(\frac{LC\ell^2}{3!}+\frac{R_S^2C^2\ell^4}{5!}\right)s^2+\left(\frac{R_S^3C^3\ell^6}{7!}+\frac{2R_SLC^2\ell^4}{5!}\right)s^3\right]. \quad (33)$$

With substituting the two terms cosh $(\theta^T\ell)$ and sinh $(\theta^T\ell)/Z_0^T$ in (31), by (32) and (33), we can obtain the linear parametric equivalent for the transfer function of (31) as

$$H(s)=\frac{1}{1+a_1s+a_2s^2+a_3s^3+a_4s^4+a_5s^5+a_6s^6} \quad (34)$$

where

$$a_1 = R_{tr}(C_{\text{out}} + C\ell + C_L) + R_{ex}(C\ell + 2C_L) + R_S\ell\left(\frac{C\ell}{2!} + C_L\right) \quad (35)$$

$$a_2 = R_{ex}R_{tr}(C_{\text{out}}C\ell + C_LC\ell + 2C_{\text{out}}C_L) + R_{tr}R_SC\ell^2\left(\frac{C_{\text{out}}C\ell}{2!} + \frac{C^2\ell^2}{3!} + C_{\text{out}}C_L + \frac{C\ell C_L}{2!}\right) \quad (35)$$

$$+R_{ex}R_SC\ell^2\left(\frac{C\ell}{3!} + C_L\right) + \frac{R_S^2C^2\ell^4}{4!} + \frac{LC\ell^2}{2!} + R_{ex}^2C\ell C_L + \frac{R_S^2C\ell^3C_L}{3!} + L\ell C_L \quad (36)$$

$$a_3 = \frac{2R_SLC^2\ell^4}{4!} + \frac{R_S^3C^3\ell^6}{6!} + \left(R_{ex} + R_{tr} + \frac{R_S}{C}C_L\right)\left(\frac{LC^2\ell^3}{3!} + \frac{R_S^2C^3\ell^5}{5!}\right) \quad (36)$$

$$+[R_{tr}(C_{\text{out}} + C_L) + 2R_{ex}C_L]\left(\frac{LC\ell^2}{2!} + \frac{R_S^2C^2\ell^4}{4!}\right) \quad (36)$$

$$+\frac{R_SC^2\ell^3}{3!}\left(R_{ex}R_{tr}(C_{\text{out}} + C_L) + R_{ex}^2C_L + \frac{L}{C}C_L + \frac{R_S}{C}R_{tr}C_{\text{out}\,L}\right) \quad (36)$$

$$+R_{ex}R_{tr}C_{\text{out}}C\ell C_L(R_{ex} + R_S\ell) + R_{tr}C_{\text{out}}L\ell C_{L'} \quad (37)$$

$$a_4 = \frac{L^2C^2\ell^4}{4!} + \frac{3R_S^2LC^3\ell^6}{6!} + \frac{R_S^4C^4\ell^8}{8!} + [R_{tr}(C_{out} + C_L) + 2R_{ex}C_L]\left(\frac{2R_SLC^2\ell^4}{4!} + \frac{R_S^3C^3\ell^6}{6!}\right)$$
$$+\left(R_{tr} + R_{ex} + \frac{R_S}{C}C_L\right)\left(\frac{2R_SLC^3\ell^5}{5!} + \frac{R_S^3C^4\ell^7}{7!}\right) + 2R_{tr}R_{ex}C_{out}C_L\left(\frac{LC\ell^2}{2!} + \frac{R_S^2C^2\ell^4}{4!}\right)$$
$$+\left(R_{tr}R_{ex}(C_{out} + C_L) + R_{ex}^2C_L + \frac{L}{C}C_L + \frac{R_S}{C}R_{tr}C_{out}C_L\right)\left(\frac{LC^2\ell^3}{3!} + \frac{R_S^2C^3\ell^5}{5!}\right)$$
$$+\frac{R_SC^2\ell^3}{3!}\left(R_{tr}R_{ex}^2C_{out}C_L + \frac{L}{C}R_{tr}C_{out}C_L\right), \quad (38)$$

$$a_5 = [R_{tr}(C_{out} + C_L) + 2R_{ex}C_L]\left(\frac{L^2C^2\ell^4}{4!} + \frac{3R_S^2LC^3\ell^6}{6!} + \frac{R_S^4C^4\ell^8}{8!}\right)$$
$$+2R_{tr}R_{ex}C_{out}C_L\left(\frac{2R_SLC^2\ell^4}{4!} + \frac{R_S^3C^3\ell^6}{6!}\right)$$
$$+\left(R_{tr}R_{ex}(C_{out} + C_L) + R_{ex}^2C_L + \frac{L}{C}C_L + \frac{R_S}{C}R_{tr}C_{out}C_L\right)\left(\frac{2R_SLC^3\ell^5}{5!} + \frac{R_S^3C^4\ell^7}{7!}\right)$$
$$+\left(R_{tr}R_{ex}^2C_{out}C_L + \frac{L}{C}R_{tr}C_{out}C_L\right)\left(\frac{LC^2\ell^3}{3!} + \frac{R_S^2C^3\ell^5}{5!}\right), \quad (39)$$

$$a_6 = 2R_{tr}R_{ex}C_{out}C_L\left(\frac{L^2C^2\ell^4}{4!} + \frac{3R_S^2LC^3\ell^6}{6!} + \frac{R_S^4C^4\ell^8}{8!}\right)$$
$$+\left(R_{tr}R_{ex}^2C_{out}C_L + \frac{L}{C}R_{tr}C_{out}C_L\right)\left(\frac{2R_SLC^3\ell^5}{5!} + \frac{R_S^3C^4\ell^7}{7!}\right). \quad (40)$$

Fig. 16 shows the step response of configuration in Fig. 15, for 32 nm technology node, using our extracted linear transfer function of (34), and HSPICE simulation.

The repeater size has been assumed 174 times larger than the minimum sized repeater, which its parameters have been extracted from ITRS 2007 [2]. Also the load capacitance has been considered equal to the input capacitance of repeater. Recall that $\lambda_{\text{CNT}} \approx 1000D$ [34] where D is the diameter of each individual CNT, the mean free path of CNT (λ_{CNT}) has been assumed 1μ m. As it is cleared from Fig. 16, there is an excellent match between the result of our extracted parametric transfer function and HSPICE simulation result. Fig. 17 shows the step response of configuration in Fig. 15, using our extracted linear parametric transfer function of (34), for the contact resistance values from 1kΩ to 50kΩ. As shown in Fig. 17, the propagation delay increases from 0.138 ns to 5.58 ns , with the increase of contact resistance value from 1kΩ to 50kΩ.

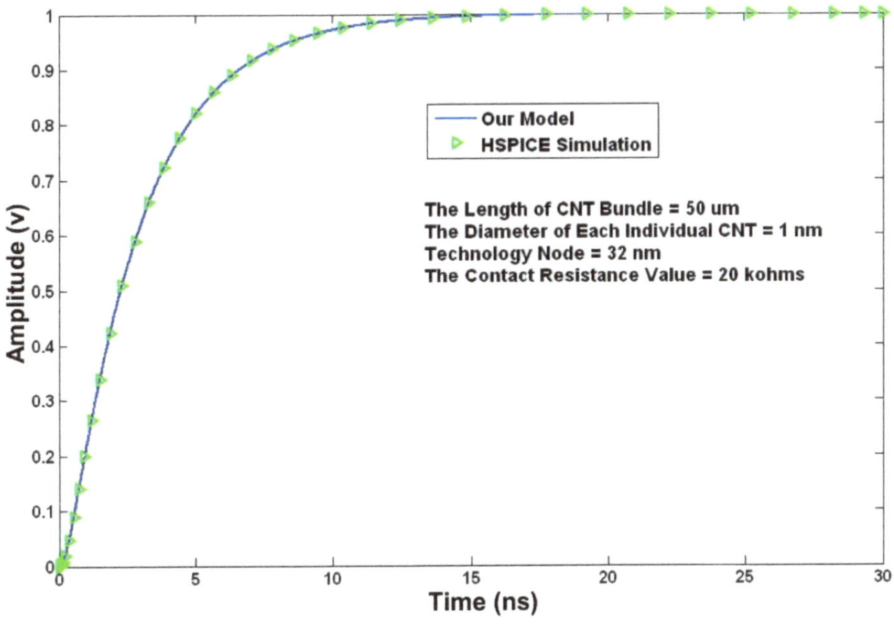

Fig. 16. The step response of configuration in Fig. 15, using our extracted linear parametric transfer function, and HSPICE simulation. In this figure, the number of transmission line model sections for HSPICE simulation, has been considered 400 [44].

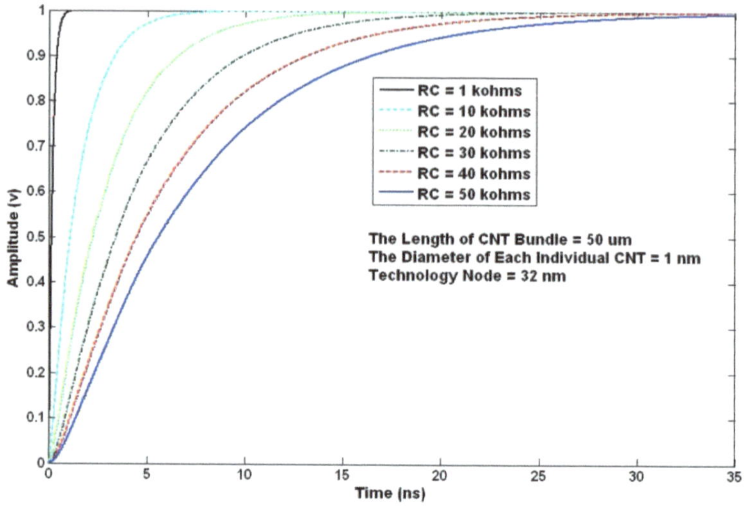

Fig. 17. The step response of configuration in Fig. 15, using our extracted linear parametric transfer function, for various values of the contact resistance [44].

Fig. 18 shows the propagation delay of configuration in Fig. 15, using our extracted linear parametric transfer function of (34), versus the contact resistance value, and for the CNT bundle lengths 50μ m, 200μ m, 500μ m and 1000μ m.

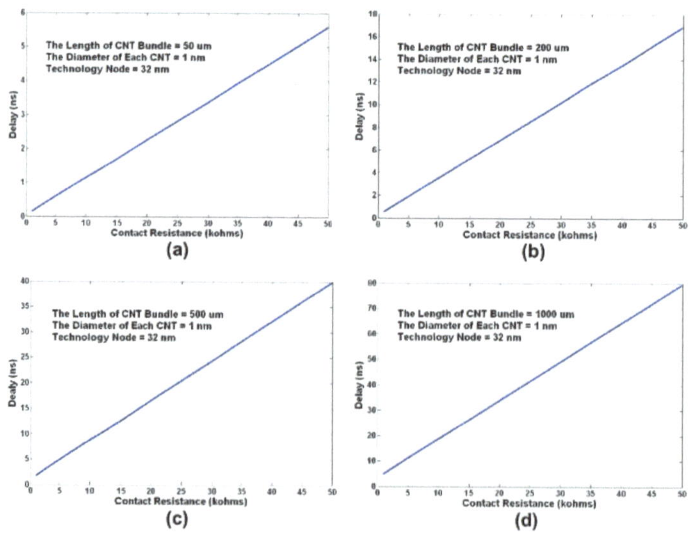

Fig. 18. The propagation delay of configuration in Fig. 15, using our new linear parametric transfer function, versus the contact resistance value, for the CNT bundle lengths: (a) 50μ m, (b) 200μ m, (c) 500μ m, (d) 1000μ m [44].

In this figure, the diameter of each individual CNT has been chosen 1 nm , and therefore as discussed before, the mean free path of CNT will be 1μ m. As shown in Fig. 18, for the length of CNT bundle equal to 50μ m, the propagation delay changes from 0.138 ns to 5.58 ns for the contact resistance values from $1\text{k}\Omega$ to $50\text{k}\Omega$, i.e. a variation range of 39.43 times the minimum value. The related delay variation ranges for the length values 200μ m, 500μm, and 1000μ m, are 31.37, 22.61 , and 15.42 times the minimum value, respectively. This means that, the impact of the contact resistance on the propagation delay, decreases with the increase of the bundle length. The reason is that, with the increase of the bundle length, the role of scattering resistance which increases with the length [29], would be more important. In Fig. 19, the nyquist diagrams for a driven CNT bundle interconnect, versus the length of CNT bundle and the diameter of each individual CNT, have been plotted using MATLAB [23].

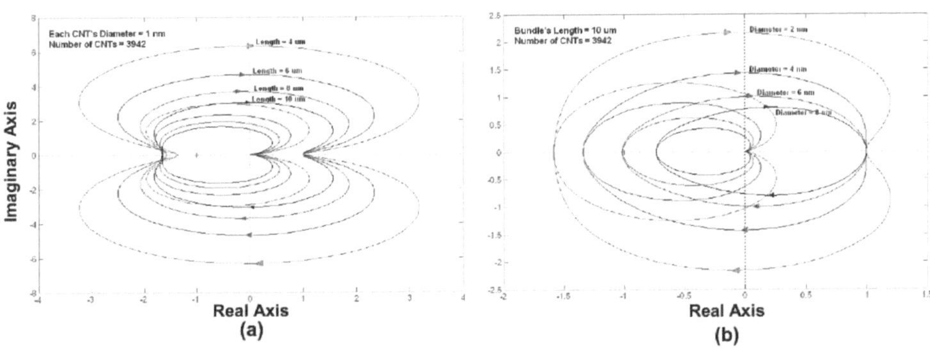

Fig. 19. The nyquist stability diagrams for a driven CNT bundle interconnect; (a) versus length; (b) versus diameter. In this figure, the space between two adjacent CNTs has been chosen as 0.34 nm , and the length, the width and the thickness of CNT bundle, have been considered 10μ m, 50 nm and 125 nm , respectively. Also the load capacitance has been assumed equal to 5 fF [44].

As shown in Fig. 19 (a), by increasing the length of CNT bundle, the complex point ($-1,0$) goes toward outside the diagram. So, by increasing the length of CNT bundle, the system becomes more stable. As shown in Fig. 19 (b), by increasing the diameter of each individual CNT, the complex point ($-1,0$) goes toward outside the diagram and then, the diagram goes farther from this point. So, by increasing the diameter of each individual CNT, the system becomes more stable. It should be noted that in simulations of Fig. 19, the driver has been considered ideal with perfect contacts, and all individual CNTs in the bundle have been assumed metallic. A more detailed discussion about the stability analysis in CNT interconnects has been presented in [48].

6. Summary

In this chapter, we have studied interconnect challenges and the behaviour of carbon nanotube (CNT) as interconnect in VLSI circuits. In this review we discussed about the two main structures of CNT, including CNT bundles and MWCNTs, which achieve good performance due to the parallel SWCNTs in a bundle or the parallel shells in a MWCNT. These optimized configurations give the better characteristics including decreased delay time in comparison with SWCNTs, which is a vital parameter for the application as interconnect. The repeater (buffer) insertion technique that is used for the reduction of delay time in the global interconnects, has been discussed. Also in this chapter, we analyzed the time domain response of CNT bundle interconnect in a driver-interconnect-load configuration, based on the formulations and discussions we have presented in the reference [44]. At the continuation, we discussed briefly about the stability concept in CNT bundle interconnects, versus the length and diameter of each CNT in a CNT bundle.

7. References

A. Naeemi, R. Sarvari, J. D. Meindl, " Performance Comparison Between Carbon Nanotube and Copper Interconnects for Gigascale Integration (GSI)", IEEE Electron Device Letters, vol. 26, No. 2, pp. 84-86, Feb. 2005.

International Technology Roadmap for Semiconductors (ITRS), 2007.

N. Srivastava, K. Banerjee, "Performance Analysis of Carbon Nanotube Interconnects for VLSI Applications", IEEE Int. Conf. on Computer-Aided Design (ICCAD), pp. 383-390, CA, Nov. 2005.

N. Srivastava, K. Banerjee, "A Comparative Scaling Analysis of Metallic and Carbon Nanotube Interconnections for Nanometer Scale VLSI Technologies", Proc. 21 st Intl. VLSI Multilevel Interconnect Conf., pp. 393-398, 2004.

K. Banerjee, N. Srivastava, "Are Carbon Nanotubes the Future of VLSI Interconnections?", ACM Design Automation Conf. (DAC), pp. 809-814, CA, Jul. 2006.

J. Li. et al., "Bottom-Up Approach for Carbon Nanotubes Interconnects", Appl. Phys. Lett., vol. 82, pp. 2491-2493, Apr. 2003.

A. Naeemi, J. D. Meindl, "Design and Performance Modeling for Single-Walled Carbon Nanotubes as Local, Semiglobal, and Global Interconnects in Gigascale Integrated Systems", IEEE Trans. Electron Devices, vol. 54, No.1, pp. 26-37, Jan. 2007.

P. L. McEuen, et al., "Single-Walled Carbon Nanotube Electronics", IEEE Trans. Nanotechnology, vol. 1, No. 1, pp. 78-85, 2002.

Min. Tang, Jun-Fa Mao, "Optimization of Global Interconnects in High Performance VLSI Circuits", Proc. 19th Int. Conf. VLSI Design, 2006.

D. Fathi, B. Forouzandeh, "Accurate Analysis of Global Interconnects in Nano FPGAs", Nano, pp. 171-176, Vol. 4, No. 3, 2009.

V. Prasad, M.P. Desai, "Interconnect delay minimization using a novel pre-mid-post buffer strategy", Proc. 16 th Int. Conf. VLSI Design, pp. 417-422, Jan. 2003.

Jan M. Rabaey, Anantha Chandrakasan, "Digital Integrated Circuit", 2 nd Edition, Chapters 4 & 9, 2006.

A.Naeemi, R.Venkatesan J.d.mendlS, "SYSTEM-ON-CHIP GLOBAL INTERCONNECT OPTIMIZATION", Proc. ASIC/SoC Conf., Sep. 2002, pp. 399-403.

M. Mui, K. Banerjee, "A Global Interconnect Optimization Scheme for Nanometer Scale VLSI With Implications for Latency, Bandwidth, and Power Dissipation", IEEE Trans. ELECTRON DEVICES, vol. 51, NO. 2, pp. 195-203, Feb. 2004.

K. Banerjee, Amit Mehrotra, "A Power-Optimal Repeater Insertion Methodology for Global Interconnects in Nanometer Designs", IEEE Trans. Electron Devices, vol. 49, No. 11, pp. 2001-2007, Nov. 2002.

R. Venkatesan, J. A. Davis, J. d. Mendil, "Compact Distributed RLC Interconnect Models-Part IV: Unified Models for Time Delay, Crosstalk, and Repeater Insertion", IEEE Trans. ELECTRON DEVICES, vol. 50, NO. 4, pp. 1094-1102, Apr. 2003.

B. Bakoglu, "Circuits, Interconnections and Packaging for VLSI", Addison-Wesley, 1990.

D. Fathi, B. Forouzandeh, A New Repeater Insertion Technique for Optimization of Global Interconnects in Nano VLSI", NANO, To be published.

Shyh-Chyi Wong, Gwo-Yann Lee, "Modeling of interconnect capacitance, delay, and crosstalk in VLSI", IEEE Trans. Semiconductor Manufacturing, vol. 13, Issue 1, pp. 108-111, Feb. 2000.

A. B. Kahng, K. Masuko, S. Muddu, "Analytical delay models for VLSI interconnects under ramp input", Proc. ACM/IEEE Int. Conf. Computer-Aided Design, pp. 30-36, Nov. 1996.

W. C. Elmore, "The Transient Analysis of Damped Linear Networks with Particular Regard to Wideband Amplifiers", J. Appl. Phys., vol. 19, No. 1, pp. 55-63, Jan. 1948.

X. Li, J. Mao, H. Huang, Y.Liu, "Global Interconnect Width and Spacing Optimization for Latency, Bandwidth and Power Dissipation," IEEE Trans. Electron Devices, vol. 52, No. 10, pp. 2272-2279, Oct. 2005.

Mathematics Laboratory (MATLAB), 2007.

N. Srivastava, R. V. Joshi, K. Banerjee, "Carbon Nanotube Interconnects: Implications for Performance, Power Dissipation and Thermal Management", IEEE Int. Electron Devices Meeting (IEDM), Washington DC, pp. 257-260, Dec. 2005.

P. J. Burke, "Luttinger Liquid Theory as a Model of the Gigahertz Electrical Properties of Carbon Nanotubes", IEEE Trans. Nanotechnology, vol. 1, No. 3, pp. 129-144, Sep. 2002.

K. Banerjee, N. Srivastava, "Are Carbon Nanotubes the Future of VLSI Interconnections?", ACM Design Automation Conf. (DAC), San Francisco, CA, pp. 809-814, Jul. 2006.

S. Datta, "Electrical Resistance: An Atomistic View", Nanotechnology, vol. 15, S433-S451, 2004.

J. Y. Park, et al., "Electron-Phonon Scattering in Metallic Single-Walled Carbon Nanotubes", Nano Letters, vol. 4, No. 3, pp. 517-520, 2004.

S. Datta, "Electronic Transport in Mesoscopic Systems", Cambridge Univ. Press, 1995.

A. Naeemi, R. Sarvari, J. D. Meindl, "Performance Comparison Between Carbon Nanotube and Copper Interconnects for Gigascale Integration (GSI)", IEEE Electron Device Letters, vol. 26, No.2, pp. 84-86, Feb. 2005.

P. J. de Pablo, et al., "Nonlinear Resistance Versus Length in Single-Walled Carbon Nanotubes", Phys. Rev. Lett., vol. 88, pp. 036804/1-036804/4, Jan. 2002.

A. Andriotis, et al., "Non-Linear Resistance Dependence on Length in Single-Wall Carbon Nanotubes", Nano Lett., vol. 3, pp. 131-134, 2003.

S. Li, et al., "Electrical Properties of 0.4 cm Long Single-Walled Carbon Nanotubes", Nano Lett., vol. 4, pp. 2003-2007, 2004.

H. Li, et al., "Circuit Modeling and Performance Analysis of Multi-Walled Carbon Nanotube Interconnects", IEEE Trans. Electron Devices, vol. 55, No. 6, pp. 1328-1337, Jun. 2008.

X. Zhou, et al., "Band Structure, Phonon Scattering, and the Performance Limit of Single-Walled Carbon Nanotube Transistors", Phys. Rev. Lett., vol. 95, No. 14, p. 146805, Sep. 2005.

J. Jiang, et al., "Universal Expression for Localization Length in Metallic Carbon Nanotubes", Phys. Rev. B, Condens. Matter, vol. 64, No. 4, p. 045409, Jul. 2001.

A. Naeemi, J. D. Meindl, "Design and Performance Modeling for Single-Walled Carbon Nanotubes as Local, Semiglobal, and Global Interconnects in Gigascale Integrated Systems", IEEE Trans. Electron Devices, vol. 54, No.1, pp. 26-37, Jan. 2007.

A. Naeemi, J. D. Meindl, "Compact Physical Models for Multiwall Carbon-Nanotube Interconnects", IEEE Electron Device Lett., vol. 27, No. 5, pp. 338-340, May 2006.

A. Naeemi, J. D. Meindl, "Mono-Layer Metallic Nanotube Interconnects: Promising Candidates for Short Local Interconnects", IEEE Electron Device Letters, vol. 26, No. 8, pp. 544-546, Aug. 2005.

A. Raychowdhury, K. Roy, "Modeling of metallic Carbon-Nanotube Interconnects for Circuit Simulations and a Comparison With Cu Interconnects for Scaled Technologies", IEEE Trans. Computer-Aided Design of Integrated Circuits and Systems, vol. 25, No. 1, pp. 58-65, Jan. 2006.

P. J. Burke, "An RF Circuit Model for Carbon Nanotubes", IEEE Trans. Nanotechnol., vol. 2, No. 1, pp. 55-58, Mar. 2003.

M. W. Bockrath, "Carbon Nanotubes: Electrons in One Dimension", Ph.D. Dissertation, Dept. Phys., Univ. California, Berkeley, 1999.

A. Nieuwoudt, Y. Massoud, "Evaluating the Impact of Resistance in Carbon Nanotube Bundles for VLSI Interconnect Using Diameter-Dependent Modeling Techniques", IEEE Trans. Electron Devices, vol. 53, No. 10, pp. 2460-2466, Oct. 2006.

D. Fathi, B. Forouzandeh, S. Mohajerzadeh, R. Sarvari, "Accurate Analysis of Carbon Nanotube Interconnects Using Transmission Line Model", Micro & Nano Letters, vol. 4, No. 2, pp. 116-121, 2009.

K. Banerjee, A. Mehrotra, "Analysis of On-Chip Inductance Effects for Distributed RLC Interconnects", IEEE Trans. Computer-Aided Design of Integ. Circuits and Systems, vol. 21, No. 8, pp. 904-915, Aug. 2002.

A. K. Palit, et al., "Reduced Order Long Interconnect Modeling", 15. ITG-GI-GMM Workshop Testmethoden und Zuverlässigkeit von Schaltungen und Systemen, Timmendorfer Strand, pp. 42-47, Mar. 2003.

D. Fathi, B. Forouzandeh, "Time Domain Analysis of Carbon Nanotube Interconnects Based on Distributed RLC Model", Nano, vol. 4, No. 1, pp. 13-21, Feb. 2009.

D. Fathi, B. Forouzandeh, "A Novel Approach for Stability Analysis in Carbon Nanotube Interconnects", IEEE Electron Device Letters, vol. 30, No. 5, pp. 475-477, May 2009.

Chapter-16

Affinity of CNT for Metal - Its Importance to Application

1. Introduction

As we know, carbon nanotube (especially single-walled carbon nanotube: SWCNT) possesses outstanding properties such as high thermal conductivity, ultra high physical strength, durability, and so forth. Therefore, we expect SWCNT to be a candidate for future material that can realize all of our hopes. However, thinking about its application, the affinity for metal and/or the bonding to metal species are inevitably important problems and those still remain to be unsolved. If there are some discontinuities between SWCNT and metal species, these discontinuities must result in thermal resistance and electrical resistance; that mean we loose the reason and advantage of using SWCNT. If SWCNT can be coated with metal, some new properties may arise, but there is no guarantee that SWCNT can maintain its outstanding properties. These kinds of new approach for application have been partially reported experimentally; however, their mechanisms are not solved. For example, some metal species were deposited on single-walled carbon nanotube forests (Ishikawa et al., 2007) and an isolated single-walled carbon nanotube was coated by metals using e-beam (Zhang et al., 2000). But to investigate the properties of functionalizedSWCNT (metal coated SWCNT) experimentally is extremely difficult. In this chapter molecular dynamics simulation (MD) easily tells us the answers for above-mentioned questions.

The classical MD is merely solving the Newtonian equation according to the forces affected among atoms, but provided the potential function is appropriate, it gives us quite proper perspectives. In this chapter I mention how to build up the appropriate potential function and the coincidence with experimental results of metal deposition on SWCNT. Furthermore, I refer to the perspectives of physical properties of the functionalized-SWCNT.

2. Computational methods

2.1 Driving the potential function

The Brenner potential was applied for carbon-carbon interaction and a classical many-body potential was applied for metal-metal and carbon-metal interactions as a function of the bond order potential. The parameter sets for nickel-nickel and carbon-nickel were derived as a function of coordination number (Shibuta & Maruyama, 2007) while those for gold-gold and carbon-gold were derived by density functional theory (DFT) calculations. Gaussian 03 was employed in this study, and Becke's three-parameter exchange functional with the Lee-Yang-Parr correlation functional (B3LYP) (Beche, 1993 and Lee et al. 1998) was applied. The Los Alamos effective core potential with DZ (LANL2DZ) (Hay et al. 1985) was used as the basis set.

Table 1. Potential parameters for MD simulations

atom	S	$\beta(\text{Å}^{-1})$	$D_{e1}(\text{eV})$	$D_{e2}(\text{eV})$	C_D	$R_{e1}(\text{Å})$	$R_{e2}(\text{Å})$	C_R	R_1 (Å)	R_2 (Å)
Fe – Fe*	1.3	1.2173	0.4155	0.8392	0.8730	2.627	0	-	2.7	3.2
Ni-Ni*	1.3	1.570	0.4217	1.0144	0.8268	2.4934	0.1096	0.3734	2.7	3.2
Ti-Ti	1.3	2.331	0.6500	0.2001	0.3700	3.819	2.411	0.2357	2.7	3.2
Au – Au	1.3	1.750	0.5290	1.3510	1.5251	3.3043	0.7573	0.2939	2.9	3.5
atom	De (eV)		S	$\beta(\text{Å}^{-1})$	$R_e(\text{Å})$	$R_1(\text{Å})$	$R_2(\text{Å})$	b		δ
Fe – C*	3.3249		1.3	1.5284	1.7304	2.7	3.0	0.0656		-0.4279

Ni-C*	2.4673		1.3	1.8706	1.7628	2.7	3.0	0.0688		-0.5351
Ti-C	3.6240		1.3	1.466	1.8000	2.7	3.0	0.0600		-0.5000
Au – C	2.1840		1.3	2.0745	1.9030	2.9	3.3	0.0970		-0.5350

Because the method for deriving parameters is similar to abovementioned paper, it is briefly explained here. First, we assume the symmetrical structure of Aux and AuCx clusters that means we do not take into account the Jahn-Teller distortion (Castro et al., 1997) and calculate their total energy. The binding energy is calculated by subtracting isolated energies from the total energy and dividing its result by the number of bonds, as shown in Eq. 1:

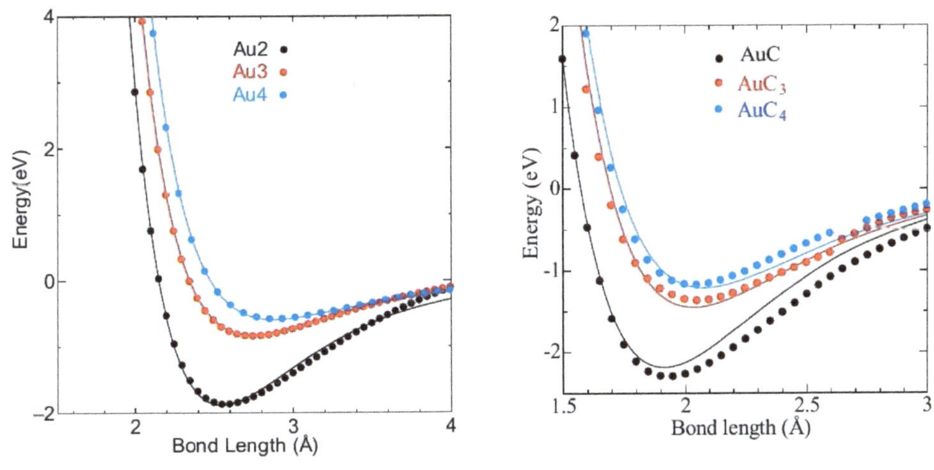

Fig. 1. Fitted potential curves from the lowest ground state derived by DFT calculation (Gaussian03, B3LYP/LANL2DZ). Dots show the binding energies, and the fitted curves are the bond order potential curves. In each case, various spin states are examined and their minimum values are employed.

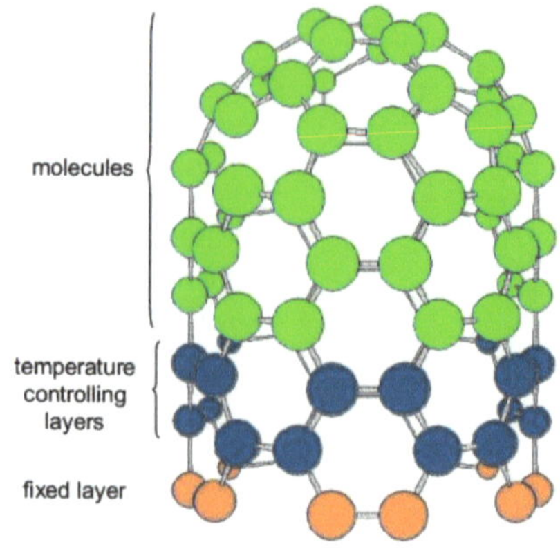

Fig. 2. Surface consisting of a capped carbon nanotube. The calculated system consists of 6 × 6 units (2880 atoms) and has dimensions of 60Å × 52Å × 60Å.

$$De = \left(E_{\text{min.}}^{\text{total}} - E_{m,\text{min.}}^{\text{isolated}} - N_c \cdot E_{c,\text{ min.}}^{\text{isolate}}\right)/N_b \tag{1}$$

where $E_{\text{min.}}^{\text{total}}$ denotes the total energy at a certain spin state, $E_{m,\text{ min.}}^{\text{isolated}}$ is the isolated energy for the metal, N_c is the coordination number of carbon, $E_{c,\text{ min}}^{\text{isolate}}$ is the isolated energy for carbon, and N_b is the bond number. In each case, several spin states have been calculated, and their minimum values corresponding to some distance are employed for the fitting. The clusters might take various spin states, but we employ the spin state with the minimum energy. Figure 1 shows the binding energies of gold-gold and carbon-gold derived from DFT calculation and fitted by the potential functions of the bond order potential shown in Eqs. (2-9).

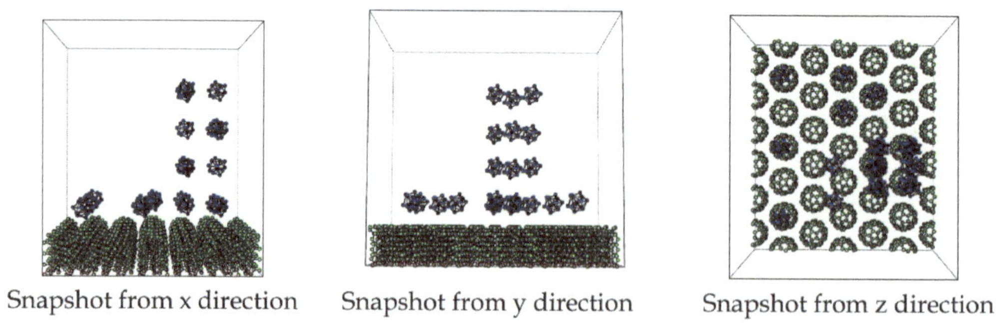

Fig. 3. Cluster configuration onto the flat CNT surface. The center of gravity of each cluster is above the center of the carbon nanotube or its hollow site. To express continuous deposition, some clusters are placed at distances of 10Å vertically.

$$E = V_R - V_A \tag{2}$$

$$V_R = \frac{D_e}{S-1}\exp\{-\beta\sqrt{2S}(r-R_e)\} \tag{3}$$

$$V_A = B^*\frac{D_eS}{S-1}\exp\{1+b(N^C-1)\} \tag{4}$$

$$D_e = D_{e1} + D_{e2}\exp\{-C_D(N^M-1)\} \tag{5}$$

$$R_e = R_{e1} - R_{e2}\exp\{-C_R(N^M-1)\} \tag{6}$$

$$f(r) = \begin{cases} 1 & (r < R_1) \\ 0.5\cdot\left(1+\cos\dfrac{r-R_1}{R_2-R_1}\pi\right) & (R_1 < r < R_2) \\ 0 & (R_2 < r) \end{cases} \tag{7}$$

$$N_i^C = 1 + \sum_{\text{carbon}} f(r_{ik}) \tag{8}$$

$$N_i^M = 1 + \sum_{\text{metal}} f(r_{ik}) \tag{9}$$

where r denotes the distance between two atoms and V_R and V_A are the Morse-type potential (Morse, 1929) repulsion and attraction terms, respectively. R_e represents the equilibrium bond length, and D_e is the potential depth at $r = R_e$. S means the ratio of the effective repulsive and attractive terms. N_i^C and N_i^M are the coordination numbers, which are derived by using the cut-off function of $f(r)$. All these parameters are shown in Table1.

2.2 Physical vapor deposition onto vertically aligned single-walled carbon nanotube

The vertically aligned single-walled carbon nanotube (VA-SWCNT) surface is expressed by a capped short carbon nanotube with a (5,5) chirality, as shown in Fig. 2. In this paper, the calculated system consists of 6 × 6 units with dimensions of 60Å × 52Å × 60Å, and their configuration resembles the FCC lattice (111). Atoms interact with each other within their cut-off distances. The first layer is a fixed layer, while the second and third layers are temperature-controlling layers. These layers maintain their temperature by velocity scaling. The system has periodic boundary conditions in each direction and the velocity Verlet algorithm is employed. Assuming a realistic metal evaporation in a vacuum chamber, physical vapor deposition (PVD) onto a VA-SWCNT surface is expressed by the collision of each M_{13} cluster with small evaporation energy (kinetic energy, K.E.) of 10.0 meV. Each metal cluster is fully annealed above its melting point (Ni: 1400 K, Au: 1000 K) and set as shown in Fig. 3. The VA-SWCNT is also annealed at 300 K. The melting point of the bulk expressed by these bond order potentials is higher than these temperatures, but the M_{13} cluster is very small and therefore its melting temperature is less than that of the bulk (Ni: 1728 K, Au: 1338 K) (Kittel, 2004).

2.3 Coating nanotube with metal

The dimensions of the calculated system are 60Å × 100Å × 100Å, and a periodic boundary condition is applied in each direction. As shown in Fig. 4, an isolated SWCNT is 60Å in length with (5,5) chirality; both its ends are fixed, and the next two layers works to control its temperature. Before coating this carbon nanotube with a metal, it is annealed at 300 K by velocity scaling technique. The molecules of the coating metal are expressed as M_{13} clusters; this expression is suitable for actual experiment because in the conventional metal evaporation process, the average metal particle size is approximately a few angstroms in diameter. This metal cluster is also fully annealed at 1200 – 1400 K; the annealing temperature is clearly higher than their melting temperature but less than their boiling temperature. The melting temperatures in the bulk form expressed by these potential parameters are approximately 1800 – 2000 K; however, in this paper, the metal clusters under study are too small to exhibit bulk properties. In order to achieve continuous metal evaporation coating, the fully annealed metal clusters are distributed at appropriate distances (15 – 20Å) and made to collide with the carbon nanotube at low evaporation energy of 10 meV or 30 meV .

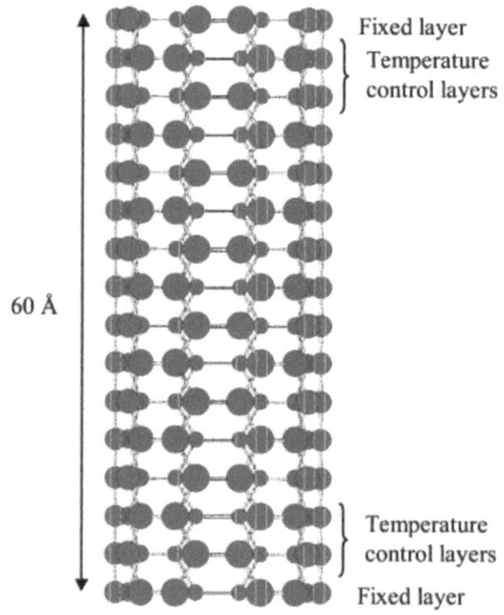

Fig. 4. Isolated single-walled carbon nanotube. The both ends are fixed layer and following two layers work to control the temperature.

3. Results and discussion

3.1 PVD on VA-SWCNT surface

The initial configuration of the metal clusters is above mentioned and shown in Fig. 3. Each cluster is placed just above the carbon nanotube, or its hollow site for expressing various collision models. In order to express continuous deposition, the clusters are placed at distances of 10Å. Figure 5 shows the result of PVD onto a VA-SWNTs surface. In both cases, some of the metal clusters collided with each other before arriving at the surface and formed a larger cluster, while others arrived at the surface without clustering. In the case of nickel deposition, irrespective of clustering, the deposited nickel atoms spread over the VASWCNT surface and formed a smooth surface; however, in the case of gold clusters, small clusters of gold could spread over the VA-SWCNT surface, but large clusters could not, and they formed a grain-like structure. On comparing the potential curves of both metals, the equilibrium bond length is slightly different, but the binding energy appears to be almost similar. The point at which they differ the most is the potential depth at which their coordination number is one. The potential depth of gold suddenly decreases as the coordination number increases. This could be a reason for the formation of the grain-like structure, because the contact angle is closely related to the wettability, which is well explained by solid-liquid affinity ($\varepsilon_{sl}/\varepsilon_l$) (Maruyama et al., 1998).

(a) Deposition of nickel clusters.

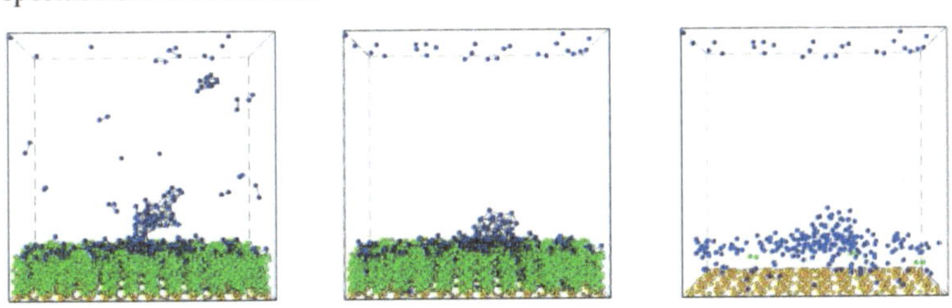

(b) Deposition of gold clusters.

Fig. 5. Metal PVD onto VA-SCNT surface. In both cases, small metal clusters collide with each other before arriving at the VA-SWCNT surface and form large clusters.

The carbon atoms except for those in the fixed layer are hidden in the third figure to identify the metal atoms.

According to their results the contact angle tends to become larger as the solid-liquid affinity becomes smaller. In the condensed phase the coordination number should be larger, so that the affinity becomes smaller comparing to that of nickel. The electron configuration of an isolated gold atom is [4f ^{14}, 5 d^{10}, 6 s^{1}] and that of a nickel atom is [3d ^{8}, 4 s^{2}]. Therefore, it is speculated that the number of electrons that greatly contribute to binding should be one for gold and more than two for nickel. The order of energy levels of 6s, 4f, and 5d are this order but they are not so far with each other; thus sometimes this order can be reversal for taking a closed shell. In case of gold the 4f-orbit and the 5d-orbit take a closed shell as they are; therefore, lone electron is one. Thus, the gold atoms have a strong binding energy in the form of a dimmer. On the other hand, the nickel atom has some free electrons, so that their binding energy is smaller than that of the gold atom. Compared to nickel deposition, gold deposition may require stringent conditions to form a flat surface; however, this can be achieved by carrying out deposition under a high vacuum condition and at a low deposition rate. Small clusters of gold can spread over the VA-SWCNT surface, while some of the gold clusters collide with each other before arriving at the surface and form a grain-like structure; it is thus essential to avoid clustering.

3.2 Coating

Figure 6 shows the simulation result of continuous metal evaporation coating ((a) Ti, (b) Fe, (c) Ni, (d) Au). In this simulation, $24M_{13}$ clusters were gently collided with one SWCNT. The impact direction was perpendicular to the axis of the carbon nanotube, and at each impact point, about four clusters collided simultaneously. In case of titanium and iron clusters (Fig. 6(a), (b)), the metal atoms seemed to firmly combine with carbon atoms; thus, the metal atoms entered the carbon nanotube and distorted its structure. The strong interaction between the metal atoms and carbon caused the structural deformation of the nanotube; moreover, the continuous impact in a short time for reducing the calculation load (10 ps /layer; usually $1 - 250$ ms /layer in evaporation method for pure gold in experiment) resulted in unrealistic phenomena. In the case of nickel clusters (Fig. 6(c)), the nickel atoms smoothly and seamlessly covered the carbon nanotube. The equilibrium position of the nickel atoms coincided with the center of the hexagonal carbon network. When the first nickel cluster collided with the carbon nanotube, a grain structure was temporarily formed; however, the nickel atoms immediately found a stable position and moved into it. Once they reached the equilibrium position, they remained stationary at this temperature. Next, the second cluster reached the surface, acquired a grain structure instantly, and then moved to the most stable site. When all the clusters reached the surface, the equilibrium position on the carbon network was saturated; therefore, the nickel clusters formed a second metal layer on the carbon nanotube. Even at this stage, the nickel atoms did not form a grain structure but formed a smooth and seamless layer.

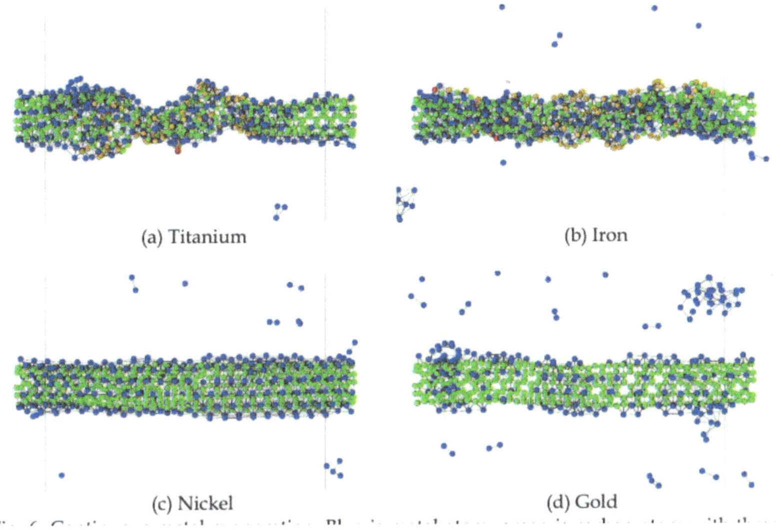

Fig. 6. Continuous metal evaporation. Blue is metal atom, green is carbon atom with three bonds (sp^2), orange is carbon atom with two bonds (one dangling bond), and red is carbon atom with one bond (two dangling bonds).

Figure 6(d) shows the result of gold cluster coating. According to Zhang et al., gold atoms form a grain structure and exhibit a highly discontinuous coating; however, according to our simulation, the gold atoms can be coated smoothly but discontinuously. The gold atoms initially formed a grain structure, as reported by Zhang et al.; this was because the gold clusters were accidentally concentrated in the same area on the carbon nanotube. This indicates that if the evaporation rate decreases or the experiment is conducted under a higher vacuum condition, the formation of a grain structure can be avoided. Zhang et al. explained this result in terms of the weak interaction between the gold atoms and the SWCNT. However, according to our DFT calculations, the binding energy of the Au-C bond is not low; on the contrary, it is slightly larger than the binding energy of the Au-Au bond. The value of the binding energy is not considered to be significant; however, the ratio of the binding energies, i.e., $\varepsilon_M - c/\varepsilon_M - M$, should be substantial. Because these condition each atom has many coordination number, the values of this ratio for titanium, iron, nickel, and gold are 2,3,1.5 (approximately), and 1 (approximately), respectively. Thus, in the case of titanium and iron coating, the carbon atoms experience an outward pull and the nanotube structure gets deformed. In the case of nickel, a very smooth coating is achieved. On the other hand, in the case of gold, though the Au-C interaction is slightly stronger than the Au-Au interaction, a grain structure is formed. In some areas, the gold atoms are deposited smoothly but not seamlessly. The reason for this phenomenon is attributed to the equilibrium bond length. As mentioned above, the most stable site for the first layer metal atoms is the center of the hexagonal carbon network, of which diameter is 2.5Å.

Coincidently, the equilibrium bond lengths of nickel and iron are comparable, but that of gold is slightly greater (2.6-3.3 Å, depending on the coordination number). The gold atoms must have a shorter bond length so that they can get accommodated in the stable site of the carbon nanotube, thereby creating a discontinuous coating that reduces the distortion energy. Thus, even gold atoms can be coated smoothly, provided the distances between them is sufficient to reduce the distortion for each fragment of coating (in our simulation, the fragment of coatings is approximately $10 - 15$ nm in length). Owing to the strong C-Ti and C-Fe interactions, the structure of the carbon nanotube is distorted under a high coating rate in our simulation. In order to confirm this, a lower coating rate is applied and the results are examined. In this case, each M_{13} cluster collides with the center of the carbon nanotube in a sequence. For the fully annealing, this sequential impact took an enough interval. Figure 7 shows this sequential impact of the metal clusters on the carbon nanotube. A full annealing duration for each impact step results in a smooth and continuous coating and no defects are observed in the carbon nanotube. A schematic diagram of the iron cluster impact is shown in Fig. 3(a). Similar to the nickel atom, the iron atom finds a stable position and moves into it. However, in this simulation, because each cluster reaches almost the same position, all the iron atoms do not spread over the surface but form a second layer. This can occur in the nickel system as well. When too many clusters arrive at the same point, each atom seeks the stable site and tends to move into it, but the temperature reduces before the diffusion is completed. As a result, the iron atoms form a second layer. A similar behavior is observed in titanium coating, as shown in Fig. 7 (b). However, in the case of titanium clusters, when the impact energy is small, a different phenomenon is observed. In our simulation, first three clusters were able to reach the surface even at a low evaporation energy, but the next cluster could notreach. This can be attributed to the difference in the equilibrium bond lengths. According to Eq. 6, if the number of coordinates is sufficient, the equilibrium bond length must be equal to R_{e1} for the metal-metal potential and the metal-carbon equilibrium bond length is simply determined as a function of R_e (Table 1). A comparison reveals that the difference in this value for the titanium coating system are larger than those for the iron coating system. This is because if there is a titanium layer on the carbon nanotube, the cluster having a small impact energy cannot overcome the repulsive energy. Thus, the cluster is reflected before it is attracted to the carbon atoms.

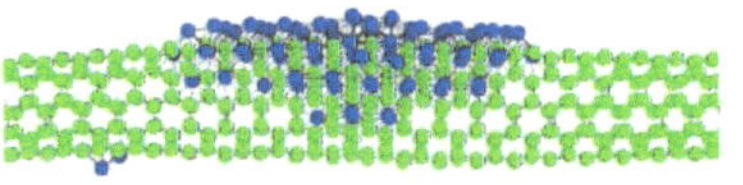

(a) Sequential Fe coating. Totally six Fe_{13} clusters collide.

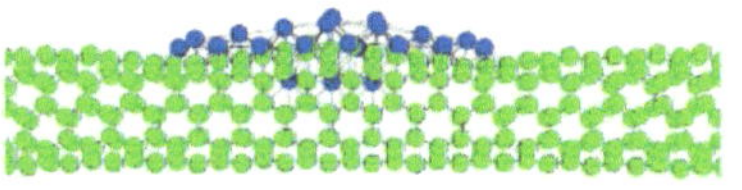

(b) Sequential Ti coating. Totally three Ti_{13} clusters collide.

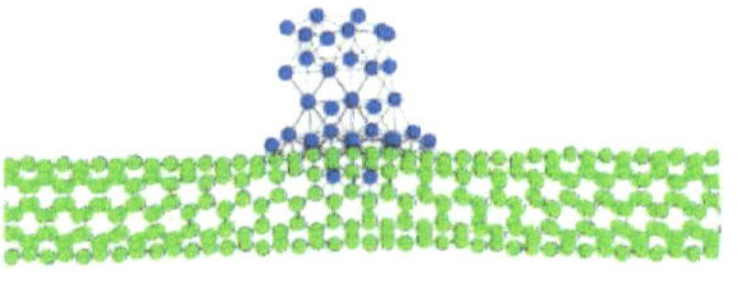

(c) Sequential Au coating. Totally three Au_{13} clusters collide.

Fig. 7. Sequential impacts of Fe, Ti, and Au cluster on the carbon nanotube surroundings with enough annealing duration after every impact.

This proves that at a high evaporation energy (e.g., 1 eV), the titanium cluster can arrive at the surface and produce a smooth coating. This might be a reason why the titanium coating is smoother than the nickel coating in the experimental results. [15] In the case of gold coating, as shown in Fig. 6(c), the first cluster was able to spread over the surface but the second and the third could not; therefore, they formed a grain structure. This indicates that if the gold cluster reaches the same place, as in the case of the experiment with a high deposition rate, the grain formation is not avoidable, as inferred from Fig. 7(d). From the point of wettability the contact angle is closely related to the wettability, which is mentioned above section explained by solid-liquid affinity ($\varepsilon_{sl}/\varepsilon_l$). According to their results the contact angle tends to become larger as the solid-liquid affinity becomes smaller that means the gold cluster tends to form a grain structure.

3.3 Physical properties

Figure 8 shows the stress-strain (force-strain) curves of coated and uncoated SWCNT. In this simulation one end of SWCNT was fixed and the other end was pulled slowly enough to be annealed that avoid unintentionally breaking of SWCNT. Owing to the disturbance of perfect sp^2 bonding of SWCNT, it might be easily expected that the physical strength becomes much weaker, but the decrease is not so large; or rather, the force constant becomes larger. The rupture is induced by metal atom that tends to take the bonds from the carbon. Because the bond length of carbon-metal is longer than that of carbon-carbon, the interaction between carbons becomes suddenly weaker than that of carbon-metal when the SWCNT pulled.

Figure 9 indicates the thermal diffusivity of coated and uncoated SWCNT. The thermal diffusivity is derived by fitting the one-dimensional non-Fourier heat conduction equation with two relaxation timescales (τ_1, τ_2) shown below equation.

$$\tau_1 \frac{\partial^2 T}{\partial t^2} + \frac{\partial T}{\partial t} = \alpha \left(\nabla^2 T + \tau_2 \frac{\partial}{\partial t} \nabla^2 T \right) \tag{10}$$

The dominant heat carrier in SWCNT is phonon and this equation model is fully described in other's paper (Shiomi & Maruyama, 2006).

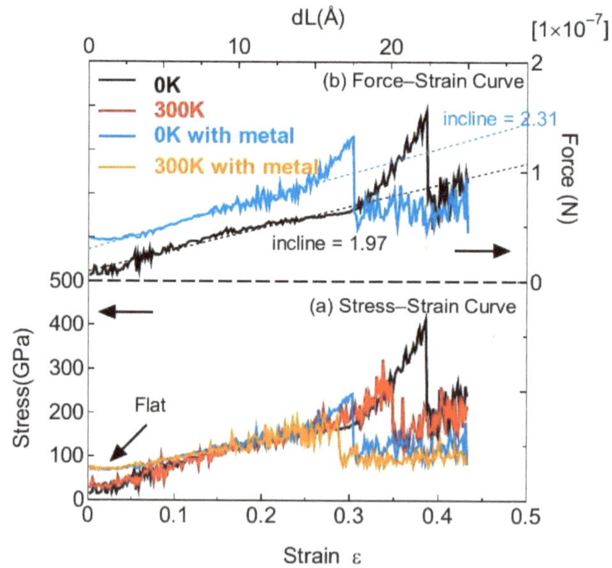

Fig. 8. The stress-strain curve (a) and force-strain curve (b). The stress is defined by assuming their cross section with 6.93Å for the uncoated SWCNT and 8.32Å for the metalcoated SWCNT in diameter. The metal-coated SWCNT meets earlier rupture point but has a larger force constant. The metal-coated SWCNT has a residual stress that makes the inclination flat in a small strain (displacement) range.

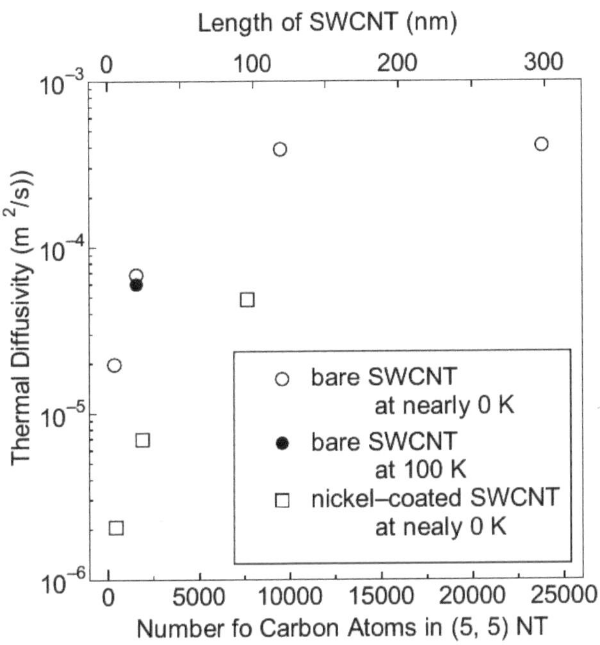

Fig. 9. Thermal diffusivity estimated in this study. The thermal diffusivity of the nickelcoated SWCNT decreases by 90%. The estimated thermal conductivity (λ) is expressed as follows: $\lambda = 1 \times 10^6\alpha - 3 \times 10^6\alpha$. It depends on the definition of the density and specific heat capacity (C_v) of the SWCNTs is assumed to be $3R$, where R is a gas constant.

4. Conclusion

In this chapter MD approach for CNT analysis was explained. The classical MD merely solves the Newtnian equation, but if the potential function is fully appropriate, the results must be convincive. In case of PVD and coating this simulation shows good agreement with the experimental results; on the other hand, for the physical properties, the adequate results are given and also the mechanism should become clear.

5. References

Beche AD. (1993) Density-Functional Thermochemistry .3. The Role of Exact Exchange. J. Chem. Phys. 98, 1993, 5648-5652.

Castro M, et al. (1997) Structure, bonding, and magnetism of small Fe-n, Co-n, and Ni-n, clusters, n <= 5. Chem. Phys. Lett. 271 1997, 133-142.

Hay PJ & Wadt WR. (1985) Abinitio Effective Core Potentials for Molecular Calculations Potentials for the Transition-Metal Atoms Sc to Hg. J. Chem. Phys. 82, 1985, 270-283.

Ishikawa K. et al. (2007) Extended abstracts, ASME-JSME Thermal Eng., (2007) HT200732783.

Kittel C. (2004)Introduction to Solid State Physics. John Wiley & Sons Inc. 2004.

Lee CT, et al. (1998) G. Development of the Colle-Salvetti Correlation-Energy Formula Into a Functional of the Electron-Density. Phys. Rev. B 37, 1998, 785-789.

Maruyama S. et al. Liquid droplet in contact with a solid surface. Micro Thermophys. Eng. 2, 1998, 49-62.

Morse PM. (1929) Diatomic Molecules According to the Wave Mechanics. II. Vibrational Levels. Phys. Rev. 34, 1929, 57-64.

Shibuta Y, Maruyama S. (2007) Bond-order potential for transition metal carbide cluster for the growth simulation of a single-walled carbon nanotube. Comput. Mat. Sci. 39, 2007, 842-848.

Shiomi J & Maruyama S. (2006) Non-Fourier heat conduction in a single-walled carbon nanotube: Classical molecular dynamics simulations. , 273 2006, 205420.

Zhang Y. et al. Phys. Lett., 331 (2000) 35.

Chapter-17

Carbon Nanotube Field Emitters

1. Introduction

Application of various one-dimensional nanostructure materials as field emission sources has attracted extensive scientific efforts. Elongated structures are suitable for achieving high field-emission-current density at a low electric field because of their high aspect ratio. Area of its application includes a wide range of field-emission-based devices such as flat-panel displays, electron microscopes, vacuum microwave amplifiers, X-ray tube sources, cathoderay lamps, nanolithography systems, gas detectors, mass spectrometers etc.

Since the discovery of carbon nanotubes (CNTs) (Iijima, 1991; Iijima & Ichihashi, 1993; Bethune et al., 1993) and experimental observations of their remarkable field emission characteristics (Rinzler et al., 1995; de Heer et al., 1995; Chernazatonskii et al., 1995), significant efforts have been devoted to the application of using CNTs for electron sources. One of the main problems for design such field emission emitter is the difficulties in estimation of the electric field on the apex of nanotubes. Only a few works considered forces acting on nanoemitters under electric field. Thus far, there is no analytical formula which provides a good approximation to the total current generated by the nanoscale field emitter. In this chapter, we theoretically consider the electric field strength, field enhancement factor, ponderomotive forces, and total current of a metallic elliptical needle in the form of hemiellipsoid in the presence of a flat anode. Also we shortly review the history CNT cold emitters and technology of their fabrication. Furthermore we consider the application areas of CNT electron sources.

2. Historical preview

Field emission is an emission of electrons from a solid surface under action of external high electric field E. Field emission was experimentally discovered in 1987 by R.W. Wood (Wood, 1897). In 1929 R.A. Millikan and C.C. Lauritsen established linear dependence of the logarithm of current density on $1/E$ (Millikan & Lauritsen, 1929). Field emission was explained by quantum tunneling of electrons through the surface potential barrier. This theory was developed by R.H. Fowler and L.W. Nordheim in 1928 (Fowler & Nordheim, 1928).

According to the Fowler-Nordheim theory, the current density of the field emission j is determined by the following expression

$$j \approx \frac{C_1 E^2}{\phi} \exp\left(-\frac{C_2 \phi^{3/2}}{E}\right) \tag{1}$$

where j denotes the emission current density in Acm^{-2}, E is local electric field at the emitting surface in Vcm^{-1}, ϕ is work function in eV , and the first and second Fowler-Nordheim constants are $C_1 = 1.56 \times 10^{-6} \mathrm{AeVV}^{-2}$, $C_2 = 6.83 \times 10^7 \mathrm{VeV}^{-3/2}\ \mathrm{cm}^{-1}$, respectively. The electric field E at the CNT tip increases compared with the average field E_0. Substituting in Eq. (1) the expression $E = \beta E_0$, where β is a field enhancement factor, we shall write the FowlerNordheim dependence in the form:

$$j \approx \frac{C_1 (\beta E_0)^2}{\phi} \exp\left(-\frac{C_2 \phi^{3/2}}{\beta E_0}\right) \tag{2}$$

Thus current-voltage characteristics of the field electron emission in the Fowler-Nordheim coordinates ($\log j/E_0^2, 1/E_0$) are presented by straight lines. It was assumed that $\varphi = 4.8\mathrm{eV}$ for nanotubes. The field enhancement factor β varied from 300 to 3000 depending on the tube size.

Theory of field emission considered in details in recent books (Fursey, 2005; Ducastelle et al., 2006).

Strong electric fields ($E \sim 10^7 \mathrm{Vcm}^{-1}$) near a surface are necessary to obtain the appreciable field emission current from pure metals. Therefore, the emitters in early investigations were produced in the form of thin spike with radiuses of curvature on the ends about 1 micron.

Development of lithographic techniques allowed fabricating so called "Spindt tips" in which the field emitters are small sharp molybdenum microcones. One of the first papers describing such technology has appeared in 1968 (Spindt, 1968). Essential efforts have been spent by several companies for development of the Spind-type field emission display, but no large-screen production has been forthcoming.

The new potential in designing field emitters and devices on their basis has appeared after discovery of carbon nanotubes.

Field emission of carbon nanotubes was for the first time reported by Fishbine (Phillips Lab.) (Fishbine et al., 1994), Gulyaev (Institute of Radio-engineering and Electronics, Russia) (Gulyaev et al., 1994), and Rinzler (Rice University) (Rinzler et al., 1994) in 1994.

Four first journal papers (Gulyaev et al., 1995; Chernozatonskii et al., 1995; Rinzler et al., 1995; de Heer et al., 1995) dedicated to this problem were published in 1995. As is known, several papers have appeared in the two subsequent years: two works (Chernozatonskii et al., 1996; Collins & Zettl, 1996) were published in 1996 and seven works (Collins & Zettl, 1997; Gulyaev et al., 1997; Sinitsyn et al., 1997; de Heer et al., 1997; Saito et al., 1997a; Saito et al., 1997b; Lee et al., 1997) were published in 1997. Starting from 1998, interest in fieldemission properties of CNT was increasing explosively all over the world. Today we can speak of thousands of published papers.

Recently, field emission from metals (Lee et al., 2002), metal oxides (Li et al., 2006; Banerjee et al., 2004; Jo et al., 2003; Seelaboyina et al., 2006), metal carbides (Charbonnier et al., 2001), and other elongated nanostructures have also been explored. It is now possible to control the diameter, height, radius of curvature of the tip, and basic form of emitters during growth. Elongated structures of different shapes such as nanotubes, nanocones, nanofibers, nanowires, nanoneedles, and nanorods have been successfully grown (Li et al., 2006; Li et al., 2007; Jang et al., 2005; Hu & Huang, 2003).

Promising new materials for field-emission sources are B- and N-doped CNTs. Terrones et al. (Terrones et al., 2004; Terrones et al., 2008) have reviewed the field emission properties of B- and N-doped CNTs and nanofibres. B-doped multi-wall CNTs could exhibit enhanced field emission (turn on voltages of $\sim$ 1.4 V/μm) when compared to pristine multi-wall CNTs (turn on voltages of $\sim$ 3 V/μm). N-doped CNTs are able to emit electrons at relatively low turn-on voltages (2 V/μm). This phenomenon arises from the presence of *B* atoms (holes) or N atoms (donors) at the nanotube tips.

3. Carbon nanotubes field emitters

3.1 Physical properties of carbon nanotubes suitable for cold emission

From the practical application point of view CNTs are preferable field emitters due to their low threshold voltage, good emission stability and long emitter lifetime.

CNTs possess these advantages due to the large aspect ratio, high electric and thermal conductivity, highest flexibility, elasticity, and Young's modulus. Their strong covalent bonding makes them chemically inert to poisoning and physically inert to sputtering during field emission. They can also carry a very high current density of order 10 ^{9}Acm^{-2} before electromigration. Nanotubes have a high melting point and preserve their high aspect ratio over time. CNTs emits electrons under conditions of technical vacuum. They are chemically inert to poisoning due to strong covalent bonding. Measuring of field emission properties (Kung et al., 2002) and theoretical ab-initio calculations (Park et al., 2001) shows that emission currents are significantly enhanced when oxygen is adsorbed at the tip of carbon nanotubes.

3.2 Manufacturing techniques for CNT-based field-emission cathodes

Many technologies for fabrication of CNT-based field-emission cathodes were offered. We shall consider only some of them.

Individual CNT field emitters have a large potential for application in electron guns for scanning electron microscopes. To investigate the emission properties of individual CNTs de Jonge et al (de Jonge & Bonard, 2004) improved the mounting method using a piezodriven nanomanipulator. For the mounting of an individual CNT on a tungsten tip, a tungsten wire was fixed by laser-welding on a titanium (or tungsten) filament.

Field emission CNT-based cathodes are manufactured either as a bulk solid containing nanotubes or as a film with thickness from hundreds of nanometers to tens of microns.

Bulk cathodes are known to be manufactured by two methods. The Alex Zettl team from California University, Berkley, USA used a technology in accordance with which the ready material of unsorted randomly aligned nanotubes is mixed into a compound, baked, and surface ground. Flexible and elastic nanotubes are not broken during the grinding. In accordance with the technology used by the Yahachi Saito team of Mie University (Japan), the graphite-electrode material processed by an electric arc is cut to pellets and glued to a stainless-steel plate by silver paste.

Film technologies are used in all other cases. Film cathodes are basically manufactured by two methods: either preliminary synthesized tubes are attached to a substrate or the tubes are grown directly on the substrate.

In the two methods, different technologies yield films of both well oriented and strongly entangled tubes.

N.I. Sinitsyn group from Institute of Radio-engineering and Electronics (Saratov, Russia) used CVD methods for synthesizing films of both regularly grown nanotubes (Fig. 1) and "felt" of entangled fibers. Strips were obtained using a catalyst deposited through a template (Zhbanov et al., 2004).

Fig. 1. Photograph of strips of oriented CNTs synthesized on a substrate. The strip width is 20μ m and the gap between the strips is 5μ m (Zhbanov et al., 2004).

The Jean-Marc Bonard team of Lousanne Polytechnical School (Switzerland) developed the technology of microcontact printing of catalytic precursor for growing oriented tubes arranged in accordance with a specified pattern on a substrate (Bonard et al., 2001b). The catalyst, the so-called "ink", was applied to the stamp surface. The ink was a solution containing from 1 to 50 mM of $Fe(NO_3)_3 \cdot 9H_2O$. The duration of contact during the printing was 3 s . Nanotube deposition was by the CVD method in a standard flow reactor at a temperature of 720°C.

In the case of low concentration of catalyst (1 mM , Fig. 2a), several single nanotubes are randomly distributed over the printing region. The catalyst-concentration growth is accompanied by formation of films of entangled tubes, as is shown in Figs. 2*b* and 2*c*. For concentrations about 50 mM , clusters of nanotubes oriented normally to the surface are formed. Figure 2*d* shows that the sides of the walls are flat, and not a single tube is hanging outward. For concentrations above 60 mM , growth of nanotubes is retarded, and the printed template is covered by amorphous carbon particles.

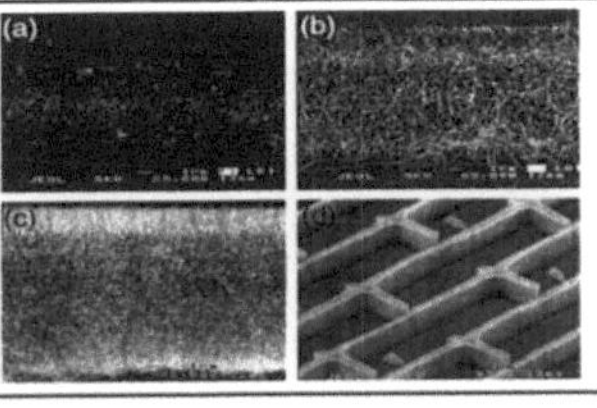

Fig. 2. Nanotube growth for various concentrations of catalytic ink used for the precursor application. Catalyst concentration in the solution was 1 mM (a), 5 mM (b), 40 mM (c), and 50 mM (d). The figure is taken from (Bonard et al., 2001b).

Hongjie Dai team of Stanford University, USA, used the following technology for obtaining arrays of well-oriented carbon nanotubes. First, porous silicon was formed on the surface of a silicon substrate by anode etching and then the ferrum film was deposited on the latter through the shadow mask by electron-beam evaporation (Fan et al., 1999). Then nanotubes were grown as a result of acetylene decomposition in argon flow at 700 °C. E. F.Kukovitsky team of the Kazan Physics-Technical Institute (Russia) developed the technology of synthesis of oriented nanotubes with conical layers (Fig. 3) (Musatov et al., 2001; Kukovitsky et al., 2003). The first stage of the process involves polyethylene pyrolysis in the first oven at a temperature of 600 °C. Then, by the helium flow, the gaseous products of pyrolysis are transferred to the second oven where nanotubes grow on the nickel foil catalyst at a temperature of 800 to 900 °C. For the obtained specimens, the current density was 10 mA/cm^2 for the electric field from 4 to 4.5 V/μm.

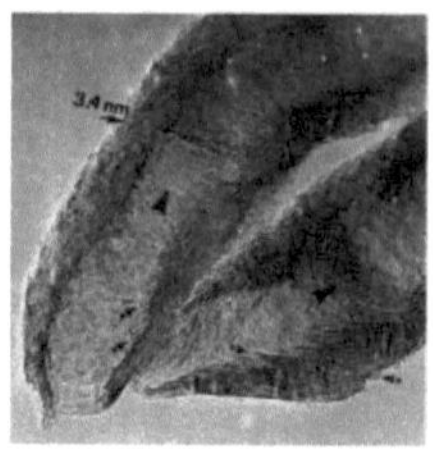

Fig. 3. High-resolution electron microscope image of nanotubes with conical layers. Graphene layers are marked by arrows with points, and the CNT growth direction is marked by large arrows (Musatov et al., 2001).

As is obvious from the literature analysis, almost all CNT-based cathodes show high emission irrespective of the fact whether the tubes are multi-wall or single-wall, welloriented or entangled. Bamboo-shaped aligned carbon nanotubes (Srivastava et al., 2006; Ghosh et al., 2008) as well as carbon nanocones (Yudasaka et al., 2008) demonstrate high field electron emission.

Let's note, that not only elongated carbon nanotubes, but also pyramids from fullerenes are used as cold cathodes. Formation and characteristics of fullerene coatings on the surface of tungsten tip field emitters and emitters with ribbed crystals formed on their surface are studied by group of Sominskii from St. Petersburg State Technical University (Russia) (Tumareva et al., 2002; Tumareva et al., 2008). Methods of creating microprotusions on the surface of the coatings that considerably enhance the electric field have been developed and tested. Emitters with a single microprotrusion demonstrated emission current densities up to $10^6 - 10^7$ A/cm^2. It was shown that single micron-sized emitters can stably operate at currents up to 100μ A.

3.3 Electric field and field enhancement factor in diode configuration

The field enhancement factor is very important parameter for characterization of CNT emitters.

The model of a hemisphere on a post for CNT emitters is widely used in analytical approximations and numerical simulations (Fig. 4). To calculate the electric-field intensity and the field enhancement factor on the nanotube tips, the following assumptions are usually done:

1. Nanotubes are regularly located on a flat substrate in a "honeycomb-like" order. A nanotube is a cylinder with height h and diameter 2ρ capped by a hemisphere of ρ radius. Total height of closed nanotube is H, the distance from cathode to anode is L, the gap between anode and nanotube tip is l, and the distance between the nearest neighbors is D.
2. A nanotube obeys the laws of continuous medium, is perfectly conducting, and the cathode potential is maintained on its entire surface.

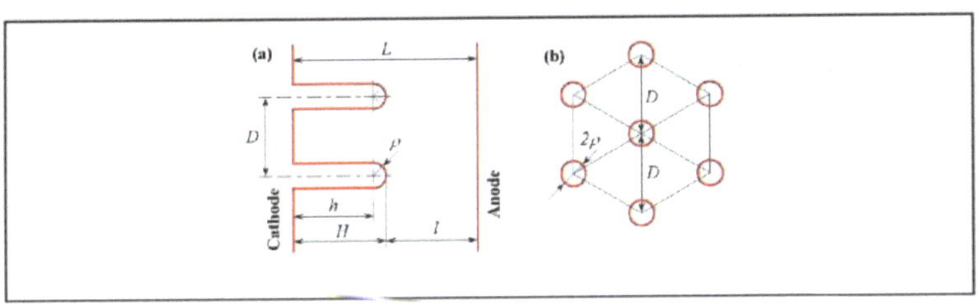

Fig. 4. Scheme of aligned nanotube film, the model of a hemisphere on a post: (a) side view; (b) top view.

Let us introduce dimensionless parameters for the geometrical characterization of model. The dimensionless height of emitter, the dimensionless gap between anode and emitter tip, and the dimensionless distance between individual emitter are the following:

$$\eta = \frac{\rho}{h}, \lambda = \frac{\rho}{l}, \delta = \frac{\rho}{D}. \quad (3)$$

Until now the analytical solution for the model of a hemisphere on a post is unknown. There is no even a solution for the individual cylindrical nanotube closed by hemispherical cap in a uniform electric field.

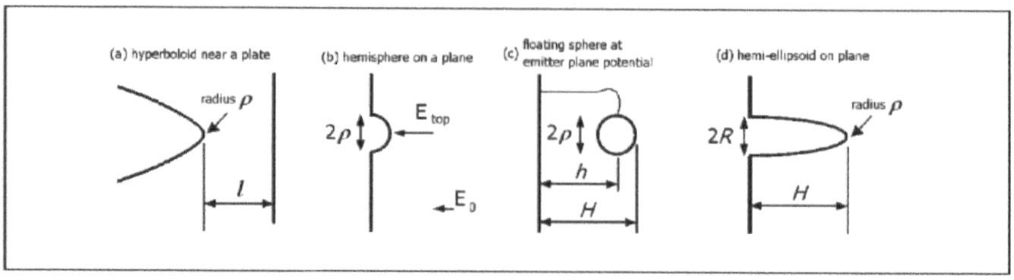

Fig. 5. Schemes of simplest models for field enhancement factor estimation: (a) hyperboloid near a plate; (b) hemisphere on a plane; (c) floating sphere at emitter-plane potential, and (d) hemi-ellipsoid on a plane.

Numerical simulations were reported in many papers (Edgcombe & Valdrè, 2001; Edgcombe & Valdrè, 2002; Read & Bowring, 2004; Musatov et al., 2001). Calculation difficulties in these numerical methods arise due to the large nanotube aspect ratio and very long distance between cathode and anode in comparison with emitter height. Usually, these numerical results were generalized and simple fitting formulas of field enhancement factor for individual nanotube (Edgcombe & Valdrè, 2001; Edgcombe & Valdrè, 2002; Read & Bowring, 2004; Shang et al., 2007), for nanotube in space between parallel cathode and anode planes (Bonard et al., 2002a; Filip et al., 2001; Nilsson et al, 2002; Smith et al., 2005), and for a nanotube surrounded by neighboring nanotubes with a screening effect (Jo et al., 2003; Glukhova et al., 2003; Nilsson et al., 2000; Read & Bowring, 2004; Wang et al., 2005) were suggested. The main problem for such algebraic fitting formulas is the lack of a definitive proof of their accuracy.

Four of the simplest models are the "hyperboloid near a plate" model, the "hemisphere on a plane" model, the "floating sphere at emitter-plane potential" model, and the "hemiellipsoid on plane" model. We follow to the classification suggested by Forbes et al. (Forbes et al., 2003).

These models allows analytical solutions, they are illustrated in Fig.5. We will use dimensionless parameters Eq. (3) to define geometry of these models.

3.3.1 Hyperboloid near a plate model

We introduce the prolate spheroidal coordinates σ and τ to consider the model of a hyperboloid near a plate (Fig. 6).

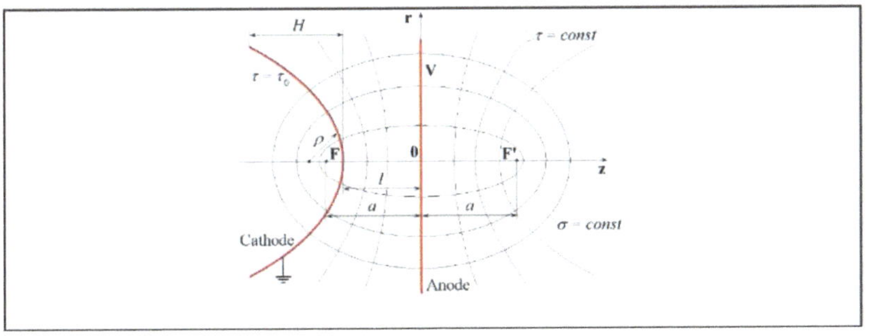

Fig. 6. Hyperboloid near a plate in prolate spheroidal coordinates.

The equation of prolate spheroid is:

$$\frac{r^2}{a^2(\sigma^2-1)}+\frac{z^2}{a^2\sigma^2}=1; \sigma \geq 1. \tag{4}$$

The equation of hyperboloid of two sheets is:

$$\frac{r^2}{a^2(\tau^2-1)}+\frac{z^2}{a^2\tau^2}=1; 1 \geq \tau \geq -1 \tag{5}$$

Points $F(0;-a)$ and $F'(0;a)$ are the foci of the hyperboloids and spheroids. The cathode represents a hyperboloid of revolution $\tau_0=\text{const}$ and the anode is a plane $\tau=0$. They are show in Fig. 6 by solid red lines. The radius of hyperboloid curvature of the tip is ρ. The electric field is calculated according to the formula:

$$E=\frac{V}{a\sqrt{(\sigma^2-\tau^2)(1-\tau^2)}\operatorname{arctanh}(l/a)}, \tag{6}$$

where function arctanh is inverse hyperbolic tangent, V is the voltage applied across a gap between anode and cathode.

The model of a hyperboloid near a plate is suitable to describe the interaction of individual CNT field emitter with surface in scanning electron microscopes. Usually in cases important for practice we have $\rho \ll l$ and $l \ll H$.

If $\rho \ll l$ the maximal value of the module of intensity is approximated by the formula (Drechsler & Müller, 1953):

$$E_{top} \approx \frac{V}{\rho \ln(4l/\rho)} \tag{7}$$

If we define the macroscopic field by $E_0 = V/l$ then we can write the field enhancement factor

$$\beta = \frac{\sqrt{1+\lambda}}{\lambda \cdot \mathrm{arctanh}\sqrt{1/(1+\lambda)}} \tag{8}$$

Let us estimate the electric force acting on the surface of ellipsoid. The electrostatic force acting on the elementary area, s of the external surface is given by

$$\vec{F} = \int_S \frac{\varepsilon_0}{2} E^2 \vec{n} ds \tag{9}$$

where ε_0 is the electric constant, $\vec{n}$ is a vector normal to the surface.

Taking into account that the infinitesimal surface element is $ds = 2\pi a^2 \sqrt{(1-\tau_0^2)(\sigma^2-\tau_0^2)} d\sigma$, we can analytically integrate the force acting on the top of a hyperboloid surface of height H (see Fig. 6). It is clear that r-component of force equals to zero, $F_r = 0$. For the z -component we have

$$F_z(H) = \frac{\pi \varepsilon_0 V^2}{2\mathrm{arctanh}^2 \sqrt{1/(1+\lambda)}} \ln\left(1 + \frac{2H}{\rho} + \frac{2H}{l} + \frac{H^2}{l\rho} + \frac{H^2}{l^2}\right) \tag{10}$$

The total current is calculated by integration of current density from Eq. (1) over a hyperboloid surface

$$J_{\text{total}} \approx \int jds \tag{11}$$

We note that the exchange and correlation effect is ignored in the basic equation (1). Thus the Fowler-Nordheim theory is suitable only for approximate calculations. Nevertheless this theory is widely used for analysis of field emission current from elongated nanostructures. After substitution of field distribution over the sphere surface and an infinitesimal surface element we have

$$\int jds = \frac{2\pi C_1 V^2}{\phi\sqrt{1-\tau_0^2}\mathrm{arctanh}^2\tau_0} \int_1^\infty \frac{\exp\left(-C_2 V^{-1} \phi^{3/2} a \sqrt{(1-\tau_0^2)(\sigma^2-\tau_0^2)}\mathrm{arctanh}\tau_0\right)}{\sqrt{\sigma^2-\tau_0^2}} d\sigma \tag{12}$$

If $\rho \ll l$ then $\tau_0 \approx 1$ and $\sqrt{\sigma^2 - \tau_0^2} \approx \sqrt{\sigma^2 - 1} \approx \sqrt{2(\sigma-1)}$. This approximation allows us to reduce our integral to another one $\int_1^\infty \frac{\exp(-a\sqrt{x-1})}{\sqrt{x-1}} dx = \frac{2}{a}$.

Thus the total field emission current is

$$J_{total} \approx \frac{2\pi C_1 V^3 \sqrt{1+\lambda}}{C_2 \phi^{5/2} \rho \operatorname{arctanh}^3 \sqrt{1/(1+\lambda)}} \tag{13}$$

where the total current, J_{total} is measured in A; the radius of curvature, ρ is measured in cm .

3.3.2 Hemisphere on a plane

The metallic sphere in a uniform electric field E_0 (Fig.5(b)) was considered in many papers (for example Refs. (Forbes et al., 2003; Wang et al., 2004; Pogorelov et al., 2009)). We can replace the sphere by point electric dipole. If the electric dipole moment is p_0 then the dipole potential is

$$\varphi_{dip} = -\frac{p_0}{4\pi\varepsilon_0} \frac{z}{(z^2 + r^2)^{3/2}} \tag{14}$$

Equation of circle is $\varphi_{\text{dip}} + zE_0 = 0$. From this equation we can find the relation between the electric dipole moment and the sphere radius: $p_0 = 4\pi\varepsilon_0 E_0 \rho^3$. The electric field on the top of hemisphere reaches $E_{\text{top}} = p_0/2\pi\varepsilon_0\rho^3 + E_0 = 3E_0$. The field enhancement factor is $\beta = E_{\text{top}}/E_0 = 3$. The field distribution over the sphere surface have the form $E = 3E_0\cos\,\theta$, where θ is polar angle.

Pogorelov et al. (Pogorelov et al., 2009) have shown that the total current emitted from the hemisphere surface is

$$J_{\text{total}} = \frac{2\pi\rho^2 C_1 \phi^{7/2} C_2^3}{3E_0} \left[\frac{E_1(C_3)}{6} + \left(\frac{1}{3C_3^3} - \frac{1}{6C_3^2} + \frac{1}{6C_3} \right) \exp\left(-C_3\right) \right], \tag{15}$$

where $C_3 = C_2\phi^{3/2}/3E_0$ and $\mathrm{E}_1(x) \equiv \int \exp\left(-xt\right)/t dt$ is the exponential integral.

Due to small β for the hemisphere we need to use very strong electrical field to produce slightly visible current in experiment.

3.3.3 Floating sphere at emitter-plane potential

The "floating sphere at emitter-plane potential" model has no "body" of the field emitter and possesses only its "head". This model gives too high estimation of electric field on the apex of nanotube but plausibly reproduce tendencies of change of the field enhancement factor. Approximate analytical solution for the "floating sphere at emitter-plane potential" model is well known (for example Refs. (Forbes et al., 2003; Wang et al., 2004)). To solve this problem the method of images (Jackson, 1999) is usually used.

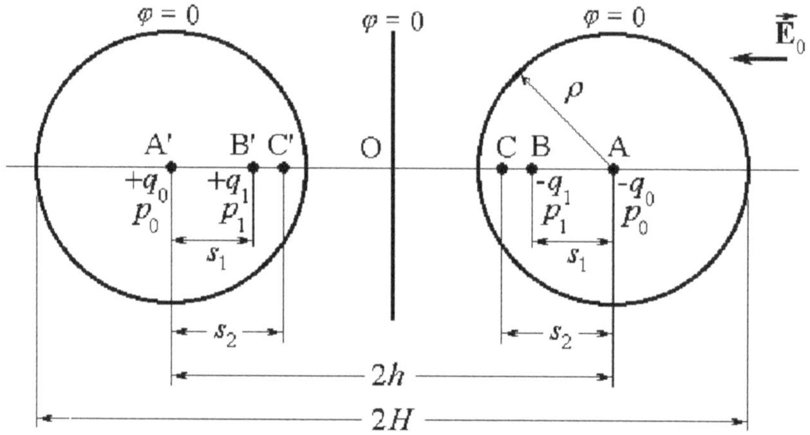

Fig. 7. Two conducting spheres of radius ρ at cathode potential in uniform electric E_0.

The charge $-q_0 = -4\pi\varepsilon_0 hE_0\rho$ and the electric dipole $p_0 = 4\pi\varepsilon_0 E_0\rho^3$ placed at point A (Fig. 7) create a sphere of radius ρ and potential $\varphi = 0$ in uniform external electric field. The charge q_0 and dipole p_0 cause a potential variation across the emitter plane. To correct this we have to place an image-charge q_0 and image-dipole p_0 at point A′ behind the emitter plane. The image-charge and image-dipole will distort the surface of sphere. To restore the shape we should place additional charge $-q_1 = -q_0\rho/2h$ and dipole $p_1 = p_0\rho^3/8h^3$ at point B on the distance $s_1 = \rho^2/2h$ from the center of sphere (see Fig. 7). Next we have to put q_1 and p_1 at point B′, after to put $-q_2$ and p_2 at C and so on. Neglecting terms of higher smallness in this series of approximation we find the electric field on the top of floating sphere

$$E_{\text{top}} = \frac{1}{4\pi\varepsilon_0}\frac{q_0}{\rho^2} + \frac{2p_0}{4\pi\varepsilon_0}\frac{1}{\rho^3} + \frac{1}{4\pi\varepsilon_0}\frac{q_1}{(s_1+\rho)^2} + E_0 \approx E_0\left(\frac{h}{\rho} + 3.5\right). \tag{16}$$

Thus the field enhancement factor is

$$\beta = \frac{1}{\eta} + 3.5 = \frac{H}{\rho} + 2.5. \tag{17}$$

We can provide more accurate calculations. Recurring formulas for the distance s_{i+1}, the charge q_{i+1}, and the dipole moment p_{i+1} through s_i, q_i, and p_i are the following $s_{i+1} =$

$\frac{\rho^2}{2h-s_i}$, $p_{i+1} = p_i \frac{\rho^3}{(2h-s_i)^3}$, and $q_{i+1} = q_i \frac{\rho}{2h-s_i} - p_i \frac{\rho}{(2h-s_i)^2}$, where the initial distance is zero: $s_0 = 0$.

Series expansion of the field enhancement factor is

$$\beta = \eta^{-1} + \frac{7}{2} - \frac{1}{2}\eta + \frac{1}{8}\eta^2 + \frac{7}{16}\eta^3 - \frac{25}{32}\eta^4 + \frac{25}{32}\eta^5 + O(\eta^6) \quad (18)$$

As the next step of approaching to CNT film, consider an assembly of floating spheres and a screening of the individual emitter by neighbors. The view from above of the sphere surrounded by another one is shown in Fig. 8. Large red circles in this picture are the floating spheres. Small black circles mark places where charges are located. Numbers " 0 " show initial charges in the center of balls. Numbers " 1 " specify image charges induced only by nearest neighbors. Numbers " 2 " concern to secondary image charges.

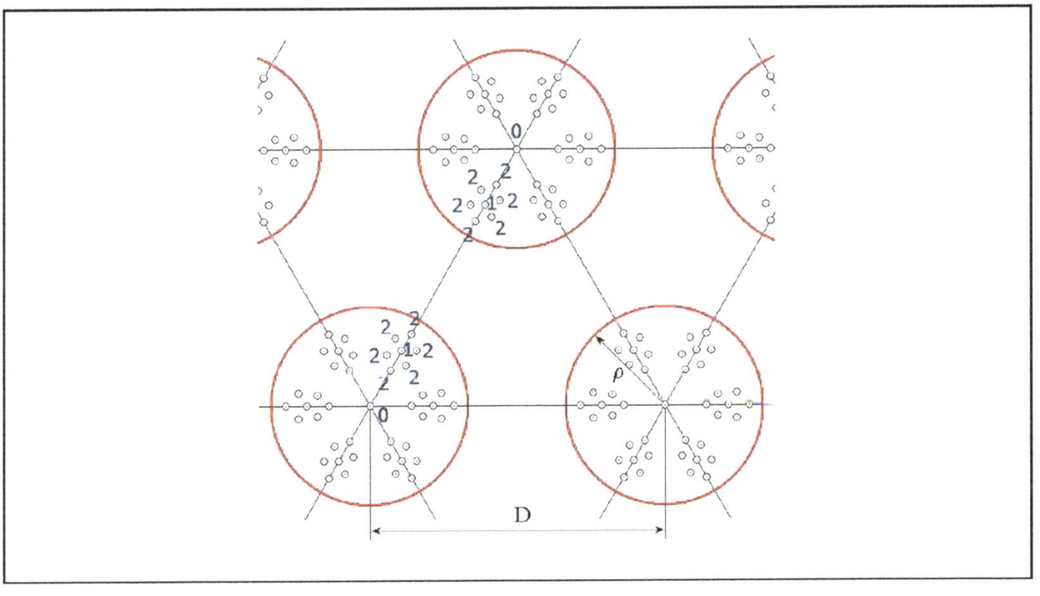

Fig. 8. Honeycomb structure, distance between spheres is *D*, sphere radius is *ρ*.

If the distance between spheres, D is large enough ($D \gg \rho$) then all image charges collect on small area around the center of sphere. In that case we can combine all charges inside the ball into its center. Also we will neglect influence of image dipoles.

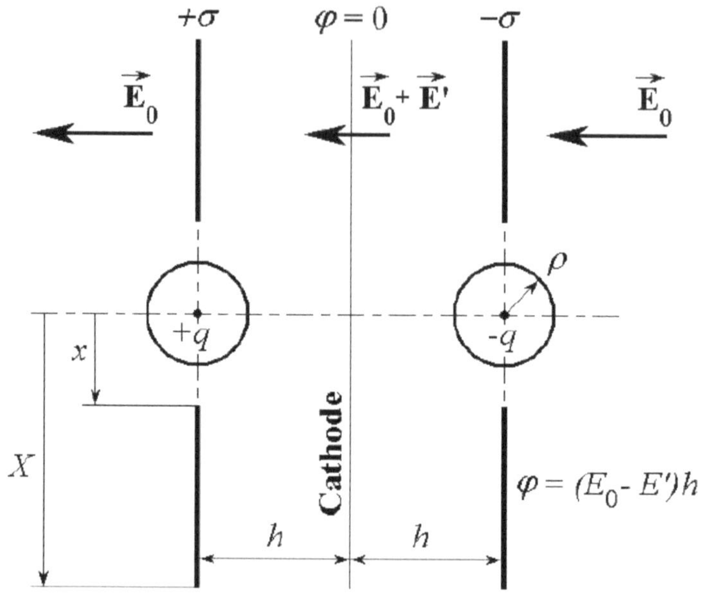

Fig. 9. Modeling of screening effect for floating spheres.

The set of floating spheres produces an idealized surface charge density $\sigma = 2q/\sqrt{3}D^2$. Positively and negatively charged surfaces form the parallel plate capacitor (Fig. 9). The electric field between two large parallel plates is given by $E' = \sigma/\varepsilon_0$.

From the equation $h\left(E_0 - \frac{2}{\sqrt{3}\varepsilon_0}\frac{q}{D^2}\right) = \frac{1}{4\pi\varepsilon_0}\frac{q}{\rho}$ we can find the total charge in the center of each sphere $q = \varepsilon_0 h E_0 \left(\frac{1}{4\pi\rho} + \frac{2h}{\sqrt{3}D^2}\right)^{-1}$.

Thus we can find the maximal field on the surface of floating sphere

$$E_{top} = \frac{hE_0}{\rho}\frac{\sqrt{3}D^2}{\left(\sqrt{3}D^2 + 8\pi\rho h\right)} + E_0 \tag{19}$$

and the field enhancement factor

$$\beta = \frac{\sqrt{3}}{\sqrt{3}\eta + 8\pi\delta^2} + 1 \tag{20}$$

More accurate approximation

$$\varphi_{\sigma^+} = 2\pi \int_x^X \frac{\sigma}{4\pi\varepsilon_0} \frac{r}{\sqrt{(2h+\rho)^2 + r^2}} dr, \varphi_{\sigma_-} = -2\pi \int_x^X \frac{\sigma}{4\pi\varepsilon_0} \frac{r}{\sqrt{\rho^2 + r^2}} dr, x = \sqrt{\frac{\sqrt{3}}{2\pi}} D \quad (21)$$

$$\varphi_{P''} = \lim_{X\to\infty} \left(\varphi_{\sigma^+} + \varphi_{\sigma_-}\right) + \frac{1}{4\pi\varepsilon_0}\left(\frac{q'}{2h+\rho} - \frac{q'}{\rho}\right) \quad (22)$$

Solving equation $\varphi_{P^n} = E_0 h$ we find

$$\beta = \frac{6(1+\eta)^2}{4\sqrt{6\pi}\delta(1+\eta)\left(\sqrt{2\pi\delta^2(2+\eta)^2 + \sqrt{3}\eta^2} - \eta\sqrt{2\pi\delta^2 + \sqrt{3}}\right) + 3\eta(2+\eta)} \quad (23)$$

Eq. (23) is transformed to Eq. (20) after neglect in values of higher order of smallness.

On the one hand the field enhancement factor and the current density on nanotube apex reach its maximum if the distance between emitters is very large. On the over hand in this case the current density on the anode will be very small. Clearly we can find optimum distance between emitters. As an approximation, assume that the emitting surface of each sphere equals $\pi\rho^2$ and that the electric field is a constant on this surface. The anode current density takes the form

$$\mathbf{j}_{\text{anode}} \approx \frac{2\pi}{\sqrt{3}}\delta^2 \frac{\mathbf{C}_1(\beta\mathbf{E}_0)^2}{\varphi} \exp\left(-\frac{\mathbf{C}_2\varphi^{3/2}}{\beta\mathbf{E}_0}\right) \quad (24)$$

If $h \gg \rho$ and $D \gg \rho$ then $\beta \approx \sqrt{3}/(\sqrt{3}\eta + 8\pi\delta^2)$. Let's use this relation for the simplicity.

Solving the equation

$$\frac{\partial \mathbf{j}_{\text{anode}}}{\partial\delta} = 0 = 24\pi\mathbf{E}_0\delta^2 - 3\sqrt{3}\mathbf{E}_0\eta + 24\pi\varphi^{3/2}\delta^2\mathbf{C}_2\eta + 64\sqrt{3}\pi^2\mathbf{C}_2\varphi^{3/2}\delta^4 \quad (25)$$

we find the optimal dimensionless distance between emitters in honeycomb structure

$$\delta_{\text{opt}} = \left(\frac{3\mathbf{E}_0\eta}{4\pi\left(\sqrt{3}\mathbf{E}_0 + \sqrt{3}\mathbf{C}_2\varphi^{3/2}\eta + \sqrt{3\mathbf{E}_0^2 + 18\mathbf{C}_2\mathbf{E}_0\varphi^{3/2}\eta + 3\mathbf{C}_2^2\varphi^3\eta^2}\right)}\right)^{1/2} \quad (26)$$

After neglect terms of higher smallness we can write the simplification

$$\delta_{\text{opt}} \approx \sqrt{\frac{\sqrt{3}\mathbf{E}_0\eta}{4\pi(\mathbf{E}_0 + 2\mathbf{C}_2\varphi^{3/2}\eta)}} \tag{27}$$

Fig. 10 illustrates the dependence of anode current density from geometrical parameters of emitter. We have assumed that the work function is $\phi = 4.8\text{eV}$, the external field is $E_0 = 60000\text{Vcm}^{-1}$, the dimensionless height is $\eta = 0.001$ (for Fig. 10a), and the dimensionless distance between emitters is $\delta = 0.002$ (for Fig. 10b).

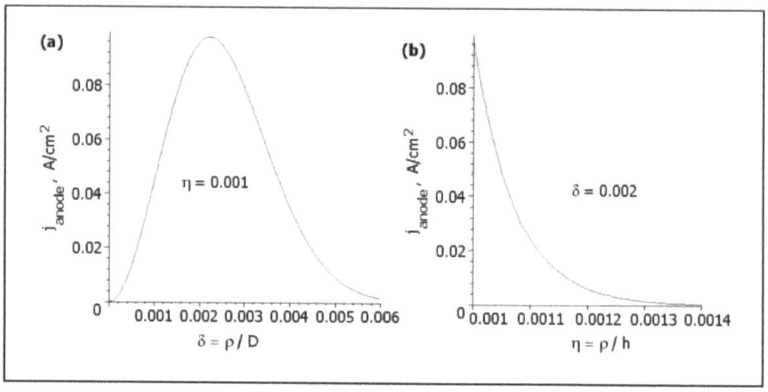

Fig. 10. Anode current density versus dimensionless sizes: (a) optimal distance between emitters if the height is fixed; (b) influence of emitter height on anode current if the density of emitting centers is constant.

Let's consider influence of the limited anode-cathode distance (Fig. 11.) on the field enhancement factor.

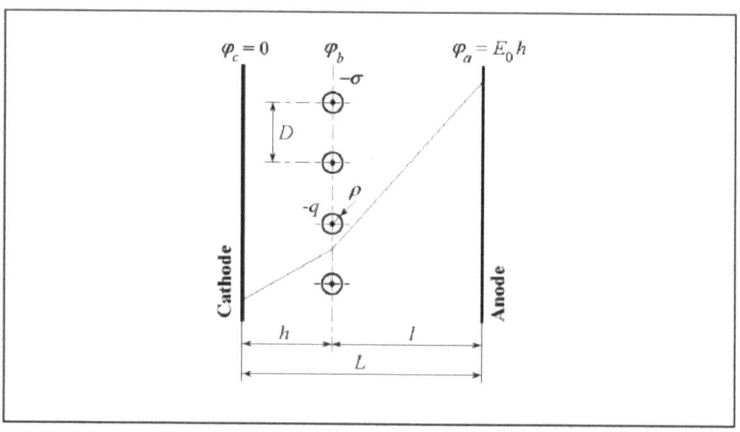

Fig. 11. Geometrical model for the limited distance, *L* between cathode and anode.

The cathode has zero potential $\varphi_c = 0$; $\varphi_a = E_0 L$ is the anode potential.

As before, assume that average charge per each conductive ball, q is concentrated at its center. Equation for average potential φ_b on plane with grounded conductive balls is

$$-\frac{\varepsilon_0}{h}\varphi_c + \left(\frac{\varepsilon_0}{h} + \frac{\varepsilon_0}{l}\right)\varphi_b - \frac{\varepsilon_0}{l}\varphi_a = -\sigma \tag{28}$$

Solving the equation $-\frac{2hl}{\sqrt{3}L\varepsilon_0 D^2}q + E_0 h = \frac{1}{4\pi\varepsilon_0\rho}q$, we find the charge in the center of ball

$$q = \frac{4\sqrt{3}\pi\varepsilon_0 E_0 h\rho D^2 L}{\sqrt{3}D^2 L + 8\pi h\rho L - 8\pi h^2\rho} \tag{29}$$

Thus the maximal field on the surface of floating sphere is

$$E_{top} = \frac{E_0 h}{\rho}\left[1 + \frac{8\pi}{\sqrt{3}}\frac{h\rho}{D^2}\left(1 - \frac{h}{L}\right)\right]^{-1} \tag{30}$$

The field enhancement factor is

$$\beta = \left[\frac{\rho}{h} + \frac{8\pi}{\sqrt{3}}\left(\frac{\rho}{D}\right)^2\left(1 - \frac{h}{L}\right)\right]^{-1} = \left[\eta + \frac{8\pi}{\sqrt{3}}\delta^2\left(1 - \frac{\lambda}{\eta}\right)\right]^{-1} \tag{31}$$

Let us note here that the model of floating sphere and the method of images allow considering field emission not only on flat anode but also on spherical anode.

3.3.4 Hemi-ellipsoid on a plane

Consider a prolate metallic spheroid in a uniform electric field. We can replace the spheroid by a linearly charged thread as we show in our recent paper (Pogorelov et al., 2009). The thread is a green line in Fig. 12 and the linear charge distribution is represented by a red line. The length of a thread is $2h$. The electrostatic potential produced by the charged thread is

$$\varphi(z, r) = -\int_{-h}^{h}\frac{1}{4\pi\varepsilon_0}\frac{\tau z' dz'}{\sqrt{(z' - z)^2 + r^2}} \tag{32}$$

where (r; z) denotes the in-plane radial and z coordinates, τz is the linear charge density at point $(0; z)$, h is half of the thread length. The solution is independent of the azimuthal angle. We assume the coefficient of linear charge density, τ to be positive.

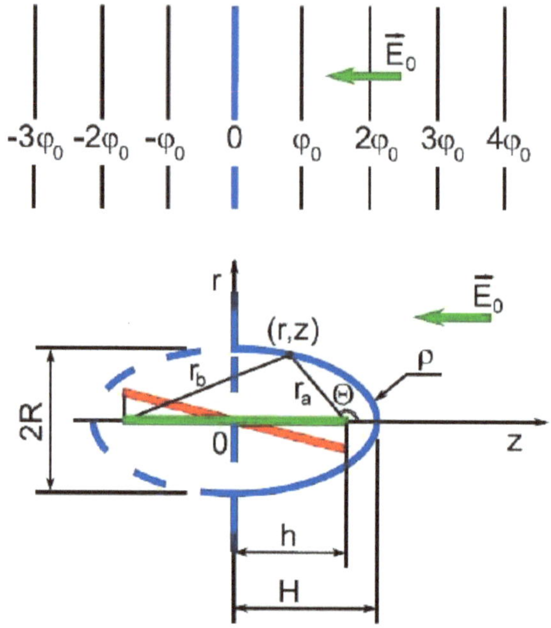

Fig. 12. Linearly charged thread in a uniform electric field along z .

The shape of metallic spheroid is given by the solution to the equation.

$$\varphi(z,r) + E_0 z = 0. \tag{33}$$

Using coordinates on the spheroid surface $r_a = \sqrt{(z-h)^2 + r^2}$ and $r_b = \sqrt{(z+h)^2 + r^2}$ and a dimensionless parameter, the eccentricity $\xi = \frac{2h}{r_a + r_b}$, $(0 < \xi < 1)$, we can rewrite Eq. (33) in the form

$$\frac{4h}{r_a + r_b} - \ln\left(\frac{r_a + r_b + 2h}{r_a + r_b - 2h}\right) = -C, \text{ where } C = 4\pi\varepsilon_0 \frac{E_0}{\tau} = \ln\left(\frac{1+\xi}{1-\xi}\right) - 2\xi. \tag{34}$$

The zero equipotential which represents the metallic hemi-ellipsoidal cathode on a plate is shown in Fig. 12 by solid blue line. Points $(0, -h)$ and (0, h) are the foci of the ellipse, r_a and r_b are distances between (r; z) and the two foci.

If ξ is close to 1 , the ellipse becomes elongated. If $\xi \to 0$ the ellipse turns into a circle. Therefore, by changing the coefficient of linear charge density, τ we may modify the shape of the ellipse.

We can also adjust other geometrical parameters of the ellipse: the length of semi-major axis or height H; the length of semi-minor axis or base radius, R at $z = 0$; and radius of curvature, ρ at point $(0, H)$ (see Fig. 13).

$$H = \frac{h}{\xi} = \frac{r_a + r_b}{2}, R = \frac{h\sqrt{1-\xi^2}}{\xi}, \rho = \frac{R^2}{H}. \tag{35}$$

We can calculate components of the electric field on the surface of the metallic spheroid:

$$E_z = -\frac{E_0}{C}\frac{h(r_b - r_a)^2}{r_a r_b (r_b + r_a)}, E_r = -\frac{E_0}{C}\frac{h(r_b - r_a)\sqrt{4h^2 - (r_b - r_a)^2}}{r_a r_b \sqrt{(r_b + r_a)^2 - 4h^2}}. \tag{36}$$

Thus the modulus of the electric field is

$$E = \sqrt{E_z^2 + E_r^2} = \frac{E_0}{C}\frac{4h^2(r_b - r_a)}{(r_b + r_a)\sqrt{r_a r_b[(r_b + r_a)^2 - 4h^2]}} \tag{37}$$

Eqs. (36), (37) allow determining the electric field strength on the surface of the half ellipsoid at an arbitrary point. The field enhancement factor at the apex of the ellipsoid is as follows:

$$\beta = \frac{2\xi^3}{(1-\xi^2)C} = \frac{2\xi^3}{(1-\xi^2)\left(\ln\frac{1+\xi}{1-\xi} - 2\xi\right)} \tag{38}$$

Analytical expressions for field strength on the z -axis and for field enhancement factor on the tip of the half ellipsoid obtained previously (Forbes et al., 2003; Kosmahl, 1991; Latham, 1981; Latham, 1995) are in agreement with our result. Here, by taking gradient of Eq. (33) we can obtain the field strength at any point we desire. In the limit $\xi \to 0$ we have a metallic half sphere and the field enhancement factor $\beta = 3$. If $\xi \to 1$ then for the elongated metallic needle, we have

$$\beta \approx 2\frac{H}{\rho}\frac{1}{\ln(4H/\rho) - 2} \tag{39}$$

Ponderomotive forces. Let us estimate the electric force acting on the surface of ellipsoid. We can calculate the force acting on the spheroid surface between circles $r_a = A$ and $r_a = B$ (see Fig. 13).

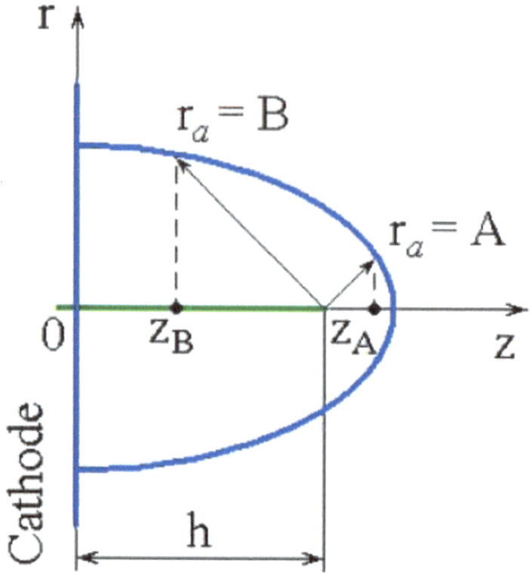

Fig. 13. Geometry for the calculation of ponderomotive force acting on the belt between $r_a = A$ and $r_a = B$.

It is obvious that r -component of force on that surface is equal to zero. For the z - component, after routine operations we obtain

$$F_z(z_A, z_B) = \frac{\pi\rho^2\varepsilon_0(\beta E_0)^2}{2\left(1-\frac{\rho}{H}\right)^2}\left[1-\left(1-\frac{\rho}{H}\right)\frac{z^2}{H^2} - \ln\left\{1-\left(1-\frac{\rho}{H}\right)\frac{z^2}{H^2}\right\}\right]\Bigg|_{z_A}^{z_B}, \quad (40)$$

where z varies from z_A to z_B (see Fig. 13). It gives us the value of net detaching force acting on the surface of the spheroid integrated over the surface between planes $z = z_A$ and $z = z_B$. The total detaching force acting on the ellipsoidal needle is

$$F_{\text{total}} = \frac{\pi\rho^2\varepsilon_0(\beta E_0)^2}{2\left(1-\frac{\rho}{H}\right)^2}\left(\frac{\rho}{H} - 1 - \ln\frac{\rho}{H}\right) \quad (41)$$

In Fig. 14a we show the relative detaching force $F_z(H,z)/F_{\text{total}}$ as a function of coordinate z, where $F_z(H,z')$ is the force acting on the surface between plane $z = z'$ and the tip of the spheroid ($z = H$). Distribution depends only on the parameter H/ρ. The major part of the detaching force is concentrated near the tip when H/ρ is large. Fig. 14b shows the total force isolines on the ($\rho, H/\rho$) plane. When we chose logarithmic scale for ρ and H/ρ with logarithmic steps we obtain a set of nearly straight isolines with equal distances in the plot.

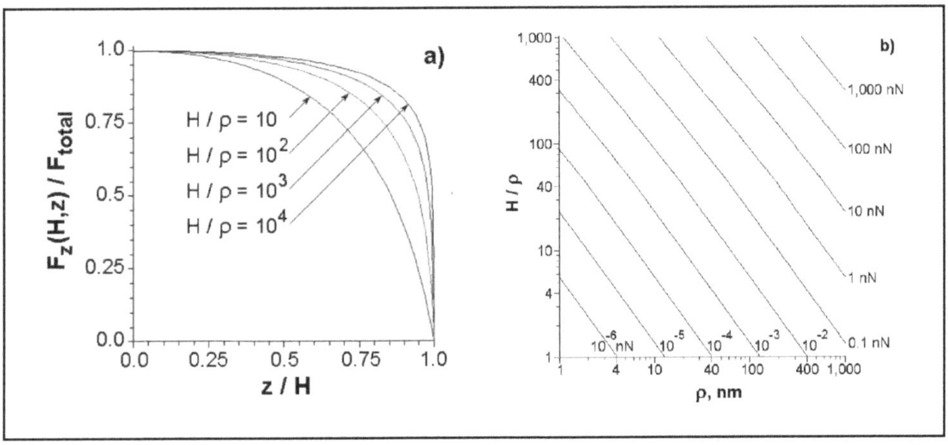

Fig. 14. Distribution of force and total force isolines. (a) Distribution of relative force $F_z(H,z)/F_{\text{total}}$ over the axis of needle. (b) Isolines of total force on the $(\rho, H/\rho)$ plane.

We think that under the action of ponderomotive forces in the external electric field, carbon nanotubes that are even chaotically located on the substrate straighten and become oriented (Musatov et al., 2001; Glukhova et al., 2003).

Field emission from individual needle. The current emitted from the surface bounded by $\theta \in (0, \Theta)$ can be expressed in terms of the E_1-function

$$J(\Theta) = \frac{2\pi\rho^2 C_1(\beta E_0)^2}{\phi} \mathrm{E}_1 \left[\frac{C_2\sqrt{2}\phi^{3/2}}{\beta E_0 \sqrt{1+\lambda}} \right] \Bigg|_{\lambda=\cos\Theta}^{1} \tag{42}$$

where $\lambda = \cos\theta, \theta$ is angle between axis z and r_a as shown in Fig. 12. By comparing with accurate numerical integration of (1) we find that formula (42) is accurate up to the first four digits for $H/\rho > 100$. So formula (42) is a good approximation for investigating the current emitted from an area of surface depending on angle Θ. The total current with similar accuracy can be written approximately as

$$J_{\text{total}} = \frac{2\pi\rho^2 C_1(\beta E_0)^2}{\phi} \mathrm{E}_1 \left[\frac{C_2\sqrt{2}\phi^{3/2}}{\beta E_0 \sqrt{1+\lambda}} \right] \Bigg|_{-\xi}^{1} \approx \frac{2\pi\rho^2 C_1(\beta E_0)^2}{\phi} \mathrm{E}_1 \left[\frac{C_2\phi^{3/2}}{\beta E_0} \right] \tag{43}$$

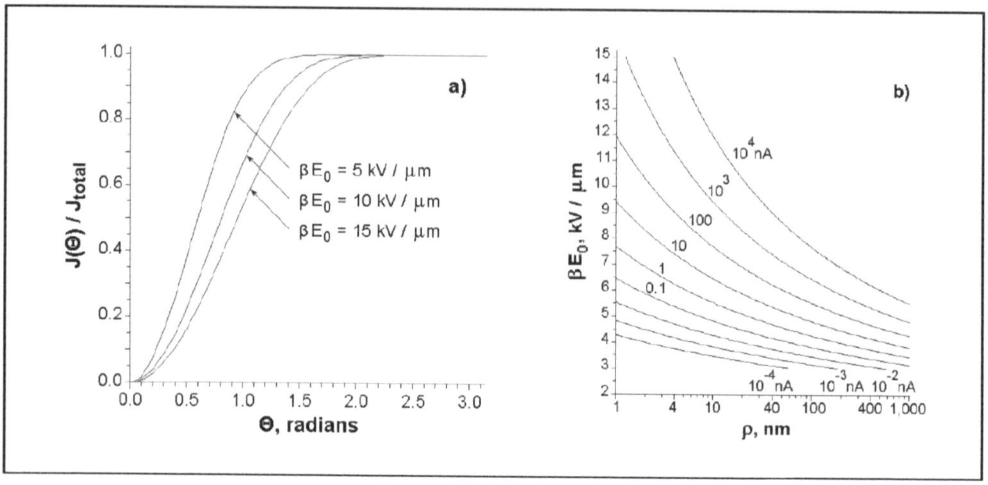

Fig. 15. Distribution of emission current and total emission current isolines for work function $\varphi = 4.8$eV. (a) Distribution of relative emission current over Θ. (b) Total emission current isolines on the ($\beta E_0, \rho$) plane.

The emission current has much sharper distribution on the tip than the detaching force due to exponential dependence of Fowler-Nordheim formula (1). In Fig. 15a we plot the relative emission current, $J(\Theta)/J_{\text{total}}$ as a function of Θ (angle between $\boldsymbol{r}_a$ and axis z), where $J(\Theta)$ is emission from the tip on the surface of the ellipsoidal needle. From (43) we see that total emission current depends only on the area of the hemisphere, $2\pi\rho^2$, work function, φ, and the electric field on the tip, βE_0. So in Fig. 15b we show the total emission current isolines within a reasonable range (from 10^{-4} to 10^4nA) with logarithmic steps on the ($\beta E_0, \rho$) plane, where the work function $\varphi = 4.8$eV is assumed. If the electric field is $E_0 \approx 1..5$ V/μm, then we have to use an enhancement factor $\beta \approx 10^3..10^4$ to get an appreciable current.

Simulation of planar anode by "image" charges. The presence of a flat anode placed at a distances comparable with the length of the nanoneedle has strong influence on the emission current and the detaching force. The basic idea of calculation is to replace the cathode by a linearly charged thread in a uniform electric field and to use a set of "image" charges to reproduce the anode as shown in Fig. 16. We put infinite set of "images" of the linearly charged thread with the same spacing, $2L(L > H)$. It is clear that planes $z = 0$ and $z = L$ are planes of symmetry for distributed charges and will have potentials 0 on cathode and $V = E_0 L$ on anode.

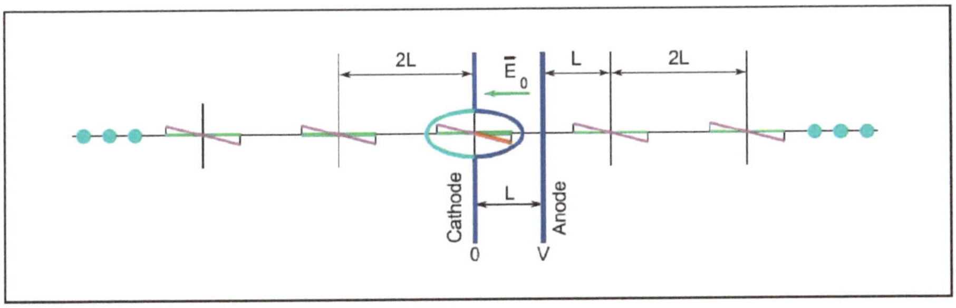

Fig. 16. Infinite set of "image" charges for the simulation of a planar anode.

With the same τ we will get different geometry of cathode due to additional potential of "images". We assume that the new form of the elongated needle will be approximately described by ellipsoid especially near the tip. On the surface of the thin ellipsoid, we have $r \ll h$ and $z < H < L$. So, we can describe the image potential φ_i as a function z.

$$\varphi_i(z) = -\frac{E_0}{C^*}\sum_{n=1}^{\infty}\left[(z-2nL)\ln\frac{2nL-z+h}{2nL-z-h} + (z+2nL)\ln\frac{2nL+z+h}{2nL+z-h}\right], \tag{44}$$

where $C^* \equiv 4\pi\varepsilon_0 \frac{E_0}{\tau^*}, L = \mu H, \mu > 1$. The thread potential is

$$\varphi = \frac{zE_0}{C}\left[\frac{4h}{r_a + r_b} - \ln\left(\frac{r_a + r_b + 2h}{r_a + r_b - 2h}\right)\right]. \tag{45}$$

Near the tip $z \approx H$ we have the following equation for describing the shape of the ellipsoid:

$$\varphi(r,z) + E_0 z + \varphi_i(z) \approx \varphi(r,z) + E_0 z + \frac{z}{H}\varphi_i(H) = 0. \tag{46}$$

So we have to change constant C on constant C^* as following

$$C^* = \ln\left(\frac{1+\xi}{1-\xi}\right) - 2\xi - P \tag{47}$$

where

$$P = \sum_{n=1}^{\infty}\left[(2n\mu - 1)\ln\frac{2n\mu - 1 + \xi}{2n\mu - 1 - \xi} - (2n\mu + 1)\ln\frac{2n\mu + 1 + \xi}{2n\mu + 1 - \xi}\right]. \tag{48}$$

Therefore we have the field emission factor in the form

$$\beta(L/H, H/\rho) = \frac{2\xi^3}{(1-\xi^2)C^*} \tag{49}$$

Finally for the force and emission current we may use the above derived formulas. Instead of (38) we should use (47), (48), and (49). So, the parameter $\mu = L/H$ in the force and emission current formulas (40), (41), (42), and (43) should be included only through the field enhancement factor $\beta = \beta(L/H, H/\rho)$.

What distance between the anode and cathode is large enough to assert that the elongated metallic spheroid is placed in a uniform electric field? In experiment we may measure distance L between the anode and cathode and the distance, $L - H$ between the anode and the needle apex. We can determine the applied electric field both by $E_0 = V/L$ and by $E_0' = V/(L-H)$. In our calculations we move the anode and increase the voltage V so as to keep E_0 or E_0' immutable. It is clear that for a large distance (in the limit $L \to \infty$) the difference between definitions of E_0 and E_0' disappears. Fig. 17 shows the total current emitted from the needle versus the anode-cathode distance.

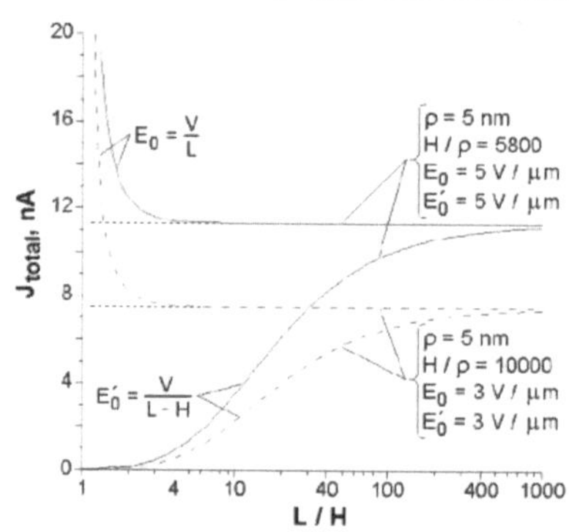

Fig. 17. Emission current as a function of the anode-cathode distance parameter $\mu = L/H$ for constant electric field $E_0 \equiv V/L$ (red line) or $E_0' \equiv V/(L-H)$ (blue line). Radius of curvature is fixed at $\rho = 5$ nm. For solid curves, $H/\rho = 5800$ and for dash curves, $H/\rho = 10000$. For red and blue solid curves $E_0 = 5$ V/μm, for red and blue dash curves $E_0 = 3$ V/μm.

We set for blue lines $E_0' = \text{const}$ and for red lines $E_0 = \text{const}$. With $E_0 = E_0' = \text{const}$ the total currents have the same limit for $\mu = L/H \to \infty$. But currents with $E_0' = \text{const}$ tend to approach the limit much slower than the corresponding currents with $E_0 = \text{constant}$. If we define the applied electric field as E_0 and the anode-cathode distance is ten times more than the needle height, then we can neglect the influence of the anode and assume that the metallic needle is placed in a uniform field. In contrast, if the applied electric field is defined as E_0', then the anode-cathode distance must exceed one thousand times of the needle height for the above statement to be valid.

3.3.5 The model of a hemisphere on a post. Numerical approximations

The model of a hemisphere on a post allows only the numerical solution. There are many numerical results obtained by various researchers which have been generalized by simple algebraic formulas of field enhancement factor for an individual nanotube and assembly of nanotubes.

From our point of view the most accurate formula belongs to Edgcombe et. al. (Edgcombe & Valdrè, 2001; Edgcombe & Valdrè, 2002; Forbes et al., 2003)

$$\beta_0 = 1.2\left(2.5 + \frac{H}{\rho}\right)^{0.9} \tag{50}$$

This formula accurately describes the field enhancement factor of individual nanotube in a uniform electric field. Comparison of field enhancement factors for the "floating sphere at emitter-plane potential" model (green line), the "hemi-ellipsoid on plane" model (red line), and fitting formula for the "hemisphere on a post" model (blue line) is shown in Fig. 18.

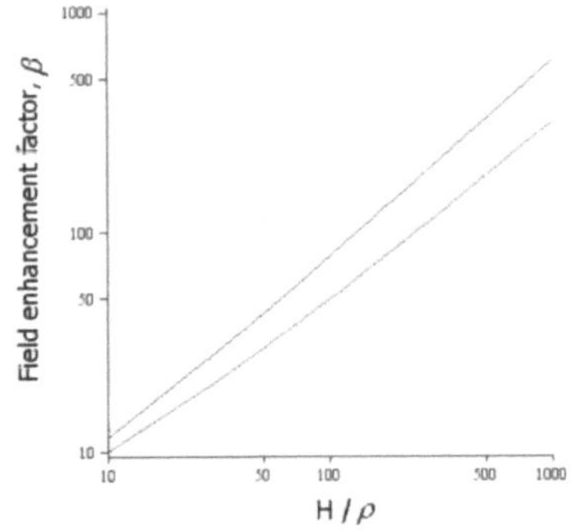

Fig. 18. Comparison the field enhancement factors for three models: floating sphere (green line), hemi-ellipsoid (red line), and hemisphere on a post (blue line).

For a nanotube in space between parallel cathode and anode planes we prefer approximation (Bonard et al., 2002)

$$\beta = \beta_0 \left[1 + 0.013 \left(\frac{L - H}{L} \right)^{-1} - 0.033 \left(\frac{L - H}{L} \right) \right] \tag{51}$$

For a nanotube surrounded by neighboring nanotubes with a screening effect we recommend (Jo et al., 2003)

$$\beta = \beta_0 \left[1 - \exp \left(-2.3172 \frac{D}{H} \right) \right] \tag{52}$$

3.4 Further investigations

Let us shortly mention the further work that should now be carried out on the fieldemission properties of CNTs.

1. Developing of field emission theory for CNT emitters
2. Temperature effects in CNTs: Joule heating; Peltier and Nottingham effects; carbon nanotube heat radiation; thermo-field emission from nanotubes
3. Action of electrostatic forces on CNT field emitters: elongation and straightening of nanotube; pulling out of nanotubes in strong electric field.
4. Reliability of CNT field emitters: degradation mechanism of field emission; lifetime of nanotube emitters.

4. Applications

Field emission is the most promising application areas of carbon nanotubes. We shall consider only a few samples of working devices.

4.1 Field emission displays

Flat-panel displays with CNT-based cathodes are proposed as alternative to other displays with film emitters. The first diode-type display consisting of 32 × 32 matrix-addressable pixels was manufactured by Wang et al. in 1998 (Wang et al., 1998). At present, flat-panel displays based on CNT field-emission cathodes are developed in hundreds of laboratories, and engineering samples have been already manufactured.

A 4.5-inch full-color CNT display developed by Choi et al. (Choi et al., 1999) from Samsung Company was demonstrated at several exhibitions.

A full-color addressable display developed jointly by the limited liability company "VolgaSvet" (Saratov, Russia) and "CopyTele Inc.} (New York, USA) (Abanshin et al., 2002) was demonstrated at the conference IveSC'02. Emission in the proposed design was from thin edges of carbon nanocluster films, which are slightly hanging from the supporting pedestals (Fig. 19). The anode covered by a phosphor layer is shown on the top in Fig. 19. Gating/blocking of emission current is performed by a metal control electrode located on the substrate between the pedestals. The figure shows the results of the trajectory analysis for an option of the structure (Zhbanov et al., 2004).

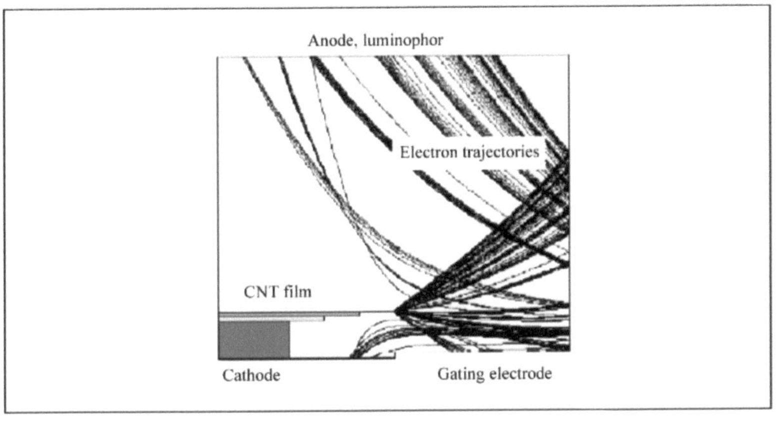

Fig. 19. Trajectory analysis for the flat-panel display produced by the limited liability company "Volga-Svet" and "CopyTele Inc." (Zhbanov et al., 2004).

4.2 X-ray tubes

A compact X-ray tube with a field emitter based on carbon nanotubes was developed by Musatov's group from Institute of Radio-engineering and Electronics (Russia) (Musatov et al., 2007). Over a long time interval, the X-ray tube maintains an anode current of 300μ A, an anode voltage of 10 kV , and the stable characteristics of the field emitter (Fig. 20).

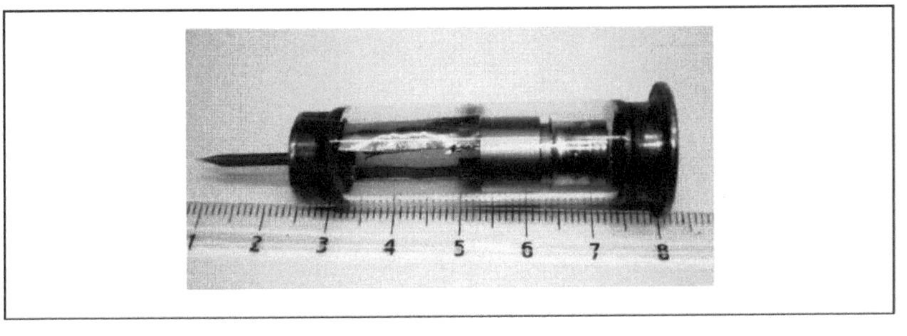

Fig. 20. Compact X-ray tube. The figure is taken from (Musatov et al., 2007).

4.3 Light elements

The Bonard team developed the cathodoluminescent light-emitting element of cylindrical geometry (Bonard et al., 2001a). The cathode in the form of a metal rod with deposited CNTs is located on the axis of a glass tube covered by phosphor from the inside. The operating voltage is 7.5 kV , the current density on the cathode is $0.25\ mA/cm^2$, the current density on the anode is $0.03\ mA/cm^2$, and the lamp luminance is $10^4\ cd/m^2$.

CNT light-emitting elements of various colors in the form of a filamentary electric lamp are proposed by Saito team in (Bonard et al., 2002b). The operating grid voltage is from 0.2 to 1.2 kV , the current density at the cathode with an area of $2\ mm^2$ is $10\ mA/cm^2$, the average electric field intensity is $1.5\ V/\mu m$, and the luminance of elements of various colors ranges from $1.5 \cdot 10^4$ to $6.3 \cdot 10^4\ cd/m^2$.

Samples of the light-emitting elements were demonstrated by A.N. Obraztsov (Obraztsov et al., 2002) and E. P. Sheshin (Leshukov et al., 2002) at the Saratov 4th International Conference on Vacuum Electrons Sources (IveSC'02).

4.4 Microwave devices

The electron gun consisting of edge or cylindrical emitters with flat faces; a grid control electrode placed above the emitters; and a three-anode focusing system which is common for all the emitters (Fig. 21) was designed by Zakharchenko's group (Zakharchenko et al., 1996). The flat face surfaces of the edge emitter are covered with a film made of carbon nanotubes of diameter $30 - 100Å$. A substantial density of emitting centers being as great as $10^8 - 10^{10} tips/cm^2$ provides a high current density at low accelerating voltages.

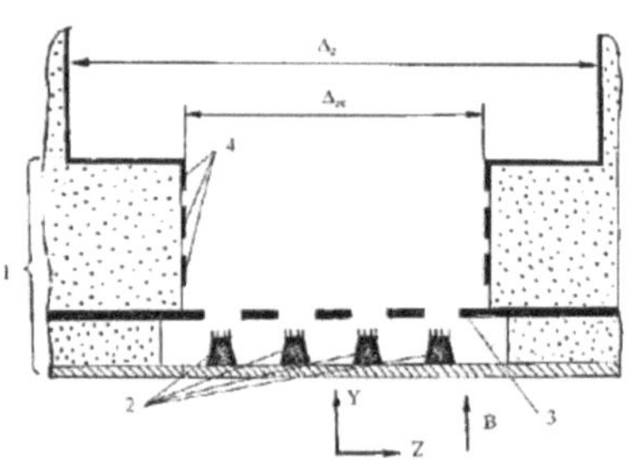

Fig. 21 Schematic of the electron gun for an amplifier. Magnetic field $B = 0.08$ T, voltage $V = 400$ V, and current $I = 1\ A/cm^2$. The electron gun (1) consists of edge or cylindrical field emitters (2) covered with CNT film, a control electrode (3), and three anodes (4) (Zakharchenko et al., 1996).

5. Conclusions and Acknowledgements

In this chapter we theoretically investigate the field emission from carbon nanotube field emitters in diode configuration between a flat anode and cathode. Exact analytical formulas of the electrical field, field enhancement factor, ponderomotive force, and field emission current are found. Applied voltage, height of the needle, radius of curvature on its top, and the work function are the parameters at our disposal. The field enhancement factor, total force and emission current, as well as their distributions on the top of the needle for a wide range of parameters have been calculated and analyzed. Also we review the technology of fabrication and the application areas of CNT electron sources.

We have right to conclude that carbon nanotubes are excellent field emitters. Now CNTs conquer appreciable positions as electron sources in compact X-ray tubes, lighting elements, scanning electron microscopy electron guns, and electron guns for microwave devices. We think CNTs have good potential in the field emission display market.

We gratefully acknowledge support through the National Science Council of Taiwan, Republic of China, through the project NSC 95-2112-M-001-068-MY3.

6. References

Abanshin, N.; Muchina, E. & Nikishin, N. et al. (2002). Proc.IVeSC'2002 Supplement, Saratov, Russia, July 15-19, p. 13.

Banerjee, D.; Jo, S. & Ren, Z. (2004). Enhanced Field Emission of ZnO Nanowires. Advanced Materials, 16, 2028.

Bethune, D.; Kiang, C. & de Vries, M. et al. (1993). Cobalt-Catalysed Growth of Carbon Nanotubes with Single-Atomic-Layer Walls. Nature, 363, 605.

Bonard, J-M.; Stöckli, Th. & Noury, O. et al. (2001a). Field emission from cylindrical carbon nanotube cathodes: Possibilities for luminescent tubes. Applied Physics Letters, Vol. 78, No. 18, P. 2775-2777.

Bonard, J.-M.; Kind, H. & Stöckli, Th. et al. (2001b). Field emission from carbon nanotube: the first five years. Solid State Electronics, Vol. 45, P. 893-914.

Bonard, J. -M.; Dean, K. & Coll, B. et al. (2002a). Field Emission of Individual Carbon Nanotubes in the Scanning Electron Microscope. Physical Review Letters, 89, 197602.

Bonard, J.-M.; Gaal, R. & Garaj, S. et al. (2002b). Field emission properties of carbon nanohorn films. Journal of Applied Physics, 91, 10107-10109.

Charbonnier, F.; Mackie, W. & Hartman, R. et al. (2001). The Journal of Vacuum Science Technology B, 19(3), 1064.

Chernozatonskii, L.; Gulyaev, Yu. & Kosakovskaja, Z. et al. (1995). Electron field emission from nanofilament carbon films. Chemical Physics Letters, Vol. 233, P. 63-68.

Chernozatonskii, L.; Kosakovskaya, Z. & Gulyaev, Yu. et al. (1996). Influence of external factors on electron field emission from thin-film nanofilament carbon structures. The Journal of Vacuum Science Technology B, Vol. 14(3), P. 2080-2082.

Choi, W.; Chung, D. & Kang, J. et al. (1999). Fully sealed, high-brightness carbon-nanotube field-emission display. Applied Physics Letters, Vol. 75, No. 20, P. 3129-3131.

Collins, P. & Zettl, A. (1996). A simple and robust electron beam source from carbon nanotubes. Applied Physics Letters, Vol. 69, No. 13, P. 1969-1971.

Collins, P. & Zettl, A. (1997). Unique characteristics of cold cathode carbon-nanotube-matrix field emitters. Physical Review B, Vol. 55, No. 15, P. 9391-9399.

Drechsler, M. & Müller, E. (1953). Zur Feldelektronenemission und Austrittsarbeit Einzelner Kristallflachen. Zeitschrift Fur Physik, Vol. 134, Issue 2, P. 208-221.

Ducastelle, F.; Blase, X. & Bonard, J.-M. et al. (2006). Electronic Structure, Lecture Notes of Physics 677, 199-276.

Edgcombe, C. & Valdrè, U. (2001). The enhancement factor and the characterization of amorphous carbon field emitters. Solid-State Electronics, Vol. 45, No. 6, pp. 857-863.

Edgcombe, C. & Valdrè, U. (2002). Experimental and computational study of field emission characteristics from amorphous carbon single nanotips grown by carbon contamination I. Experiments and computation. Philosophical Magazine B, Vol. 82, No. 9, P. 987-1007.

Fan, Sh.; Chapline, M. & Franklin, N. et al. (1999). Self-Oriented Regular Arrays of Carbon Nanotubes and Their Field Emission Properties. Science, Vol. 283, P. 512-514.

Filip, V.; Nicolaescu, D. & Tanemura, M. et al. (2001). Modeling the electron field emission from carbon nanotube films. Ultramicroscopy, Vol. 89, P. 39-49.

Fishbine, B.; Miglionico, C. & Hackett, K. et al. (1994). Graphene nanotubule cold field emission electron sources. Materials Research Society Symposium - Proceedings, Vol. 349, pp. 319-324.

Forbes, R.; Edgcombe, C. & Valdrè, U. (2003). Some comments on models for field enhancement. Ultramicroscopy, Vol. 95, pp. 57-65.

Fowler, R. & Nordheim, L. (1928). Electron Emission in Intense Electric Fields. Proceedings of the Royal Society (London) A, 119, pp. 173-181.

Fursey, G. (2005). Field emission in vacuum microelectronics. Kluwer Academic / Plenum Publishers, New York, ISBN: 0-306-47450-6.

Ghosh, P.; Tanemura, M. & Soga, T. et al. (2008). Field emission property of N-doped aligned carbon nanotubes grown by pyrolysis of monoethanolamine. Solid-State Communications, Vol. 147, Issue 1-2, pp. 15-19.

Glukhova, O.; Zhbanov, A. & Torgashov, I. et al. (2003). Ponderomotive forces effect on the field emission of carbon nanotube films. Applied Surface Science, Vol. 215, pp. 149159.

Gulyaev, Yu.; Nefyodov, I. & Sinitsyn, N. et al. (1994). Simulation of the emission characteristics of field emitter arrays based on nanofilament carbon structures. 7-th International Vacuum Microelectronics Conference. ATRIA, World Trade Center, Grenoble, France, July 4-15, pp. 319-321.

Gulyaev, Yu.; Chernozatonskii, L. & Kosakovskaja, Z. et al. (1995). Field emitter arrays on nanotube carbon structure films. The Journal of Vacuum Science Technology B, Vol. 13(2), pp. 435-436.

Gulyaev, Yu.; Sinitsyn, N. & Torgashov, G. et al. (1997). Work function estimate for electrons emitted from nanotube carbon cluster films. The Journal of Vacuum Science Technology B, Vol. 15(2), pp. 422-424.

de Heer, W. ; Chátelain, A. & Ugarte, D. (1995). A Carbon Nanotube Field-Emission Electron Source. Science, Vol. 270, No. 5239, pp. 1179-1180.

de Heer, W.; Bonard, J.-M. & Fauth K. et al. (1997). Electron Field Emitters Based on Carbon Nanotube Films. Advanced Materials, Vol. 9, No 1, pp. 87-89.

Hu, Y. & Huang, C.-H. (2003). Computer simulation of the field emission properties of multiwalled carbon nanotubes for flat panel displays. The Journal of Vacuum Science Technology B, Vol. 21(4), pp. 1648-1654.

Iijima, S. (1991). Helical microtubules of graphitic carbon. Nature, Vol. 354, pp. 56-58.

Iijima, S. & Ichihashi, T. (1993). Single shell carbon nanotubes of one nanometer diameter. Nature, Vol. 363, pp. 603-605.

Jackson, J. (1999). Classical Electrodynamics. John Wiley & Sons, Inc. - 808 p. ISBN 0-471-30932-X.

Jang, H.; Kim, D.-H. & Lee, H.-R. et al. (2005). Field emission from cone-like single crystalline indium tin oxide nanorods. Materials Letters, Vol. 59, pp. 1526-1529.

Jo, S.; Tu, Y. & Huang, Z. et al. (2003a). Effect of length and spacing of vertically aligned carbon nanotubes on field emission properties. Applied Physics Letters, Vol. 82, No. 20, pp. 3520-3522.

Jo, S.; Lao, J. & Ren, Z. et al. (2003b). Field-emission studies on thin films of zinc oxide nanowires. Applied Physics Letters, Vol. 83, No. 23, pp. 4821-4823.

de Jonge, N. & Bonard, J.-M. (2004). Carbon nanotube electron sources and applications. Philosophical Transactions of the Royal Society (London) A, Vol. 362, pp. 2239-2266.

Kosmahl, H. (1991). Analytic Evaluation of Field Emission Enhancement Factors for Ellipsoidal Cones and Elliptic Cross-Section Wedges. IEEE Trans. Electron Devices, Vol. 38, No. 6, pp. 1534-1537.

Kukovitsky, E.; L'vov, S. & Sainov, N. et al. (2003). Applied Surface Science, Vol. 215, pp. 201208.

Kung, S.; Hwang, K. & Lin, I. (2002). Oxygen and ozone oxidation-enhanced field emission of carbon nanotubes. Applied Physics Letters, Vol. 80, pp. 4819-4821.

Latham, R. (1981). High Voltage Vacuum Insulation: The Physical Basis, Academic Press, London.

Latham, R. (1995). High Voltage Vacuum Insulation: Basic Concepts and Technological Practice, Academic Press, London.

Lee, Y.; Kim, S. & Tománek, D. (1997). Field-induced unraveling of carbon nanotubes. Chemical Physics Letters, Vol. 265, pp. 667-672.

Lee, Y.-H.; Choi, Ch.-H. & Jang, Y.-T. et al. (2002). Tungsten nanowires and their field electron emission properties. Applied Physics Letters, Vol. 81, No. 4, pp. 745-747.

Leshukov, M.; Baturin, A. & Chadaev, N. et al. (2002). Proc. IVeSC'2002 Supplement, Saratov, Russia, July 15-19, 2002, p. 246.

Li, S.; Lee, Ch. & Lin, P. et al. (2006). Gate-controlled ZnO nanowires for field-emission device application. The Journal of Vacuum Science Technology B, Vol. 24(1), pp. 147151.

Li, J.; Wang, Q. & Gu, C. (2007). Growth and field emission properties of tubular carbon cones. Ultramicroscopy, Vol. 107, pp. 861-864.

Millikan, R. & Lauritsen, C. (1929). Dependence of Electron Emission from Metals upon Field Strengths and Temperatures. Physical Review, Vol. 33, pp. 598-604.

Musatov, A.; Kiselev, N. & Zakharov, D. et al. (2001). Field electron emission from nanotube carbon layers grown by CVD process. Applied Surface Science, Vol. 183(1-2), pp. 111119.

Musatov, A.; Gulyaev, Yu. & Izrael'yants, K. et al. (2007). A Compact X-ray Tube with a Field Emitter Based on Carbon Nanotubes. Journal of Communications Technology and Electronics, Vol. 52, No. 6, pp. 714-716.

Nilsson, L.; Groening, O. & Emmenegger, C. et al. (2000). Scanning field emission from patterned carbon nanotube films. Applied Physics Letters, Vol. 76, pp. 2071-2073.

Nilsson, L.; Groening, O. & Kuettel, O. et al. (2002). Microscopic characterization of electron field emission. The Journal of Vacuum Science Technology B, Vol. 20(1), pp. 326-337.

Obraztsov, A.; Petrushenko, Yu. & Satanovskaya, O. et al. (2002). Proc. IVeSC'2002 Supplement, Saratov, Russia, July 15-19, p. 254.

Park, N.; Han, S. & Ihm, J. (2001). Effects of oxygen adsorption on carbon nanotube field emitters. Physical Review B, Vol. 64, Issue 12, No. 125401.

Pogorelov, E.; Zhbanov, A. & Chang, Y. (2009). Field enhancement factor and field emission from hemi-ellipsoidal metallic needle. Ultramicroscopy, Vol. 109, Issue 4, pp. 373378.

Read, F. & Bowring, N. (2004). Field enhancement factors of random arrays of carbon nanotubes. Nucl. Instr. Meth. Phys. Res. A, Vol. 519, pp. 305-314.

Read, F. & Bowring, N. (2004). Field enhancement factors of one-dimensional and twodimensional arrays of nanotubes. Microelectronic Engineering, Vol. 73-74, pp. 679685.

Rinzler, A.; Hafler, J. & Nikolaev, P. et al. (1994). Field Emission and Growth of Fullerene Nanotubes. MRS proceeding, Vol. 359, pp. 61-68 (December 1994).

Rinzler, A.; Hafner, J. & Nikolaev, P. et al. (1995). Unraveling Nanotubes: Field Emission from an Atomic Wire. Science, Vol. 269, pp. 1550-1553.

Saito, Y.; Hamaguchi, K. & Nishino, T. et al. (1997a). Field Emission Patterns from SingleWalled Carbon Nanotubes. Jpn. J. Appl. Phys., Vol. 36, Part 2, No. 10A, P. L1340L1342.

Saito, Y.; Hamaguchi, K. & Hata, K. et al. (1997b). Conical beams from open nanotubes. Nature, Vol. 389, No. 6651, pp. 554-555.

Seelaboyina, R.; Huang, J. & Park, J. et al. (2006). Multistage field enhancement of tungsten oxide nanowires and its field emission in various vacuum conditions. Nanotechnology, Vol. 17, pp. 4840-4844.

Shang, X. ; Wang, M. & Qu, S. et al. (2007). A model calculation of the tip field distribution for a single carbon nanotube. The Journal of Appied. Physics, Vol. 102, 054301.

Sinitsyn, N.; Gulyaev, Yu. & Torgashov, G. et al. (1997). Thin films consisting of carbon nanotubes as a new material for emission electronics. Applied Surface Science, Vol. 111, pp. 145-150.

Smith, R.; Carey, J. & Forrest, R. et al. (2005). Effect of aspect ratio and anode location on the field emission properties of a single tip based emitter. J. Vac. Sci. Technol. B, Vol. 23(2), pp. 632-635.

Spindt, C. (1968). A Thin-Film Field-Emission Cathode. The Journal of Applied Physics, Vol. 39, Issue 7, 3504.

Srivastava, S.; Vankar, V. & Rao, D. et al. (2006). Enhanced field emission characteristics of nitrogen-doped carbon nanotube films grown by microwave plasma enhanced chemical vapor deposition process. Thin Solid Films, Vol. 515, pp. 1851-1856.

Terrones, M.; Jorio, A. & Endo, M. et al. (2004). New direction in nanotube science. Materials Today, Vol. 7, No. 10, pp. 30-45.

Terrones, M; Souza, A. & Rao, A. (2008). Doped carbon nanotubes: Synthesis, characterization and applications. Carbon Nanotubes Book, Series: Topics in Appl. Phys., vol. 111, pp. 531-566.

Tumareva, T.; Sominskiĭ, G. & Efremov, A. et al. (2002). Tip Field Emitters Coated with Fullerenes. Technical Physics, Vol. 47, No. 2, pp. 244-249.

Tumareva, T.; Sominski, G. & Svetlov, I. et al. (2008). Fullerene-Coated Field Emitters Activated by a Potassium Ion Flux in High Electric Fields. Technical Physics, Vol. 53, No. 11, pp. 1504-1507.

Wang, Q.; Setlur, A. & Lauerhaas, J. et al. (1998). A nanotube-based field-emission flat panel display. Applied Physics Letters, Vol. 72, No. 22, pp. 2912-2913.

Wang, X.; Wang, M. & He, P. et al. (2004). Model calculation for the field enhancement factor of carbon nanotube. J. Appl. Phys., Vol. 96, No. 11, pp. 6752-6755.

Wang, X.; Wang, M. & Ge, H. et al. (2005). Modeling and simulation for the field emission of carbon nanotubes array. Physica E, Vol. 30, pp. 101-106.

Wood, R. (1897). A New Form of Cathode Discharge and the Production of X-Rays, together with Some Notes on Diffraction. Preliminary Communication" Phys. Rev. (Series I), Vol. 5, pp. 1-10.

Yudasaka, M.; Iijima, S. & Crespi, V. (2008). Single-wall carbon nanohorns and nanocones. Carbon Nanotubes Book, Series: Topics in Appl. Phys., vol. 111, pp. 605-629.

Zakharchenko, Y.; Torgashov, G. & Gulyaev Y. et al. (1996). Two-stage distributed amplifier on field emitter arrays. J. Vac. Sci. Technol. B, Vol. 14(3), pp. 1982-1985.

Zhbanov, A.; Sinitsyn, N. & Torgashov, G. (2004). Nanoelectronic Devices Based on Carbon Nanotubes. Radiophysics and Quantum Electronics, Vol. 47, pp. 435-452.

Chapter-18

Sensing Mechanisms of Carbon Nanotube Based NH_3 Gas Detectors

1. Introduction

It is well known that the electrical performance of carbon nanotube (CNT) devices is extremely sensitive to chemical environment. Studying the interaction between CNTs and chemical gases could, from fundamental point of view, significantly deepen our understanding on nanoscale device physics. Meanwhile, as large scale integration of CNT devices is still challenging at the current stage, individual electronic devices containing CNTs as their key elements for sensing purpose would be more feasible for practical applications of CNTs. In addition to a small diameter, an extremely large surface to volume ratio makes CNT a suitable material for nanoscale chemical sensing. Since the first CNT gas sensor reported in 2000 (Kong, et al., 2000), many types of CNT-based chemical sensors have been demonstrated. CNT networks (Li, et al., 2003), functionalized CNTs (Qi, et al., 2003), CNT and polymer composites (Wei, et al., 2006), etc, are used as sensing elements. Although tremendous progresses have been achieved in this area, the underlying sensing mechanism still remains unclear. NH_3 gas detection represents the most typical argument in CNT based gas sensing area. Previously proposed mechanisms include the indirect interaction through the hydroxyl group on SiO_2 substrates (Kong, et al., 2000) or pre-adsorbed water layer (Bradley, et al., 2003), adsorption of gas molecules at the interstitial sites in the CNT bundle (Zhao, et al., 2002), direct charge transfer from the adsorbed gas molecules to the CNT channel (Chang, et al., 2001), and modulation of the SB at CNT/metal contacts (Yamada, 2006), etc. Till now, there is no well recognized sensing mechanisms. Hence in this chapter, we will first review the development of CNT based NH_3 gas sensors. After that, we will present a systematic study on the sensing mechanisms through selective Si_3N_4 passivation, which enables us to truly distinguish the sensing signal from the CNT channel and CNT/metal contacts. By comparing the strikingly distinct sensing performance at various testing conditions, we clearly show that the Schottky barrier modulation at the CNT/metal contacts dominates the sensing performance at room temperature. At higher temperatures, say 150°C or above, NH_3 molecules start to adsorb on the CNT wall and the charge transfer from the adsorbed NH_3 molecules to the CNTs contributes to the sensing signal. Next, we will demonstrate tunable real-time NH_3 sensors with three-terminal CNT field-effect transistors. The room-temperature sensitivity and reversibility of such sensors can be greatly enchaned by proper gate voltages. Finally, we will conclude the signicance achieved so far and give an outlook of the future development of CNT based gas sensors.

2. Literature reviews: Contact Vs Channel

It is well know that most of CNTFETs exhibited p-type characteristics in air, which is commonly attributed the contact SB modulation by environmental oxygen (Derycke, et al., 2002, Heinze, et al., 2002). Will this argument still holds for NH_3 gas? On one hand, theoretical calculations predicted limited interaction between NH_3 gas and CNT at room temperature (Kong, et al., 2000, Bauschlicher & Ricca, 2004, Peng & Cho, 1999). On the other hand, CNT based sensors showed high sensitivity to chemical gas. It is natural for researchers to consider the role of metal-CNT contacts. In order to differentiate, or separate the effect from CNT channel and metal-CNT contacts, the design of experiments is to protect or cover either of them and then compare the sensing performance against NH_3 gas. The first effort belongs to Bradley et al. He and his coworkers used thermally evaporated SiO to passivate a short-channel CNTFETs, as shown Fig. 1. They found a good sensitivity to NH_3 after contact passivation. (Bradley, et al., 2003).

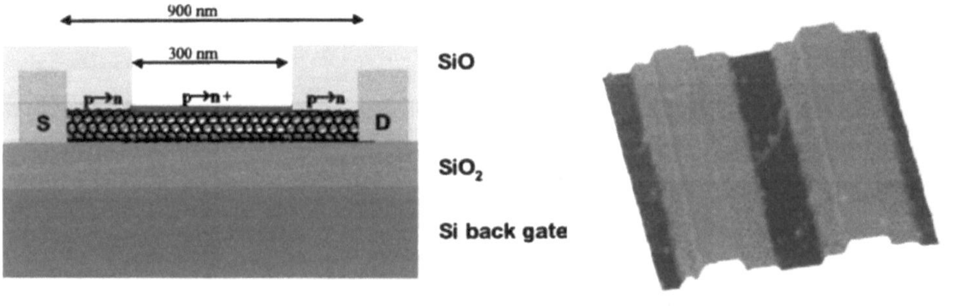

Fig. 1. The device strucutre in Bradley et al's work

Zhang et al. argued that when the passivation length was comparable to the depletion length in the CNT, the contacts could be indirectly affected, which could be the case in the work above. In their work, polymethylmethacrylate (PMMA) was applied to protect the CNT/metal contacts from NO2 exposure and their devices became insensitive after contact passivation, as shown in Fig. 2. (Zhang, et al., 2006)

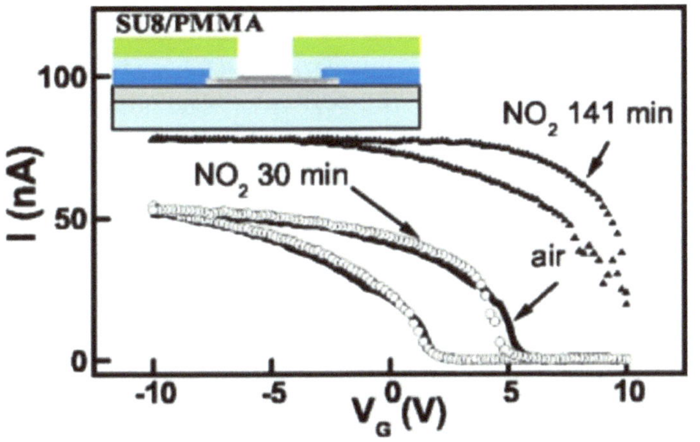

Fig. 2. Zhang et al's sensing results after contact passivation.

Interestingly, Liu et al. also employed PMMA as a passivation layer. They observed changes in the transfer characteristics upon exposure to NH_3 and NO_2 for both contact-passivated device and channel passivated devices, suggesting that both the CNT channel and the CNT/metal contacts play a role in the detection process (Liu, et al., 2005). The obvious ambiguity in those reports could arise from the permeable passivation materials used. Moreover, as the experiments were carried out at room temperature and air ambient only, exclusive identification of the sensing mechanisms is not possible. Next, we will introduce a selective $Si_3 N_4$ passivation technique, with which the sensing signals from the CNT channel and CNT/metal contacts are truly distinguished. Strikingly distinct sensing performance at various testing conditions is observed.

3. Experimental details

CNTs were aligned between Ti/Au source and drain electrodes predefined on a p-type silicon wafer using an ac DEP technique (Li, et al., 2005, Peng, et al., 2006), which is simple and cost effective, suitable for CNT sensor fabrications. The CNT suspension is introduced onto the electrodes with an AC voltage across with a frequency of 2 – 10MHz. The peak-topeak voltage ranged from 6 V to 16 V , depending on the electrodes separation. Due to the induced DEP force, the CNTs in the suspension align with the electric field direction and move towards the electrodes surface. The AC voltage was turned off immediately once SWNTs bridged the electrodes, confirmed by monitoring the resistance across electrodes. The number of the SWNTs between the electrodes can be controlled by adjusting the SWNT concentration and the manipulation time. A heavily doped Si with a 200 nm thick thermally grown SiO_2 top layer was used as the backgate. Note that the CNTs in this work are on top of Au electrodes and the contact regions are fully accessible to the ambient.

These devices are typically the SB-CNTFETs. As illustrated in Fig. 3, three device structures were employed in our experiments: (1) an as-prepared CNTFET with the exposed CNT channel and CNT/Au contacts; (2) only the contacts passivated with 500 nm$Si_3 N_4$ layer and (3) only the channel passivated with 500 nm$Si_3 N_4$ layer. Dry air was used as the background gas with a flow rate of 500 sccm in the following experiments unless otherwise stated. NH_3 gas was selected as the detecting species to study the sensing mechanisms of the CNT sensors.

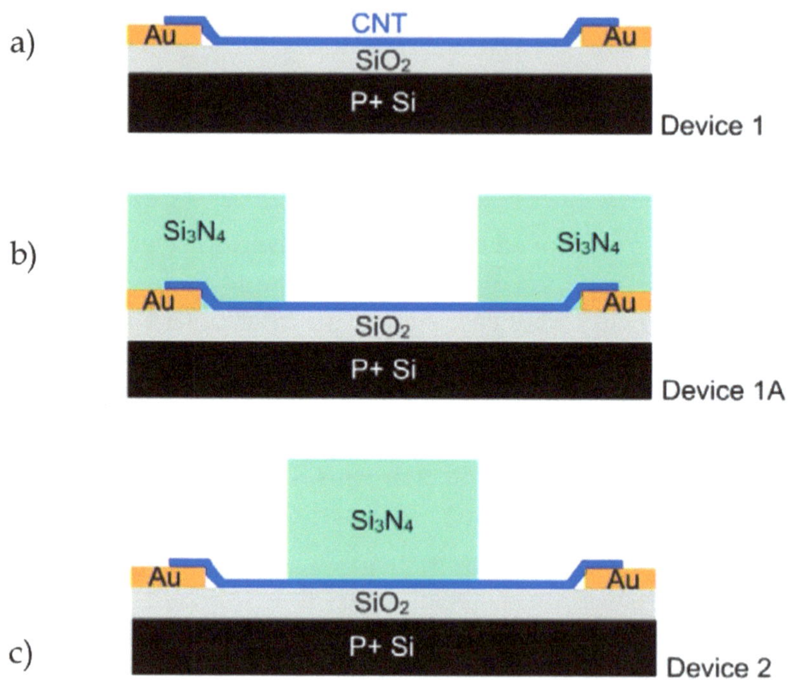

Fig. 3. Schematics for (a) Device 1: As-prepared CNTFET; (b) Device 1A: the contacts passivated by $Si_3 N_4$ and (c) Device 2: the central CNT channel passivated by $Si_3 N_4$.

4. Results and Discussion

4. 1NH_3 sensing at room-temperature

An as-prepared CNTFET (Device 1) showed a sensitive response to small concentrations of NH_3 at room temperature (see Fig. 4 (a)). It is seen that, under a positive gate voltage, both the sensitivity and reversibility were much higher than those under a negative one.

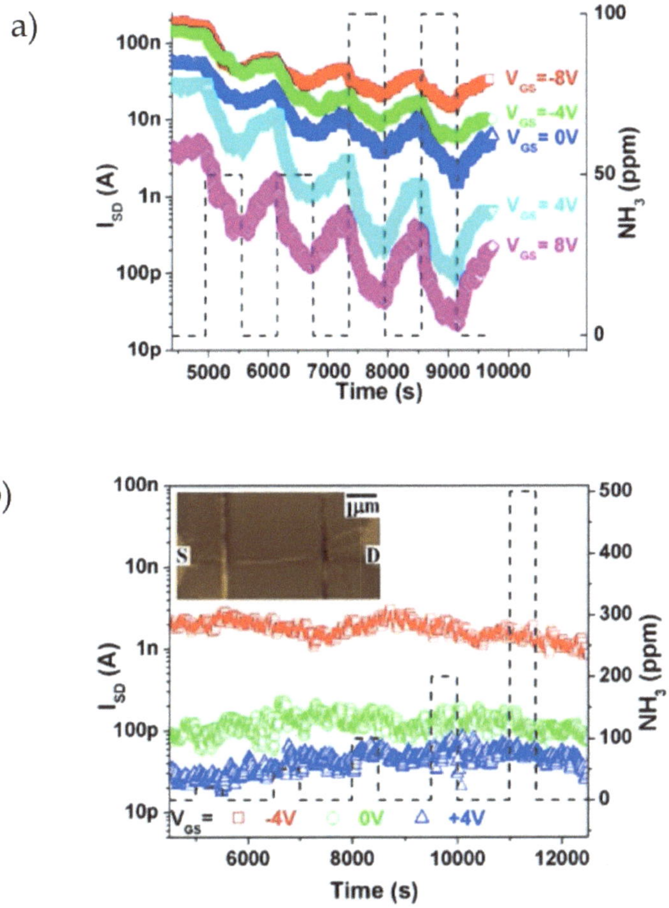

Fig. 4. Real-time detection of NH_3 at room temperature under various gate voltages (a) before (Device 1) and (b) after the contacts passivation (Device 1A), respectively. Inset: an atomic force microscope (AFM) image of Device 1A after the passivation.

In order to experimentally differentiate whether the sensing responses are from the CNT channel and/or the CNT/Au contacts, we passivated the CNT/Au contacts of Device 1 with a Si_3 N_4 thin film, leaving the CNT channel open. After the passivation, we found that the device (Device 1A) did not respond to NH_3 at room temperature, even at a concentration up to 500 ppm , as shown in Fig. 4 (b). For comparison, we only passivated the CNT channel with Si_3 N_4 thin film in another CNTFET (Device 2), but uncovering the CNT/Au contacts. Interestingly, Device 2 showed a high sensitivity at room temperature (see Fig. 5). Therefore, we can unambiguously conclude that NH_3 gas induced SB modulation is a dominant mechanism for our CNT gas sensors at room temperature.

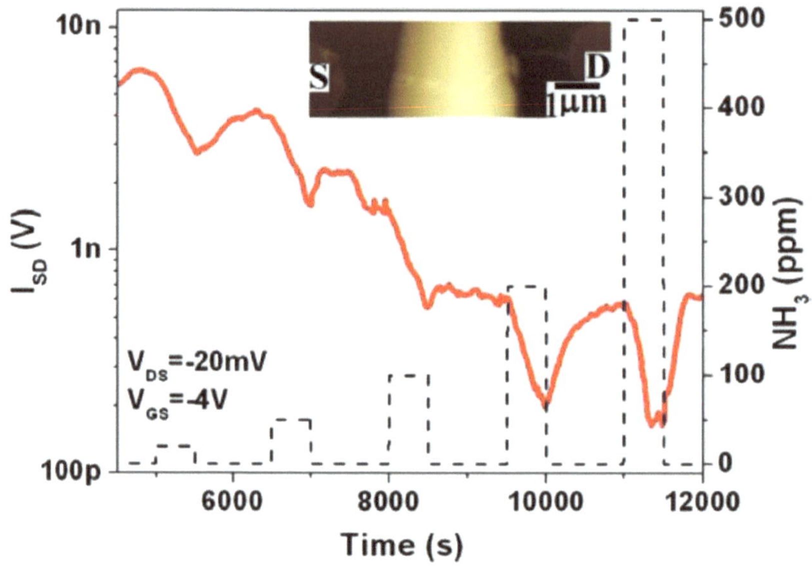

Fig. 5. Sensing response of Device 2 at room temperature. Inset: AFM image of Device 2 with central channel passivated.

Actually, PMMA was widely employed as a passivation material to protect the CNT/metal contact regions for gas (Zhang, et al., 2006, Liu, et al., 2005) and protein sensing (Heller, et al., 2008). However, two major problems exist due to the polymer nature of PMMA. Firstly, PMMA is not dense enough to fully passivate the contacts. For example, NO_2 was found to penetrate the 2.2μ m thick SU-8/PMMA layer in 30 mins (Zhang, et al., 2006). Thus, the CNT/metal contacts are inevitably affected by the gradual diffusion of the detecting species through the PMMA layer, so that the role of the contact in the detection could not be eliminated. Secondly, PMMA is thermally unstable above 100°C. This is a critical limitation as the adsorptions of some biomolecules and gas molecules on CNTs are enhanced at high temperatures. In contrast, Si_3 N_4 is much denser and it can completely insulate the contacts from chemical environment (Kaminishi, et al., 2005). Meanwhile, its thermal stability allows for high temperature sensing experiments, as shown later.

4.2 NH_3 sensing at elevated temperature

The transfer curves of Device 1A before and after exposure to 500ppmNH_3 for 1000 s are monitored at T = 25°C, T = 50°C and T = 100°C in Fig. 6 (a), (b) and (c), respectively. No significant change in V_{TH} curves was observed. We suggest that, the adsorption of NH_3 on the CNT is not favored at this temperature range.

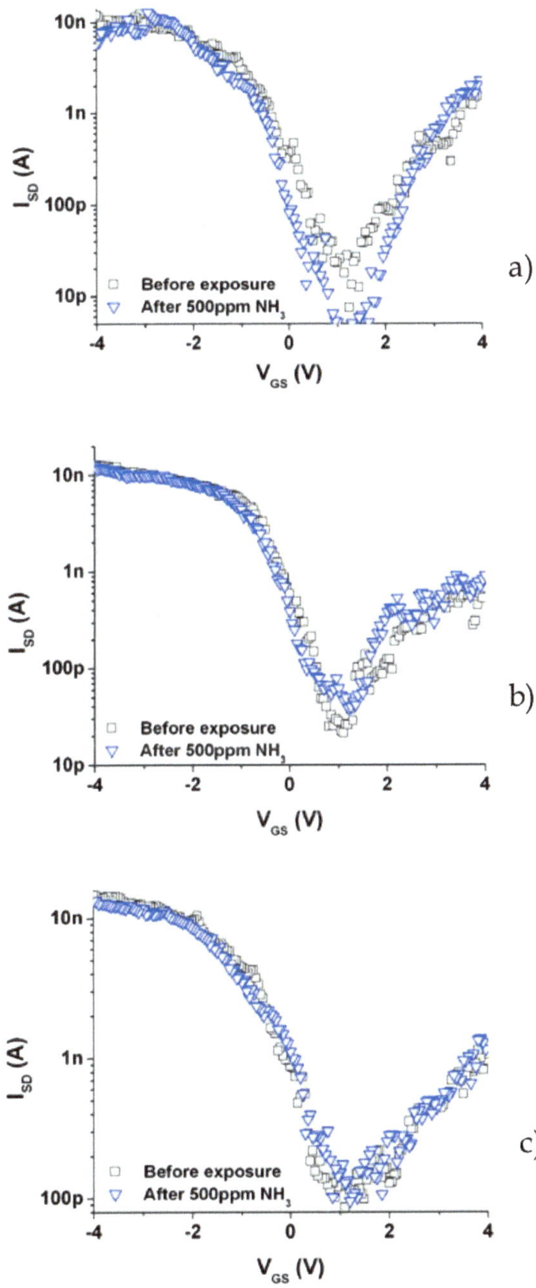

Fig. 6. Transfer characteristics of Device 1A with contacts passivated before and after exposure to NH_3 at (a) T = 25°C, (b) T = 50°C and (c) T = 100°C respectively.

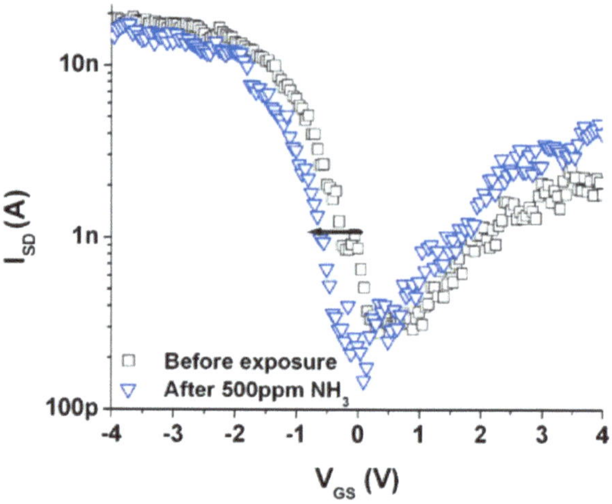

a)

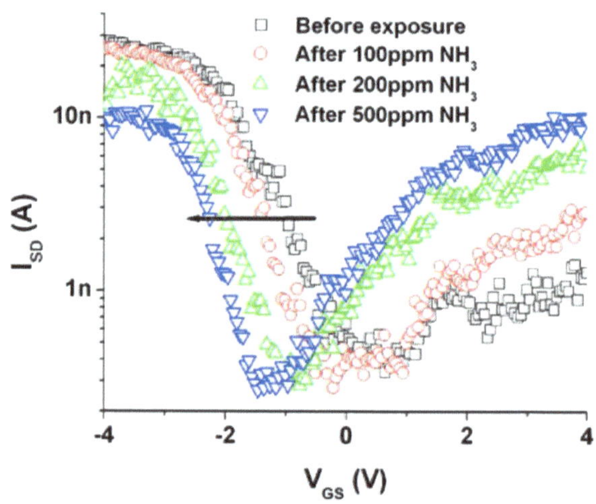

b)

Fig. 7. The transfer characteristics for Device 1A with the contacts passivated before and after exposure to NH_3 at (a) T = 150°C and (b) T = 200°C, respectively.

The transfer curve started to shift towards negative gate voltage after NH_3 exposure at 150°C and above, see Fig. 7 (a). Since the contacts were fully isolated from NH_3, this parallel shift in the transfer curve suggests that NH_3 could adsorb on the CNT wall and donate electrons to the CNT. Consequently, the Fermi level of the CNT moves towards the conduction band edge so that the threshold voltage V_{TH} becomes more negative. When the testing temperature reached 200°C, this phenomenon became more prominent. Progressive shift of the transfer curve in accordance with NH_3 concentrations is shown in Fig. 7 (b).

Under a first order estimation, the total charge transferred $\Delta Q = C_G \Delta V_{TH}$, where the gate capacitance $C_G = \frac{2\pi\varepsilon\varepsilon_0 L}{\cosh^{-1}(h/r)}$, the SiO_2 dielectric constant $\varepsilon = 3.9$ and thickness $h =$ 200 nm. For a SWNT bundle with a length $L \sim 5\mu$ m and a radius $r \sim 5$ nm, $C_G \approx 0.25$fF. Thus, ΔQ is approximately 0.625 fC or about 4000 electrons, at 200°C with $\Delta V_{TH} = 2.5$ V for 500 ppm NH_3. If the cross-sectional area of $NH_3 A \sim 0.13$ nm^2, the length of the exposed CNT channel L′ $\sim 3\mu$ m, the coverage $\theta \sim 0.07$ (interpolated from the Langmuir plot for 500 ppm NH 3 on CNT(Bradley, et al., 2003)), the charge transfer rate $f \approx \Delta QA/q\theta\pi rL'$ is about 0.02 electron. This value is reasonably consistent with typical theoretical predictions.

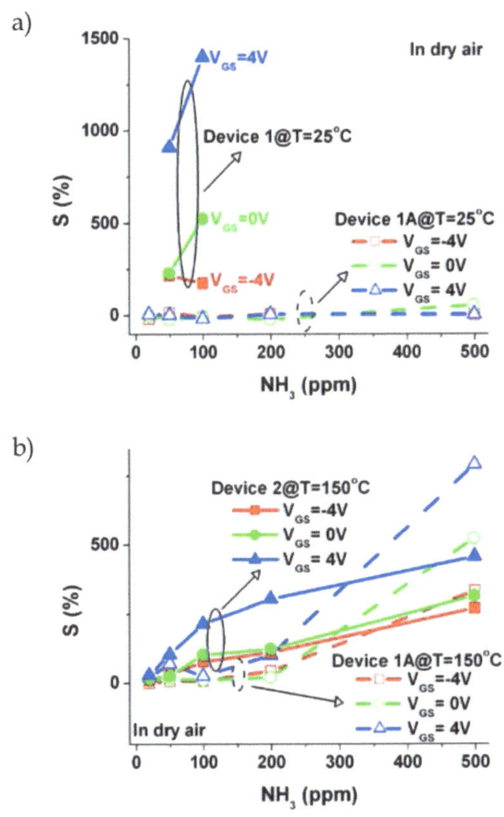

Fig. 8. Extracted sensitivities for (a) Device 1 and Device 1A at T = 25°C; (b) Device 2 and Device 1A at T = 150°C, respectively.

The extracted sensitivity $S = \Delta R/R_0$ of Device 1 and Device 1A under three gate voltages at T = 25°C are shown in Fig. 8 (a). For Device 1, a very high sensitivity and significant gate modulation were observed. When the gate voltage is varied from negative to positive, the dominant carrier injection process changes from tunneling to thermionic emission, and the source-drain current becomes very sensitive to the SB height. A small change in the contact SB height due to NH_3 adsorption will be prominently reflected in the source-drain current. Our device structure with a CNT on top of metal electrodes could also enhance the SB modulation effect. Once the contacts passivation is carried out in Device 1A, it essentially does not respond to NH_3, implying that the CNT channel is not active to NH_3 at room temperature. Fig. 8 (b) compares the sensitivities for Device 1A and Device 2 at $T = 150°C$. At small NH_3 concentrations, a low coverage of NH_3 on the CNT channel and poor charge transfer efficiency result in a small sensitivity in Device 1A. When the NH_3 concentration is increased, the sensitivity for Device 2 becomes saturated, probably due to limited interaction area in the CNT/Au contacts.

4.3 Effect of oxygen on NH_3 sensing

Theoretical studies suggest that, NH_3 interacts weakly with pristine CNTs with little charge transfer (Kong, et al., 2000, Bauschlicher & Ricca, 2004, Peng & Cho, 1999). Existence of a large activation barrier prevents adsorption of NH_3 on perfect CNTs even at high temperatures. However, the adsorption of the gas molecules on defective CNTs could be much easier (Feng, et al., 2005, Robinson, et al., 2007). In addition, the adsorption barrier of NH_3 on a defective CNT can be further lowered by pre-dissociated oxygen atoms, leading to an enchanted charge transfer rate, as pointed out by Andzelm et al (Andzelm, et al., 2006). In order to study the influences of oxygen on NH_3 adsorption onto the CNT wall, we changed the background gas from dry air to N_2. Device 1A was first annealed in N_2 environment at 350°C for 2 hrs to degas the adsorbed oxygen. Note that during the hightemperature annealing, remaining oxygen molecules at the contacts can be further desorbed and the device became more n -type. The transfer curves before and after exposure to 500 ppm NH 3 for 1000s from $T = 25°C$ to $T = 150°C$ were shown in Fig. 9, and real-time sensing results at $T = 200°C$ are shown in Fig. 10. No detectable changes due to NH_3 exposure were observed. Comparing with the sensing response observed in dry air environment, we can confirm that the adsorption of NH_3 is facilitated by environmental oxygen. It was also found that the sensitivity was restored after the background was changed to dry air again. Our results are consistent with Andzelm et al's predictions. NH_3 could preferentially attach to the defect sites on CNT with pre-dissociated oxygen, as illustrated in Fig. 11.

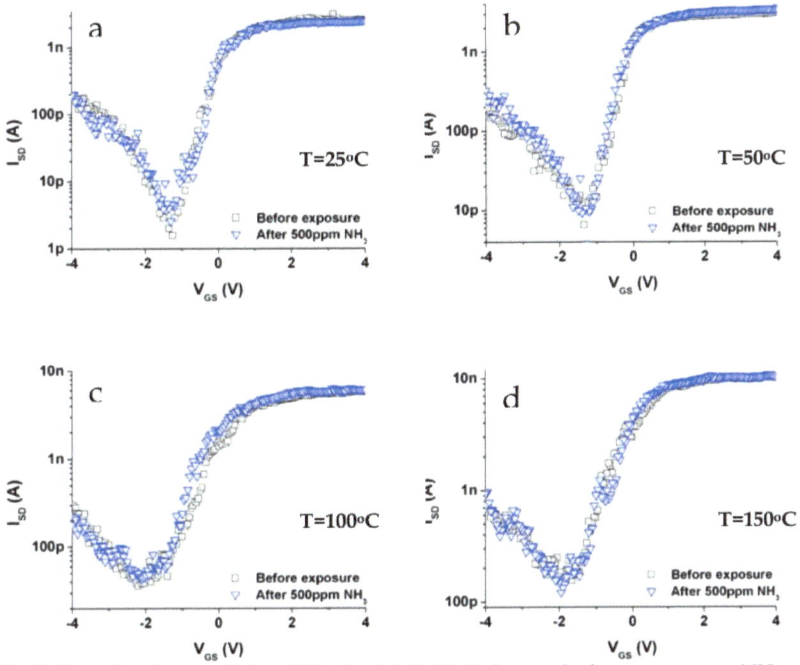

Fig. 9. Transfer characteristics of Device 1A in N_2 before and after exposure to NH_3 at (a) T = 25°C, (b) T = 50°C, (c) T = 100°C and (d) T = 150°C respectively.

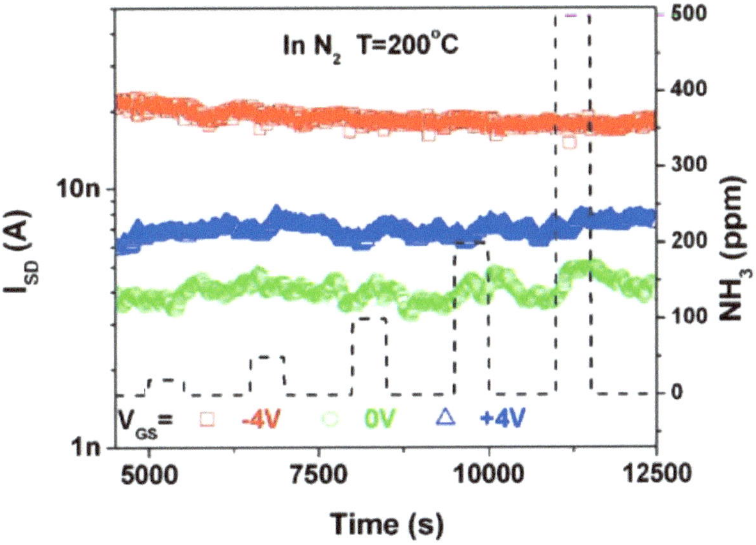

Fig. 10. Response of I_{SD} to various concentrations of NH_3 in N_2 at T = 200°C for Device 1A.

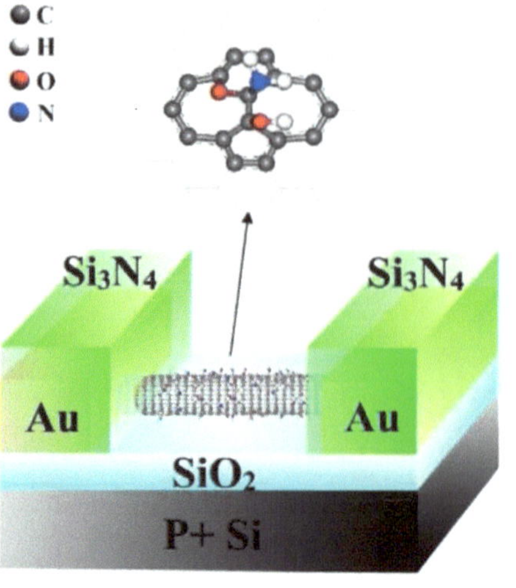

Fig. 11. Schematic of NH_3 adsorption on Device 1A. Here, we illustrate a NH_3 molecule adsorbs on a CNT with Stone-Wales defect with pre-dissociated oxygen atoms, as suggested in by Andzelm et al.

4.4 Comparisons of the sensing mechanisms in CNT based gas sensors

From our results, we are able to rule out several possibilities of indirect interactions between NH_3 and CNT. Firstly, as the testing environment was totally dry, NH_3 adsorption through water layer is not applicable here. Secondly, If NH_3 could interact through the SiO_2 substrate or adsorb inside the CNT bundles, a reduced sensitivity should have been observed after contacts passivation. However, our observation that Device 1A is totally insensitive to NH_3 at room temperature does not support this hypothesis. In fact, charge transfer and the SB modulation are the two mechanisms in our CNT sensors.

At room temperature, the weak adsorption of NH_3 on the CNT wall does not induce any measurable effect on the source-drain current. The sensing signal mainly arises from the contacts. When NH_3 molecules are adsorbed on the CNT/Au interface, the work function of the Au electrode is reduced (Bilic, et al., 2002) and/or the electrostatic charge balance between the CNT and Au is disturbed by the dipoles of NH_3 molecules, (Yamada, 2006) leading to an increased SB for hole injection. The sensitivity reflected in the source-drain current are, however, gate voltage dependant. As illustrated in Figure 12 (b), when a negative gate voltage bends the energy band of the CNT upwards, the SB width becomes very narrow and holes could tunnel through the barrier, even when the SB height is increased by NH_3. In contrast, at a positive gate voltage, the SB width is too thick for tunneling process. Thus, hole injection is only through thermionic emission over the SB height. The source-drain current is then expressed as:(Appenzeller, et al., 2004) $I_{DS} \sim T^2 \exp(-q\phi_B/K_BT)$, where ϕ_B is the Schottky barrier height, K_B is the Boltzmann's constant and T is the temperature in Kelvin. In this case, the sensitivity for the SB modulation is $S = \Delta R/R_0 \approx \exp(q\Delta\phi_B/K_BT) - 1$. Owing to the exponential dependence on the SB height change $\Delta\phi_B$, extremely high sensitivity can be achieved. Our results are consistent with Yamada's theoretical prediction that the SB modulation is most significant when CNT is operating in the depletion mode.(Yamada, 2006)

One disadvantage of the SB dominated sensors is a typical long recovery process at room temperature. Once the operating temperature is increased, the sensitivity degrades shapely due to its exponential dependence with 1/T. The nanoscale CNT/metal contacts could also be prone to subtle environmental noises, causing fluctuations during detection. In contrast, using the CNT channel as a sensing element can avoid the problems. In addition, the large surface to volume ratio of CNT offers another merit. At $T > 150°C$, when the adsorption barrier is further lowered by oxygen, NH_3 adsorption on the defect sites of the CNT becomes favored. Upon the consequent charge transfer, the Fermi level of the exposed central CNT channel moves upwards and the energy band shifts downwards with respect to those passivated parts, a potential barrier is therefore created and impedes the current flow, see Fig. 12 (d). Here, we would like to emphasize that, the charge transfer effect can be studied only when the contacts are fully protected. Otherwise, as temperature increases, the sensitivity enhancement from the charge transfer effect and the degradation due to SB modulation counteract each other.

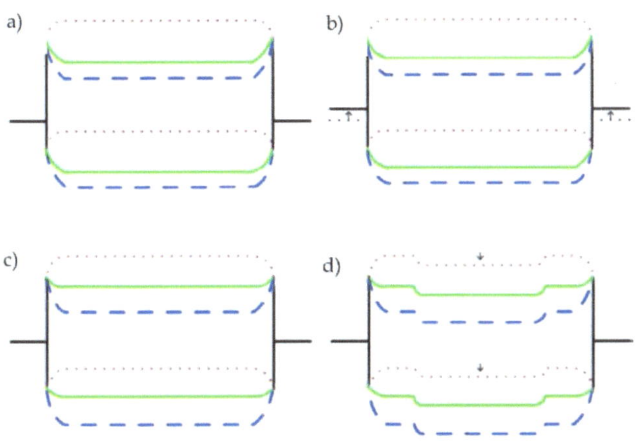

Fig. 12. Schematic energy band diagram for: Device 1 (a) before and (b) after the NH_3 exposure (An intrinsic CNT is considered. The work function of the source/drain electrodes is initially near the valence band edge of the CNT and is reduced after NH_3 exposure); Device 1A (c) before and (d) after the NH_3 exposure

(After passivation, the work function of electrodes aligns near the midgap of CNT. The Fermi level of the exposed central CNT channel shifts upwards due to electron-doping from NH_3). Legend: Red dotted for $V_{GS} < 0$; Green solid for $V_{GS} = 0$; Blue dashed for $V_{GS} > 0$.

Table 1 compares the SB modulation and charge transfer mechanisms. The sensors with the SB modulation usually demonstrate a high sensitivity at room temperature. Their sensing performance can be adjusted through the gate voltages. In contrast, the sensors under the charge transfer mechanism require high working temperatures, showing a good reversibility.

Table 1. Comparisons between sensors operated by SB modulation and charge transfer.

Mechanism	SB modulation	Charge transfer
Sensitivity	Extremely high in depletion mode	Low
Gate dependence	Strong	Weak
Operating Temperature	Room temperature	T > 150°C
Reversibility	Low, can be improved at positive gate voltage	Good

5. Conclusions and Future Work

In this chapter, we discussed in details the sensing mechanisms of CNT based NH_3 detection. By selective Si_3 N_4 passivation, We clearly show that the SB modulation at the CNT/metal contacts dominates the sensing performance at room temperature, and the sensor exhibits high sensitivity and good tunability under appropriate gate voltages. At higher temperatures, say 150°C or above, NH_3 molecules start to adsorb on the CNT wall and the charge transfer process from the adsorbed NH_3 molecules to the CNTs contributes to the sensing signal. As the mechanisms are identified, the next step is how to improve the sensing performance, and more chanllegingly, how to differentiate between different gas species. One promising way could be functionalizaton method (Feng, et al., 2005, Qi, et al., 2003). For example, using metal catalysts in the form of nanoparticles to decorate CNT, which promotes the interaction with speicific gas speicies. With a combination of several metal nanoparticle, the sensing performance to various gas molecules could be compared and contrasted.

6. Reference

Andzelm, J.; Govind, N. & Maiti, A. (2006). Nanotube-based gas sensors - Role of structural defects. Chemical Physics Letters, Vol. 421, No. 1-3: 58-62, 0009-2614

Appenzeller, J.; Radosavljevic, M.; Knoch, J. & Avouris, P. (2004). Tunneling versus thermionic emission in one-dimensional semiconductors. Physical Review Letters, Vol. 92, No. 4, 0031-9007

Bauschlicher, C. W. & Ricca, A. (2004). Binding of NH_3 to graphite and to a (9,0) carbon nanotube. Physical Review B, Vol. 70, No. 11, 1098-0121

Bilic, A.; Reimers, J. R.; Hush, N. S. & Hafner, J. (2002). Adsorption of ammonia on the gold(111) surface. Journal of Chemical Physics, Vol. 116, No. 20: 8981-8987, 0021-9606

Bradley, K.; Gabriel, J. C. P.; Briman, M.; Star, A. & Gruner, G. (2003). Charge transfer from ammonia physisorbed on nanotubes. Physical Review Letters, Vol. 91, No. 21, 00319007

Bradley, K.; Gabriel, J. C. P.; Star, A. & Gruner, G. (2003). Short-channel effects in contactpassivated nanotube chemical sensors. Applied Physics Letters, Vol. 83, No. 18: 38213823, 0003-6951

Chang, H.; Lee, J. D.; Lee, S. M. & Lee, Y. H. (2001). Adsorption of NH_3 and NO_2 molecules on carbon nanotubes. Applied Physics Letters, Vol. 79, No. 23: 3863-3865, 0003-6951

Derycke, V.; Martel, R.; Appenzeller, J. & Avouris, P. (2002). Controlling doping and carrier injection in carbon nanotube transistors. Applied Physics Letters, Vol. 80, No. 15: 2773-2775, 0003-6951

Feng, X.; Irle, S.; Witek, H.; Morokuma, K.; Vidic, R. & Borguet, E. (2005). Sensitivity of ammonia interaction with single-walled carbon nanotube bundles to the presence of defect sites and functionalities. Journal of the American Chemical Society, Vol. 127, No. 30: 10533-10538, 0002-7863

Heinze, S.; Tersoff, J.; Martel, R.; Derycke, V.; Appenzeller, J. & Avouris, P. (2002). Carbon nanotubes as Schottky barrier transistors. Physical Review Letters, Vol. 89, No. 10, 0031-9007

Heller, I.; Janssens, A. M.; Mannik, J.; Minot, E. D.; Lemay, S. G. & Dekker, C. (2008). Identifying the mechanism of biosensing with carbon nanotube transistors. Nano Letters, Vol. 8, No. 2: 591-595, 1530-6984

Kaminishi, D.; Ozaki, H.; Ohno, Y.; Maehashi, K.; Inoue, K.; Matsumoto, K.; Seri, Y.; Masuda, A. & Matsumura, H. (2005). Air-stable n-type carbon nanotube field-effect transistors with $Si_3 N_4$ passivation films fabricated by catalytic chemical vapor deposition. Applied Physics Letters, Vol. 86, No. 11, 0003-6951

Kong, J.; Franklin, N. R.; Zhou, C. W.; Chapline, M. G.; Peng, S.; Cho, K. J. & Dai, H. J. (2000). Nanotube molecular wires as chemical sensors. Science, Vol. 287, No. 5453: 622-625, 0036-8075

Li, J. Q.; Zhang, Q.; Peng, N. & Zhu, Q. (2005). Manipulation of carbon nanotubes using AC dielectrophoresis. Applied Physics Letters, Vol. 86, No. 15, 0003-6951

Li, J.; Lu, Y. J.; Ye, Q.; Cinke, M.; Han, J. & Meyyappan, M. (2003). Carbon nanotube sensors for gas and organic vapor detection. Nano Letters, Vol. 3, No. 7: 929-933, 1530-6984

Liu, X. L.; Luo, Z. C.; Han, S.; Tang, T.; Zhang, D. H. & Zhou, C. W. (2005). Band engineering of carbon nanotube field-effect transistors via selected area chemical gating. Applied Physics Letters, Vol. 86, No. 24, 0003-6951

Peng, N.; Zhang, Q.; Li, J. Q. & Liu, N. Y. (2006). Influences of ac electric field on the spatial distribution of carbon nanotubes formed between electrodes. Journal of Applied Physics, Vol. 100, No. 2, 0021-8979

Peng, S. & Cho, K. J. (1999). Chemical control of nanotube electronics. 7th Annual Foresight Conference on Molecular Nanotechnology, 57-60, Santa Clara, California.

Qi, P.; Vermesh, O.; Grecu, M.; Javey, A.; Wang, O.; Dai, H. J.; Peng, S. & Cho, K. J. (2003). Toward large arrays of multiplex functionalized carbon nanotube sensors for highly sensitive and selective molecular detection. Nano Letters, Vol. 3, No. 3: 347351, 1530-6984

Robinson, J. A.; Snow, E. S. & Perkins, F. K. (2007). Improved chemical detection using single-walled carbon nanotube network capacitors. Sensors and Actuators a-Physical, Vol. 135, No. 2: 309-314, 0924-4247

Wei, C.; Dai, L. M.; Roy, A. & Tolle, T. B. (2006). Multifunctional chemical vapor sensors of aligned carbon nanotube and polymer composites. Journal of the American Chemical Society, Vol. 128, No. 5: 1412-1413, 0002-7863

Yamada, T. (2006). Equivalent circuit model for carbon nanotube Schottky barrier: Influence of neutral polarized gas molecules. Applied Physics Letters, Vol. 88, No. 8, 0003-6951

Yamada, T. (2006). Equivalent circuit model for carbon nanotube Schottky barrier: Influence of neutral polarized gas molecules. Applied Physics Letters, Vol. 88, No. 8: 083106, 0003-6951

Zhang, J.; Boyd, A.; Tselev, A.; Paranjape, M. & Barbara, P. (2006). Mechanism of NO_2 detection in carbon nanotube field effect transistor chemical sensors. Applied Physics Letters, Vol. 88, No. 12, 0003-6951

Zhao, J. J.; Buldum, A.; Han, J. & Lu, J. P. (2002). Gas molecule adsorption in carbon nanotubes and nanotube bundles. Nanotechnology, Vol. 13, No. 2: 195-200, 09574484.

Index

www.ingramcontent.com/pod-product-compliance
Lightning Source LLC
LaVergne TN
LVHW070849220926
848486LV00032B/359
* 9 7 8 1 7 8 7 1 5 6 4 5 6 *